Advanced Structured Materials

Volume 230

Common engineering materials are reaching their limits in many applications, and new developments are required to meet the increasing demands on engineering materials. The performance of materials can be improved by combining different materials to achieve better properties than with a single constituent, or by shaping the material or constituents into a specific structure. The interaction between material and structure can occur at different length scales, such as the micro, meso, or macro scale, and offers potential applications in very different fields.

This book series addresses the fundamental relationships between materials and their structure on overall properties (e.g., mechanical, thermal, chemical, electrical, or magnetic properties, etc.). Experimental data and procedures are presented, as well as methods for modeling structures and materials using numerical and analytical approaches. In addition, the series shows how these materials engineering and design processes are implemented and how new technologies can be used to optimize materials and processes.

Advanced Structured Materials is indexed in Google Scholar and Scopus.

Andreas Öchsner · Holm Altenbach
Editors

Engineering Design Applications VII

Springer

Editors
Andreas Öchsner
Faculty of Mechanical and Systems Engineering
Esslingen University of Applied Sciences
Esslingen, Baden-Württemberg, Germany

Holm Altenbach
Institut für Werkstoffe, Technologie und Mechanik
Fakultät für Maschinenbau Otto von Guericke University Magdeburg
Magdeburg, Germany

ISSN 1869-8433 ISSN 1869-8441 (electronic)
Advanced Structured Materials
ISBN 978-3-031-84345-7 ISBN 978-3-031-84346-4 (eBook)
https://doi.org/10.1007/978-3-031-84346-4

This Springer imprint is published by the registered company Springer Nature Switzerland AG
The registered company address is: Gewerbestrasse 11, 6330 Cham, Switzerland

Preface

Different engineering disciplines such as mechanical, materials, computer and process engineering provide the foundation for the design and development of improved structures, materials and processes. The modern design cycle is characterized by an interaction of different disciplines and a strong shift to computer-based approaches where only a few experiments are performed for verification purposes. A major driver for this development is the increased demand for cost reduction and higher efficiency, which is also connected to environmental demands. One way to fulfil such requirements is lighter structures and/or new composite materials and structures. Another emerging area is the interaction of classical engineering with the health, medical and environmental sectors. This further volume in this series gives an update on recent developments in the mentioned areas of modern engineering design application.

Esslingen, Germany — Andreas Öchsner
Magdeburg, Germany — Holm Altenbach

Contents

Chapter 1
Refinement of AZ91 Alloy for Quality Casting Production

Vadim Shalomeev, Galyna Tabunshchyk, Kinga Korniejenko, and Marek Nykiel

Abstract Magnesium alloys are a promising material for bioresorbable implants in osteosynthesis manufacture. Magnesium is contained in bone and muscle tissue and it participates in various metabolic processes of the human body. Magnesium biocorrosion products have good biocompatibility. Magnesium alloys positively affect the human body, but their properties and quality do not always meet the requirements for using them as implants. It is possible to improve the magnesium alloy's properties by improving the refining processes of the liquid alloy. The aim of our research is to investigate and test different materials for refining magnesium melt. Optical microscopy, standard methods for determining the mechanical properties of magnesium alloy, and special methods for determining the refining ability of various materials were implemented. The structure and properties of the cast magnesium alloy AZ91 in its refined state were investigated. Filtering materials that provide melt refining for obtaining high-quality casting with an increased set of properties are proposed. It has been shown that the usage of a carbon-containing filter provides not only effective refining of the alloy but also its additional modification. At the same time, an increased amount of intermetallic phase is observed in the alloy structure, which improves its structure and physical and mechanical characteristics. The developed complex technology for processing the AZ91 liquid melt allows improvement of the yield of suitable castings for the manufacture of implants and improves their quality. Preclinical testing of implants, which were manufactured using the experimental technology, was carried out and a positive result of their usage was obtained.

V. Shalomeev · G. Tabunshchyk (✉)
National University Zaporizhzhia Polytechnic, Zaporizhzhia, Ukraine
e-mail: galina.tabunshchik@gmail.com

V. Shalomeev
e-mail: shalomeev@zp.edu.ua

K. Korniejenko · M. Nykiel
Faculty of Materials Engineering and Physics, Cracow University of Technology, Cracow, Poland
e-mail: kinga.korniejenko@pk.edu.pl

M. Nykiel
e-mail: marek.nykiel@pk.edu.pl

A. Öchsner and H. Altenbach (eds.), *Engineering Design Applications VII*, Advanced Structured Materials 230, https://doi.org/10.1007/978-3-031-84346-4_1

1.1 Introduction

Titanium or steel implants are widely used in medicine to treat bone fractures with the osteosynthesis method. However, in the case of intensive bone regeneration, rigid titanium and steel implants are unproductive: they slow down bone growth (Yadroitsava et al. 2019). Inflammatory processes could also start around them, leading to further bone loosening (Borys et al. 2019). In addition, such implants should be removed from the bone to stabilize the fracture after joining the fragments, which requires a second surgery.

The solution to this problem is the usage of bioresorbable materials (Moshood et al. 2022). The most common biosoluble material for these types of implants are various polyacids, the usage of which eliminates the need for repeated removal surgery. However, the usage of polyacids for osteosynthesis is complicated due to the insufficient mechanical properties of this material and its high cost.

Magnesium-based alloys can be a promising biosoluble material for implant manufacturing (He et al. 2023; Shalomeev et al. 2019). Medical studies have shown the positive effects of magnesium on the human body. It has been established that magnesium is a natural element of the body, so it is well absorbed (Jahnen-Dechent and Ketteler 2012). This element optimally matches the properties of the skeleton, it ensures good fusion of bone fragments and quickly restores them at the site of a fracture. However, metal magnesium does not provide the required level of mechanical properties and biosolubility for manufacturing from its structures. Therefore, most of the current research activities are devoted to the improvement of the physical, mechanical, and special properties of magnesium alloys by variation of the existing compositions and developing new ones (Prasad et al. 2021; Shalomeev et al. 2022a). At the same time, metallurgical processes during alloy smelting play a significant role in shaping the structure and properties of the metal (Jiang 2017). Combined with smelting, refining, and modification determine the nature of crystallization, the degree of granularity of the structure, and the set of alloy properties (Zhang et al. 2016; Shalomeev et al. 2022b).

During the production of magnesium alloys, non-metallic inclusions of exogenous and endogenous nature accumulate in the liquid metal, which is associated with the high chemical activity of magnesium and alloying elements, as well as the usage of fluxes. Therefore, the technology of manufacturing magnesium-based alloys involves refining processes. The most advanced way to purify liquid metal is to filter it through granular solid filters, which allows simultaneously conducting refining and alloy modification.

Explanation of the sorbent action mechanism in contact with the liquid metal and, based on its analysis, choosing the most appropriate and rational filtration material is a rather important task.

1.2 State of the Art

Melting easily oxidizable magnesium alloys has several difficulties. This could be explained by the fact that a film is formed on the surface of the melt, which does not protect the melt from further oxidation and combustion. At melting temperatures, magnesium interacts with nitrogen and oxygen to form nitrides and oxides, and also absorbs hydrogen intensely (Chunmiao et al. 2013). The intensive interaction of magnesium melts with furnace gases, lining and charge moisture leads to the formation of compounds that are removed by refining. For the magnesium alloys are used such types of refining as sedimentation, flux treatment, and filtration (Cao et al. 2019).

Flux treatment is based on the removal of suspended inclusions from the melt as a result of adhesion, dissolution, or chemical interaction with molten salts. Surface phenomena play a major role in the flux refining process. Fluxes reduce the adhesion of inclusions to the metal and absorb them under the action of interfacial forces. To prevent contact of the melt with the atmosphere, melting of magnesium alloys is carried out under fluxes consisting of chloride and fluoride salts of alkali and alkaline earth metals or in the environment of protective gases.

Magnesium melts always contain impurities in the form of Prasad et al. (2022):

(a) elements, which are present in the melt in a dissolved state;
(b) oxides of magnesium and alloying elements, flux residues;
(c) gases.

Fluxes for melting magnesium alloys are used by Cao et al. (2019):

- protect the charge from oxidation and ignition during melting;
- insulate the melt from contact with air;
- refine the alloy from non-metallic inclusions.

The composition of fluxes for melting magnesium alloys is determined by the chemical composition of the alloy and can vary. The basis of most fluxes is carnallite ($MgCl_2$. KCl). Barium chloride is added to fluxes to increase their specific gravity and facilitate flux separation from the melt. Calcium fluoride increases the viscosity of the flux. Magnesium oxide is added as a thickener, which ensures the formation of a crust when the metal overheats.

At smelting of magnesium alloy AZ91 flux VI-2 (38…46% $MgCl_2$, 32…43% KCl, 8…10% $CaCl_2$, 5…9% $BaCl_2$, 3…5% CaF_2, < 1,5% MgO) has proved itself well, showing good protective and refining properties. Due to its increased density, it easily separates from the melt surface and forms a strong dense film, which protects the alloy from oxidation. However, when the chemical composition of the alloy changes, additional flux components are required (Kumar et al. 2018). Additional sedimentation of the melt, which leads to a significant reduction in its gas saturation, dramatically reduces the reject rate of castings.

The refining ability of the flux is determined by the complex physical and chemical properties, not only by the chemical activity of its components to magnesium oxide.

The impurities removal will be more effective, the better the flux wets magnesium oxide, the lower the surface activity of the flux to magnesium, and the higher the density of the flux (Perera et al. 2024).

Melting of magnesium alloys under a layer of flux also has several disadvantages: the penetration of flux into the castings, the formation of intense corrosion centers (flux hygroscopicity), and the need to add fresh flux to protect the melted mirror (Esmaily et al. 2017).

One of the effective ways of melt purification from large inclusions and films is filtration, when liquid metal is passed through filters, which are installed in transfer crucibles, casting funnels, and casters. In this case, the efficiency of non-metallic removal of the inclusions depends on the nature of the filter material, the thickness of the filter layer, and the value of the metallostatic pressure, which determines the filtration rate. The best effect of melt purification is obtained with the usage of materials that are characterized by high work of adhesion to non-metallic inclusions (Wagner et al. 2024).

The following requirements are imposed on filtration materials (Raabe et al. 2022; Shalomeev et al. 2022c): the possibility of easy crushing and sieving, sufficient mechanical strength, inertness of the filter to the chemical composition of the alloy, and stable effect of refining.

Limestone, magnesite, and graphite, which are widely used in metallurgy, satisfy these requirements. When these materials interact with a magnesium alloy, simultaneously with melt refining, the process of its modification with carbon, which is part of the filter materials, is possible. Obviously, different carbon-containing materials will have different refining and modifying capacities, due to their chemical composition and physicochemical properties (Long et al. 2013). Therefore, the correct choice of filter material, which provides increased refining capacity, maximum refinement of metal grains, and, as a consequence, an increased complex of properties of the alloy, is an important task.

1.3 Methods of Research

Experimental meltings were carried out in a IPM-500 crucible furnace, into which preheated charge materials were loaded and after melting were poured into extraction crucibles at 650…730°C. Extraction crucibles were installed in distribution furnaces, in which the alloy was refined according to its chemical composition and refined with VI-2 flux at 740…760°C. Efficiency of filtration through different materials was investigated on magnesium alloy, after refining with flux (% wt.: 38…46 MgCl; 32…43 KCl; 8…10 $CaCl_2$; 5…9 $BaCl_2$; 3…5 CaF_2). Preheated to the temperature of 500°C filter materials with the granularity of 10…50 mm were alternately poured onto the grid of removable casting bowl with the height of 100 mm, installed above the riser of the casting mold, and cast samples were poured for determination of mechanical properties and metallographic control.

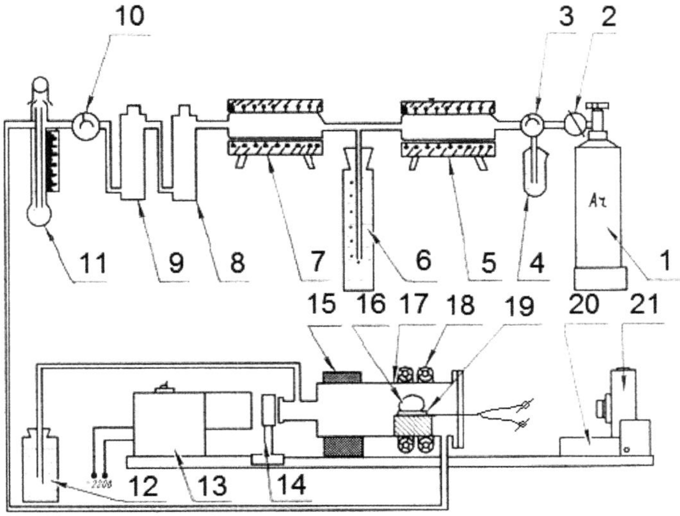

Fig. 1.1 Schematic diagram of the installation for determining the surface tension of liquid metal by the "lying drop" method: 1- argon cylinder; 12- flask; 2- reducer; 13- projection lamp; 3- three-way faucet; 14- lens; 4- drexel; 15- furnace mounting post; 5- furnace with reduced copper; 16- drop; 6- monostat; 17- quartz glass furnace; 7- furnace with palladium catalyst; 18- inductor; 8–9- silicogel and zeolite columns; 19- substrate with thermocouple; 10- two-way faucet; 20- chamber support; 11- rheometer; 21- photo camera

The surface tension of magnesium alloy melt was studied by the "lying drop" method (Fig. 1.1). The research technique consists of melting a cylindrical sample (Ø 7.3 ± 0.1 mm; $h = 7.3 \pm 0.1$ mm) on a horizontal substrate in an induction furnace in an argon atmosphere. By fixing a magnified image of the droplet, the geometric parameters of the droplet projection were measured and the experimental data were mathematically processed.

Flux samples of appropriate sizes were made by withdrawing the molten flux with a quartz tube. To study the interaction of AZ91 melt and liquid flux with substrate materials during the research process, a special graphite heater was designed and manufactured, which was located in a quartz glass furnace in the middle of the inductor. The substrates representing the filter material (magnesite, limestone, and graphite) and oxides (MgO, Al_2O_3, and SiO_2) were located in the middle of the graphite heater.

The operating temperatures during the studies were determined using thermocouples. The change in the substrate's temperature was recorded in time depending on the operation mode of the high-frequency generator.

The surface interaction characteristics (Fig. 1.2) of the investigated materials were determined as:

(a) Surface tension ($\sigma_{p.g.}$) is the interaction at the melt-inert gas interface, i.e., it is the excess energy that is attributed to the surface unit and is determined by the

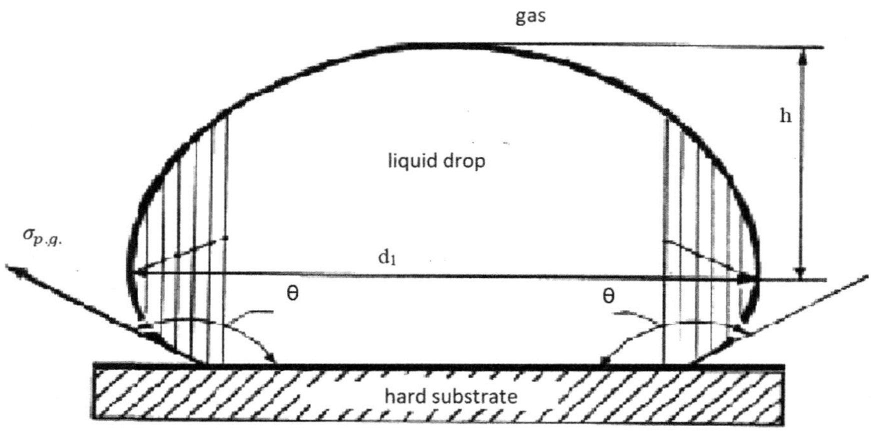

Fig. 1.2 Schematic of a metal drop on a solid substrate: θ–marginal wetting angle; d_1 –equatorial diameter of the drop; h–distance from the top to the equatorial diameter of the drop; $\sigma_{p.g}$–surface tension

formula:

$$\sigma_{p.g.} = d_e^2\,(1/H)\,(\rho_{\mathrm{Pr}} - \rho_{Ar})\,\mathrm{g},\ \mathrm{mJ}/m^2, \tag{1.1}$$

where de is the equatorial diameter of the drop, m;
ρ_{Pr} is the density of the sample under study, kg/m^3;
ρ_{Ar} is the density of argon, kg/m^3; ρ_{Ar}= 0.0178 kg/m^3;
g is the acceleration of free fall, m/s^2; g = 9.81 m/s^2;
1/H = f (d_1/2 h) is the ratio of the drop diameter and height.

(b) cohesion (A_k) is the work that is spent to overcome the forces of attraction between the molecules of a physical body:

$$\mathrm{A_k} = 2 - \sigma_{\mathrm{p.g.}},\ \mathrm{mJ/m^2} \tag{1.2}$$

(c) adhesion (Aa) is the work required for phase separation (drop-substrate):

$$\mathrm{A}_a = \sigma_{r.g.} - (1 + \cos\theta),\ mJ/m^2, \tag{1.3}$$

where θ is the marginal angle of wetting of the solid surface by the liquid, deg.

(d) spreading coefficient (K_r) - used to clarify the conditions of adhesion:

$$K_r = A_a - A_k = \sigma_{r.g} - (1 + \cos\theta) - 2 - \sigma_{p.g} = \sigma_{r.g} - (\cos\theta - 1),\ mJ/m^2. \tag{1.4}$$

Flux refining is based on the removal of suspended inclusions from melts as a result of adhesion, dissolution or chemical interaction with molten salts. The refining properties of the flux are characterized by the work of adhesion of the inclusions to the metal in the flux medium:

$$W_{fl} = \sigma_m \cos\theta_{M-B} - \sigma_{fl} \cos\theta_{fl-B} + A_a \tag{1.5}$$

where W_{fl}- adhesion of inclusions to metal in flux environment, mJ/m2;

σ_m is the surface tension of metal melt, mJ/m^2;

σ_{fl} is the surface tension of flux, mJ/m^2;

θ_{m-B} is the edge angle of inclusions wetting by metal melt, deg;

θ_{fl-B} is the marginal angle of wetting of inclusions with liquid flux, deg;

A_a is the interphase tension (adhesion) at the boundary between liquid metal and flux, mJ/m^2.

To determine the characteristics of the surface interaction, substrates were made of periclase (MgO), alundum (Al_2O_3), and quartzite (SiO_2), to which the surface characteristics of the magnesium melt and liquid flux were studied. The interaction of the flux with non-metallic inclusions was determined by formula (1.5). The efficiency of inclusions removal during filtration was determined by the formula:

$$\mathrm{W}_f^{fl-f} = \mathrm{Aa} + \mathrm{A}_a^{fl-f} + \mathrm{A}_a^{m-f} \tag{1.6}$$

where W_f^{m-fl} is the joint energy of sorbent-inclusion bond in the system: solid filter (sorbent) - metallic melt (AZ91) - inclusions (flux), mJ/m^2;

Aa is the interfacial tension (adhesion) in the system metal (AZ91) - inclusion (flux), mJ/m^2;

A_a^{m-f} is the interfacial tension (adhesion) in the system inclusion (flux) - sorbent (filter), mJ/m^2;

A_a^{m-f} is the interfacial tension (adhesion) in the system metal (AZ91) - sorbent (filter), mJ/m^2.

The hardened drops of metal were cut in half and polished sections were made. The microstructure of the metal on the metal-substrate contact surface was studied by optical microscopy after etching with a 7% alcohol solution of nitric acid. The microhardness of the metal was determined on a "Buehler" microhardness tester with an indenter load of 10 g.

Metal research was carried out after heat treatment in a thermal mine furnace of the Bellevue type, with a capacity of 112 kW and a productivity of 95 kg/h, as well as in a thermal furnace of the PAP-4 M type, with a productivity of 50 kg/h, according to the mode: heating 540 ± 5°C, aging for 8 h, cooling in air and aging at 200 ± 5 °C, aging for 15 h, cooling in air.

The mechanical properties of magnesium alloy samples were tested on the "INSTRUN 2801" universal testing machine. The mechanical properties were determined both on standard magnesium alloy samples and on samples after their aging in artificial blood substitutes - gelofusin, venofundin, and physiological saline, for 1, 2, and 3 months at a temperature of 36 ± 1.0 °C using an ultra-thermostat UT- 15. The macro- and microstructure of the samples after mechanical tests was studied using the methods of light microscopy with the devices "Neophot 32" and "OLYMPUS IX 70". Phase analysis of the structural components of magnesium alloys was studied using an electron microscope - a microanalyzer with an energy dispersive microanalysis system "SELMI PEM-106Й".

1.4 Experimental Part

The efficiency of melt purification by filtration materials was compared by determining the characteristics of surface interaction in the systems: alloy AZ91 - filter, flux - filter, alloy AZ91 - inclusions, flux - inclusions. For this purpose, the method of "lying drop" was used, when the molten sample of metal or flux was placed on horizontal substrates of the materials under study. There were used substrates made of graphite, magnesite, and limestone. Substrates from SiO_2, Al_2O_3, and MgO were used as materials that simulated inclusions.

Analysis of the research results (Table 1.1) showed that surface tension ($\sigma_{p.g.}$) at a constant temperature at the alloy-gas interface in the systems AZ91 alloy - carbonate substrate and AZ91 alloy - oxide was at the level of 70…78 mJ/m^2. Flux VI-2 spread on carbonate materials and practically did not wet oxide materials, which caused different values of surface tension ($\sigma_{p.g.}$) on oxide and carbonate substrates (see Table 4.3). Adhesion (A_a) at the interface between flux and substrate of $CaCO_3$, $MgCO_3$ and graphite was 130 mJ/m^2, 126 mJ/m^2 and 125 mJ/m^2 respectively, which is 2…3 times higher than the adhesion at the interface between flux and oxide substrate.

In systems, AZ91 alloy - carbonate substrate and AZ91 alloy - oxide the smaller wetting angle and accordingly the greater work of adhesion was for $CaCO_3$ material.

Table 1.1 Characteristics of the surface interaction between the AZ91 alloy, VI-2 flux and the substrate material

Substrate material	$\sigma_{p.g.}$, mJ/m^2		θ, deg		A_a, mJ/m^2		$A_к$, mJ/m^2		K_p, mJ/m^2	
	AZ91	VI2	AZ91	VI2	AZ91	VI2	AZ91	VI2	AZ91	VI2
$CaCO_3$	78	77	127	47	31	130	156	154	– 125	– 24
$MgCO_3$	70	78	140	52	16	126	140	156	– 124	– 30
Graphite	72	75	154	48	7	125	144	150	– 137	– 25
SiO_2	69	101	132	132	23	33	138	202	– 115	– 169
Al_2O_3	74	99	137	143	20	19	148	198	– 128	– 180
MgO	70	100	148	137	11	27	140	200	– 129	– 173

Note The table shows average values

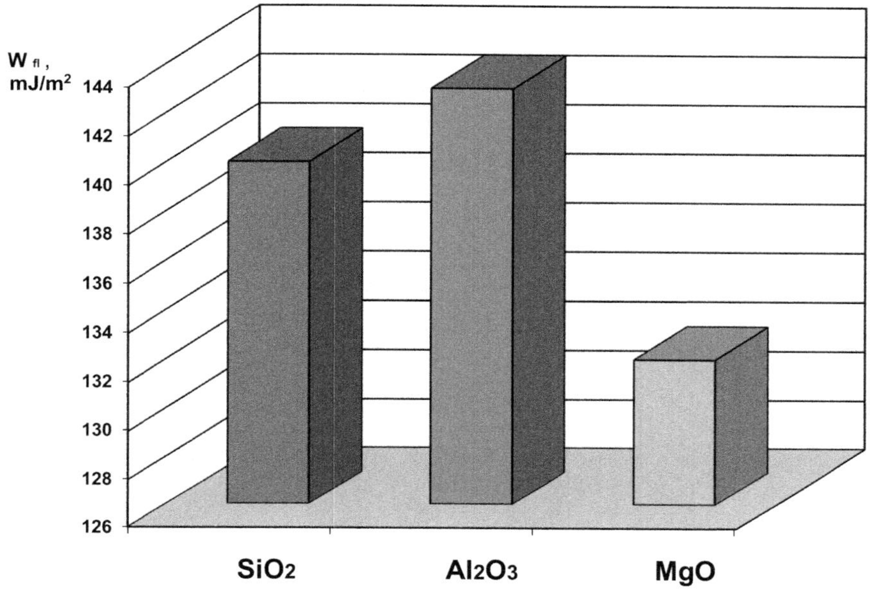

Fig. 1.3 Refining ability of flux to inclusions

The refining flux property is characterized by the work of adhesion to the alloy AZ91 (W_{fl}) (2.5). Interfacial tension (adhesion) at the boundary of alloy AZ91 - flux VI-2 is: $\sigma_{r.g.} = 71$ mJ/m^2; $\theta = 49°$. The adhesion of inclusions to metal in a flux medium was for SiO_2, Al_2O_3, and MgO, respectively 140 mJ/m^2, 143 mJ/m^{2}and 132 mJ/m^2. Considering that the refining property of flux (W_{fl}) increases with the decrease of work of adhesion we can conclude that flux VI-2 adsorbs these inclusions, but more effectively - magnesium oxides (Fig. 1.3).

Refining efficiency is characterized by the total energy of the sorbent-inclusion bond (Wϕ) in the system: solid filter (sorbent) - metal melt (AZ91) - inclusion (flux). Therefore, the completeness of inclusion removal in the melt decreased during filtration from limestone to magnesite and graphite - 279 mJ/m^2, 259 mJ/m^{2,} and 250 mJ/m^2, respectively (Fig. 1.4).

The depth of interaction between the alloy and the filter material was evaluated by metallographic analysis. It was found that the AZ91 alloy penetrated the CaCO3 filter material to a depth of 175 µm, which significantly exceeded the refining effect of graphite (6 mm) and magnesite (10 mm) (Fig. 1.5).

In the surface contact zone of the investigated droplets with the substrate material, an increased (compared to the rest of the droplet volume) amount of intermetallides was observed. It was found that the greater amount of intermetallides was detected in the surface zone of the drop at the contact with magnesite substrate. The size of intermetallides reached 25 microns, which is 3…4 times more than in drops at contact with limestone and graphite.

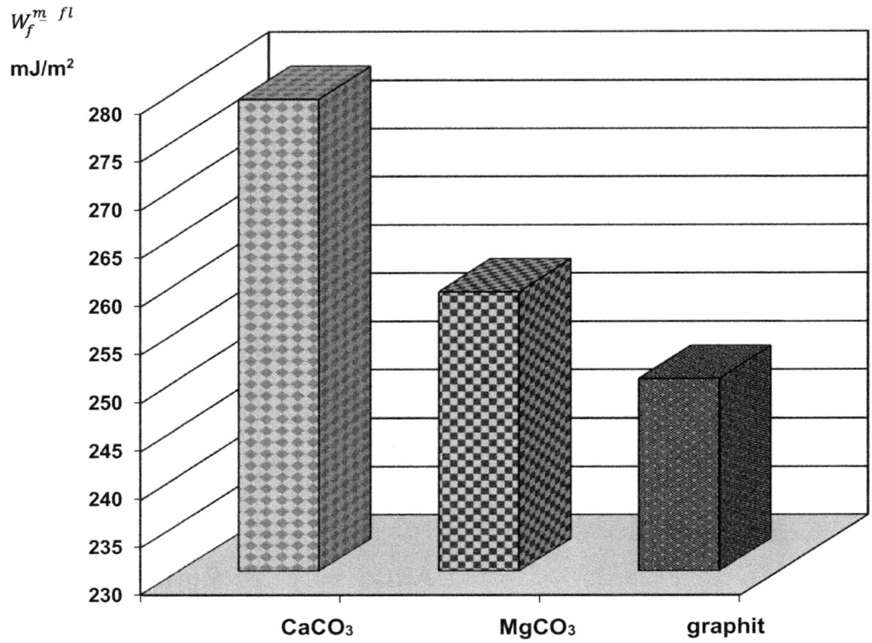

Fig. 1.4 Refining efficiency of filter materials

The micrograin size in the material of all investigated drops was at the same level (Table 1.2).

Microhardness δ -solid solution in the surface zones of drops of all variants was slightly higher than in the center (Table 1.3), higher values of microhardness of matrix, eutectic, and intermetallide were observed in the drop located on the magnesite substrate, which is most likely associated with the processes of diffusion of substrate elements into the metal melt.

Quantitative analysis of structural components in the studied samples from AZ91 alloy showed that both single intermetallides and their clusters formed during metal interaction with substrate materials. Their index and average size increased from magnesite to graphite and limestone (Table 1.4). The topography and morphology of inclusions changed after metal contact with substrate materials.

Filtration of the alloy through magnesite, limestone, and graphite promoted a noticeable refinement of the metal micrograin, especially when using a filter containing equal parts of magnesite, graphite, and limestone.

Mechanical tests showed that melt filtration enhanced the strength (σ_v) and plastic (δ) properties of the AZ91 alloy. Higher values of mechanical properties and density were obtained on the samples, which were fabricated using a complex filter of magnesite, graphite, and limestone in the ratio of 1:1:1 (Table 1.5).

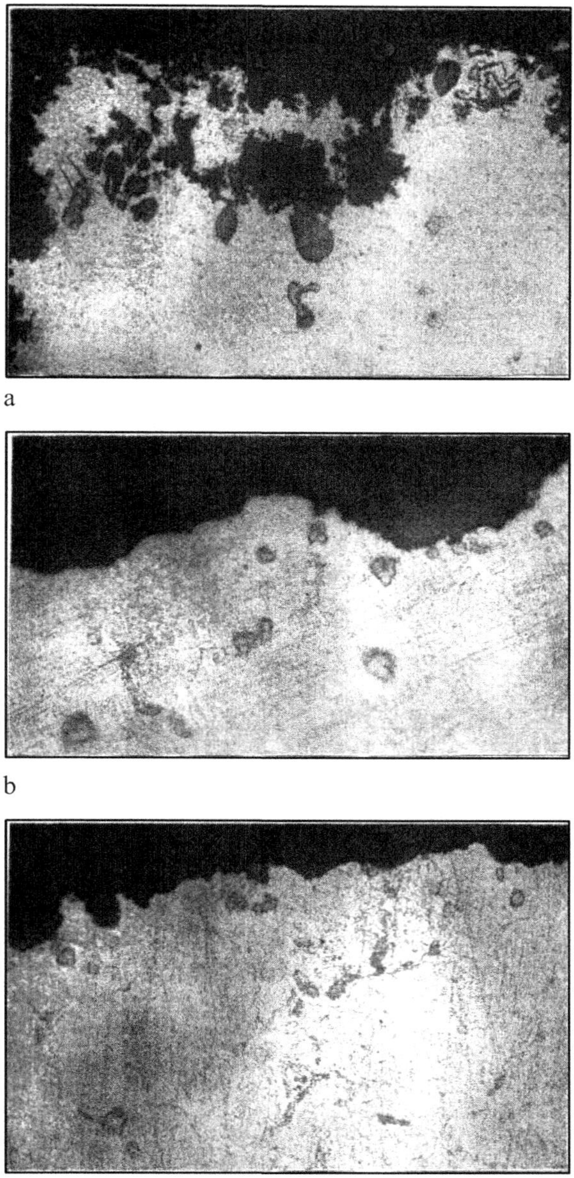

Fig. 1.5 Microstructure of the interaction boundary of AZ91 alloy with the filter, × 500: a—rom limestone; b—from graphite; c—from magnesite

Table 1.2 Structural components of AZ91 alloy droplets in contact with different substrates

Substrate material	Interaction depth, μm	Micro grain size, μm	Size of γ-phase, μm
Limestone	Until 175	80…150	2.0 … 6.0
Graphite	Until 10	80…175	3.0 … 8.0
Magnesite	Until 6	75…150	3.0 … 25.0

Table 1.3 Microhardness of alloy droplets AZ91

Substrate material	Microhardness, HV, MPa			
	Matrix		(δ + γ)-phase	γ - phase
	Border	Center		
Limestone	858.0…973.5	733.4…932.5	1188.4…1368.9	2825.8
Graphite	792.0…894.1	761.8…792.0	1225.8…1891.6	2825.8
Magnesite	894.1…1167.8	824.0…1017.3	1225.8…2288.9	2825.8…5150.0

Table 1.4 Quantitative assessment of inclusions in alloy test samples AZ91

Filter material	Single inclusions		Clusters of inclusions	
	Index, I	Average size d, μm	Index, I	Average size d, μm m
Magnesite	0,00,701	3,895	0,00,140	9,990
Graphite	0,00,907	4,491	0,00,125	8,910
Limestone	0,01,181	5,670	0,00,272	17,10

Table 1.5 Physico-mechanical properties of heat-treated AZ91 alloy samples after various filtration options (average values)

Filtering optionи	Physico-mechanical properties		
	σ_V, MPa	δ, %	Physical density, g/sm^3
Without filtration	234.0	3.0	1.6858
Magnesite	242.0	4.2	1.6980
Graphite	240.0	4.3	1.6753
Limestone	266.0	5.4	1.6876
Complex	275.0	6.2	1.7067

Thus, the best properties of AZ91 alloy are achieved when it is filtered through a complex filter containing limestone, magnesite, and graphite.

For preclinical testing, prototypes in the form of maleolar screws of various configurations were made of refined AZ91 alloy (Fig. 1.6). The macro- and microstructure of the developed alloy had a finely dispersed structure with a high degree of homogeneity (Fig. 1.7). The average grain size of the alloy was 57 μm. The grain boundaries were clean, and no eutectic emissions were observed along the grain boundaries. A large number of small spherical clusters of the secondary hardening phase were present, evenly distributed throughout the microstructure.

The tensile strength of the new alloy was 302…315 MPa, and the relative elongation was 7.2…8.5%, which fully meets the requirements.

Studies of the biodegradation rate (Table 1.6) showed that after 3 months of exposure to artificial blood substitutes, the developed alloy lost up to 43% of its

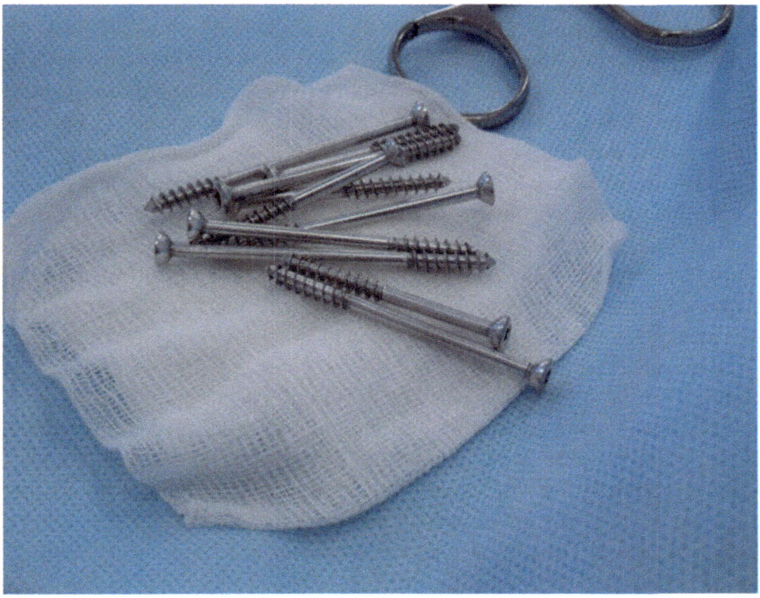

Fig. 1.6 Experimental maleolar screws for osteosynthesis made of AZ91 alloy

a b

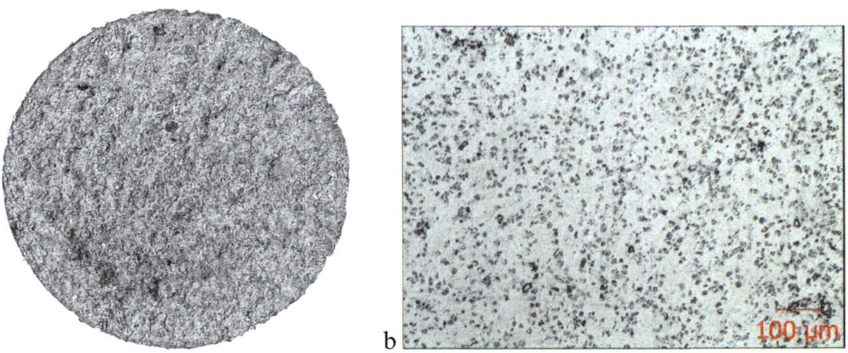

Fig. 1.7 Macro- (a, × 5) and microstructure (b, × 100) of the experimental AZ91 alloy

strength and up to 40% of its ductility. At the same time, the set of properties exceeded the corresponding properties of bone tissue, which ensured reliable fracture growth with subsequent biodegradation of the implant.

To implement the developed alloy into the medical industry, it is necessary to conduct a number of preclinical studies to identify the possible negative impact of its biodegradation products on bone tissue repair processes and the living organism as a whole.

The investigation of the toxic effects of biodegradation products of the experimental magnesium alloy AZ91 was carried out on rats using the "open field" test, which can detect minimal deviations in the animal body caused by toxic agents. As control parameters, the coefficients of horizontal and vertical motor activity, grooming (brushing of wool), examination of orifices, and defecation were used.

Registration of weight dynamics in the study by the "open field" method showed a decrease in this indicator by 7% during the first 2 weeks after surgery (Fig. 1.8). Subsequently, there was a significant weight gain and a significant improvement in the appetite of animals from the experimental group, which suggests that the initial decrease was the result of the impact of surgical trauma and pain at the site of intervention, rather than toxic effects. In addition, regular monitoring of the condition of the hair coat and mucous membranes did not reveal any abnormalities. Urine tests of the experimental animals also showed no signs of toxic effects of implant biodegradation products.

The study of the tentative experimental activity of animals in the "open field" (Fig. 1.9) showed a decrease in the total indicator by 60.6% on day 2 after surgery. On day 14, the index was 35.3% higher but did not exceed the index in the control group. After 1 month, the results did not differ significantly. The results from 2 to 6 months, when the greatest toxic effect of biodegradation products is expected, are especially indicative. Thus, the dynamics of motor and exploratory activity of the animals indicated the absence of toxic effects of the implant, and the decrease in the index in the early stages of the study was caused by the influence of surgical trauma.

The analysis of the indicators of tentative research activity of animals (Table 1.7) showed a positive trend for the animals of the experimental group, which is due

Table 1.6 Mechanical properties of the experimental AZ91 alloy after aging in different environments

Basys		1 month		2 month		3 month	
σ_B, MPa	δ, %	σ_B, MPa	δ, %	σ_B, MPa	δ, %	σ_B, MPa	δ, %
Gelofusin							
310	7.9	273	7.1	240	6.2	198	5.1
Venofundin							
315	7.2	269	6.3	236	5.2	182	4.3
Physiological solution							
302	8.5	284	7.9	261	7.2	230	6.1

Note average values

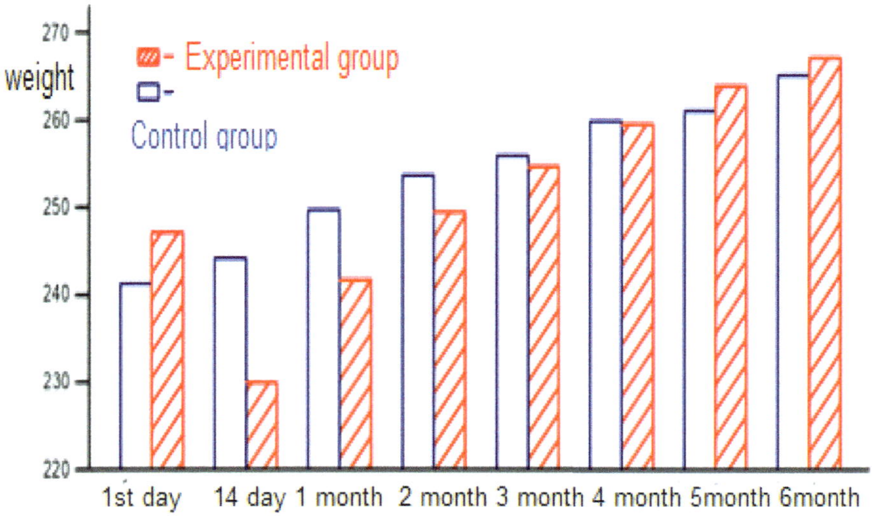

Fig. 1.8 Weight dynamics of animals of the experimental and control groups after surgery

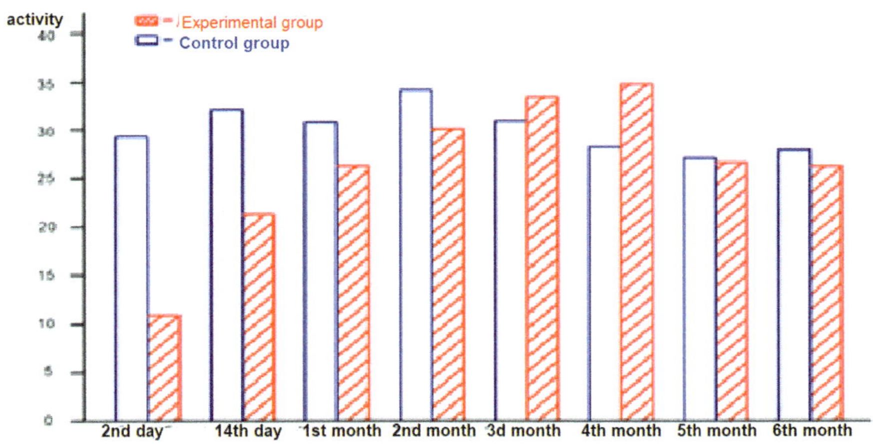

Fig. 1.9 Dynamics of total and motor and research activity of experimental animals

to a gradual increase in the motor and tentative research activity of the animals of the experimental group starting from 2 months after surgery. By almost all criteria, no significant differences were found between the control and experimental groups. The high level of general neurological status of animals confirms the absence of neurotoxicity of magnesium alloy biodegradation products.

The processes of bone regeneration with the usage of implants made of the developed alloy were investigated. For this purpose, fractures of the upper third of both

Table 1.7 Indicative experimental behavior and motor activity of animals in "open field" conditions

Sample	Horizontal motor activity	Vertical motor activity	Grooming	Inspection of holes	The number of acts of defecation
2nd day after surgery					
Experimental, N = 14	5.07 ± 0.32*	1.45 ± 0.22*	3.13 ± 0.33	2.1 ± 0.29*	3 ± 0.28
Control, N = 6	13.33 ± 0.71	3.27 ± 0.46	3.12 ± 0.33	5.5 ± 0.4	2.35 ± 0.31
14th day after surgery					
Experimental, N = 14	6.19 ± 0.48*	4.54 ± 0.31*	4.31 ± 0.38	4 ± 0.25*	2.74 ± 0.18
Control, N = 6	13.27 ± 0.92	5.33 ± 0.44	6.31 ± 0.48	5 ± 0.38	2.83 ± 0.3
1st month after surgery					
Experimental, N = 14	10.23 ± 0.58	5,51 ± 0.28	4 ± 0,29	4.24 ± 0.24*	2,9 ± 0,11
Control, N = 6	12 ± 0.55	5.54 ± 0.52	6,47 ± 0,51	6 ± 0.38	2,07 ± 0,3
2nd month after surgery					
Experimental, N = 14	13.8 ± 0.2	5.94 ± 0.48	5.1 ± 0.32	4.11 ± 0.21*	2.41 ± 0.98
Control, N = 6	14.21 ± 0.58	6.1 ± 0.28	5.76 ± 0.5	6.13 ± 0.48	2.23 ± 0.14
3rd month after surgery					
Experimental, N = 14	13.48 ± 0.52*	6.54 ± 0.28	5.56 ± 0.18	5.8 ± 0.87	2.10 ± 0.8
Control, N = 6	10.68 ± 0.78	8 ± 0.14	6 ± 0.98	4.14 ± 0.58	2 ± 0.4
4th month after surgery					
Experimental,N = 14	13.8 ± 0.47*	7.1 ± 0.48	5.04 ± 0,47	6.58 ± 0.68*	1.88 ± 0.25
Control, N = 6	10.4 ± 0.48	6 ± 0.85	5.14 ± 0.52	4.25 ± 0.46	1.79 ± 0.13
5th month after surgery					
Experimental, N = 14	11 ± 0.64	5.96 ± 0.14	4.25 ± 0.67	3.14 ± 0.64	3.64 ± 0.33
Control, N = 6	12.41 ± 0.57	5.64 ± 0.85	4.58 ± 0.34	4.47 ± 0.46	2.57 ± 0.14
6th month after surgery					
Experimental, N = 14	9.44 ± 0.24	5 ± 0.57	4.5 ± 0.13	3.54 ± 0.53	1.59 ± 0.34
Control, N = 6	10.48 ± 0.59	5.64 ± 0.57	4.94 ± 0.34	4.14 ± 0.67	1.47 ± 0.11

femurs were modeled in animals of the control and experimental groups. Animals in the experimental group were implanted with fixators made of magnesium alloy AZ91, and those in the control group were implanted with stainless steel 12X18H10T (Fig. 1.10).

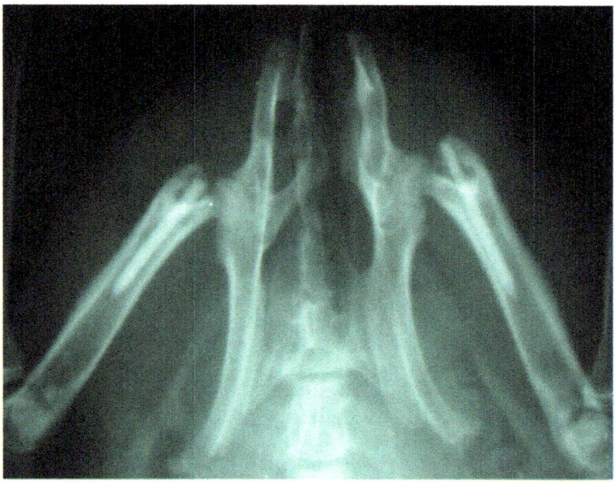

Fig. 1.10 Radiograph of a rabbit after osteosynthesis with implants made of the experimental magnesium alloy

Animals were withdrawn from the experiment at 2 weeks, 1, 4, and 6 months after surgery, after which the bone was sawn together with the implant through the fracture zone (Fig. 1.11).

The study showed that the use of implants made of a new magnesium alloy did not lead to a violation of the processes of growth of femoral fractures in animals. The fracture healing time in the control and experimental groups did not differ significantly, and no pathological changes occurred. The use of implants made of a new

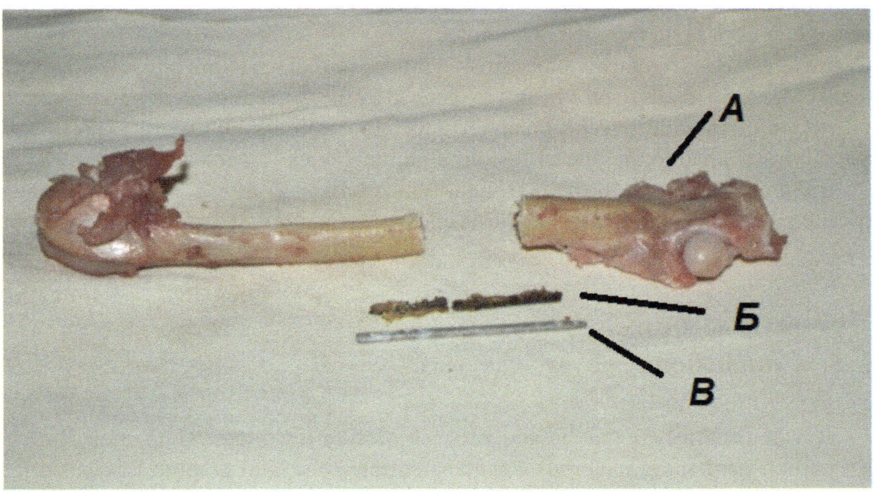

Fig. 1.11 A - femur of a rabbit; B—working implant; B—view of the implant before surgery Fi

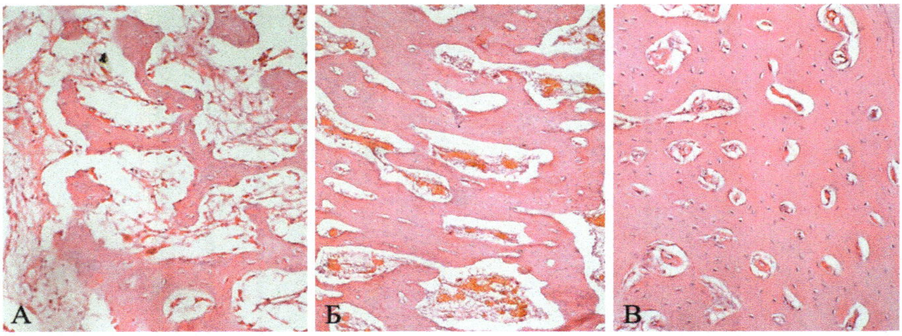

Fig. 1.12 Dynamics of the reparative process after an experimental injury with the use of implants made of a new magnesium alloy: a—2 weeks; b—1 month; in—4 months, × 100

magnesium alloy did not disrupt the formation of new blood vessels in the fracture areas. At all stages of observation, a widespread network of microvessels was found, the density of which increased in direct proportion to the intensity and duration of the regeneration process. For comparison, at these stages of treatment, implants made of stainless steel caused a disturbance in blood supply and death of bone cells, followed by their dissolution and displacement by connective tissue fibers.

Normal blood supply ensured a gradual and growing process of regenerative changes with minimal deviations from the normal course of reparative osteogenesis (Fig. 1.12).

At the late stages of treatment (6 months), the structure of the bone tissue at the site of implantation did not differ from the normal structure.

At the late stages of treatment (6 months), the structure of the bone tissue at the site of implantation did not differ from the normal structure. Only some disorderly arrangement of bone crossbars was noted.

It was established that the biocorrosion products of magnesium alloy do not have a toxic effect on body tissues, which is evidenced by the absence of signs of endogenous intoxication. In addition, the behavior and rate of biodissolution of experimental implants are equal to the rate of fusion of animal bones.

The preclinical tests of the high-quality bisoluble AZ alloy showed the promise of its use as implants for osteosynthesis and can be recommended for further clinical tests.

1.5 Conclusions

1. It was established that in the process of filtering magnesium melt through carbon-containing materials (magnesite, limestone, and graphite), it was effectively refined. The usage of a filter, which contains equal amounts of magnesite, graphite, and limestone when pouring a magnesium alloy provided a reduction

in the size of the structural components of the metal by 40…50%, increased its strength by ~20%, and plasticity–almost twice.

2. It was established that when the filter materials interact with the magnesium melt, it is effectively modified. At the same time, an increased amount of intermetallic phase is observed in the structure of the alloy, which provides an increase in the microhardness of the structural components of the alloy and contributes to an increase in its physical and mechanical characteristics.
3. The usage of carbon-containing materials for the modification and refining of the AZ91 alloy is quite effective as for improving the casting quality as for improving its physical and mechanical properties.
4. The aging of test samples in artificial blood substitutes showed that after 3 months, the AZ91 alloy has a level of mechanical properties that exceed the properties of bone tissue: $\sigma B = 182–230$ MPa, $\delta = 4.3–6.1\%$, and provides reliable fracture growth followed by complete biodegradation.
5. Research using the "open field" method showed the absence of any significant violations in the physiological manifestations of experimental animals, and the biocorrosion products of the new magnesium alloy do not cause endogenous intoxication of body tissues and do not increase cellular destruction. Also, no adverse effect on the general physical condition of the experimental animals and significant changes in their behavior were found. The results indicate the absence of toxic effects of biodegradation products of the new alloy on living organisms.
6. As a result of an experimental morphological study, it was established that the usage of magnesium alloy AZ91 for osteosynthesis of fractures did not disrupt the processes of reparative regeneration of bone tissue.
7. The results of preclinical tests showed that the AZ91 alloy does not have a toxic or carcinogenic effect on the living organism, has a positive effect on osteogenesis and reparative processes, and can be recommended for further clinical tests.

References

Borys J, Maciejczyk M, Antonowicz B, Sidun J, Świderska M, Zalewska A (2019) Free Radical Production, Inflammation and Apoptosis in Patients Treated With Titanium Mandibular Fixations—An Observational Study. Front Immunol 10:2662. https://doi.org/10.3389/fimmu.2019.02662

Cao H, Huang M, Wang C, Long S, Zha J, You G (2019) Research status and prospects of melt refining and purification technology of magnesium alloys. J Magnes Alloy 7:370–380. https://doi.org/10.1016/j.jma.2019.07.002

Chunmiao Y, Lifu Y, Chang L, Gang L, Shengjun Z (2013) Thermal analysis of magnesium reactions with nitrogen/oxygen gas mixtures. J Hazard Mater 260:707–714. https://doi.org/10.1016/j.jhazmat.2013.06.047

Esmaily M, Svensson JE, Fajardo S, Birbilis N, Frankel GS, Virtanen S, Arrabal R, Thomas S, Johansson LG (2017) Fundamentals and advances in magnesium alloy corrosion. Prog Mater Sci 89:92–193. https://doi.org/10.1016/j.pmatsci.2017.04.011

He M, Chen L, Yin M, Xu S, Liang Z (2023) Review on magnesium and magnesium-based alloys as biomaterials for bone immobilization. J Market Res 23:4396–4419. https://doi.org/10.1016/j.jmrt.2023.02.037

Jahnen-Dechent W, Ketteler M (2012) Magnesium basics. Clin Kidney J 5:i3–i14. https://doi.org/10.1093/ndtplus/sfr163

Jiang LY (2017) Metallurgy: importance, processes, and development status. Membrane-based separations in metallurgy, pp 3–18. https://doi.org/10.1016/b978-0-12-803410-1.00001-3

Kumar A, Kumar S, Mukhopadhyay NK (2018) Introduction to magnesium alloy processing technology and development of low-cost stir casting process for magnesium alloy and its composites. J Magnes Alloy 6:245–254. https://doi.org/10.1016/j.jma.2018.05.006

Long CM, Nascarella MA, Valberg PA (2013) Carbon black vs. black carbon and other airborne materials containing elemental carbon: Physical and chemical distinctions. Environ Pollut 181: 271–286. https://doi.org/10.1016/j.envpol.2013.06.009

Moshood TD, Nawanir G, Mahmud F, Mohamad F, Ahmad MH, AbdulGhani A (2022) Sustainability of biodegradable plastics: new problem or solution to solve the global plastic pollution? Curr Res Green Sustain Chem 5:100273. https://doi.org/10.1016/j.crgsc.2022.100273

Perera HCS, Gurunanthanan V, Singh A, Mantilaka MMMGPG, Das G, Arya S (2024) Magnesium oxide (MgO) nanoadsorbents in wastewater treatment: a comprehensive review. J Magnes Alloy. https://doi.org/10.1016/j.jma.2024.05.00

Prasad SVS, Prasad SB, Verma K, Mishra RK, Kumar V, Singh S (2022) The role and significance of magnesium in modern day research-a review. J Mag Alloy. 10:1–61. https://doi.org/10.1016/j.jma.2021.05.012

Prasad SVS, Prasad SB, Verma K, Mishra RK, Kumar V, Singh S (2021) The role and significance of magnesium in modern day research-a review. J Magnes Alloy 10. https://doi.org/10.1016/j.jma.2021.05.012

Raabe D, Ponge D, Uggowitzer P, Roscher M, Paolantonio M, Liu C, Antrekowitsch H, Kozeschnik E, Seidmann D, Gault B, De Geuser F, Dechamps A, Hutchinson C, Liu C, Li Z, Prangnell P, Robson J, Shanthraj P, Vakili S, Sinclair C (2022) Making sustainable aluminum by recycling scrap: the science of "dirty" alloys. Prog Mater Sci 128:100947. https://doi.org/10.1016/j.pmatsci.2022.100947

Shalomeev V, Tabunshchyk G, Greshta V, Nykiel M, Korniejenko K (2022a) Influence of alkaline earth metals on structure formation and magnesium alloy properties. Materials 15:4341–4341. https://doi.org/10.3390/ma15124341

Shalomeev V, Tabunshchyk G, Greshta V, Korniejenko K, Guigou MD, Parzych S (2022c) Casting welding from magnesium alloy using filler materials that contain scandium. Materials 15:4213–4213. https://doi.org/10.3390/ma15124213

Shalomeev V, Aikin N, Chorniy V, Naumik V (2019) Design and examination of the new biosoluble casting alloy of the system Mg–Zr–Nd for osteosynthesis. East-Eur J Enterp Technol 1:40–48. https://doi.org/10.15587/1729-4061.2019.157495

Shalomeev V, Greshta V, Liutova O, Bovkun S (2022) Devising resource-saving technologies for the production of casting from magnesium alloys using waste of metallurgical enterprises. East-Eur J Enterp Technol 3:6–12. https://doi.org/10.15587/1729-4061.2022.260190

Wagner R, Schmiedel A, Döring P, Storti E, Dudczig S, Henkel S, Aneziris CG, Biermann H, Weidner A (2024) Relation of nonmetallic inclusions to the cyclic properties of steel 42CrMo4 after steel melt cleaning by filtration and related processes. Mater Sci Eng, A 892:146070–146070. https://doi.org/10.1016/j.msea.2024.146070

Yadroitsava II, du Plessis A, Yadroitsev I (2019) Bone regeneration on implants of titanium alloys produced by laser powder bed fusion: a review, pp 197–233. https://doi.org/10.1016/b978-0-12-815820-3.00016-2

Zhang M, Guo B-H, Xu J (2016) A review on polymer crystallization theories. Crystals 7:4. https://doi.org/10.3390/cryst7010004

Chapter 2
Structural Optimization of Mechanical Elements Based on the Lattice Method Applying Additive Manufacturing with a Sustainable Approach

Juan Cruz-Castro, Yunuén López-Grijalba, Jesús Mares-Carreño, Griselda Stephany Abarca-Jiménez, José Luis Calderón-Osorno, Hugo Ernesto Gaspariano-Mendoza, and Luis Héctor Hernández-Gómez

Abstract Technological development and progress in various areas such as manufacturing and materials have resulted in various components becoming leaner, reducing material consumption, thus minimizing waste, and contributing to the environment. In addition, the manufacturing time of mechanical elements is reduced, reducing costs. By using the finite element method, a lattice structure optimization was performed in an original model. Three lattices were applied in the optimized model. This method uses optimization algorithms to generate geometries with an

---

J. Cruz-Castro · Y. López-Grijalba (✉) · J. Mares-Carreño · G. S. Abarca-Jiménez · J. L. Calderón-Osorno · H. E. Gaspariano-Mendoza
Instituto Politécnico Nacional, Unidad Profesional Interdisciplinaria de Ingeniería campus Hidalgo, carretera Pachuca-Actopan km 1+500, Distrito de Educación, Salud, Ciencia, Tecnología e Innovación, 42162 San Agustín Tlaxiaca, Hidalgo, Mexico
e-mail: ylopezg@ipn.mx

J. Cruz-Castro
e-mail: jucruzc@ipn.mx

J. Mares-Carreño
e-mail: jmaresc@ipn.mx

G. S. Abarca-Jiménez
e-mail: gabarcaj@ipn.mx

J. L. Calderón-Osorno
e-mail: jcalderono@ipn.mx

H. E. Gaspariano-Mendoza
e-mail: hgasparianom2000@alumno.ipn.mx

L. H. Hernández-Gómez
Instituto Politécnico Nacional, SEPI, ESIME Zacatenco, Unidad Profesional Adolfo López Mateos, Edificio 5, Segundo piso, Colonia Lindavista, Alcaldía Gustavo A. Madero 07738, Ciudad de México, Mexico
e-mail: luishector56@hotmail.com

A. Öchsner and H. Altenbach (eds.), *Engineering Design Applications VII*, Advanced Structured Materials 230, https://doi.org/10.1007/978-3-031-84346-4_2

optimal distribution of material, maximizing the strength and minimizing the weight of the component. From the lattices: regular cubic, double pyramid and double transversal pyramid a reduction between 52.63–60.53% of the mass was achieved with respect to the model with an infill of 100%. Experimental tests were performed on the 3D printed models to verify their mechanical strength compared to the spur gear model with the 100% filling. All models were subjected to the maximum torque and speed in the test bench, none of the spur gears showed any damage when inspected.

Keywords Spur gear · Finite element method · Lattice structure optimization method · Topology optimization

2.1 Introduction

Gears are mechanical elements that have been widely used to power transmission in diverse equipment, components or machinery. These are made of several materials in accordance with their applications. The most common materials used in manufacturing gears are carbon steel, alloy steel, iron and bronze. Gears can be different by their teeth geometry and named as: spur gear, helical gears, bevel gears and crown worm gears (Petrescu, et al. 2017).

Gears are robust elements due to the mechanical resistance that they must present to transmit the torque of the drive shaft. Failures in gears can occur principally for the following conditions: (1) bending fatigue, (2) contact fatigue, (3) wear and (4) scuffing (Errichello 2002).

Structural optimization of gears involves improving the design and material selection of these elements to improve different characteristics such as performance, durability and efficiency. One of the main objectives of the optimization process is to minimize component weight while maintaining or improving strength and reducing factors such as noise and vibration. Computational analyses based on numerical methods such as the finite element method (FEM) are often employed to simulate the behavior of gears under different loads and conditions to identify areas of improvement and refine the gear design accordingly (Tyflopoulos and Steinert 2022). In addition, advanced materials and manufacturing techniques such as additive manufacturing can be used to further improve the structural integrity and performance of the gears.

Optimization in areas such as engineering has had a wide application due to the importance of improving the design, making the use of materials more efficient. This is without compromising the structural integrity of the element. In industry, it is implemented in order to find solutions to obstacles such as dynamics and buckling problems, stress limitations, pressure loads, nonlinear geometric problems, compatible mechanisms, support and material design, etc.

A structural optimization problem is considered when certain conditions are established in the quest to improve structural design. According to Mares-Carreño

(Mares-Carreño, et al. 2021), the methods for structural optimization can be classified in:

1. Shape optimization. It aims to vary some contours of the model to be designed to improve its mechanical behavior.
2. Parametric optimization. It depends on the dimensions of the cross section.
3. Topology optimization. Used to achieve an acceptable initial design of the structure, subsequently refined with a shape optimization tool. It is usually the most used because it allows obtaining the desired configuration starting from an initial morphology.

The most used materials for plastic gears are polymers including nylon, polyurethane and phenolic laminates from injection molding (Landi et al. 2021). These have shown several significant advantages, for example: lower weight, lower noise during operation, low sliding friction, lower inertia, low displacement friction, chemical resistance and ability to operate in corrosive environments, ability to operate under vibration conditions, adaptation to higher tolerances, among others. However, it should be noted that plastic gears have lower mechanical properties and poor heat conduction compared to metal gears. These characteristics depend mainly on the manufacturing parameters and the material selected (Muminovic et al. 2022). This means that the failure mechanism of plastic gears is completely different from that of metal gears (Jain et al. 2019).

Furthermore, when structural optimization is implemented, it can be obtained complex shapes. In these cases, traditional manufacturing process cannot be applied, so additive manufacturing (AM) is used. Nowadays there is a large variety of equipment focused on AM. The more common are 3D printers based on fused deposition modeling (FDM) technology. This technology uses a plastic filament made of polylactic acid (PLA), polyethylene terephthalate glycol (PETG) or acrylonitrile butadiene styrene (ABS) (Bhatia and Sehgal 2023).

Several research projects on optimization the mechanical elements such as gears were reviewed. Kulangara (Kulangara et al. 2018) has implemented a structural optimization by honeycomb mesh in spur gears based on the application of the tangential force in the tooth. The optimization was made by applying a finite element study. A reduction of 19% of the volume in the spur gear was achieved, later a comparison between the optimized and original model was carried out. It was noticed a similitude in the strains and stresses. In (Malakova and Sivak 2023), Malakova used a topology optimization to improve gears by finite element techniques. It was considered in this case repeated loads in each tooth and a criteria of volume reduction in the body. Another goal was to preserve the stiffness of geometry. The results showed a mass reduction from 23.41% to 41.75%. Concli and Molinaro (Concli and Molinaro 2022) chose three mechanical components: a hip replacement, a die for plastic bottles and a car part. These were analyzed to review their resistance and other mechanical properties. By using FEM, it was achieved an optimization by applying a multi-object genetic algorithm. Body-centered cubic (BCC), body-centered cubic-z (BCCZ) and face-centered cubic lattice cell topologies were applied in each element. A reduction in the average mass was between 21–23% for each element.

In this paper, a topology optimization is applied to a spur gear to reduce the mass by implementing a lattice structure optimization method in the AM process. It proposed to use a spur gear with the presence of a hub where it is necessary for the implementation of a seat key in the transmission assembly.

2.2 Statement of the Problem

As was presented before, there are several advances in the process related to the mass reduction in spur gears. However, in all cases there has not been included spur gears with hub which is needed to comply with the assembly by using a key seat as a union with the shaft.

In most of the cases, the implementation of spur gears in a transmission system implies higher costs due to the use of more robust elements, the hub is an added mass that provides stiffness to the assembly and sufficient space to locate the seat key. When implementing lattice method in the optimization process, it is necessary to keep the stiffness necessary in the elements to assure their mechanical performance.

While a structural evaluation is performed in spur gears it is common to find high stresses reflected in located areas or like in these specific cases in the seat key, the tooth root and the contact zone between teeth. Spur gears are commonly used in a torsional condition that leads to shear stresses above all in the seat key. This leads to another important point to consider in the optimization process, to monitor these compromised areas.

2.3 Materials and Methods

In order to verify the methodology that is presented in this paper, it was considered to establish a standard geometry for a pair of spur gears. The geometries were in accordance with specifications of Martin® (Martin and Gears 2023), a commercial brand. These elements can be mounted in a test bench PTD® for the experimental tests. The bench test can reach a speed of 274.817 rpm and a torque of 12.956 Nm (114.6703 lbf.in). The pair of spur gears have a diametral pitch of 12 and a pressure angle of 14.5° with a reduction ratio of 12.68:1. For the assembly test, bearing housings self-lubricant and shafts of 15.875 mm (5/8 in) were implemented in the test bench (Fig. 2.1).

For the comparison in the experimental tests, original and optimized models were manufactured by AM using FDM in PLA material. The material properties of the PLA filament are shown in Table 2.1.

The methodology developed for the structural optimization of spur gears is shown in Fig. 2.2.

For the analyzes, it needed to calculate the torque applied in the test bench. This is transmitted by the reduced shaft to the drive gear from an input power of 372.85 W

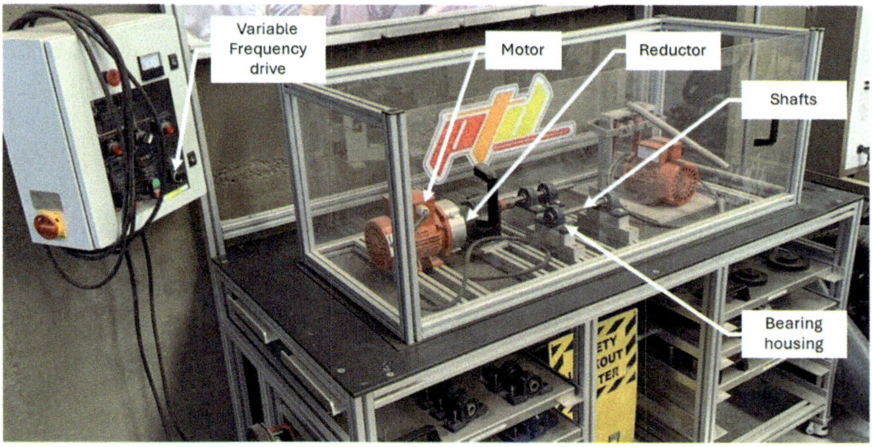

Fig. 2.1 PTD® test bench

Table 2.1 Mechanical and physical properties of PLA filament material

Property	PLA (Ansys® Academic Research Mechanical, Granta, Release R2022 xxxx)
Density (kg/m^3)	1255
Tensile strength, yield (MPa)	52.44
Tensile strength, ultimate (MPa)	62.93
Modulus of elasticity (GPa)	3.447
Poisson's ratio	0.3899

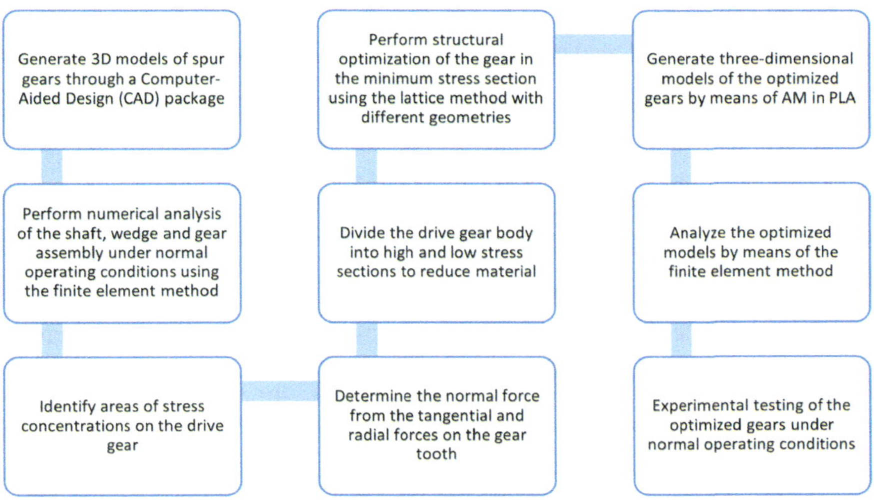

Fig. 2.2 Proposed methodology for the optimization of spur gear

(1/2 Hp) with an outlet speed of 28,779 rad/s. The torque is required to simplify the number of components used in the numerical analysis. It was determined a torque of 12.956 N.m (114.6703 lbf.in) by Eq. (2.1).

$$T = \frac{P}{\omega} \tag{2.1}$$

where T is the torque supplied by the motor to the shaft in N.m; P is the motor power in W; ω is the angular velocity of the shaft mounted on the reducer rad/s.

Likewise, the tangential and radial forces are determined (Mott 2018; Budynas and Nisbett 2011), from Eq. (2.2) and Eq. (2.3)

$$F_t = \frac{60000P}{\pi dn} \tag{2.2}$$

$$F_r = F_t tan\phi \tag{2.3}$$

where F_t is the tangential force in N; P is the motor power in kW; d is pitch diameter in [mm]; n is rotational speed in [rpm]; F_r is radial force in [N]; ϕ is the pressure angle [°].

From Eq. (2.2) and Eq. (2.3), it was determined a tangential, radial and total force of 611.959 N (137.5739 lbf), 158.263 N (35.57894 lbf) y 632.093 N (142.1002 lbf), respectively.

2.3.1 Numerical Analysis

For the first numerical analysis were considered in the assembly the pair of spur gears, the key seat and a shaft section (Fig. 2.3). The dimensions correspond to the elements used in the test bench. For the spur gears a geometry of 30 teeth for the driven gear and 20 teeth for the pinion were considered. The material assigned is PLA with the properties mentioned in Table 2.1. The assembly shown in Fig. 2.3 was performed to locate the compromised areas generated by the operational conditions of the test bench. In order to simplify the study, a two-dimensional analysis was performed (Fig. 2.4).

The components were discretized with higher order 2D elements. A total of nodes and elements were 69 854 and 26 660, respectively (Fig. 4a). Nonlinear contacts of frictionless type were considered between each element of the system. Besides, boundary and load conditions were established (Fig. 4b) in the assembly, a fixed support in the gear bore and a torque at the shaft.

Under these considerations, the stress field in the pinion spur gear was obtained. It identified the zones with minimum stresses where a reduction of mass can be applied by using the lattice structure optimization method. Figure 2.5 shows the total deformation and equivalent von Mises stress.

Fig. 2.3 3D Model assembly of the spur gears for the first analysis

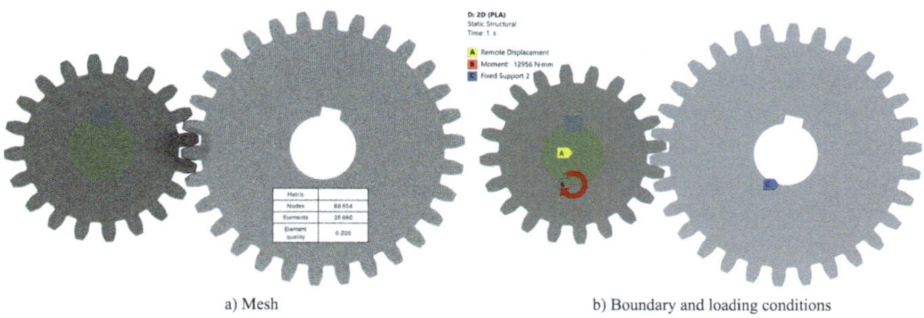

Fig. 2.4 2D model for structural analysis

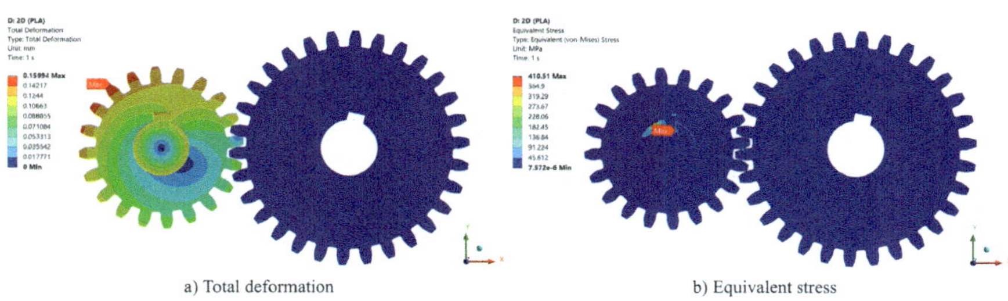

Fig. 2.5 Deformation and stress in gear assembly

In Fig. 2.6 the stress field is observed in the spur gear made of PLA. In this step, it was identified through circles in the middle zone with minimum stresses. This zone is the objective for the reduction of mass and in it, the lattice method will be applied.

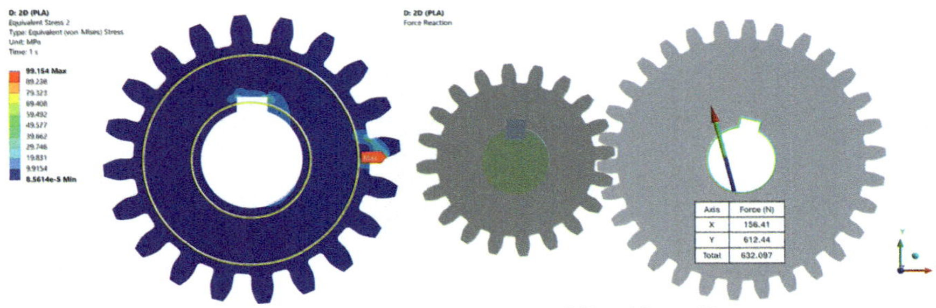

a) Equivalent stress in the PLA spur gear   b) Normal force and its components

Fig. 2.6 Normal force and stress in the PLA spur gear pinion

From these, the normal force and its components (radial and tangential forces) can be determined on the tooth of the spur gear pinion. Figure 6b shows these values.

The 3D model was divided in three sections, la interior part where the key seat can be found united to the shaft, the middle section and the external section where the gear tooth crown is situated (Fig. 2.7).

From the model shown in Fig. 2.7, lattice structure optimization was made. The boundary and loading conditions were the force applied on the spur gear tooth, a fixed support in the shaft and seat key (Fig. 8a). Then, it was identified the exclusion zone as the internal and external faces (Fig. 8b). The objective of the optimization process was maximizing the stiffness through a response type compliance. It was applied a response constraint of mass. Figures 8c, d show the lattice density tracker and the resultant optimized with a cubic lattice, respectively.

Diverse types of lattices were generated by SpaceClaim ANSYS® (Ansys® Academic Research Mechanical, Granta, Release R2022 xxxx) in the middle zone to verify the mechanical behavior of the spur gears. The lattices implemented were

Fig. 2.7 Sections in the PLA spur gear pinion

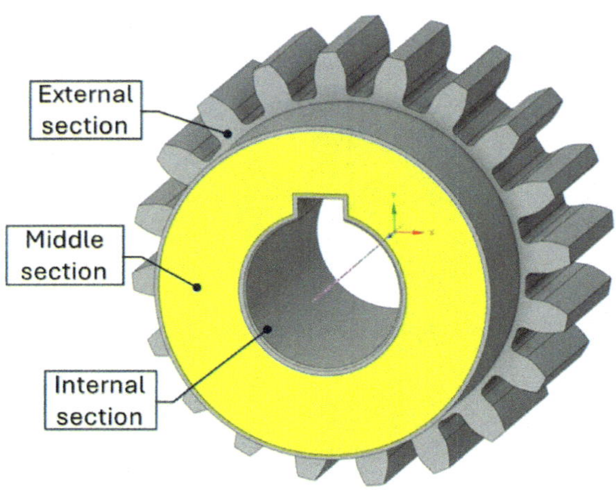

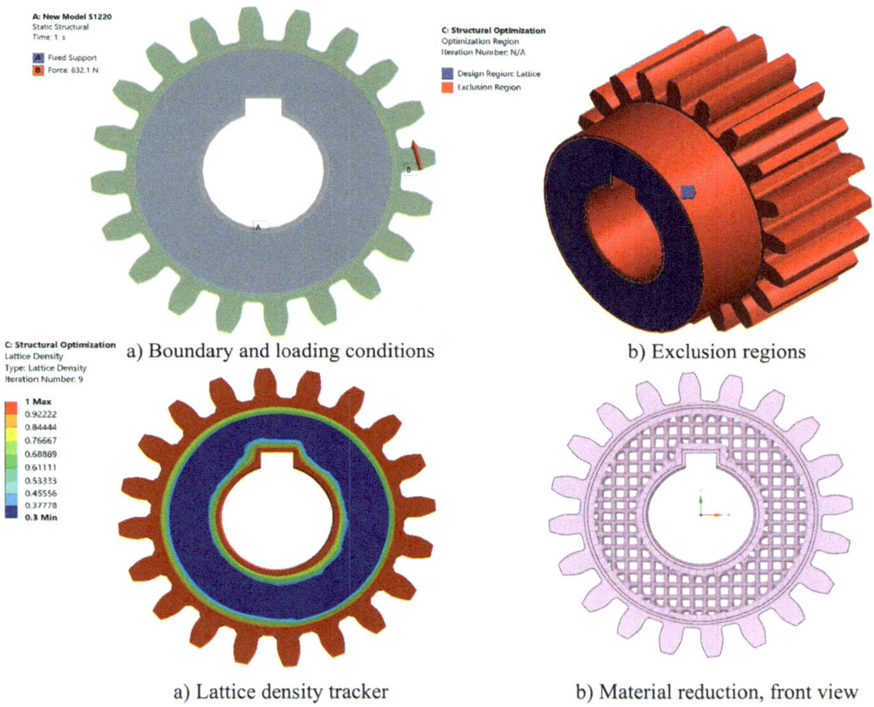

Fig. 2.8 Lattice type optimization of the spur gear pinion

(a) regular cubic, (b) double pyramid and (c) double transversal pyramid. Figure 2.9 shows the spur gears with the lattices proposed and their schematic representation.

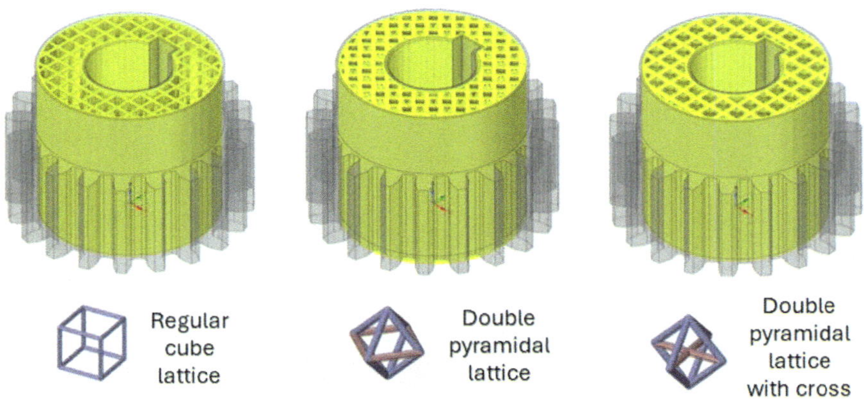

Fig. 2.9 Digital models of the optimized gear

2.3.2 *Experimental Analysis*

From the optimization, three spur gears were made by AM using FDM in PLA material. In Table 2.2 can be observed the printing models in white and the original steel model. It presents the mass for each model. It can be noticed that there is an important reduction in the mass against the original model made in steel.

The PLA printed gears were mounted on the PTD® test bench. These have been subjected to various tests by varying the speed from zero to the maximum speed delivered by the gearbox (274.817 rpm). Speed measure of the drive spur gear was made by using a digital photo reflector tachometer. Each speed was maintained for about 20 s. Figure 2.10 shows the speed increase in the drive shaft.

Likewise, the metal gears were interchanged, varying their size. In order to subject the printed gears to different conditions. This is shown in Fig. 2.11.

Table 2.2 Specifications of the original spur gear versus the proposal 3D printing models

Component	Mass (g)	Material	Image	Component	Mass (g)	Material	Image
Spur gear 30 teeth	722.0	Steel	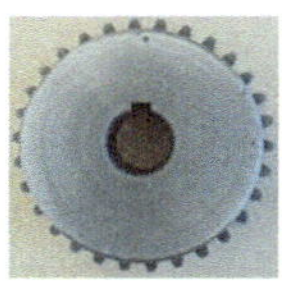	Regular cube lattice	15.0	PLA white	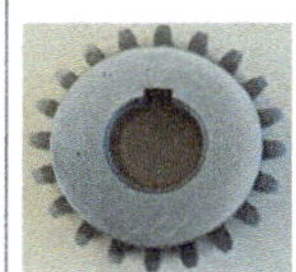
Spur gear pinion 20 teeth	247.0	Steel		Double pyramidal lattice	18.0	PLA white	
Spur gear (w/100% filling)	38.0	PLA black		Double pyramidal lattice with cross	18.0	PLA white	

Fig. 2.10 Speed variation applied in the test bench

a) *Drive spur gear 30 teeth*

b) *Drive spur gear 40 teeth*

c) *Drive spur gear 48 teeth*

d) *Drive spur gear 64 teeth*

Fig. 2.11 Tests on optimized PLA spur gear

For each test favorable results were obtained in the different spur gears considering the different types of lattices. The only element that presents a failure during the test was the double pyramidal lattice. However, it was observed that inadequate assembly and deficiencies in the material were present in the printed element. Figure 2.12 shows the failure of the failure in the spur gear.

2.4 Results

The models optimized were analyzed by simulation considering a static analysis with the conditions established in part 3 of this paper. The normal load applied to the tooth of each model was determined from the first analysis where the complete assembly was used, this load was validated analytically.

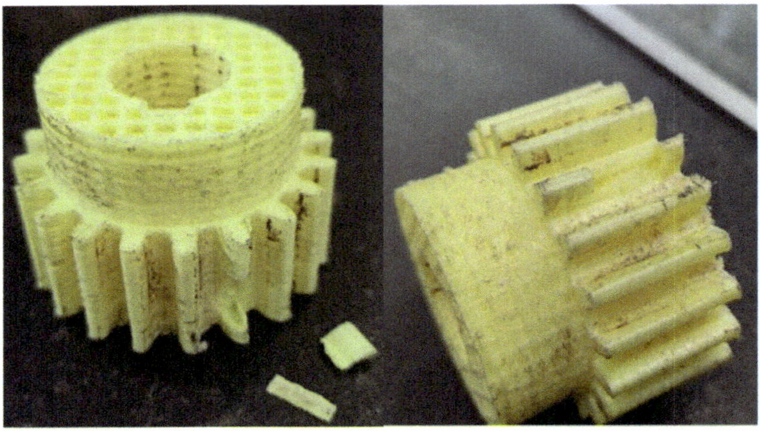

Fig. 2.12 Optimized spur gear failure

Figure 2.13 shows the corresponding results from the printed spur gear pinion with infill at 100%, total deformation and equivalent von Mises stress were obtained for the comparison with the optimized designs.

According to the ANSYS Mechanical user guide (Ansys® Academic Research Mechanical: Structural Optimization Analysis. ANSYS Mechanical user guide 2022), a homogeneous model is used for lattice validation in order to reduce the solution time. So, the steps to create a downstream validation are slightly different than for the other optimization methods available in ANSYS®. In Fig. 2.14 the results of the total deformation and equivalent von Mises stresses can be observed for each optimized spur gear pinion applying the different lattice geometries proposed.

It is observed the increase in the values of stresses in the optimized models. Also, it is noted a growth in the distribution of the area deformed around the contact tooth that implies a reduction in the stiffness of the spur gear. The highest values,

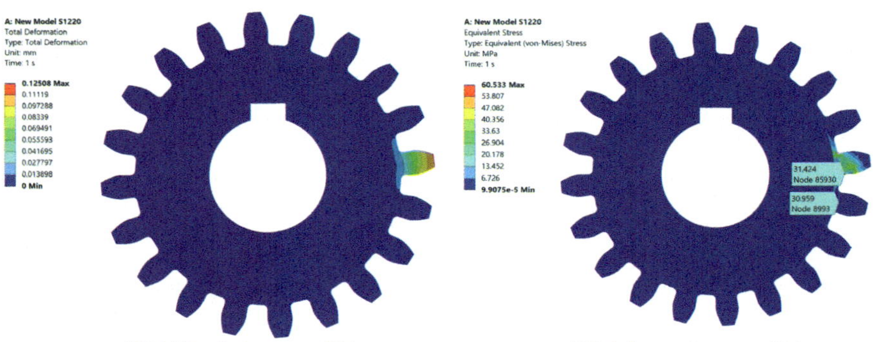

a) Total deformation in spur gear of PLA b) Equivalent stress in spur gear of PLA

Fig. 2.13 Results of the PLA gear under normal force

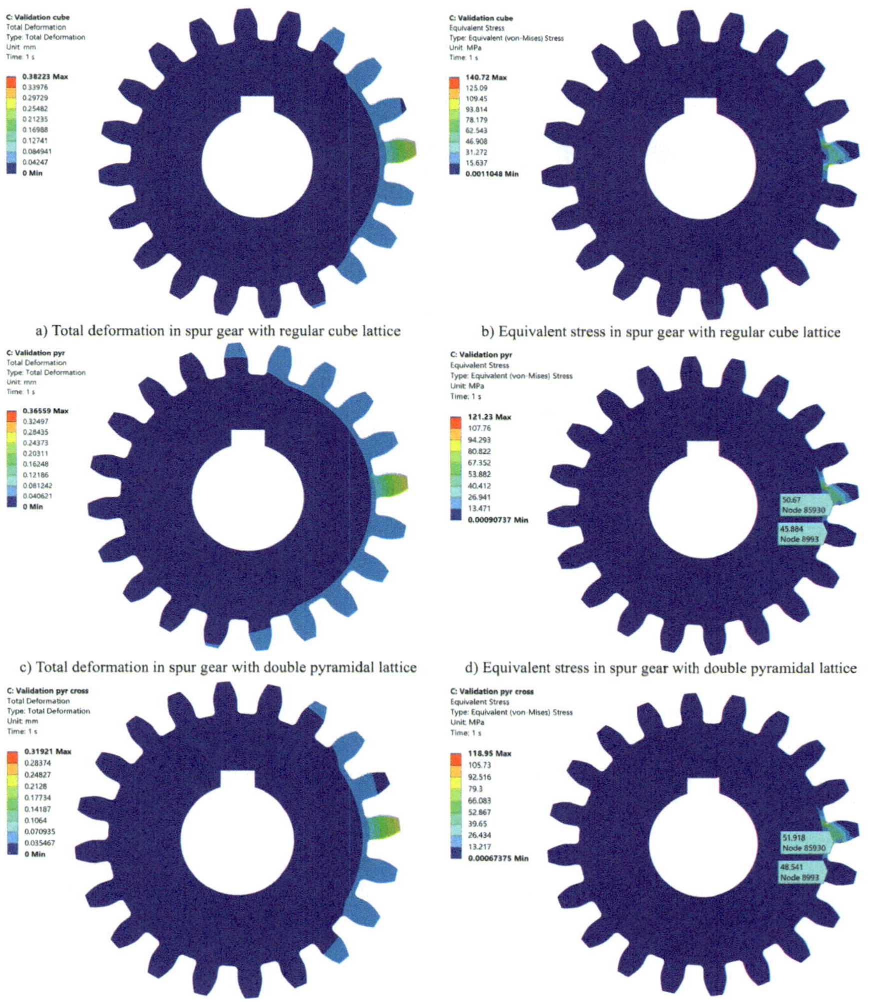

a) Total deformation in spur gear with regular cube lattice

b) Equivalent stress in spur gear with regular cube lattice

c) Total deformation in spur gear with double pyramidal lattice

d) Equivalent stress in spur gear with double pyramidal lattice

e) Total deformation in spur gear with double pyramidal lattice with cross

f) Equivalent stress in spur gear with double pyramidal lattice with cross

Fig. 2.14 Results of the optimized PLA spur gear under normal force

as expected, are located at the fillet radius near the root diameter of the spur gear (stress concentration). The points where stress concentration is located can be studied by considering fatigue loads and projecting their remaining life. According to the experimental tests, none of the specimens presented visual damage when they were subjected to variation in speed or torque. The printed models fulfill structural and mechanical with its purpose in the test bench.

2.5 Conclusion

Spur gears that present a hub in their design, including a seat key for their assembly, can be optimized by reducing their mass without compromising their structural integrity. The lattice structure optimization method has proved to be an alternative in the application of the manufacturing process. These techniques entail to less material consumption, reduction of manufacturing times, etc.

A lattice structure optimization method was applied to a spur gear made by PLA. It was considered standard and commercial dimensions applicable to a test bench for transmission systems. A reduction in mass was established as an objective of the optimization process. The optimization was focused on a spur gear pinion with a hub, it was of interest since it presents several zones with stress concentrations due to the presence of a seat key. Since the hub is a mass added to the spur gear it has the potential to be optimized by reducing its mass.

By using the finite element method, a lattice structure optimization was performed in an original model. Three lattices were applied in the optimized model. From the lattices: regular cubic, double pyramid and double transversal pyramid a reduction between 52.63 and 60.53% of the mass was achieved with respect to the model with an infill of 100%.

In accordance with the numerical results changes in the stiffness of the models are affected, it was deduced from observing the increased distribution area of deformations near the contact tooth. However, when experimental tests were performed it was noticed an appropriate structural behavior in the printed optimized spur gears and no damage was registered.

It is also noted that a better reduction of mass was achieved in the double pyramidal and double pyramidal with cross lattices than the regular cubic lattice. In terms of mechanical performance, it was observed that applying the double pyramidal with cross conduce to a reduction in the stress compared to regular cubic and double pyramidal lattices. It can be concluded that infill percentage and type lattice can improve the mechanical properties. It is necessary to establish the load operational conditions at which the element of interest will be subject.

Both numerical and experimental studies have only been conducted for a static problem. However, fatigue tests on these optimized elements are required in order to better understand the behavior of polymers such as PLA.

Acknowledgements The grant for the development of the Projects SIP20242357 and SIP20242641 was awarded by the National Polytechnic Institute (IPN). It is kindly acknowledged.

References

Ansys® Academic Research Mechanical, Granta, Release R2022

Ansys® Academic research mechanical: structural optimization analysis. ANSYS mechanical user guide, release 2022R1 (2022)

Bhatia A, Sehgal AK (2023) Additive manufacturing materials, methods and applications: a review. Mater Tod: Proceed 81(2):1060–1067. https://doi.org/10.1016/j.matpr.2021.04.379

Budynas RG, Nisbett JKS (2011) Shigley´s mechanical engineering design, Mc Graw Hill, USA, pp 673–720

Concli F, Molinaro M (2022) Design for additive manufacturing–material characterization and geometrical optimization. Int J Comput Methods Exp Meas 10(2):146–157. https://doi.org/10.2495/CMEM-V10-N2-146-157

Errichello R, Muller J (2002) How to analyze gear failures. J Fail Anal Prev 2(6):8–16. https://doi.org/10.1007/BF02715492

Jain M, Patil S, Ghosh SS (2019) A review on failure characteristics of polymeric gears. AIP Conf Proc 2148:1. https://doi.org/10.1063/1.5123979

Kulangara AJ et al (2018) Generation and optimization of lattice structure on a spur gear. Mater Tod: Proceed 5(2):5068–5073. https://doi.org/10.1016/j.matpr.2017.12.085

Landi L et al (2021) Analytical procedure for the optimization of plastic gear tooth root. Mech Mach Theory 166:104496. https://doi.org/10.1016/j.mechmachtheory.2021.104496

Martin Sprocket & Gears, Inc.: Sprocket (2023). https://es.martinsprocket.com/documents/catalogs/engineering. Accessed 01 Aug 2023

Malakova S, Sivak S (2023) Design optimization of gear wheel bodies in order to reduce weight, MM Sci J 6453–6459. https://doi.org/10.17973/MMSJ.2023_06_2023010

Mares-Carreño J, et al (2021) La optimización estructural y sus aplicaciones (The structural optimization and its applications), CyT UPIITA-IPN, 82, 1–19. https://www.boletin.upiita.ipn.mx/index.php/ciencia/916-cyt-numero-82/1889-la-optimizacion-estructural-y-sus-aplicaciones. Accessed 25 May 2024

Mott RL (2018) Spur gear design in machine elements in mechanical design. Pearson, USA, pp 304–423

Muminovic AJ et al (2022) Failure analysis of nylon gears made by additive manufacturing. Eng Fail Anal 137:106272. https://doi.org/10.1016/j.engfailanal.2022.106272

Petrescu RV, et al (2017) Gears-part I, Analytical procedure for the optimization of plastic gear tooth root. Am J Eng Appl Sci 10(2):457–472. https://doi.org/ https://doi.org/10.3844/ajeassp.2017.457.472

Tyflopoulos E, Steinert M (2022) A comparative study of the application of different commercial software for topology optimization. Appl Sci 12(2): 611. https://doi.org/10.3390/app12020611

Chapter 3
Study of the Effects on Shear Behavior Due to the Variation in Printing Parameters for 3D Printed Parts

Jesús Mares-Carreño, Griselda Stephany Abarca-Jiménez, Juan Cruz-Castro, Yunuén López-Grijalba, Manuel Vladimir Vega-Blanco, Eli Ignacio-Juárez, and Juan Alfonso Beltrán-Fernández

Abstract Additive manufacturing, also known as 3D printing, has experienced a surge in popularity in recent years. Currently, the additive manufacturing market is valued at approximately twenty billion dollars, with a projected annual growth rate of twenty percent for the period between 2023 and 2030. This technology is fundamentally altering the approach of product designers towards prototyping, and increasingly, 3D printed components are being integrated as functional parts in consumer products. Given the significance of additive manufacturing, it becomes imperative to characterize the mechanical behavior of 3D printed parts to ensure they possess the required rigidity and resistance for their intended applications. However,

J. Mares-Carreño · G. S. Abarca-Jiménez (✉) · J. Cruz-Castro · Y. López-Grijalba · M. V. Vega-Blanco · E. Ignacio-Juárez
Instituto Politécnico Nacional, Unidad Profesional Interdisciplinaria de Ingeniería campus Hidalgo, Carretera Pachuca-Actopan km 1+500, Distrito de Educación, Salud, Ciencia, Tecnología e Innovación, 42162, San Agustín Tlaxiaca, Hidalgo, Mexico
e-mail: gabarcaj@ipn.mx

J. Mares-Carreño
e-mail: jmaresc@ipn.mx

J. Cruz-Castro
e-mail: jucruzc@ipn.mx

Y. López-Grijalba
e-mail: ylopezg@ipn.mx

M. V. Vega-Blanco
e-mail: mvvegab@ipn.mx

E. Ignacio-Juárez
e-mail: eignacioj2000@alumno.ipn.mx

J. A. Beltrán-Fernández
Instituto Politécnico Nacional, SEPI, ESIME Zacatenco, Unidad Profesional Adolfo López Mateos, Alcaldía Gustavo A. Madero, Edificio 5, segundo piso, colonia Lindavista, 07738, Ciudad de México, Mexico
e-mail: jbeltranf.ipn@gmail.com

A. Öchsner and H. Altenbach (eds.), *Engineering Design Applications VII*, Advanced Structured Materials 230, https://doi.org/10.1007/978-3-031-84346-4_3

recent efforts in characterizing the mechanical behavior of 3D printed parts have primarily focused on tension behavior, neglecting the crucial aspect of shear stress behavior, which is highly relevant in various applications such as axles, pins, and rods. This study investigates the effects of variations in 3D printing parameters on shear behavior through experimental torsional testing. Test specimens with a diameter of 9 mm, according to ASTM E143-02 standards, were subjected to testing, varying parameters such as infill pattern, infill percentage, and printing orientation. Results indicate that printing orientation has the most significant influence on shear behavior during torsional testing, while infill percentage and infill pattern exhibit minor effects. The findings highlight the significant influence of printing parameters on the mechanical behavior of 3D printed parts. Further experimentation, including tests involving variations in parameters such as layer height, printing speed, and material composition, is essential to comprehensively understand their impact on mechanical behavior.

Keywords Additive manufacturing · Infill pattern · Infill percentage · Printing orientation · Torsion test

3.1 Introduction

Additive manufacturing (AM) has had an upgrowing use in recent years, especially the 3D printing of polymers, this is due to the ease of access and relatively low cost of the hardware and software necessary to the process in comparison to other manufacturing processes, along with the increased capability of reproduced complex geometries. This technology is fundamentally altering the approach of product designers towards prototyping, and increasingly, 3D printed components are being integrated as functional parts in consumer products.

In this context, investigation and technical development in AM has also increased, as shown in Kumar and Sathiya (2021); Jemghili et al. 2021; Kunovjanek et al. 2022; Ponis et al. 2021), being the most popular research topics the "Additive manufacture methods" and "Recycling and reuse" (Ponis et al. 2021) where the most studied additive manufacturing methods are the powder bed fusion techniques (Jemghili et al. 2021). Another relevant topic which has been under investigation is the further integration of AM in supply chains, with focus in product design and sustainability (Kunovjanek et al. 2022). While the clear tendency is to integrate more and more the AM to the production of consumer goods, there is a lack of test standards and structured methods for design and materials selection in AM (Kumar and Sathiya 2021), for these methods and standards to be set, the mechanical behavior of the 3D printed parts needs to be studied in more depth, as there are several printing parameters that affect the final performance of the printed part.

Several studies involving the evaluation of the mechanical performance of 3D printed parts have been conducted, some are mainly focused on the performance related to the materials (Sepahi et al. 2021; Rodríguez et al. 2001; Palanisamy et al. 1286; Lakshman Sri et al. 2024), while others are focused on the effect of the printing parameters over the mechanical properties (Chacón et al. 2017; Balderrama-Armendariz et al. 2018; Khosravani and Reinicke 2020; Gonabadi et al. 2022; Ganeshkumar et al. 2022), and some others are proposed from an optimization approach (Kadhum et al. 2023; Valvez et al. 2022). While all these studies contribute to a better understanding of the relation between printing parameters and mechanical performance, most of them are focused on tension behavior (as shown in Gao et al. (2022)), but as 3D printing application has broaden a deeper characterization of the mechanical behavior of the printed parts is necessary. Given the above, this work is focused on the characterization of the shear behavior of the 3D printed parts, by means of torsion tests according to ASTM E143-02 standard, and the effects of the variation of the printing parameters, such as infill pattern, infill percentage, and printing orientation, over the shear performance of 3D printed PLA (polylactic acid) test specimens.

3.2 Statement of the Problem

The problem to be addressed is that of the characterization of the shear behavior of the 3D printed PLA parts, and how this is affected by the variation of the printing parameters such as the infill pattern, infill percentage, and printing orientation. To have consistent results, all tests must be performed under similar conditions and following a standard test guideline. Furthermore, test specimens also need to be manufactured in accordance with some standards and under similar conditions.

Once the tests have been carried out, data must be processed to identify the variations in the shear behavior of the test specimens and correlate them with the variations of the printing parameters, to establish relations between the printing parameters and the performance of a 3D printed part subject to shear stress.

3.3 Materials and Methods

For the development of this paper, it was proposed to study the change of the mechanical behavior in shear, of 3D printed PLA parts due to the variation of the manufacturing process, namely printing parameters. To fulfill this purpose, the following methodology was established:

(1) The geometrical characteristics of the test specimens and the method for the experimental torsion tests were defined in accordance with ASTM E143-02 (ASTM 2012).
(2) Varying printing parameters for 3D printed PLA test specimens were established, being the infill pattern, infill percentage, and printing orientation (see Table 3.1).

Table 3.1 Varying printing parameters for 3D printed PLA test specimens

Infilling pattern	Infilling percentage (%)	Printing orientation
Tri-hexagonal	25	Horizontal
Cubic	50	Vertical
Gyroid	75	45 degrees

(3) FDM (fused deposition material) was selected as the manufacturing process for all the test specimens.
(4) An inspection previous to the tests was realized for each test specimen.
(5) Experimental torsion tests were carried out in accordance with ASTM E-143–02 (ASTM 2012).
(6) A comparison was performed using the collected data.

Tests specimens were printed using the FDM process with a 1.75 mm white Creality Hyper PLA filament under the considerations showed in Table 3.1. The printing pattern can be observed in Fig. 3.1 in accordance with the orientation established in Table 3.1. A printing velocity of 300 mm/s for the infill and 200 mm/s for the outer wall, with a layer height of 0.2 mm were established for all specimens. A total of 27 specimens was used for the test.

The dimensions of the test specimens are shown in Fig. 3.2. The ends of the test specimens were made in hexagonal shape due to the housings of the machine test, driver connection, and gauge tree, respectively.

The equipment used in the tests was a Gunt® Hamburg WP510 torsion test machine with a 200 N.m (1770.15 lb.in) motor drive serial number 229829. It allows the acquisition of data as torque and angle twist. It performs four velocities of deformation: 50 rpm/min, 100 rpm/min, 200 rpm/min, and 500 rpm/min. The measures of torque are performed through tensile gauges with a range of 0 to 199.9 N.m (1769.26 lb.in) and indication resolution of 0.1 N.m (0.88 lb.in). Meanwhile, the absolute angle twist is measured with an optoelectronic sensor with a range of 0 to

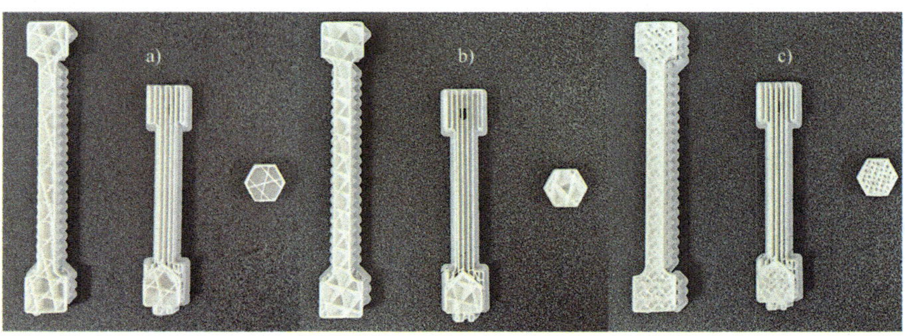

Fig. 3.1 Infill pattern with varying orientation form left to right, horizontal, 45 degrees and vertical. **a** Cubic pattern **b** tri-hexagonal pattern **c** gyroid pattern

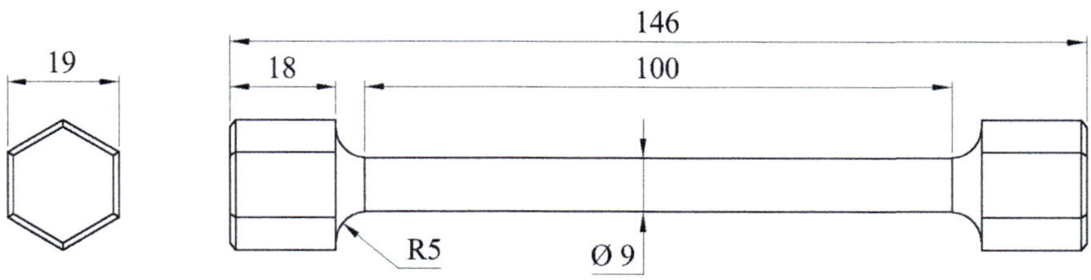

Fig. 3.2 Dimensions in millimeters, of the test specimens used in the experimental torsion test

3200° and indication resolution of 0.1°. The direction of the torque considered in the test was clockwise and the speed established was 50 rpm/min.

For the selection and preparation of the test specimens, a visual inspection was performed for each one in order to verify the absence of superficial imperfections. In the case of additive manufacturing, it is expected the existence of uniformity in the diameter of the transversal section, and in the length of the test specimens. However, measures were reviewed, obtaining an average polar moment of inertia, J, of 12.79 × 10–9 m^4 (0.031 in^4), assuming a constant area section over the useful length of the test specimens. The free length fulfills with at least four diameters, and the ratio of free length to diameter is minor than helical twisting of the axis of the specimen as required for the ASTM E-143–02 (ASTM 2012).

Figure 3.3 shows the fixture test preparation used for each test, all tests were conducted under conditions of ambient temperature and pressure. An incremental load applied by an external torque was applied for each test, changes in torque and the corresponding changes in angle of twist were determined, then the plot results were obtained for each case of study (see Fig. 3.4).

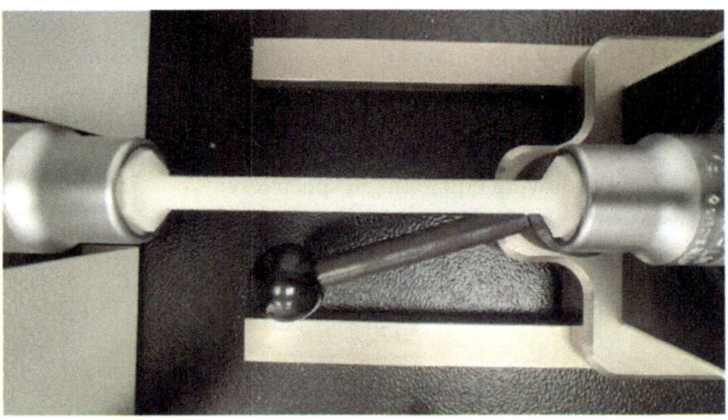

Fig. 3.3 Torsion test fixture preparation

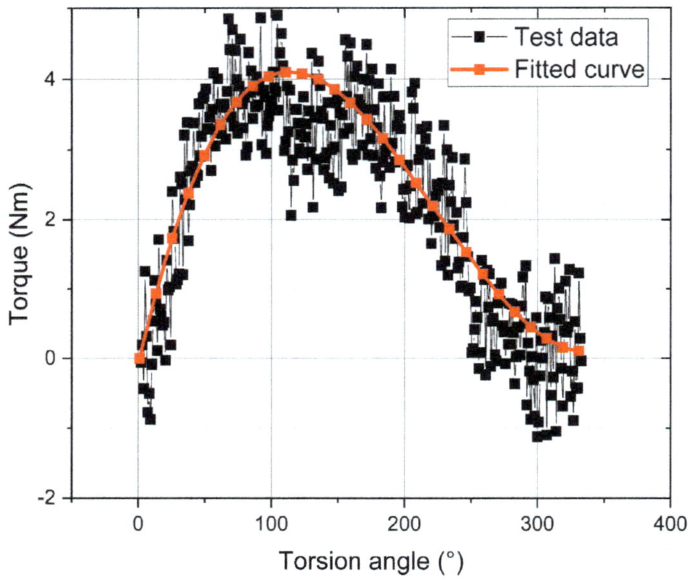

Fig. 3.4 Torsion test data and fitted curve for the horizontal cubic pattern with 50% infill

Once the torsion test data were obtained for all the test specimens, a third order polynomial curve fit was applied to each case, to obtain appropriate curves for analysis of the response variation. For this test method, random variations in the data are considered as variations in the angle of twist θ. Due to possible small offsets at zero torque and small variations in establishing the load path in the specimens during the first small increment of torque, the readings at zero torque and the first small increment of torque are typically not included in the calculations.

It was not expected that there can exist variations in the tests that can compromise the results obtained due to the care about them. It was monitored that the parameters of alignment of the specimen, speed of testing, temperature, and errors in torque and twist values were the same for the tests. The error due to torque ΔT, and angle twist Δθ are given by the following equations, respectively:

$$\frac{\Delta G}{G} = \frac{\Delta T}{T} \tag{3.1}$$

$$\frac{\Delta G}{G} = \frac{\Delta \theta}{\theta} \tag{3.2}$$

3.4 Results

Figures 3.5, 3.6, and 3.7 show the results of the torsion tests carried out for this work. Figure 3.5 shows the results for the cubic infill pattern, in a) shows the three different infill percentages with a horizontal printing direction where the blue graph corresponding to 75 infill percentage shows an unexpected low performance. Comparing this with the other graphs for the horizontal printing direction it can be concluded that for this particular case, we are facing a defect in the test specimen. A similar conclusion can be made for the cubic vertical at 50% infill (see Fig. 5b), the cubic vertical at 50% infill (see Fig. 5c), and the gyroid horizontal at 50% infill (see Fig. 6a) test specimens.

From the results obtained in the gyroid infill pattern, it can be observed a higher resistance when the specimens were printed at 45-degree, however, the strain energy under the curves is lower in comparison with the horizontal printing specimens.

In the case of the three hexagonal infill patterns, it can be observed a persistence of higher resistance when the specimens were printed at 45-degree, so the difference in the areas under the curve.

As expected, in most cases, the better mechanical performance was in specimens with a 75-infilling percentage. In Fig. 3.8 are shown the plot results from these tests comparing the printing orientation. It is noticed that the maximum torque supported by the specimens was obtained when is applied in printing orientation

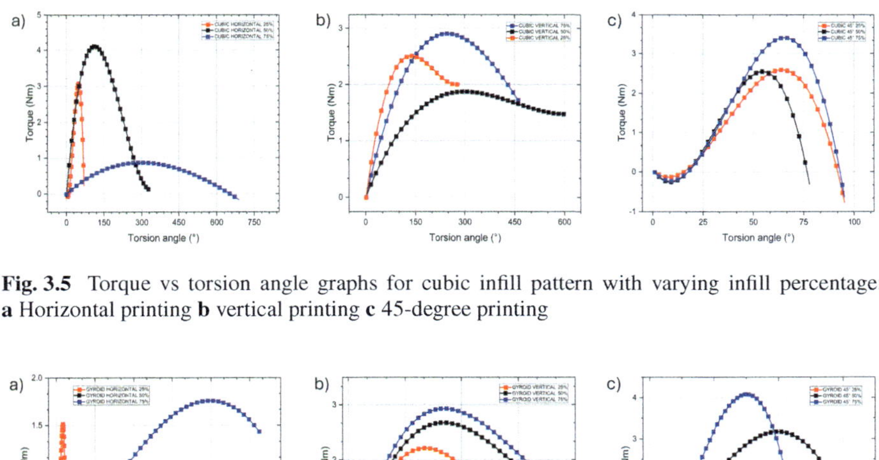

Fig. 3.5 Torque vs torsion angle graphs for cubic infill pattern with varying infill percentage: **a** Horizontal printing **b** vertical printing **c** 45-degree printing

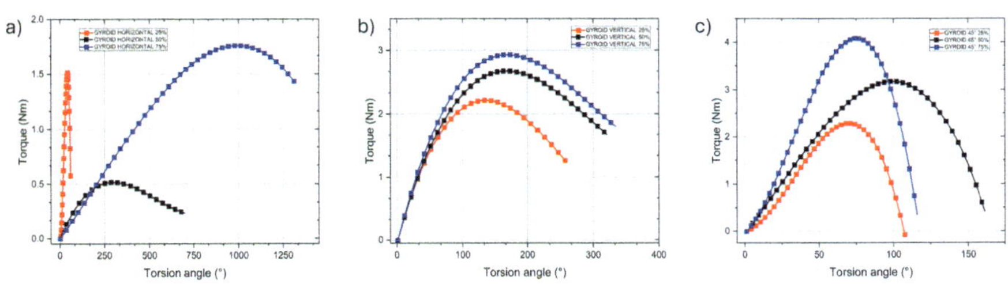

Fig. 3.6 Torque vs torsion angle graphs for gyroid infill pattern with varying infill percentage: **a** Horizontal printing **b** vertical printing **c** 45-degree printing

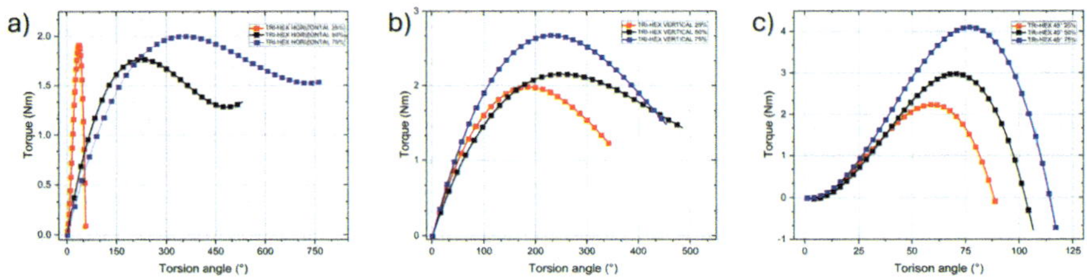

Fig. 3.7 Torque vs torsion angle graphs for tri-hexagonal infill pattern with varying infill percentage: **a** Horizontal printing **b** vertical printing **c** 45-degree printing

of 45°. However, a brittle behavior is observed, and a maximum torque is achieved when a gyroid and tri-hexagonal infilling pattern is used. The difference between these patterns is the area under the curve.

Besides, when a comparison of the result tests was performed using as a comparison variable the infilling pattern (Fig. 3.9) it was noted a similar performance when is used a 45° printing orientation. Though, it is minimum the torque supported in the specimens with a horizontal printing orientation, the angular deformation achieved is major. Finally, the vertical printed test specimens show a mid-response, both in torque and angular deformation, in comparison to the other two printing directions.

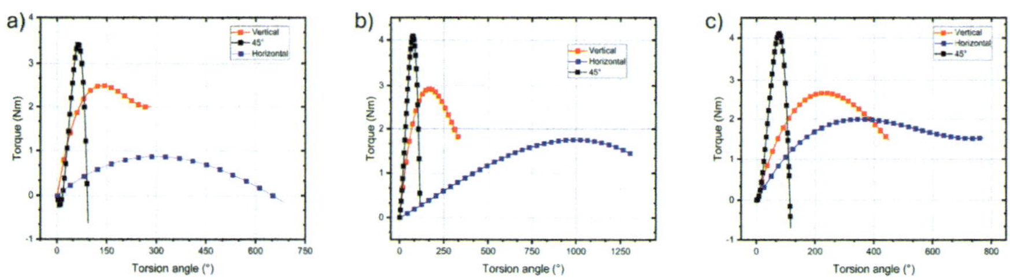

Fig. 3.8 Torque vs torsion angle graphs for a 75% infill density: **a** Cubic infilling pattern, **b** gyroid infilling pattern, and **c** tri-hexagonal infilling pattern

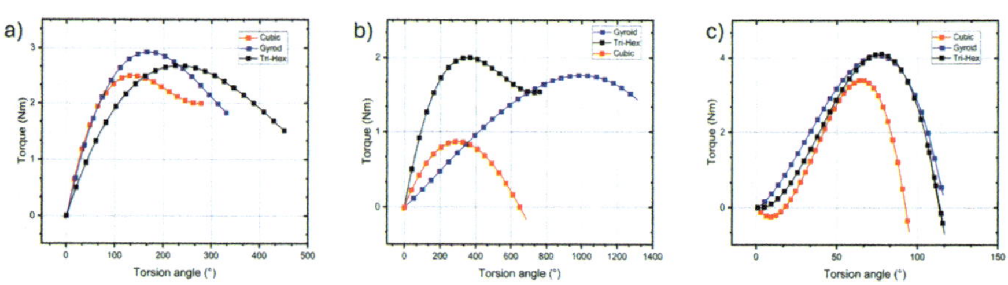

Fig. 3.9 Torque vs torsion angle graphs for a 75% infill density: **a** Vertical, **b** horizontal, and **c** 45° printing orientation

When examining the graphs in Figs. 3.5, 3.6, and 3.7, all the graphs show a linear section from which we were able to calculate the shear modulus G by means of the classical strength of materials formulae along with an approximate shear yield strength and the maximum shear stress for each case. For these calculations an equivalent polar moment of inertia J was calculated for every specimen, assuming a constant annular cross section in accordance with the infill percentage along the useful length of the test specimens. Table 3.2 summarizes the results for all 27 test specimens.

3.5 Conclusion

Several torsion tests were performance in specimens made by AM. The tests were carried out using a standard methodology. It was obtained their mechanical properties for each specimen. A brittle behavior was observed in relation to the infill percentage printing in almost all the cases, and therefore a reduction in the strain energy. The linear behavior of these cases conducts to a higher estimated shear modulus. In these cases, the printing pattern plays an important role in the performance of the specimen, either to have higher shear resistance or to increase strain energy.

It is concluded that printing orientation has the most significant influence on shear behavior during torsional testing, while infill percentage and infill pattern exhibit minor effects. It was also noted that the proposed method for printing the test specimens turned out to be correct as most of the specimens had a good performance, however, more experiments are needed to ascertain repeatability.

Further experimentation, including tests involving variations in parameters such as layer height, printing speed, and material composition, is essential to comprehensively understand their impact on mechanical behavior. It is worth noting that comparison with results for other materials under similar conditions will be also valuable in order to fully characterize the mechanical behavior of the 3D printed parts despite the used materials.

Table 3.2 Estimated values for shear modulus, shear yield strength, and maximum shear stress

Test specimen (%)	Cross section area (mm^2)	Polar moment of inertia (mm^4)	Estimated shear modulus G (MPa)	Estimated shear yield strength (MPa)	Estimated maximum shear stress (MPa)	Maximum torsion angle (°)
Cubic H 25	63.61	2260.72	161.89	11.54	12.26	69.8
Cubic H 50	127.23	3867.95	97.27	7.56	9.51	333
Cubic H 75	190.85	4830.93	6.86	0.56	1.80	781
Cubic V 25	63.61	2260.72	59.25	7.96	9.95	464
Cubic V 50	127.23	3867.95	19.087	2.9	4.33	601
Cubic V 75	190.85	4830.93	41.97	4.278	5.39	284
Cubic 45° 25	63.61	2260.72	50.42	9.154	10.26	95
Cubic 45° 50	127.23	3867.95	30.18	5.57	5.86	77.9
Cubic 45° 75	190.85	4830.93	24.53	5.76	6.324	95
Gyroid H 25	63.61	2260.72	85.03	5.97	6.00	57.7
Gyroid H 50	127.23	3867.95	4.7	0.93	1.16	698
Gyroid H 75	190.85	4830.93	2.9	2.976	3.25	1311
Gyroid V 25	63.61	2260.72	76.84	6.37	8.75	258.9
Gyroid V 50	127.23	3867.95	45.5	4.176	6.17	321
Gyroid V 75	190.85	4830.93	40.47	4.092	5.44	334
Gyroid 45° 25	63.61	2260.72	67.02	7.96	9.11	107.4
Gyroid 45° 50	127.23	3867.95	60.78	6.264	7.33	161.3
Gyroid 45° 75	190.85	4830.93	67.77	6.88	7.58	115.7
Tri-Hex H 25	63.61	2260.72	131.47	7.164	7.60	57.7
Tri-Hex H 50	127.23	3867.95	21.18	3.48	4.03	532

(continued)

Table 3.2 (continued)

Test specimen (%)	Cross section area (mm^2)	Polar moment of inertia (mm^4)	Estimated shear modulus G (MPa)	Estimated shear yield strength (MPa)	Estimated maximum shear stress (MPa)	Maximum torsion angle (°)
Tri-Hex H 75	190.85	4830.93	12.84	3.16	3.72	768.8
Tri-Hex V 25	63.61	2260.72	48.99	6.37	7.88	344
Tri-Hex V 50	127.23	3867.95	23.97	3.712	4.98	485.7
Tri-Hex V 75	190.85	4830.93	26.53	3.99	4.96	453.7
Tri-Hex 45° 25	63.61	2260.72	94.67	7.96	8.91	88.9
Tri-Hex 45° 50	127.23	3867.95	54.33	6.5	6.89	106.4
Tri-Hex 45° 75	190.85	4830.93	49.57	6.98	7.62	116.8

Acknowledgements The grant for the development of the Project SIP20241688 was awarded by the National Polytechnic Institute (IPN). It is kindly acknowledged.

References

ASTM (American Society for Testing and Materials) (2012) E143–02: Standard Test Method for Shear Modulus at Room Temperature. USA

Balderrama-Armendariz CO, MacDonald E, Espalin D, Cortes-Saenz D, Wicker R, Maldonado-Macias A (2018) Torsion analysis of the anisotropic behavior of FDM technology. JAMT 96:307–317. https://doi.org/10.1007/s00170-018-1602-0

Chacón JM, Caminero MA, García-Plaza E, Núñez PJ (2017) Additive manufacturing of PLA structures using fused deposition modelling: effect of process parameters on mechanical properties and their optimal selection. Maters des 124:143–157. https://doi.org/10.1016/j.matdes.2017.03.065

Ganeshkumar S, Kumar S, Magarajan U, Rajkumar S, Arulmurugan B, Sharma S, Li C, Ilyas R, Badran M (2022) Investigation of tensile properties of different infill pattern structures of 3D-Printed PLA Polymers: analysis and validation using finite element analysis in ANSYS. Materials 15:5142. https://doi.org/10.3390/ma15155142

Gao G, Xu F, Xu J, Tang G, Liu Z (2022) A survey of the influence of process parameters on mechanical properties of fused deposition modeling parts. Micromachines 13(4):553. https://doi.org/10.3390/MI13040553

Gonabadi H, Chen Y, Yadav A, Bull S (2022) Investigation of the effect of raster angle, build orientation, and infill density on the elastic response of 3D printed parts using finite element microstructural modeling and homogenization techniques. Int J Adv Manuf Technol 118:1485–1510. https://doi.org/10.1007/s00170-021-07940-4

Jemghili R, Ait Taleb A, Khalifa M (2021) A bibliometric indicators analysis of additive manufacturing research trends from 2010 to 2020. Rapid Prototyp J 27(7):1432–1454. https://doi.org/10.1108/RPJ-11-2020-0274

Kadhum AH, Al-Zubaidi S, Alkareem SSA (2023) Optimization of mechanical properties and surface characteristics of PLA+ 3D printing materials. Int J Chem Eng 2023:8887905. https://doi.org/10.1155/2023/8887905

Khosravani MR, Reinicke T (2020) Effects of raster layup and printing speed on strength of 3D-printed structural components. Procedia Struct Integr 28:720–725. https://doi.org/10.1016/j.prostr.2020.10.083

Kumar MB, Sathiya P (2021) Methods and materials for additive manufacturing: a critical review on advancements and challenges. Thin-Walled Struct. 159:107228. https://doi.org/10.1016/j.tws.2020.107228

Kunovjanek M, Knofius N, Reiner G (2022) Additive manufacturing and supply chains–a systematic review. Prod plan control 33(13):1231–1251. https://doi.org/ https://doi.org/10.1080/09537287.2020.1857874

Lakshman Sri SV, Karthick A, Dinesh C (2024) Evaluation of mechanical properties of 3D printed PETG and polyamide (6) polymers. Chem Phys 8:100491. https://doi.org/ https://doi.org/10.1016/j.chphi.2024.100491

Palanisamy C, Aaron Tay Hong Kiat HK (2021) Experimental study and modeling the tensile strength of 3D-printed aluminium polylactic acid (PLA) parts using artificial neural networks. F1000Res. 10:1286. https://doi.org/10.12688/f1000research.73796.1

Ponis S, Aretoulaki E, Maroutas TN, Plakas G, Dimogiorgi K (2021) A systematic literature review on additive manufacturing in the context of circular economy. Sustainability 13(11):6007. https://doi.org/10.3390/su13116007

Rodríguez JF, Thomas JP, Renaud JE (2001) Mechanical behavior of acrylonitrile butadiene styrene (ABS) fused deposition materials. Experimental Investigation Rapid Prototyp J 7(3):148–158. https://doi.org/10.1108/13552540110395547

Sepahi MT, Abusalma H, Jovanovic V, Eisazadeh H (2021) Mechanical Properties of 3D-printed parts made of polyethylene terephthalate glycol. J Mater Eng Perform 30(9):6851–6861. https://doi.org/10.1007/s11665-021-06032-4

Valvez S, Silva AP, Reis PNB (2022) Optimization of printing parameters to maximize the mechanical properties of 3D-Printed PETG-based parts. Polymers 14:2564. https://doi.org/10.3390/polym14132564

Chapter 4
Exploration of Microfluidic Technology for Additive Manufacturing: Devices for Precise Material Dispensing

Erik Omar Alvarado-Alcántara, Juan Alfonso Beltrán-Fernández, Alejandro Tonatiu Velázquez-Sánchez, Karen Pamela Vázquez-Thierry, Carolina Alvarado Moreno, and Verónica Guzmán-Mercado

Abstract The application of the microfluidic technology within the field of additive manufacturing is investigated, with a specific focus on devices designed for precise material dosing. An in-depth examination is conducted to explore how microfluidic technology can be integrated into Cartesian 3D printing systems, enabling precise and controlled material dispensing throughout the manufacturing process. A novel methodology was developed to assess these devices, evaluating their feasibility and effectiveness in improving the quality and efficiency of manufactured products. Experimental analysis was performed to examine various microfluidic device configurations and their impacts on material dosing precision, along with their potential for reducing waste. Promising advancements in additive manufacturing were demonstrated by the findings, which offer new insights and opportunities in fields such as medicine, electronics, and engineering. The foundation for enhancing material dosing control in additive manufacturing was laid by this study.

E. O. Alvarado-Alcántara · J. A. Beltrán-Fernández (✉) · A. T. Velázquez-Sánchez · K. P. Vázquez-Thierry · C. A. Moreno · V. Guzmán-Mercado
Sección de Estudios de Posgrado E Investigación, Instituto Politécnico Nacional, ESIME Unidad Zacatenco, Unidad Profesional Adolfo López Mateos, Ciudad de México, México
e-mail: jbeltran@ipn.mx

E. O. Alvarado-Alcántara
e-mail: eomar.alv.alc@gmail.com

A. T. Velázquez-Sánchez
e-mail: atvelazquezs@ipn.mx

K. P. Vázquez-Thierry
e-mail: kappa.thierry@gmail.com

C. A. Moreno
e-mail: calvaradom1800@alumno.ipn.mx

V. Guzmán-Mercado
e-mail: vero_guzmanm94@hotmail.com

A. Öchsner and H. Altenbach (eds.), *Engineering Design Applications VII*, Advanced Structured Materials 230, https://doi.org/10.1007/978-3-031-84346-4_4

Keywords Microfluidic technology · Additive manufacturing · Precise material dosing · Synthetic material

4.1 Introduction

The 3D printing market has experienced substantial growth within the scientific community, particularly in fused deposition modeling (FDM). However, it is questioned why a similar surge has not been observed in other technologies, such as binder jetting, whose patents have now entered the public domain and were initially conceived for materializing digital models.

The FDM technology is recognized for offering significant technological and economic advantages compared to other methods, such as binder jetting. Binder jetting involves the controlled deposition of a liquid binder into a powder bed to form a physical model from a digital design. Despite the expiration of key patents (Alvarado-Alcántara et al. 2023), the widespread popularity of binder jetting has not been achieved, especially among enthusiasts of rapid prototyping devices (Magdassi 2009).

Microfluidic technologies have been successfully integrated into desktop printers by companies such as EPSON®, Brother®, and HP®, revolutionizing the speed and precision of liquid deposition (Rapp 2016). This has prompted exploration into the fields of micro electro mechanical systems (MEMS) and microfluidics. Microfluidics is defined by Rivas (Rivas, 2008) as a field that exploits the dominant properties of fluids at scales smaller than one millimeter (Crane et al. 2013; Squires and Quake 2005; De menech 2008; Wang et al. 2020).

The study of microfluidics, as described by Rennie (2019), involves the examination of fluid mechanics at the submillimeter scale, where physical properties like surface tension are more pronounced than at larger scales. Conventional fluid mechanics principles have been found insufficient for explaining the functioning of inkjet printer systems. The application of macroscopic fluid properties to microscale fluids does not yield equivalent results, as the governing laws of microfluidics (Berthier and Brakke 2012; Bingcheng 2011) are markedly different from those at the macro scale, as illustrated in Fig. 4.1.

Whitesides (2006) highlights that the potential of microfluidics extends beyond its traditional applications in chemical synthesis and biological analysis. It is proposed in this study that microfluidics could also be utilized in the field of additive manufacturing. Although microfluidic devices have been intrinsically used in binder injection technology, their broader application within additive manufacturing has not been extensively explored since its inception. The importance of examining this perspective is emphasized by the broad range of opportunities it presents. This is evidenced by the steady increase in related publications, as illustrated in Fig. 4.2.

However, the application of microfluidics in other fields has also been investigated, as illustrated in Fig. 4.3. Out of a total of 334 publications containing the

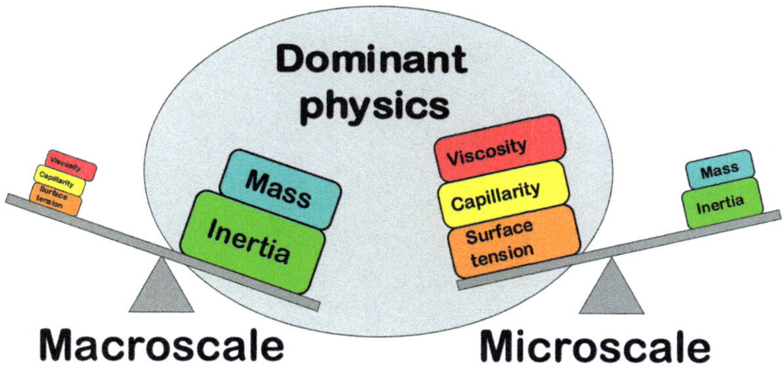

Fig. 4.1 Dominant physical properties at the macroscale versus the microscale (Rennie and Law 2019; Bingcheng 2011)

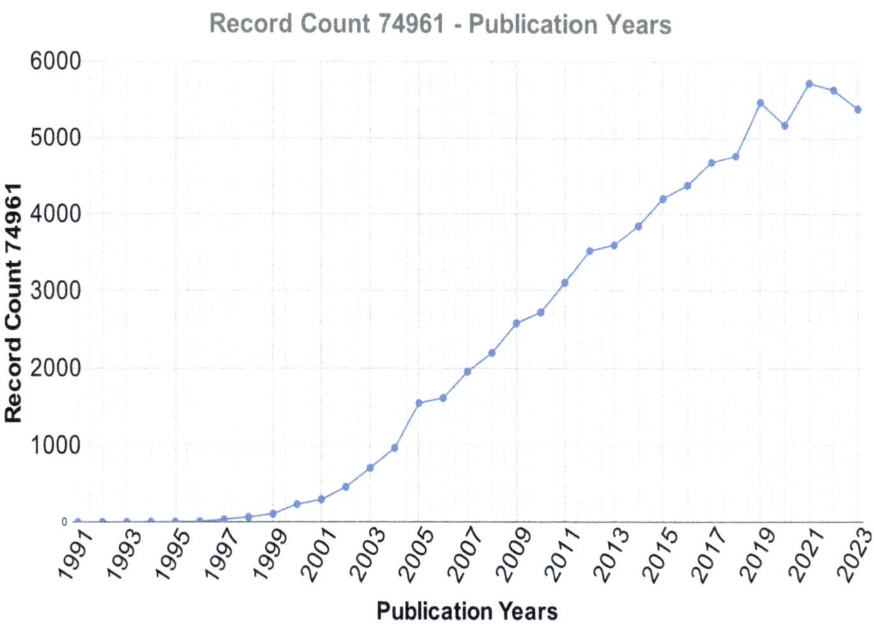

Fig. 4.2 Number of publications in the field of microfluidics since 1991 (Dongqing 2015)

keywords "microfluidics" and "additive manufacturing," only 44 specifically address the approach examined in this research.

The expiration of patents, along with the simultaneous technological advancements in microfluidics, has allowed for a new approach to additive manufacturing. It has been indicated that 3D printing techniques have facilitated the development

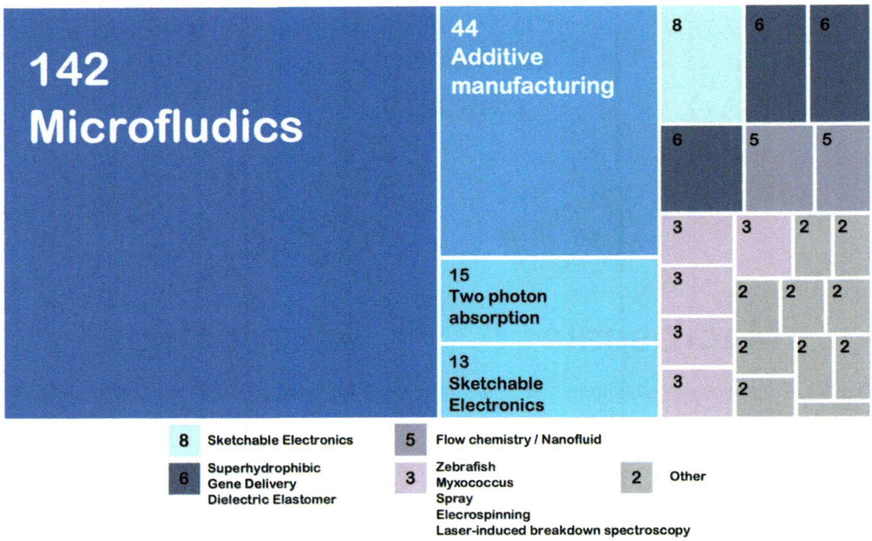

Fig. 4.3 Publications on microfluidics applied to additive manufacturing

of devices for microfluidics, particularly with the use of light-curing resins (Naderi et al. 2019; Rosero 2021; Xie et al. 2024).

In order to achieve controlled dosing, similar to binder injection technology, and transition from a macro to a micro regime in terms of droplet generation, it is necessary to consider capillarity, surface tension, and the viscosity of the liquid (van der Waals forces). If these properties are disregarded, and only microtubes are used, the following effects will occur:

1. Liquid particles will cohere.
2. Surface tension will determine the morphology of the liquid (Prakash and Yeom 2014).
3. Viscosity will cause the particles to adhere to the microtube.

The accumulation of molecules at the end of the tube will result in the formation of droplets. Due to the influence of gravity (Rayleigh 1878) on the fluid's mass, as shown in Fig. 4.4, the droplets will eventually grow in size, break the van der Waals forces, and detach.

This procedure does not closely achieve the desired outcome for 3D printing, unlike binder injection technology, which provides lower and more controlled dosage. Although dripping occurs, it does not exhibit the same characteristics, particularly in terms of droplet size, as demonstrated in Fig. 4.5.

In the experiment, only a peristaltic pump was used, as illustrated in Fig. 4.6, to replace the extruder motor of the 3D printer. Its function was to supply water through the tubes and maintain a controlled flow at the outlet.

Fig. 4.4 Formation of a water droplet

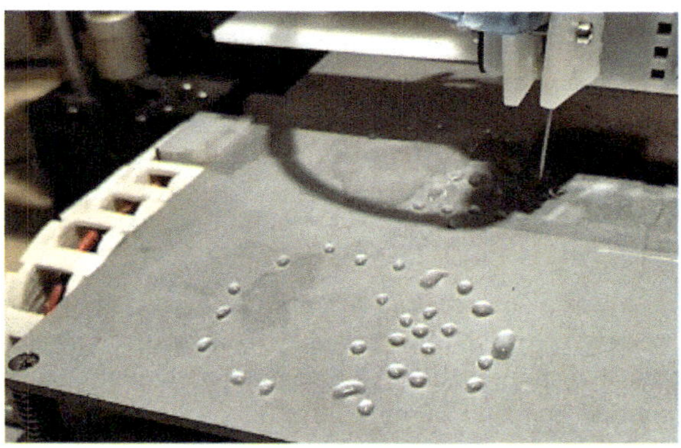

Fig. 4.5 Macro drip attached to a Cartesian motion system

As previously mentioned, the experiment did not achieve fluid deposition with droplet sizes smaller than 1 mm, which would have ensured better definition in the trajectory indicated by the Cartesian system. Binder injection technology utilizes a piezoelectric component (Ham et al. 2010; Brünahl and Grishin 2002) for liquid dosing, one of several methods for generating microdroplets (Wei et al. 2019). This

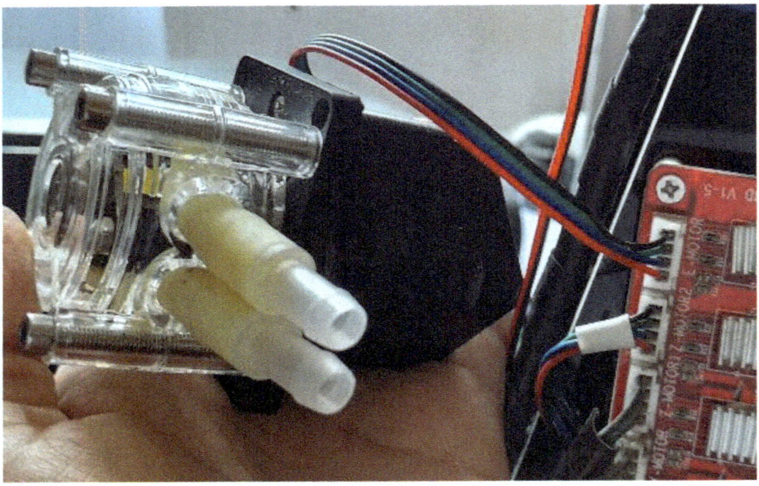

Fig. 4.6 Peristaltic pump

component complements a system that operates from the macro to the micro scale and can be integrated into the Cartesian system.

4.2 Materials and Methods

To achieve micro-dripping in the Cartesian system, a piezoelectric disc, typically used for water atomization or cold steam generation, was employed, as shown in Fig. 4.7.

The operation of this device was controlled using the corresponding controller, as shown in Fig. 4.8. The main features of this device include its compact size and low power consumption, requiring only 5 V, which can be supplied through a type C cable.

It was also necessary to improve the peristaltic pump, as commercially available models have high flow rates due to the 6 mm internal diameter of the hose. To address this, a custom version was designed and manufactured using silicone hoses with an internal diameter ranging from 0.5 to 2 mm, as shown in Fig. 4.9. This adjustment was made to ensure that the material flow would be comparable to the flow rate of materials in a standard commercial FDM printer, which uses a 1.75 mm filament.

The redesign of the pump was necessary due to the change in hose dimensions, which required adjustments to the rest of the components. Screws, bearings, and spacers were incorporated in the construction of the pump, as shown in Fig. 4.10.

Fig. 4.7 Piezoelectric for micro-drip generation

4.3 Development

Several acrylic elements of varying thicknesses, cutting using a CNC laser machine, were also required and will be detailed in the development section.

The methodology used was as follows:

- Design, manufacturing, and adjustments of the peristaltic pump.
- Design and adaptation of the elements of the piezoelectric system with the peristaltic pump

The peristaltic pump was the first element to undergo redesign. As shown in Fig. 4.11, the dimensions of the components (highlighted in magenta) and their tolerances were determined based on the bearings (indicated in cyan) that were available on the market.

The specifications of the shaft, where the set of bearings would be attached, which is part of the stepper motor of the commercial peristaltic pump, were also considered.

To determine the tolerance of the planetary bearings and the piece shown in magenta, the silicone hose was first obtained. Due to its characteristics, it was the least resistant to mechanical deformation, allowing for successful peristalsis.

Fig. 4.8 Controllers for micro-drip piezoelectric generators

Fig. 4.9 Silicone hoses

Fig. 4.10 Elements for the construction of the peristaltic pump

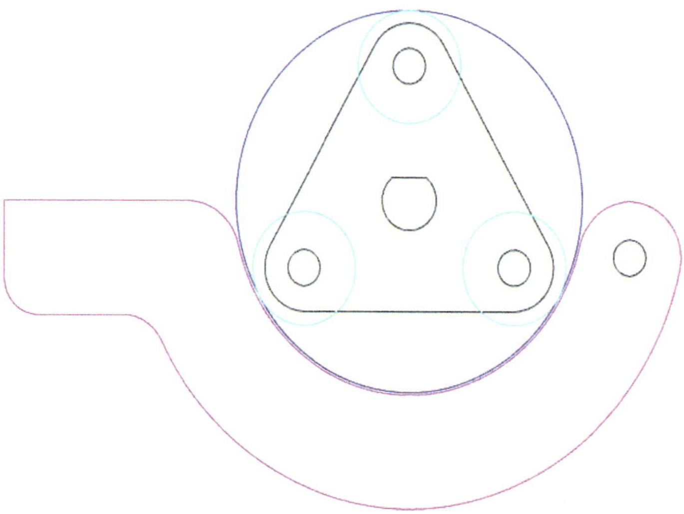

Fig. 4.11 Design of planetary bearings

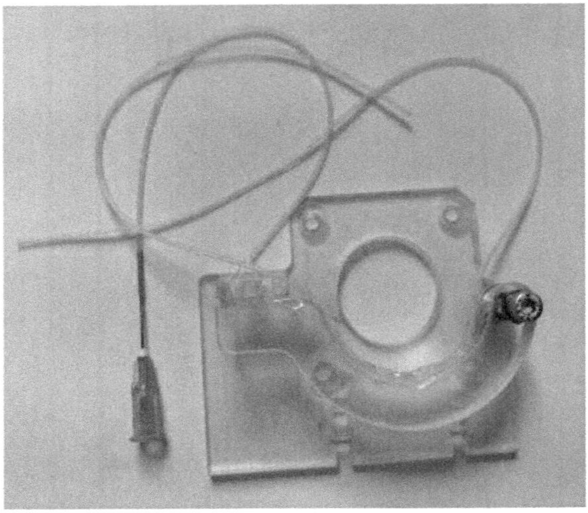

Fig. 4.12 0.5 mm inner diameter hose

The piece in magenta features a fixed point at one end, functioning as a pivot, which allows for adjustment on the opposite side to accommodate hoses of varying external diameters, ranging from 1 to 4 mm, as shown in Fig. 4.12.

The planetary bearings will perform as rollers were constructed using the previously described materials, as shown partially assembled in Fig. 4.13. This subassembly of the peristalsis system is displayed in Fig. 4.14 (Fig. 4.15).

Through this modelling method, the correct assembly of components, tolerances, and integration with other systems were ensured. The individual drawings of each component, as shown in the design of Fig. 4.16, were extracted to facilitate their manufacture using CNC laser cutting.

The result of these drawings is shown in Fig. 4.17, where the components were fabricated using different thicknesses of acrylic, produced by CNC laser cutting.

The pump parts were chemically bonded and eventually assembled by being bolted to the stepper motor, as shown in Fig. 4.18.

A peristaltic pump, capable of accommodating hoses ranging from 1 to 4 mm in diameter, was thus developed as part of a macro drip system.

Following the construction of the pump, the integration of the micro-drip system, comprising the piezoelectric pump and its controller, was undertaken. The components were subsequently modeled, and configurations for their arrangement were made, as illustrated in Fig. 4.19.

The same procedure used for manufacturing the previous parts was applied to this assembly. The system was physically validated, and design improvements were made, as shown in Fig. 4.20.

This redesign enables integration with the peristaltic pump. The assembled system is illustrated in Fig. 4.21.

Fig. 4.13 Construction of planetary bearings

Fig. 4.14 Assembled planetary bearings

Fig. 4.15 3D modeling of the peristaltic pump

The integration of this system was designed to utilize the extruder space in the 3D printer, allowing for its implementation in the FDM printer by taking advantage of the stepper motor control of the extruder. This adaptation was intended as a modification of binder jetting technology for use with this type of printer.

4.4 Results

The installation of the newly developed 'micro monodisperse module' in the FDM 3D printer is shown in Fig. 4.22.

The stepper motor of the extruder was replaced with the stepper motor of the peristaltic pump, as both motors share identical electrical characteristics, connectors, and operating specifications.

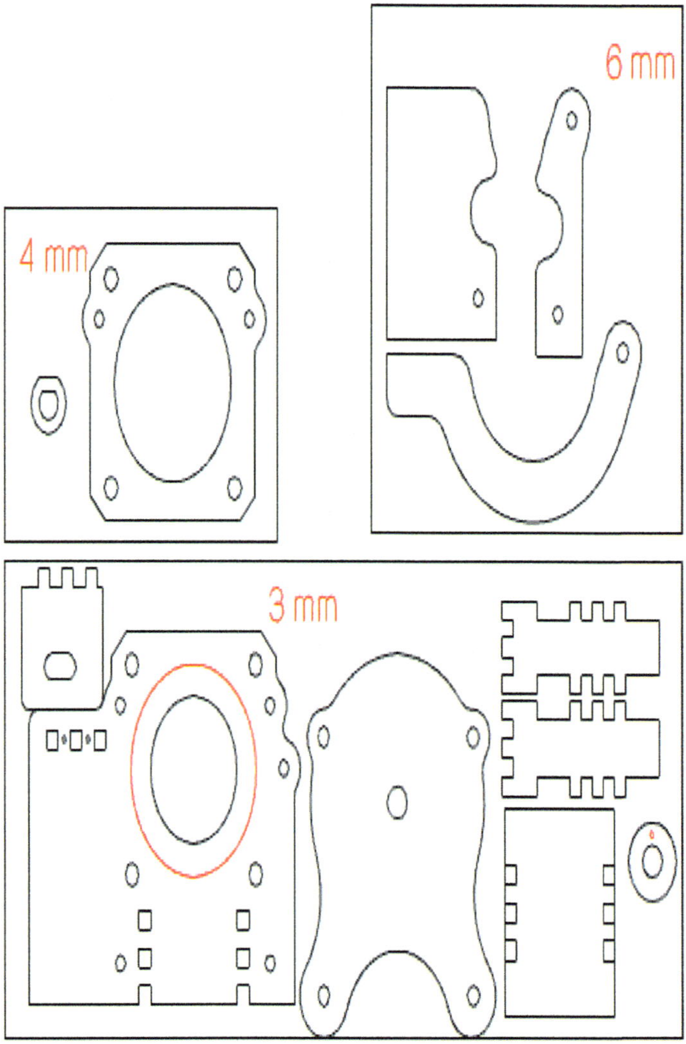

Fig. 4.16 Components of the peristaltic pump designed for CNC laser cutting

4.5 Discussion

The results of this study highlight the potential of integrating microfluidic technology into additive manufacturing systems, particularly in achieving precise material dosing. The successful replacement of the extruder motor with a peristaltic pump, coupled with the integration of a piezoelectric micro-drip system, demonstrates a significant advancement in controlling material deposition. The observed reduction

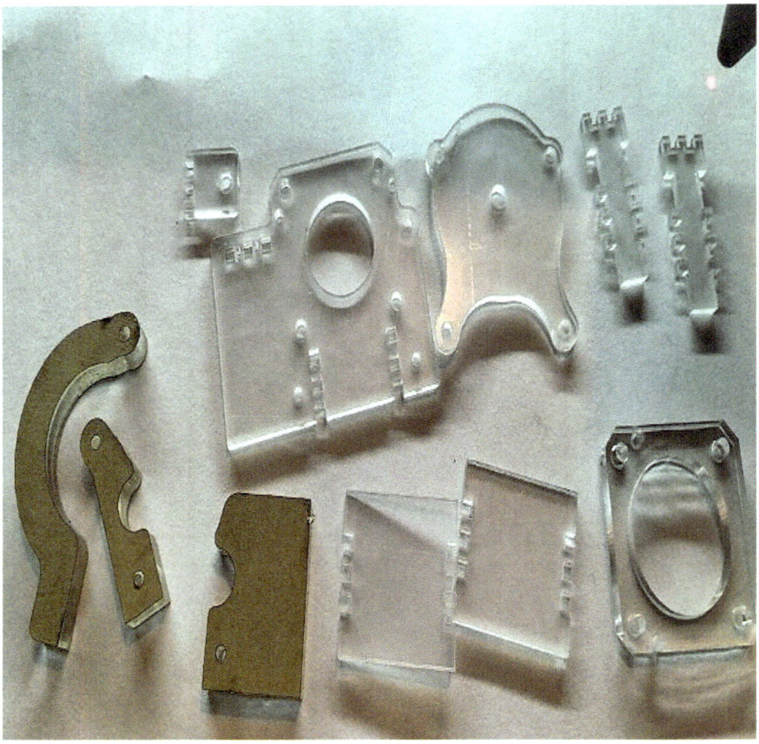

Fig. 4.17 Drawings materialized with a CNC laser cutter

in droplet size and improved control over material flow validate the effectiveness of the proposed modifications. However, the study also revealed several challenges, including the need for further optimization of the micro-drip system to address issues related to droplet consistency and flow stability. These findings suggest that while the integration of microfluidics offers promising enhancements, additional research and development are necessary to fully realize its potential in additive manufacturing. Future work should focus on refining the system to mitigate the challenges identified and exploring its application across various materials and printing processes.

4.6 Conclusions

The generation of micro-droplets relies on a comprehensive understanding of physical properties such as surface tension, capillarity, and viscosity. These properties play a crucial role in achieving precise droplet formation.

Fig. 4.18 Assembly of all components of the peristaltic pump

Fig. 4.19 Piezoelectric system parts

Fig. 4.20 Piezoelectric system assembly

Fig. 4.21 Integration of peristaltic pump with piezoelectric system

The exploration of these properties and their applications is relatively novel and offers opportunities for new approaches in additive manufacturing with compatible materials.

While the micro monodisperse module is theoretically feasible, new and unpredictable challenges have emerged with each advancement, which needed to be

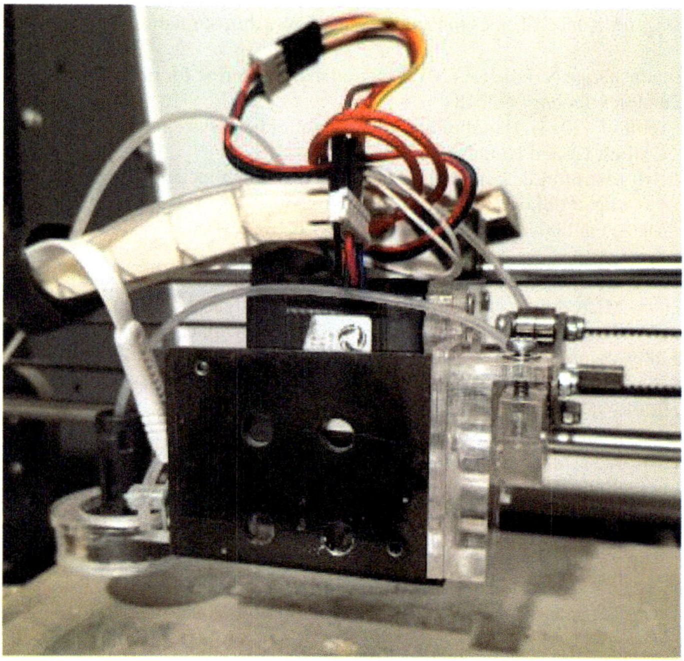

Fig. 4.22 Monodisperse micro module installed in FDM printer

addressed to advance the research. Therefore, time and resources must be appropriately allocated to overcome these challenges effectively.

References

Alvarado-Alcántara EO, Beltrán-Fernández JA, Hermida-Ochoa JC, Andrade JLC, Hernández-Gómez LH, Moreno-Garibaldi P, González MRY, González ARY, Guzmán-Mercado V, Uribe-Cortés TB (2023) Design of a system to produce rapid biomedical prototypes with synthetic materials: state of the art. In: Advanced structured materials. Springer Science and Business Media Deutschland GmbH, pp 173–179

Berthier J, Brakke KA (2012) The physics of microdroplets. In: The physics of microdroplets. Wiley

Bingcheng L (2011) Microfluidics technologies and applications. Springer

Brünahl J, Grishin AM (2002) Piezoelectric shear mode drop-on-demand inkjet actuator

Crane NB, Onen O, Carballo J, Ni Q, Guldiken R (2013) Fluidic assembly at the microscale: progress and prospects

De menech M, Garstecki P, Jousse F, Stone HA (2008) Transition from squeezing to dripping in a microfluidic t-shaped junction. J Fluid Mech 595:141–161. https://doi.org/10.1017/S002211200700910X

Dongqing L (2015) Encyclopedia of microfluidics and nanofluidics (P). Springer, New York

Ham YB, Seo WS, Oh SJ, Park JH, Yun SN (2010) A piezoelectric-driven single-nozzle dispenser for liquid droplet jetting. J Korean Phys Soc 57:877–881. https://doi.org/10.3938/jkps.57.877

Magdassi (2009) Advanced Topics in Printing Techniques. https://piams.org/inkjet-printing-technologies/
Naderi A, Bhattacharjee N, Folch A (2019) Digital manufacturing for microfluidics. https://doi.org/10.1146/annurev-bioeng-092618
Prakash S, Yeom J (2014) Nanofluidics and microfluidics: systems and applications. William Andrew, Oxford, United Kingdom
Rapp BE (2016) Microfluidics: modeling, mechanics and mathematics. Microfluid: Model, Mech Math 1–766. https://doi.org/10.1016/C2012-0-02230-2
Rayleigh L (1878) On the instability of jets. In: Proceedings of the London mathematical society. pp 1–10, 4–13. https://doi.org/10.1112/PLMS/S1-10.1.4
Rennie R, Law J (2019) A dictionary of physics. Oxford University Press
Rivas DF (2008) Microfluidos: ¿cuánto hay de nuevo? Divulgación-Física
Rosero G (2021) Dispositivos de Microfluídica utilizados en aplicaciones biotecnológicas. https://www.youtube.com/watch?v=Hav8KhToTgw&t=1039s
Squires TM, Quake SR (2005) Microfluidics: fluid physics at the nanoliter scale. 77. https://doi.org/10.1103/RevModPhys.77.977
Wang J, Shao C, Wang Y, Sun L, Zhao Y (2020) Microfluidics for medical additive manufacturing. Engi 6:1244–1257. https://doi.org/10.1016/j.eng.2020.10.001
Wei YY, Sun ZQ, Ren HH, Li L (2019) Advances in microdroplet generation methods. Chin J Anal Chem 47:795–804. https://doi.org/10.1016/S1872-2040(19)61162-X
Whitesides GM (2006) The origins and the future of microfluidics. Nature 442:368–373. https://doi.org/10.1038/nature05058
Xie M, Zhan Z, Li Y, Zhao J, Zhang C, Wang Z, Wang Z (2024) Functional microfluidics: theory, microfabrication, and applications. Int J Extrem Manuf 6 (2024). https://doi.org/10.1088/2631-7990/ad2c5f

Chapter 5
Computation Simulations of Basic and Specific Experiments of Composites

Jan Krmela, Vladimíra Krmelová, Martina Fusková, Marek Makúch, and Nadiia Artyukhova

Abstract This article describes computational simulations of tensile tests of polymeric composites with the determination of material parameters. Composites consist of an elastomeric rubber matrix and textile or steel cords as reinforcement. Stress–strain analyses are performed. The ANSYS APDL (ANSYS Parametric Design Language) procedure is used to create computational models simply and quickly. Single- and two-layer models are created for simulations. A two-parameter Mooney–Rivlin material model is used for the rubber matrix. Comparisons are made with experimental data attention is paid to comparing the data with the video-extensometer. Therefore, the tests are called basic tensile tests if the dependence of tensile force–displacement between gauges is measured, but if the video-extensometer is used, the outputs from tests are dependencies between true stress and true strain in the work area of specimens. Tests with video-extensometer are called specific tests. The dependence of true strain–transverse strain for the two-layer specimen is done as a control curve to verify that the video-extensometer correctly measured the deformations at the measuring points. The data obtained from tests are tensile strength, ductility, contraction, Poisson's ratio, and tensile modulus of elasticity. The computational models take this into account. The difference between the computational simulation results and the experimental data is approximately 10%.

J. Krmela (✉) · V. Krmelová · M. Fusková · M. Makúch
Faculty of Industrial Technologies in Púchov, Alexander Dubček University of Trenčín, Púchov, SK, Slovak Republic
e-mail: jan.krmela@tnuni.sk; vladimira.krmelova@tnuni.sk

M. Fusková
e-mail: martina.fuskova@tnuni.sk

M. Makúch
e-mail: marek.makuch@tnuni.sk

N. Artyukhova
Department of Marketing, Academic and Research Institute of Business, Economics and Management, Sumy State University, Sumy, UA, Ukraine
e-mail: n.artyukhova@pohnp.sumdu.edu.ua

A. Öchsner and H. Altenbach (eds.), *Engineering Design Applications VII*, Advanced Structured Materials 230, https://doi.org/10.1007/978-3-031-84346-4_5

Keywords Composite · Rubber · Simulation · ANSYS · Mooney–Rivlin · Video-extensometer

5.1 Introduction

In practice, it is necessary to have knowledge about the mechanical behavior of composites (Krmela 2021, 2017; Rao et al. 2004). This knowledge can be obtained by mechanical tests such as static tensile tests, bending tests, etc. The mechanical properties of a multilayer composite depend on the orientation of the cords of the respective individual layers.

In this article, the authors used the finite element software ANSYS Mechanical for the computational simulation of composites. The APDL (ANSYS Parametric Design Language) in ANSYS (Krmela 2021; Krmela et al. 2023) is used to create 3D computational models.

It is necessary to have knowledge about geometry, material parameters, structure, and other parameters (see Fig. 5.1) for the creation of computational models of a tire casing for the strain–stress analysis. Outputs from computational simulations, e.g., strain, can be verified with experimental data.

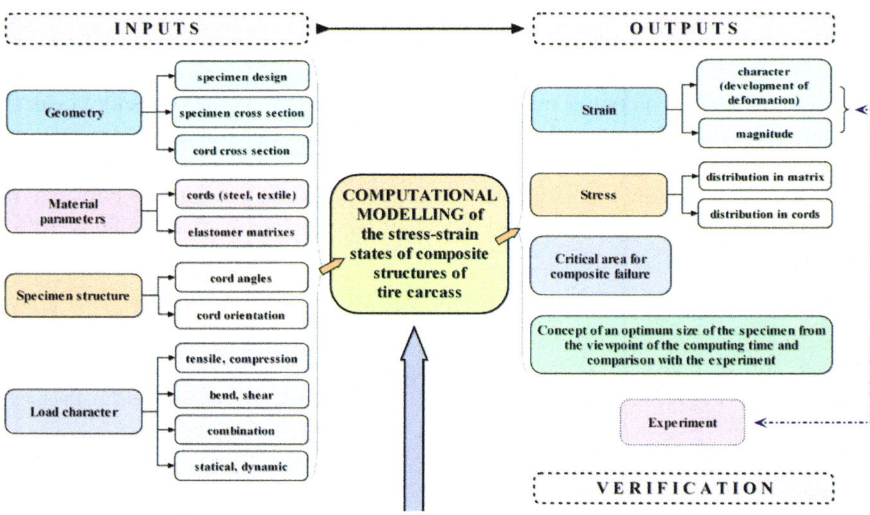

Fig. 5.1 Necessary inputs for computational models of composites and outputs

Table 5.1 Geometry parameters of composite test specimens

Specimen number	Width [mm]	Thickness [mm]	Number of layers	Cord angle
CTT33	14.6	0.95	1	23°
B16/2	14.1	1.2	2	± 24°

5.2 Materials and Methods and Computational Models

5.2.1 Materials Used

For experiments, two configurations of long fiber composites marked CTT33 and B16/2 with a rubber matrix are chosen. These composite specimens of single-layer (CCT33) and two-layer (B16/2) consist of 2 × 0.32 steel cords as reinforcement with a diameter of 0.32 mm and an EPM of 560 m^{-1}. The EPM means that the distance between the cords is approx. 1.79 mm. These composites are used as a part of the Barum passenger tire casing (manufactured by Continental company), such as a steel-cord belt (Krmela 2017).

The composite specimens have the geometry parameters in Table 5.1. The intention was to produce samples with the same geometric parameters. The samples were made in such a way that the blanks of the layers were taken directly from the tire casing factory, and in the university laboratory, the layers were inserted over each other in the case of a two-layer sample. The process of vulcanization in the vulcanizing press and manual production (insertion of layers) caused that the angles and thickness of the layers are not the same. The specimens were cut with a water jet. Therefore, the width of the specimens was not the same. The thickness of the single-layer specimen is 0.95 mm and the thickness of the two-layer specimen is 1.2 mm (0.6 mm per layer). The total length of the test samples is 135 mm.

5.2.2 Methods Used

The experiments of static tensile loading of the composite specimens are performed at a standard temperature of 20 °C. Autograph AG–X test machine plus 5 kN-Shimadzu with a video-extensometer as non-contact optical systems, (Popelar and Liechti 2003; Sharpe 2008) is used. The tensile test specimens of rectangle shape were clamped in the testing device gauges so that the initial gauge length was 80 mm. The loading speed was 25 mm/min.

The working area on the surface of the specimens was measured with a video-extensometer. The change of length–displacement [mm] and relative elongation–true strain [%] was measured by the video-extensometer at a distance of ± 25 mm from the center of the test specimen (i.e., at a distance of 50 mm between the measure points-recommended gauge length for the video-extensometer), called a working area of specimens, and the change of width–contraction (width decreases) [mm] and relative contraction–transverse strain [%] were also measured by the video-extensometer. The following data were recorded: tensile force [N], displacement between the gauges (displacement) [mm], relative elongation between the gauges (strain) [%], displacement in the working area (extensometer displacement) [mm], true strain in the working area [%], contraction [mm], and transverse strain [%]. The following data were calculated: tensile engineering stress (eng. stress) [MPa] and true tensile stress [MPa] based on the actual tensile force value and the actual cross section [mm^2] during the experiments.

In the case of a single-layer specimen CCT33, the signal from the video-extensometer was lost, therefore the true strain values are given only up to a certain value of the deformation of the specimen, and as for the measurement of the width of the specimen—here the signal was lost right at the beginning of the measurement.

The data obtained in terms of material parameters are tensile strength [MPa], ductility (value of the strain when the test specimen breaks for specimen CCT33 and value of the true strain in working area when the test specimen breaks for specimen B16/2) [%], contraction (value of the transverse strain at test specimen B16/2 break) [%], Poisson's ratio as negative ratio of transverse strain to true strain [-] for specimen B16/2, and tensile modulus of elasticity [MPa].

The tensile test, according to ISO 527–4 and ISO 527–5, is used to determine the tensile modulus and Poisson's ratio as elastic characteristic values. Poisson's ratio and the tensile modulus of elasticity are important because these material parameters can be used as input material parameters for the computational modeling of tire casings. Finally, the dependence of true strain–transverse strain for the B16/2 specimen is done as a control curve to verify that the video-extensometer correctly measured the deformations at the measuring points.

5.2.3 Computational Models

It is important to experimentally determine material parameters used for the description of each part of a tire casing for the tire computational simulation (Krmela 2021).

At first, the two-layer model is created based on the geometric parameters in Table 5.1 for the B16/2 specimen using the APDL procedure. The SOLID186 volume element is used. The material parameters as modulus of elasticity and Poisson's ratio describe the steel cords with modulus of elasticity of 210 GPa and Poisson's ratio of 0.3. The Mooney–Rivlin two-parametric hyperelastic material model (Krmela et al. 1199) with parameters $C_{10} = 0.55$ MPa and $C_{01} = 0.11$ MPa is used for the definition of the elastomeric matrix, (Krmela 2017). It corresponds to the modulus of elasticity of 4 MPa. The Mooney–Rivlin parameters are determined on the basis of data from a static tensile test of the rubber matrix. The incompressibility parameter d (Krmela 2021) is 0.14 MPa^{-1} (for Poisson's ratio of 0.49). The final width of the computational model of the specimen is 14.6 mm, and the final length is 135.1 mm. The model consists of 11,328 elements and 46,551 nodes.

Then, a single-layer model is created, which presents the CCT33 specimen. The model consists of 5,664 elements and 24,823 nodes. This model was created for comparison with the CCT33 and B16/2 specimens.

The computational model B16/2 is reverse loaded, the displacement of 12.47 mm (based on the experimental data—ductility of the B16/2 specimen) is defined in the loading *z*-axis, and the summary reaction forces in the *z*-axis are searched. The computational model CCT33 is reverse loaded by a displacement of 33.83 mm (based on the experimental data—ductility of the CCT33 specimen).

The computational models with details of cords and boundary conditions are presented in Figs. 5.2 and 5.3.

5.3 Results and Discussion

5.3.1 Experimental Results

The tensile force–displacement dependences obtained on the tensile tests are shown in Fig. 5.4. Forces at break are 595.2 N for the B16/2 specimen and 236.3 N for the CCT33 specimen. The tensile force–contraction dependence for the B16/2 specimen is shown in Fig. 5.5. The engineering stress–engineering longitudinal strain and true/engineering stress–true longitudinal strain dependencies are in Figs. 5.6 and 5.7. The true/engineering stress–transverse strain dependence for the B16/2 specimen is shown in Fig. 5.8.

The dependence of true strain—transverse strain for the B16/2 specimen is shown in Fig. 5.9.

For the two-layer B16/2 specimen, these results are obtained: true strength of 56 MPa, strength of 40.5 MPa (based on engineering approach), ductility (value of the true strain in working area when the test specimen breaks) of 12.2%, contraction of 27.6%, Poisson's ratio of 2.2, and tensile modulus of elasticity of 460.8 MPa (based on the true stress). The width of the specimen at break was 10.1 mm.

For the single-layer CCT33 specimen, these results are obtained: strength of 16.2 MPa, ductility of 41.7%, and modulus of elasticity of 55.3 MPa.

The modulus of elasticity of the two-layer specimen is 8.3 times greater than that of the single-layer specimen.

5.3.2 Computational Simulation Results

The results of the computational simulations of the strain–stress state of the B16/2 specimen are represented as a summary of the displacement of the matrix and

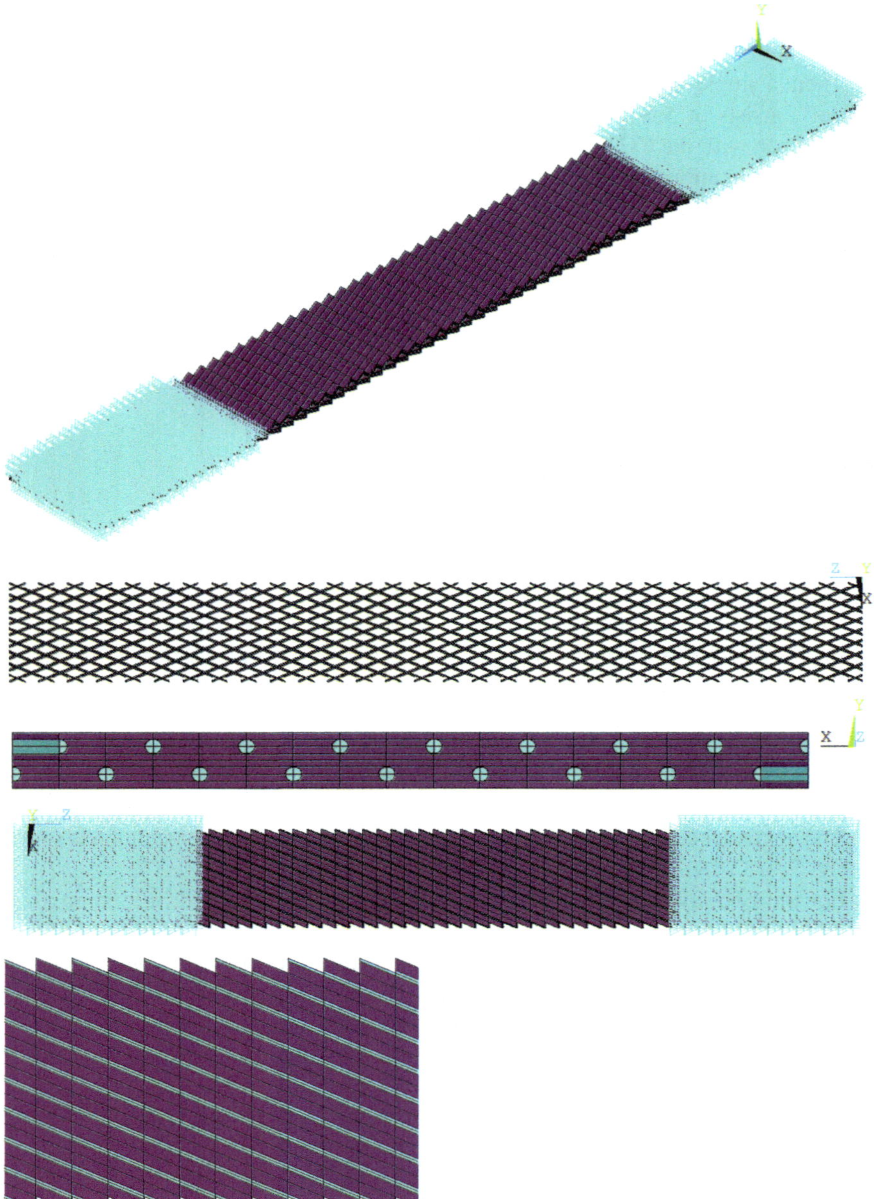

Fig. 5.2 Two-layer computational model B16/2 with details of cords and boundary conditions

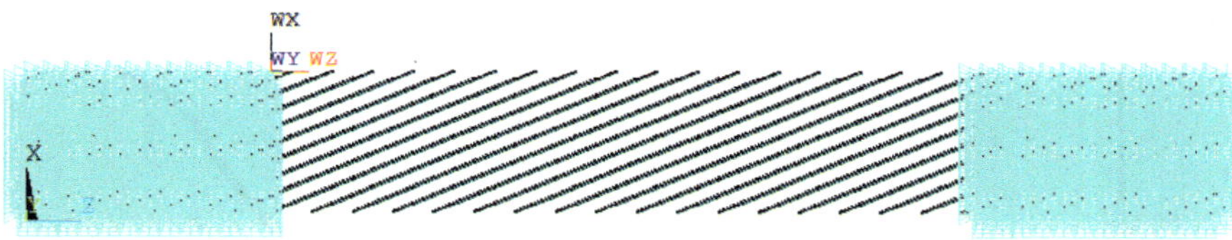

Fig. 5.3 Single-layer computational model CCT33

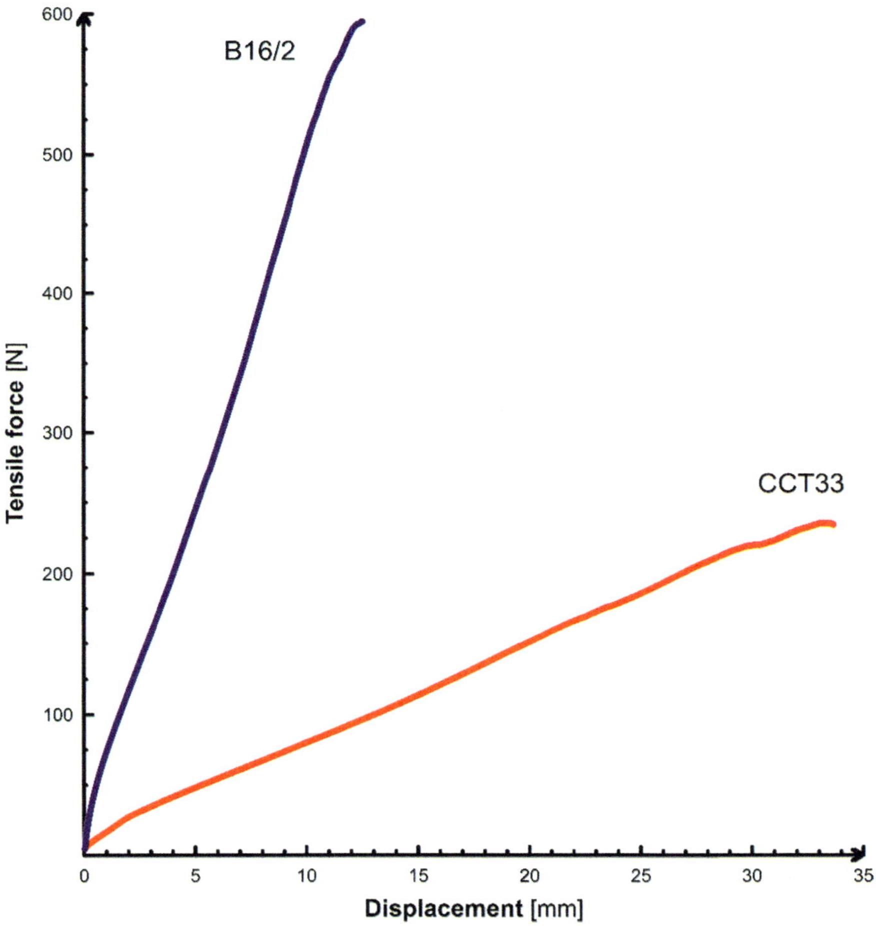

Fig. 5.4 Tensile force–displacement test dependencies

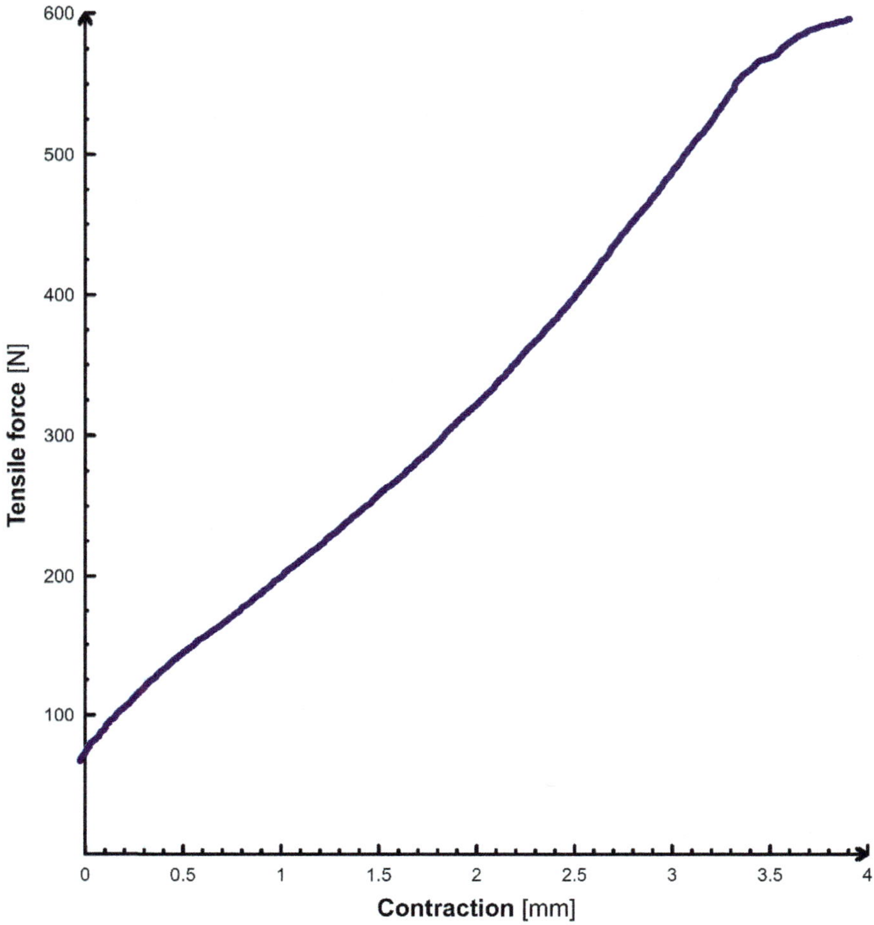

Fig. 5.5 Tensile force–contraction test dependence for B16/2

steel cords (Figs. 5.10 and 5.11), and the displacement of the cords on the *x*-axis (Fig. 5.12).

The reaction force–displacement dependence of the simulation B16/2 compared to the experimental dependence is shown in Fig. 5.13. The force–true longitudinal/transverse strain dependences of the simulation B16/2 compared to the experimental dependences are shown in Fig. 5.14.

The contraction (change of width)–displacement dependence of the simulation B16/2 compared to the experimental dependence is shown in Fig. 5.15.

The results of the computational simulations of the strain–stress state of the CCT33 specimen are represented as a summary of the displacement of the matrix and steel cords (Figs. 5.16 and 5.17), and the displacement of the cords on the *x*-axis (Fig. 5.18).

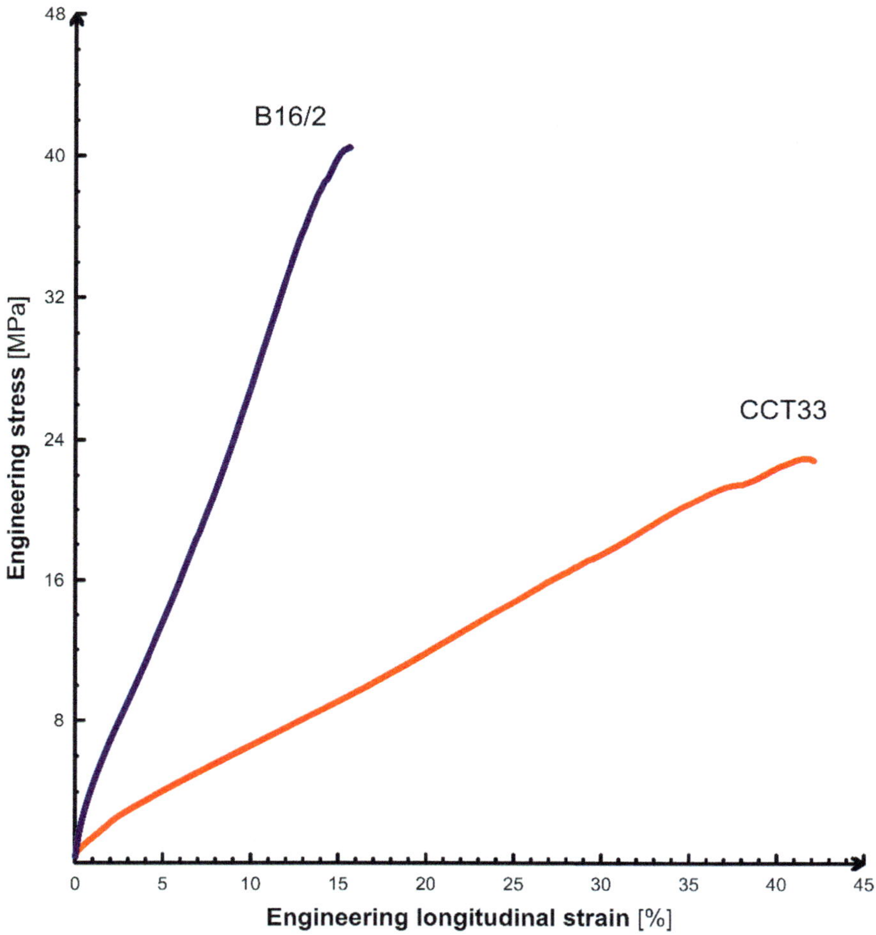

Fig. 5.6 Engineering stress–engineering longitudinal strain test dependencies

The reaction force–displacement dependence of the simulation CCT33 compared to the experimental dependence is shown in Fig. 5.19. The force–true longitudinal/transverse strain dependences of the simulation CCT33 compared to the experimental force–true longitudinal strain dependence are shown in Fig. 5.20.

The contraction–displacement dependence of the simulation CCT33 is shown in Fig. 5.21.

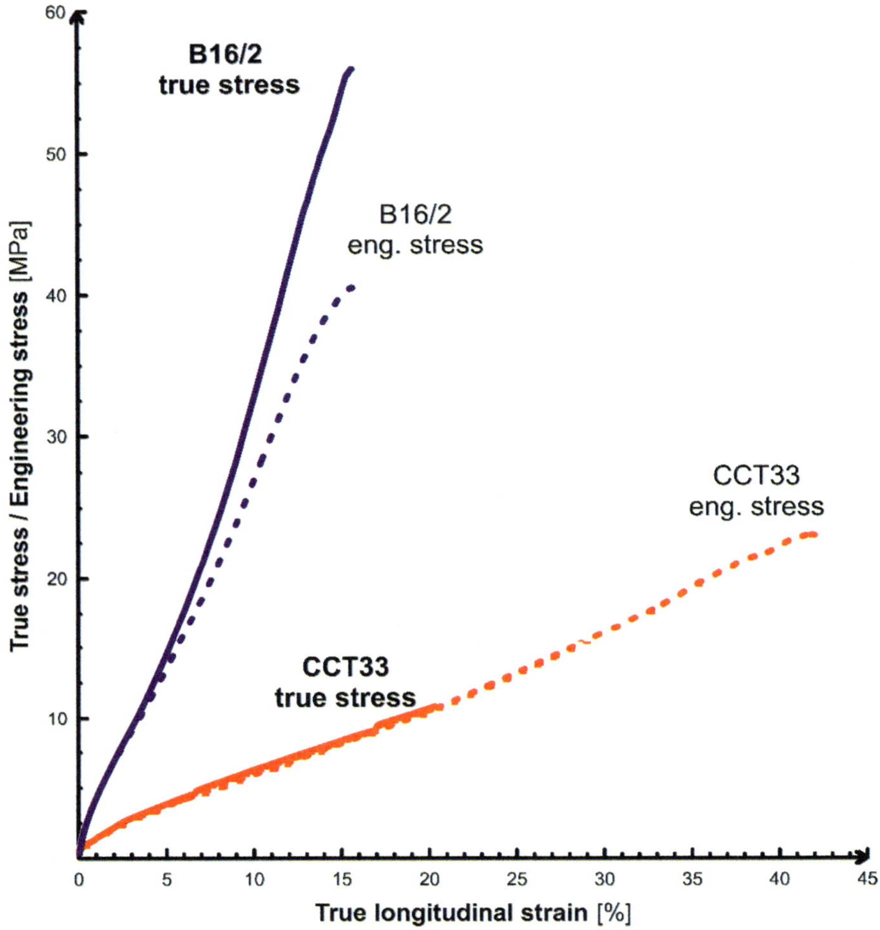

Fig. 5.7 True/engineering stress–true longitudinal strain test dependencies

5.4 Conclusions

On the basis of experiments, the material parameters of polymer composites are obtained. The experimental results were used for comparison with the results from the computational simulation. The difference in tensile force at the same deformation between the simulations and the experiments of the composite samples reached a maximum difference of 9.2%. The results of the computational simulation provide a better understanding of the mechanical properties of composites with steel cords with specific angles. Composite simulations are important for practice because the results can be used to simulate passenger car tire casings since the tire casing consists

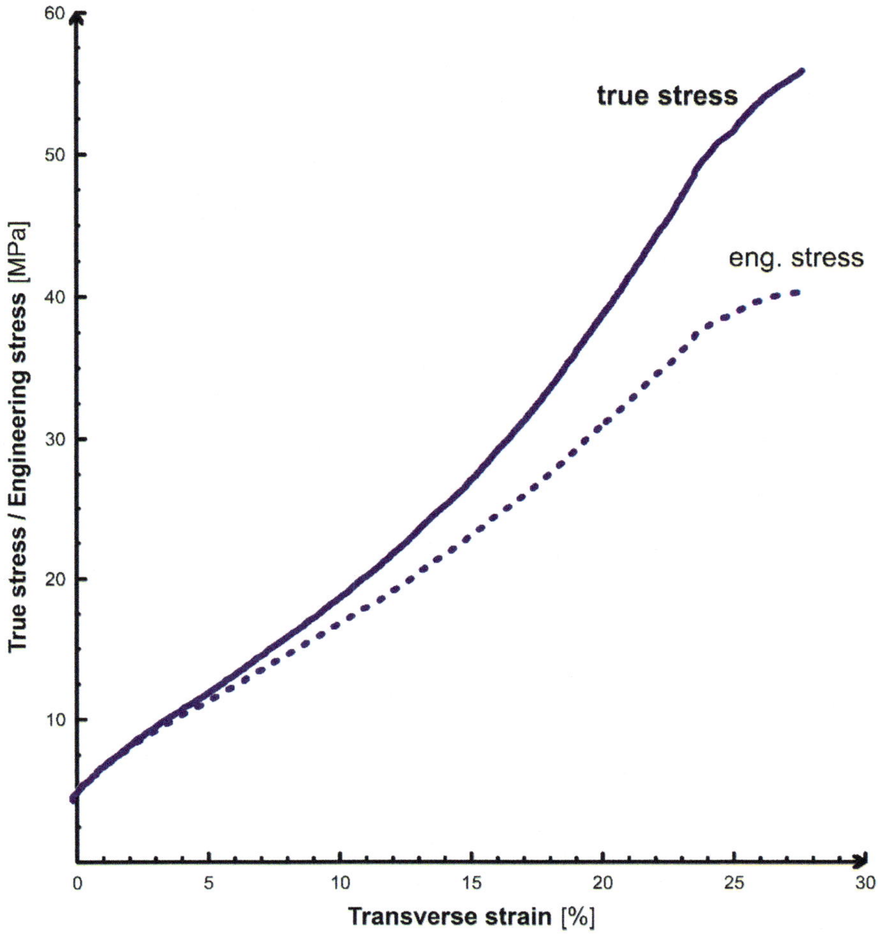

Fig. 5.8 True/engineering stress–transverse strain test dependence for B16/2

of elastomer parts and steel cords in the tire crown as reinforcement. The fully automatic ANSYS APDL procedure is proposed for the creation of computational models based on geometric and material parameters. Therefore, the models are prepared quickly and the calculation setup guarantees fast convergence. The Shimadzu test machine can be equipped with a temperature chamber for a temperature range of -60–180 °C. Other results from computational simulations and experiments of composites and polymers will be continuously published on the website www.laborator.sk, which is an output of the grant project the Cultural and Educational Grant Agency of the Slovak Republic project KEGA 003TnUAD-4/2022 on the topic "Simulations

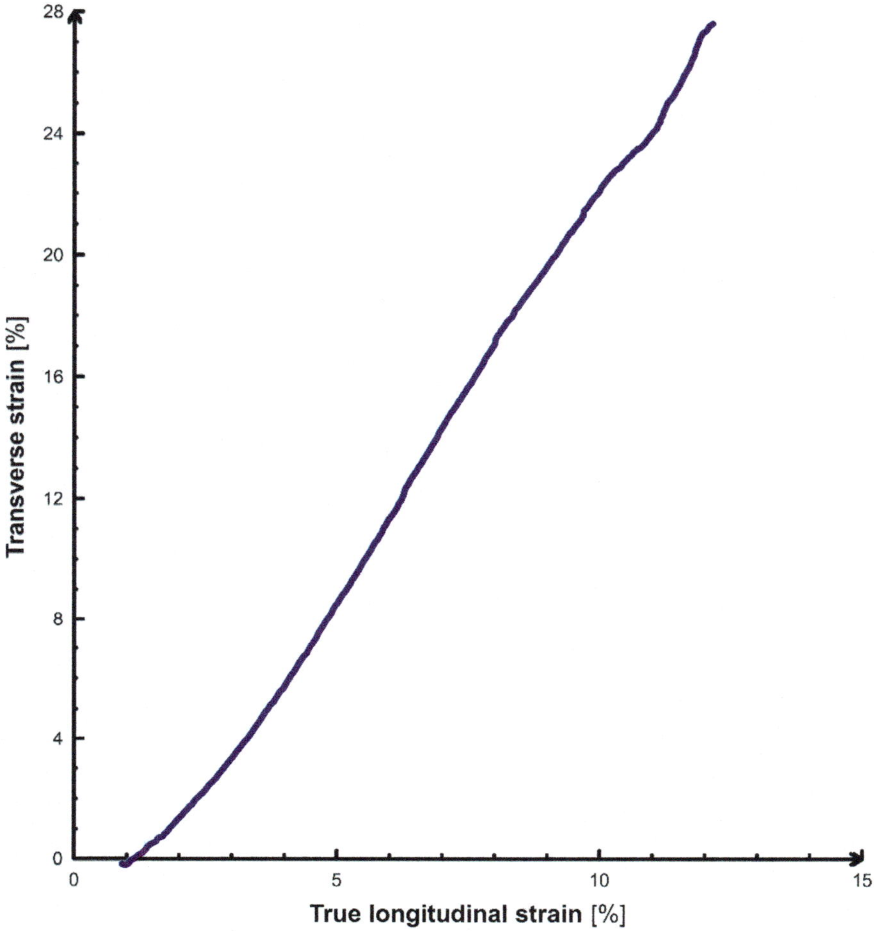

Fig. 5.9 True strain–transverse strain test dependence for B16/2

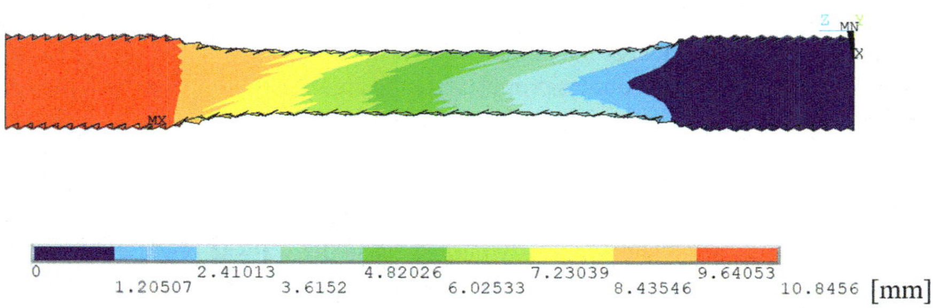

Fig. 5.10 Summary of the displacement of matrix of simulation B16/2

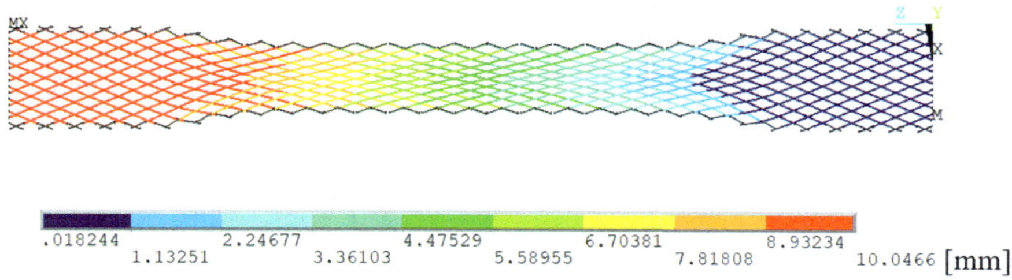

Fig. 5.11 Summary of the displacement of cord of simulation B16/2

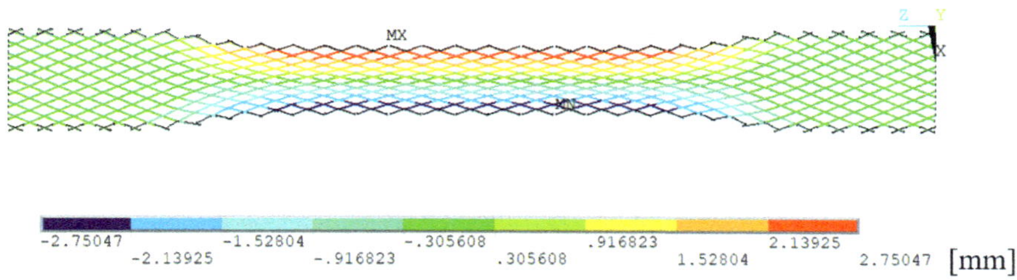

Fig. 5.12 Displacement of the cords on the x-axis of simulation B16/2

of basic and specific experiments of polymers and composites based on experimental data in order to create a virtual computational–experimental laboratory for mechanical testing."

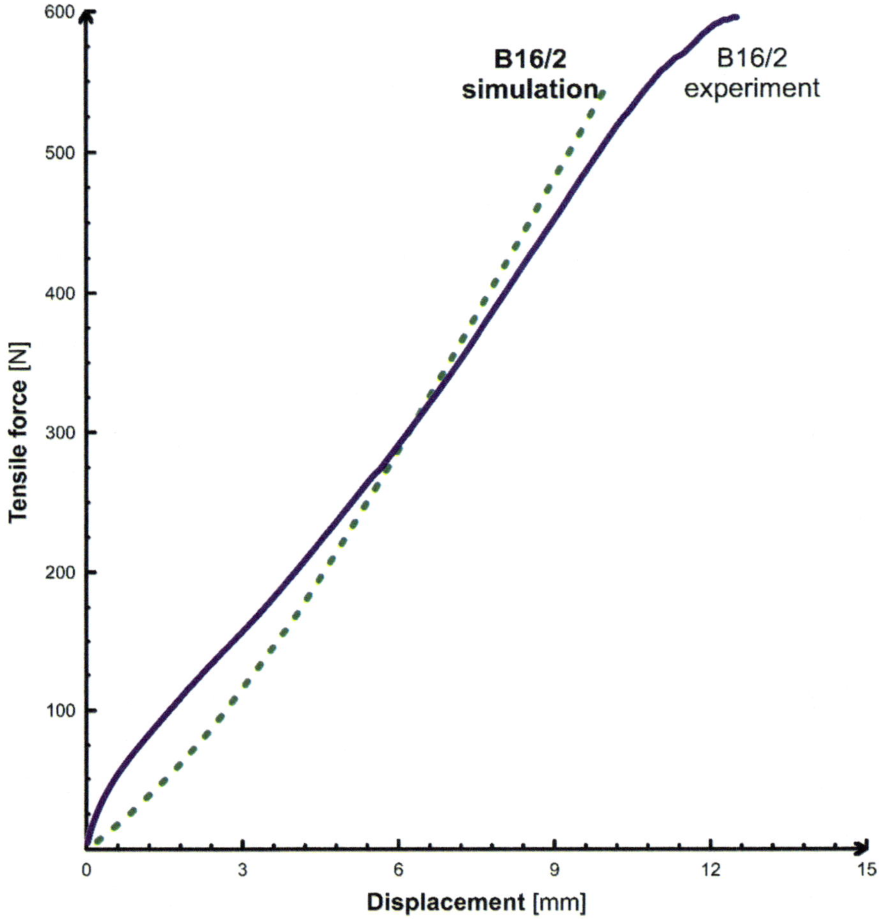

Fig. 5.13 Reaction force–displacement dependence of simulation B16/2 and comparison with experiment

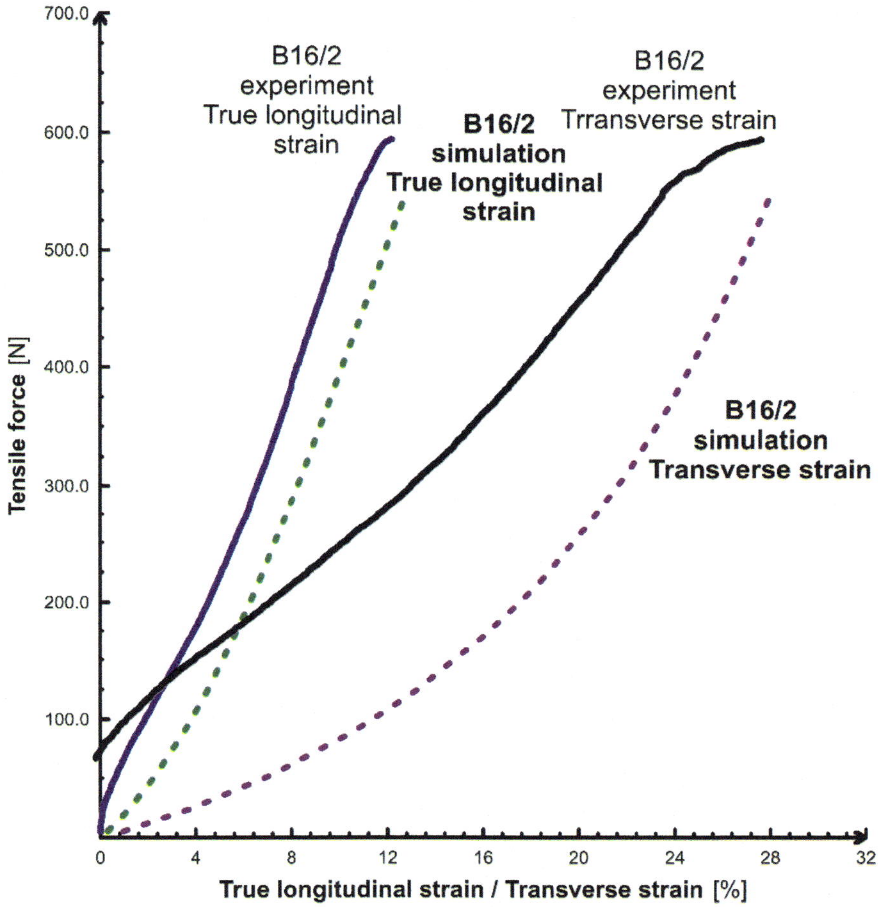

Fig. 5.14 Force–true longitudinal/transverse strain dependences of simulation B16/2 and comparison with experiment

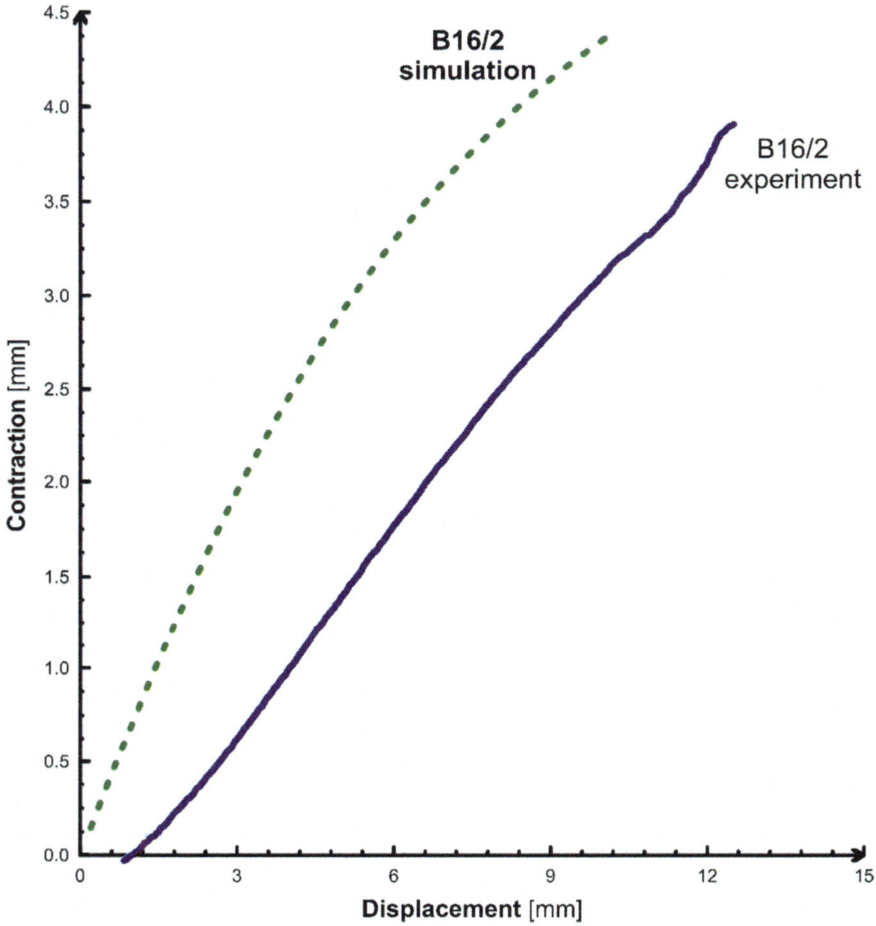

Fig. 5.15 Contraction–displacement dependence of simulation B16/2 and comparison with experiment

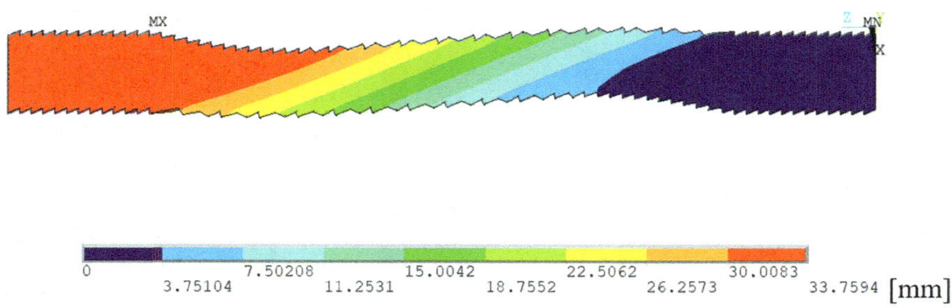

Fig. 5.16 Summary of the displacement of matrix of simulation CCT33

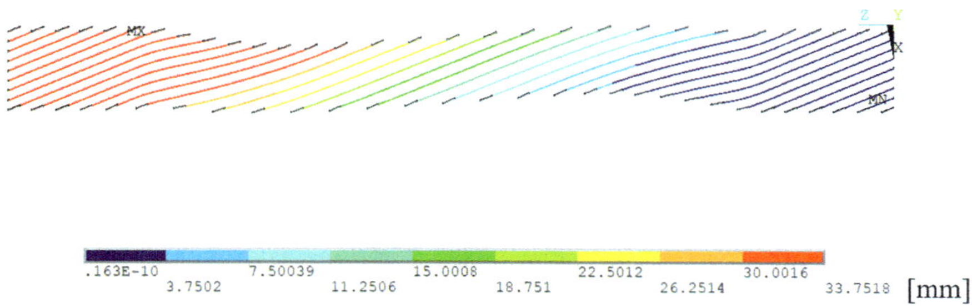

Fig. 5.17 Summary of the displacement of cord of simulation CCT33

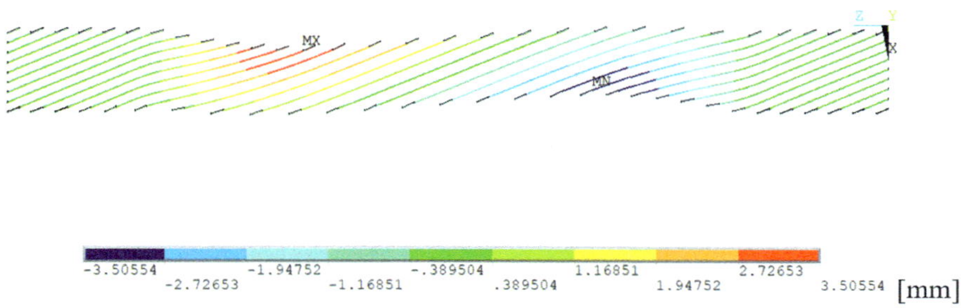

Fig. 5.18 Displacement of the cords on the x-axis of simulation CCT33

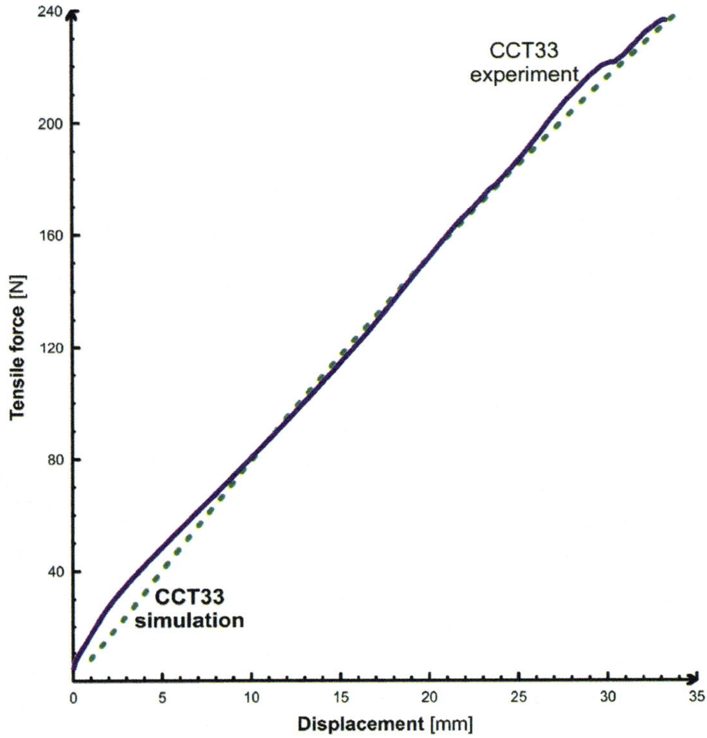

Fig. 5.19 Reaction force–displacement dependence of simulation CCT33 and comparison with experiment

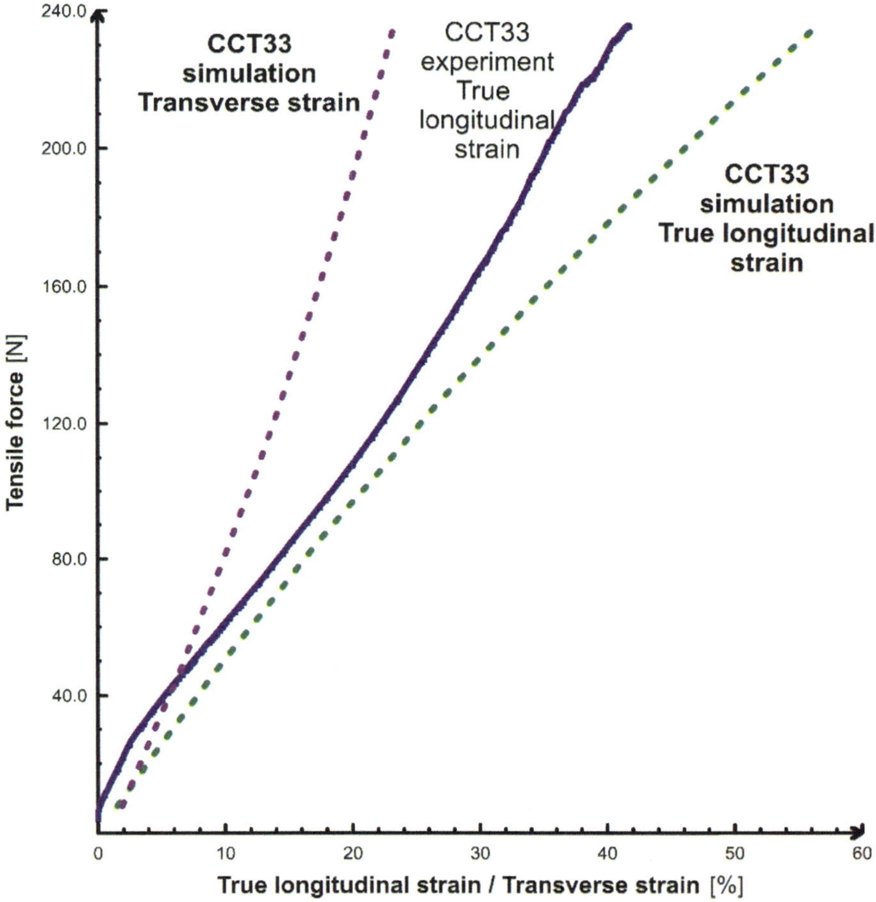

Fig. 5.20 Force–true longitudinal/transverse strain dependences of simulation CCT33 and comparison with experiment

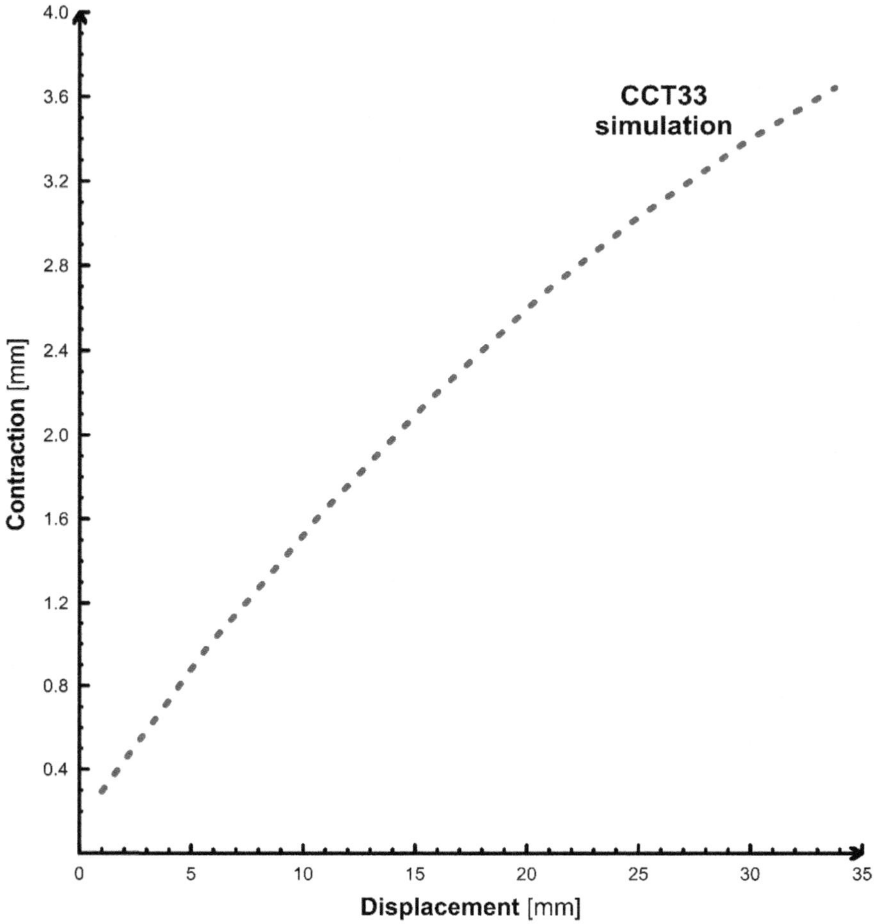

Fig. 5.21 Contraction–displacement dependence of simulation CCT33

Acknowledgements This research work has been supported by the Cultural and Educational Grant Agency of the Slovak Republic (KEGA), project No. 003TnUAD-4/2022 "Simulations of basic and specific experiments of polymers and composites based on experimental data in order to create a virtual computational-experimental laboratory for mechanical testing," and the project APVV-23-0342.

References

Krmela J (2017) Tire casings and their material characteristics for computational modeling. Oficyna wydawnicza slowarzyszenia menadzerów jakošci i produkcji, Czestochowa

Krmela J (2021) The influence of temperature and other parameters on the tensile properties of polymer composites and polymers under cyclic loading. Oficyna wydawnicza slowarzyszenia menadzerów jakošci i produkcji, Czestochowa (2021)

Krmela J, Krmelová V, Artyukhov A, Sadjiep S, Bakošová A (2021) Computational simulation of the shear test of a multi-layered long-fibre composite with a polymer matrix. In: IOP conference series: materials science and engineering, vol 1199. p 012075. https://doi.org/10.1088/1757-899X/1199/1/012075

Krmela J, Krmelová V, Artyukhov A, Lex C, Ondrušová D (2023) Simulations of tests of polymeric composites based on experimental data. In: Öchsner A, Altenbach H (eds) Engineering design applications V. Advanced Structured Materials, vol 171, pp 53–57. Springer, Cham (2023). https://doi.org/10.1007/978-3-031-26466-5_4

Popelar CF, Liechti KM (2003) A distortion-modified free volume theory for nonlinear viscoelastic behavior. v 7:89–141. https://doi.org/10.1023/A:1025625430093

Rao S, Daniel IM, Gdoutos EE (2004) Mechanical properties and failure behaviour of cord/rubber composites. Appl Compos Mater 11(6):353–375. https://doi.org/10.1023/B:ACMA.0000045312.61921.1f

Jr Sharpe WN (ed) (2008) Springer handbook of experimental solid mechanics. Springer, New York. https://doi.org/10.1007/978-0-387-30877-7

Chapter 6
Modeling Air Purification with Porous Materials

Tatiana Kudryashova, Sergey Polyakov, and Nikita Tarasov

Abstract The work discusses an important environmental problem—air purification from nanoscale solid pollutants. One of the most widely used filtration methods is passing the air through a granular sorbent, which forms a porous insert in the flow area. Effective operation of this technology requires the creation of digital twins, including the use of mathematical models to generate recommendations for replacing the active substance, its regeneration or disposal. The main difficulty in modeling air purification with porous materials is the correct description of the movement of the gaseous medium and impurities in the entire filter area. It is necessary to describe the flow in the zone free from granules and in the zone of active capture of impurities by the porous material. The paper proposes one of the approaches to open modeling "free flow"—"porous medium" system. It based on the quasi-hydrodynamic system of equations for gas flow and the convection–diffusion equation for impurities. For the model, a computational algorithm has been developed based on the finite volume method. For calculations, the staggered grids are applied, based on distribution of flow parameters into nodes and cells of the computational grid. Using the program implementation of the algorithm, a series of computational experiments were carried out to describe the flow of a medium through a region containing a porous insert. Comparison of the results of the calculations with the data obtained by the finite element software Ansys Fluent showed the consistency of the numerical approach and the developed application.

Keywords Mathematical modeling · Porous media · Free flow · Quasi-hydrodynamic system of equations · Ecology · Air purification

---

T. Kudryashova (✉) · S. Polyakov · N. Tarasov
Keldysh Institute of Applied Mathematics, RAS, Moscow, Russia
e-mail: kudryashova@imamod.ru

S. Polyakov
e-mail: polyakov@imamod.ru

N. Tarasov
e-mail: nikita_tarasov@imamod.ru

A. Öchsner and H. Altenbach (eds.), *Engineering Design Applications VII*, Advanced Structured Materials 230, https://doi.org/10.1007/978-3-031-84346-4_6

6.1 Introduction

Environmental issues are one of the most important in the modern world. One of its main objectives is to reduce the anthropogenic impact on the atmosphere. Indeed, many researchers associate the global climate changes observed today with pollution from industrial and agricultural activities (Afifa and K., Hussain, N., Ashraf, M.H., Saleem, M.Z. 2024). In addition, various toxic aerosols are concentrated in residential areas and cause inflammation of the mucous membranes, and with prolonged exposure, serious diseases of the eyes and lungs (Yoda et al. 2020). In this regard, many countries are developing programs to limit and standardize industrial emissions into the environment (Jafari et al. 2021). The main method of implementing these restrictions is the installation of cleaning devices in front of the outlet channels and pipes. The use of air filters is not limited to controlling emissions into the environment. There are industries, such as medicine, microelectronics, etc., in which the use of dust-free rooms is the main factor in product quality. The presence of even a millionth part of some impurities can cause significant damage to the technological process of such a facility (Baturina et al. 2018). In addition, air purification in public and residential premises has become especially important in the fight against the spread of COVID-19 (Mallakpour et al. 2022; Bu et al. 2021).

Today, there are many different filtration methods based on various physicochemical and electromagnetic effects that ensure the removal of harmful impurities from the air (Han et al. 2021). However, it is not possible to completely solve the purification problems without continuing detailed scientific research, most of which is currently being carried out using mathematical and computer modeling methods. This work is in line with such researches. It proposes a complex methodology for calculating the flow of polluted air through porous materials. It continues the cycle of our previous studies (Polyakov et al. 2020, 2021) and is devoted to the development of new computational technologies.

6.2 Problem Formulation

Our works (Polyakov et al. 2020, 2021) consider a multiscale approach formed with the use of different spatial levels for the description of purification process description. Such separate research allows us to use significantly different mathematical models. For example, continuum mechanics models are used at the macroscopic level. Mesoscale processes corresponding to granule sizes can be described by the large-particle method. Microscopic processes in the pores of granules (absorption of pollutants on the internal surfaces of the pores and chemical transformations) require the use of an atomistic approach, for example, molecular dynamics methods. The transition from one level to another is carried out using balance relationships and criteria adopted within the framework of the calculation.

The practice of using a multiscale approach has revealed the need for additional correction of the macroscopic level model. Previously, we described the granular structure of sorption elements by directly specifying their geometric dimensions and position. However, when modeling not only laboratory, but also real household and industrial filters, this description technique leads to the need to use overly detailed grids. Indeed, the overall dimensions of filtration devices can reach meters, with an average characteristic granule size of the order of several millimeters (Bessonov et al. 2015).

The solution of this situation lies in the additional averaging of parameters and macroscopic equations over a representative elementary volume (REV) and the introduction of medium porosity (Das et al. 2017). This method eliminates the need to consider each sorbent granule separately. The granules form some kind of skeleton that affects the air flow in the filter. The skeleton is characterized by two main parameters—porosity and permeability. Using these parameters, the equations describing the flow and convection–diffusion of the impurity are modified.

The introduction of a porous medium leads to certain difficulties. Let us explain this with an example. Figure 6.1 shows the simplest case of a porous plug located in the center of the flow region. In this situation, a unified (end-to-end) description of the gas flow in the free and porous regions is required.

The methods used within the end-to-end description can be divided into two groups. Within the first group, in the free flow region, the standard system of Navier–Stokes equations is solved, and in the porous zone, certain laws of motion through the porous medium are applied (Darcy's law, Forchheimer, etc.) (Rybak et al. 2021; Costa et al. 2004). In this case, special exchange conditions are set at the interface of the two zones. The second group of methods is based on the formulation of a general system of equations that describe flow in both free and porous regions. These approaches

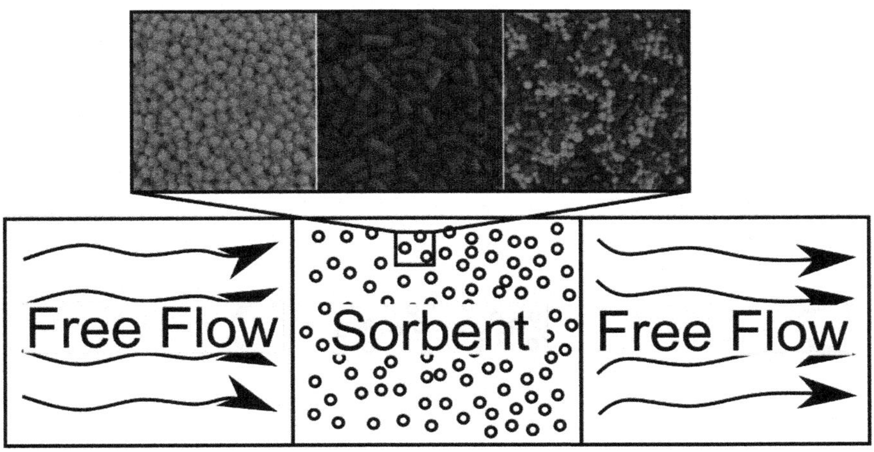

Fig. 6.1 Scheme of sorption purification

are known as volume-averaged Navier–Stokes (VANS) (Duzel and Martin 2020) or non-Darcy models (Das et al. 2017).

Each of the mentioned groups of methods has its own advantages. The approach based on the VANS model seems to be the most suitable and flexible in the case of modeling air filtration, which we consider below.

6.3 Mathematical Model

The VANS system of equations with a stationary porous skeleton under the condition of temperature equilibrium can be written in the following form (Duzel and Martin 2020):

$$\frac{\partial(\varepsilon\rho)}{\partial t} + \nabla\left(\varepsilon\rho \boldsymbol{u}_g\right) = 0, \tag{6.1}$$

$$\frac{\partial(\varepsilon\rho \boldsymbol{u_g})}{\partial t} + \nabla\left(\varepsilon\rho \boldsymbol{u}_g \otimes \boldsymbol{u}_g\right) = -\varepsilon\nabla p + \nabla(\varepsilon \boldsymbol{\Pi}_{NS}) - \overline{F}_p \boldsymbol{u_g}, \tag{6.2}$$

$$\frac{\partial E}{\partial t} + \nabla\left(\varepsilon\rho \boldsymbol{u}_g H\right) = \nabla(\varepsilon\chi\nabla T) + \nabla(\varepsilon \boldsymbol{\Pi}_{NS} \boldsymbol{u}_g) - \overline{F}_p {\boldsymbol{u_g}}^2, \tag{6.3}$$

where ε is the porosity, ρ is the gas density, $\boldsymbol{u_g} = \frac{\boldsymbol{u}}{\varepsilon}$ is the gas velocity, $\boldsymbol{u}$ is the Darcy velocity, p is the pressure, $\boldsymbol{\Pi}_{NS}$ is the Navier–Stokes tensor, $\overline{F}_p$ is the momentum drain associated with gas movement through the porous medium, E is the total energy, H is the enthalpy, χ is the coefficient of thermal conductivity, T is the temperature. Also, system (1)–(3) is supplemented with an equation of state (the relationship between temperature, density and pressure), as well as initial and boundary conditions.

The system of Eqs. (6.1)–(6.3) can be significantly simplified using the following characteristics of the granular air filtration process. Since the gas flow speed, for its effective purification is much less than the speed of sound, this allows us to consider this process in a weakly compressible approximation. In addition, since heat is not released during the sorption capture of impurities, and its operation occurs at normal ambient temperature, an isothermal model can be used. Also, when constructing the model, we use a quasi-hydrodynamic system of equations (QHD) (Elizarova 2009), which has a number of useful computational properties, including increased stability. Then, in a dimensionless formulation, system (1)–(3) is transformed to the following form (the values given below are dimensionless):

$$\nabla(\boldsymbol{u} - \varepsilon \boldsymbol{w}) = 0, \tag{6.4}$$

$$\frac{\partial \boldsymbol{u}}{\partial t} = \nabla\left(\frac{1}{Re}\left(\nabla \otimes \boldsymbol{u} + \nabla \otimes \boldsymbol{u}^T\right) + (\boldsymbol{u} \otimes \boldsymbol{w})\right) - \nabla\left(\frac{1}{\varepsilon}\boldsymbol{u} \otimes \boldsymbol{u} - \boldsymbol{w} \otimes \boldsymbol{u}\right) - \varepsilon\nabla p - \varepsilon F_p \boldsymbol{u}, \tag{6.5}$$

$$w = \tau\left[(\boldsymbol{u}, \nabla)\left(\frac{1}{\varepsilon}\boldsymbol{u}\right) + \varepsilon\nabla p + \varepsilon F_p \boldsymbol{u}\right], \tag{6.6}$$

where $\boldsymbol{w}$ is the regularization correction, Re is the Reynolds number, τ is a small relaxation parameter. The expression for calculating the pressure is obtained by substituting Eq. (6.6) into (4). The Poisson equation is formed as a result of the substitution. The character of the F_p function depends on the parameters of the medium and the shape of the skeleton of the porous plug. The function F_p in dimensionless form, with spherical pores and sufficiently large Reynolds numbers has the following form (Das et al. 2017):

$$F_p = \frac{1}{ReDa} - \frac{f}{\sqrt{Da}}|\boldsymbol{u}|. \tag{6.7}$$

Here $Da = \frac{K}{d^2}$ is the Darcy number, K is the permeability coefficient, d is the characteristic pore size, $f = \frac{1.8}{\sqrt{180\varepsilon^5}}\varepsilon$ is the empirical coefficient depending on the skeletal structure (given here for uniform spherical pores).

Further modeling of the evolution of impurities, due to their low concentration and weak influence on the flow, can be carried out using convection–diffusion equations (Polyakov et al. 2020, 2021). The system of convection–diffusion equations in the presence of a porous medium, have the following form (Whitaker 1997):

$$\frac{\partial(\varepsilon C_k)}{\partial t} = \nabla(\varepsilon D_k \nabla C_k) - \nabla\left(\frac{\boldsymbol{u}}{\varepsilon}\varepsilon C_k\right) - Q_k C_k, \tag{6.8}$$

where C_k is the average concentration of type k impurity, D_k is the diffusion coefficient of type k impurity, Q_k is the coefficient of volumetric absorption of type k impurities by a porous medium.

Thus, the "free flow"—"porous body" flow model is formed by a system of Eqs. (6.4)–(6.6), supplemented by boundary and initial conditions. The evolution of the concentration of impurities of various types is described by Eq. (6.6). If there is no influence of impurity concentration on the flow, then the flow calculation can be carried out separately. In Eq. 6.8, pre-calculated flow parameters are used. Equations (6.4)–(6.6) and (6.8) must be calculated together at each time step, for example, when the porosity and mobility coefficient change during the accumulation of impurities.

6.4 Numerical Algorithm

The computational algorithm is obtained for time-explicit discretization of the time derivatives. The spatial derivatives were approximated by the finite volume method (Eymard et al. 2000). For the correct discretization of the constructed model, it became necessary to separate the flow parameters. The velocities are determined in the nodes, and pressures are in the cells of the computational grid. This necessity

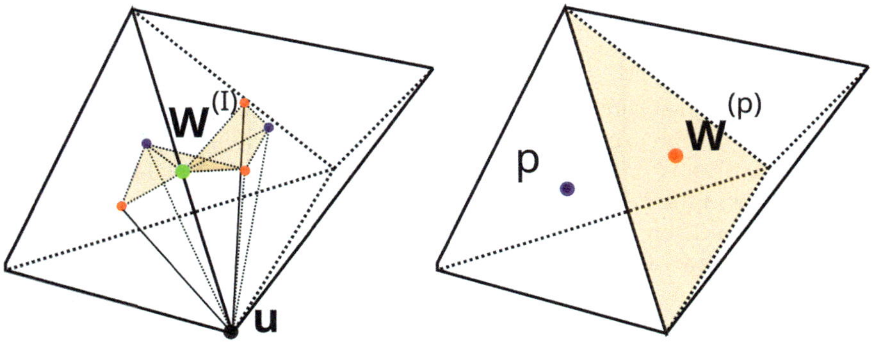

Fig. 6.2 Control volume for velocities (left) and pressure (right)

arises from the peculiarity of constructing the discrete analogue (7), in which the values of porosity and permeability are set for the cells as a whole. This leads to a discontinuity in the momentum drain in Eq. (6.5) when approximating the velocity at the cell centers and the appearance of non-physical oscillations at the interface of the free region and the porous insert. Figure 6.2 shows the generation of control volumes for velocity (left) and pressure (right). The symbols $\boldsymbol{W}^{(I)}$ and $\boldsymbol{W}^{(p)}$ denote the flows of the discrete analogue of Eq. (6.5) and the Poisson equation for pressure, accordingly. The blue dots represent the centers of cells, the red dots represent the centers of edges, and the green dots represent the centers of edges of the grid model.

The general computational algorithm consists of an approximate solution of the pressure equation by the preconditioned conjugate gradient method (Milyukova 2016), and subsequent calculation of the velocity vector on a new time layer. These actions are repeated cyclically until the required time value is reached.

The program implementation of the specified algorithm was performed in the C + + programming language (Stroustrup 2013). To implement grid operators in matrix–vector form, the Eigen3 linear algebra library (Eigen xxxx) was used. In order to speed up calculations, the program was parallelized using the geometric parallelism method. The OpenMPI standard (Gabriel et al. 2004) was used as a parallel programming technology.

6.5 Modeling Results

In this section, we present test calculations for three different Reynolds numbers in the case of two-dimensional plane geometry. The computational domain and its dimensional parameters are shown in Fig. 6.3 (the porous insert is marked in orange). On the left boundary of the computational domain, a Poiseuille profile normalized to unity was set. At the upper and lower boundaries, there are adhesion conditions. Constant pressure is saved on the right border. In addition, the following parameters

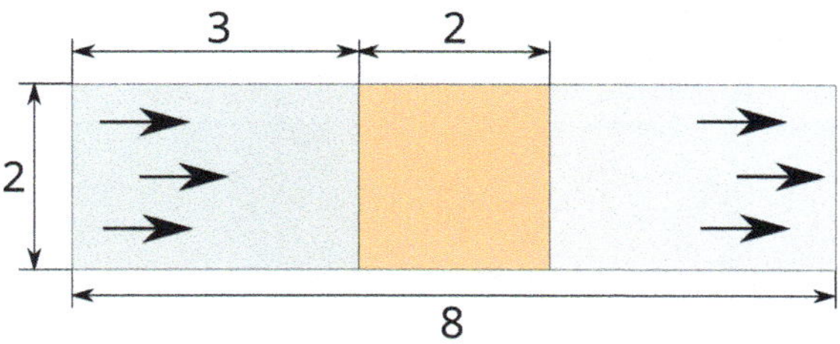

Fig. 6.3 Computational domain

of flow and porous domain were used: $Re = 100, 500, 1000$, $Da = 0.01$, $\gamma = 0.1$. The computational grid consisted of 40,501 nodes and 40,000 quadrangles.

The pressure distributions for the performed computational experiments have a similar nature. Figure 6.4 shows the pressure distribution for calculation with $Re = 1000$. The maximum pressure values obtained at the inlet are equal to 47.64, 41.71, 40.04 for $Re = 100, 500, 1000$, respectively. From Fig. 6.4 it is clear that the main decrease in pressure occurs in the region of the porous medium. This is consistent with the ideas about the nature of the process (Costa et al. 2004).

Figure 6.5 shows the distributions of the velocity modulus for three Reynolds numbers. They allow us to note the following features of the flow. Up to the porous plug, the flow has a formed Poiseuille profile. Then, when approaching the porous medium, a transverse component is formed, distributing the longitudinal flow uniformly across the cross-section. Inside the porous medium, with the exception of the small near-wall zone, the longitudinal velocity component is distributed almost uniformly. The main differences in the calculations are observed when leaving the insert—the higher the Reynolds number, the slower the Poiseuille profile is formed. This flow behavior is confirmed in the works of other authors (Michael et al. 2004).

To verify the results obtained, similar computational experiments were carried out in the Ansys Fluent software (ANSYS Fluent User's Guide 2021). Figure 6.6 shows the normalized distributions of the velocity magnitude along the centerline of the considered geometry, obtained using our original development (QHD) and the Ansys Fluent package. The results obtained using both methods are in good agreement with

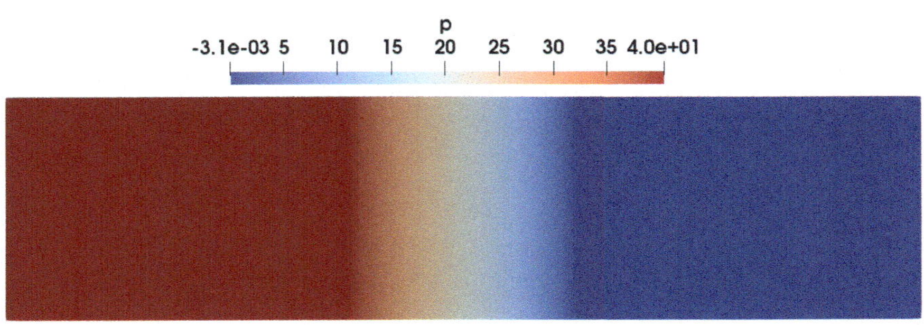

Fig. 6.4 Pressure distributions for $Re = 1000$

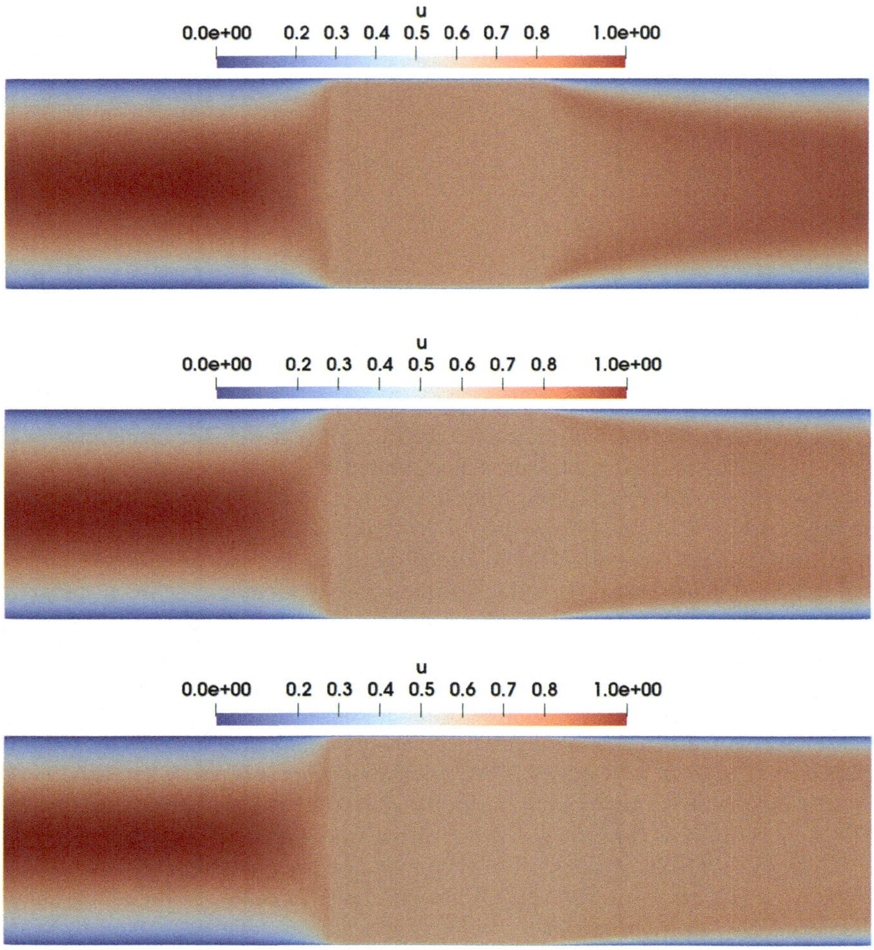

Fig. 6.5 Velocity modulus distributions for $Re = 100, 500, 1000$ from top to bottom

each other. It confirms the consistency of the constructed computational algorithm and its program implementation.

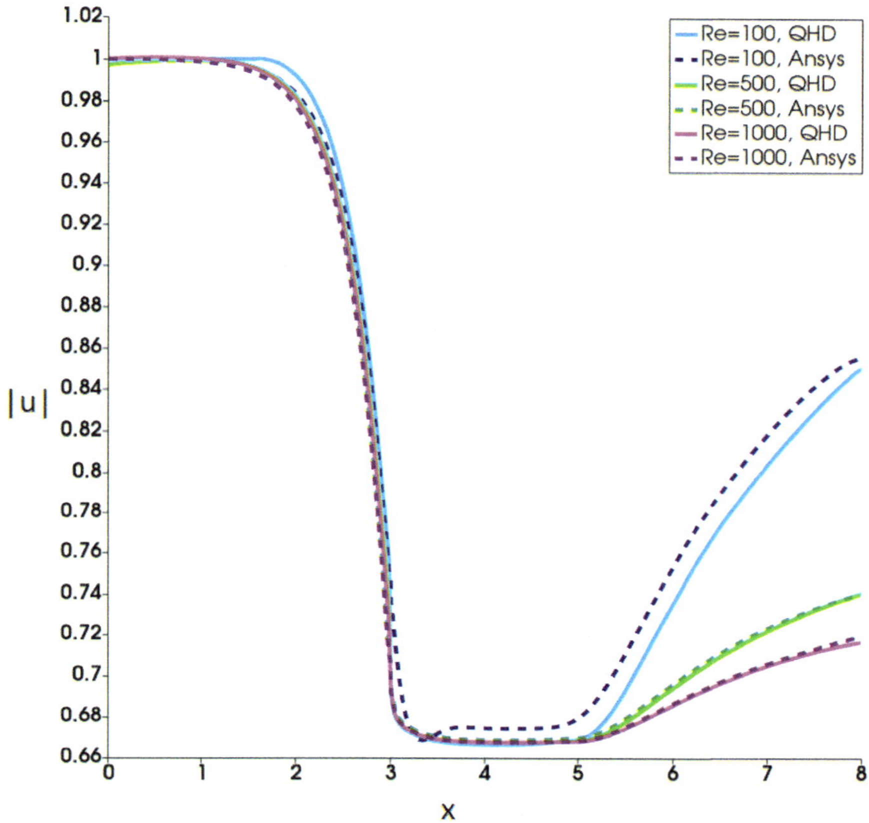

Fig. 6.6 Velocity modulus ($|\boldsymbol{u}|$) distributions along the centerline ($y = 0$) (dimensionless variables)

6.6 Conclusion

As part of a numerical analysis of the process of air filtration by granular sorbent, a new unsteady mathematical model of flow in the "free flow"—"porous medium" system was constructed. The resulting system of equations is based on a quasi-hydrodynamic approach applied under conditions of porosity and the presence of an additional source of momentum changes. Some of the computational complexity of the constructed model was the implementation of end-to-end calculation. This was overcome by applying the finite volume method on staggered grids. For this purpose, the components of the velocity vector are distributed at the grid nodes, and the pressures are distributed at the centers of the cells. Based on the numerical methodology, a parallel program was developed and test experiments were carried out to calculate flow parameters for several values of the Reynolds number. In order to verify the developed code, a series of calculations were carried out in the Ansys Fluent software. Comparison of the obtained data allows us to conclude that the proposed computational algorithm and its software implementation are correct. In the future, we intend to use the developed code for the purpose of numerical analysis of filtration processes under conditions close to reality, for example, for purification systems of real three-dimensional geometry, taking into account the properties of sorbents.

References

Afifa, Arshad K, Hussain N, Ashraf MH, Saleem MZ (2024) Air pollution and climate change as grand challenges to sustainability. Sci Total Environ 928:172370. https://doi.org/10.1016/j.scitotenv.2024.172370

ANSYS fluent user's guide (2021) Release 2021R2. Inc., ANSYS

Baturina OA, Dyatkin B, Reshetenko TV (2018) Influence of air impurities on the performance of nanostructured PEMFC catalysts. In: Li F, Bashir S, Liu J (eds) Nanostructured materials for next-generation energy storage and conversion, pp 407–441. Springer, Berlin, Heidelberg. https://doi.org/10.1007/978-3-662-56364-9_14

Bessonov AA, Kulyukhin SA, Konovalova NA, Mizina LV, Rumer IA (2015) Granular sorbents for passive environment protection system during severe accidents with total loss of power supply at NPPs. JPEE 3(4):29–34. https://doi.org/10.4236/jpee.2015.34005

Bu Y, Ooka R, Kikumoto H, Oh W (2021) Recent research on expiratory particles in respiratory viral infection and control strategies: a review. Sustain Cities Soc 73:103106. https://doi.org/10.1016/j.scs.2021.103106

Costa VAF, Oliveira LA, Baliga BR, Sousa ACM (2004) Simulation of coupled flows in adjacent porous and open domains using acontrol-volume finite-element method. Numerical Heat Transfer, Part a: Applications 45(7):675–697. https://doi.org/10.1080/10407780490424839

Das MK, Mukherjee PP, Muralidhar K (2017) Modeling transport phenomena in porous media with applications. Springer Cham. https://doi.org/10.1007/978-3-319-69866-3

Duzel U, Martin A (2020) Modeling high velocity flow through porous media. AIAA 2020–0486. AIAA Scitech 2020 Forum. https://doi.org/10.2514/6.2020-0486

Eigen 3. URL https://eigen.tuxfamily.org/

Elizarova TG (2009) Quasi-gas dynamic equations. Springer, Verlag, New York. https://doi.org/10.1007/978-3-642-00292-2

Eymard R, Gallouet T, Herbin R (2000) Finite volume methods. In: Handbook of Numerical Analysis 7, 713–1018. https://doi.org/10.1016/S1570-8659(00)07005-8

Gabriel E, Fagg GE, Bosilca G, Angskun T, Dongarra JJ, Squyres JM, Sahay V, Kambadur P, Barrett B, Lumsdaine A, Castain RH, Daniel DJ, Graham RL, Woodall TS (2004) Open MPI: Goals, concept, and design of a next generation MPI implementation. In: Kranzlmuller D, Kacsuk P, Dongarra J (eds) Recent advances in parallel virtual machine and message passing interface. EuroPVM/MPI 2004. Lecture Notes in Computer Science, 3241, 97–104. Springer, Berlin, Heidelberg. https://doi.org/10.1007/978-3-540-30218-6_19

Han S, Kim J, Ko SH (2021) Advances in air filtration technologies: structure-based and interaction-based approaches. Materials Today Advances. 9:100134. https://doi.org/10.1016/j.mtadv.2021.100134

Jafari JA, Charkhloo E, Pasalari H (2021) Urban air pollution control policies and strategies: a systematic review. J Environ Health Sci Eng 19(2):1911–1940. https://doi.org/10.1007/s40201-021-00744-4

Mallakpour S, Azadi E, Hussain CM (2022) Fabrication of air filters with advanced filtration performance for removal of viral aerosols and control the spread of COVID-19. Adv Coll Interface Sci 303:102653. https://doi.org/10.1016/j.cis.2022.102653

Michael R, Delphine R, Pierre-Henri C, Alain B (2020) Modelling of flow through spatially varying porous media with application to topology optimization. arXiv:2004.10712 [physics.flu-dyn]. https://doi.org/10.48550/arXiv.2004.10712

Milyukova OY (2016) Combination of numerical and structured approaches to the construction of a second-order incomplete triangular factorization in parallel preconditioning methods. J Comput Math Math Phys 56:699–716. https://doi.org/10.1134/S096554251605016X

Polyakov SV, Karamzin YN, Kudryashova TA, Podryga VO, Puzyrkov DV, Tarasov NI (2020) Multiscale simulation of gas cleaning processes. J Math Models Comput Simul 12:302–315. https://doi.org/10.1134/S2070048220030151

Polyakov SV, Kudryashova TA, Tarasov NI (2021) Application of the multiscale approach to simulation of air sorbent filtration. Dokl Math 104:297–300. https://doi.org/10.1134/S1064562421050094

Rybak I, Schwarzmeier C, Eggenweiler E, Rude U (2021) Validation and calibration of coupled porous-medium and free-flow problems using pore-scale resolved models. Comput Geosci 25:621–635. https://doi.org/10.1007/s10596-020-09994-x

Stroustrup B (2013) The C++ programming language. Fourth edition. Upper Saddle River, NJ, Addison-Wesley

Whitaker S (1997) Volume averaging of transport equations. In: Book: fluid transport in porous media, 1–60

Yoda Y, Tamura K, Adachi S, Otani N, Nakayama SF, Shima M (2020) Effects of the use of air purifier on indoor environment and respiratory system among healthy adults. IJERPH 17(10):3687. https://doi.org/10.3390/ijerph17103687

Chapter 7
Bentonite Microwave Modification and Its Impact on Polymer Materials

Zuzana Mičicová, Mariana Pajtášová, Slavomíra Božeková, Darina Ondrušová, Petra Skalková, and Maroš Dedinský

Abstract Microwave heating functions by converting electromagnetic energy into heat, which uniformly permeates the entire volume of the material undergoing heating. This process allows for the consistent and thorough removal of moisture from clay materials, even at lower temperatures, thereby reducing both the time and energy expenditure involved. In this study, we harnessed microwave heating to physically modify bentonite. Subsequently, we incorporated the modified bentonite as a partial replacement for carbon black content in polymer materials. Characterization of the physically modified bentonite was conducted using Fourier transform infrared spectroscopy (FTIR) and thermal analysis. Furthermore, the rheological and physical and mechanical properties of the polymeric materials were analyzed to assessment the influence of microwave modification of bentonite on the polymer matrix. In the assessment of polymeric materials, the QUV Accelerated Weathering Tester was also employed to evaluate their durability under various environmental conditions. The results indicated that substituting conventional fillers with bentonite microwave modification in polymer materials resulted in enhanced strength and ductility when compared to polymer materials containing unmodified bentonite.

Z. Mičicová (✉) · M. Pajtášová · S. Božeková · D. Ondrušová · P. Skalková · M. Dedinský
Faculty of Industrial Technologies, Alexander Dubček University of Trenčín, Púchov, Slovak Republic
e-mail: zuzana.micicova@tnuni.sk

M. Pajtášová
e-mail: mariana.pajtasova@tnuni.sk

S. Božeková
e-mail: slavomira.bozekova@tnuni.sk

D. Ondrušová
e-mail: darina.ondrusova@tnuni.sk

P. Skalková
e-mail: petra.skalkova@tnuni.sk

M. Dedinský
e-mail: maros.dedinsky@tnuni.sk

A. Öchsner and H. Altenbach (eds.), *Engineering Design Applications VII*, Advanced Structured Materials 230, https://doi.org/10.1007/978-3-031-84346-4_7

Keywords Bentonite · Microwave modification · Polymer materials · Physical and mechanical properties

7.1 Introduction

Rubber technology has made remarkable progress, particularly in the development of materials with improved tensile strength, hardness and elongation at break (Arawande et al. 2023; Bokobza 2019). To address the limitations of natural rubber-such as its low modulus and strength-and to adapt it for diverse applications in military, engineering, and biomedical fields, it is often compounded with various fillers (Chen et al. 2022; Hakimi et al. 2021). These fillers are critical to enhancing the performance of natural rubber by improving its strength, durability, electrical and thermal conductivity and barrier properties (Arawande et al. 2023; Bokobza 2019; Chen et al. 2022; Hakimi et al. 2021; Medupin et al. 2019). Derived from petroleum, carbon black is the most widely used filler in the rubber industry as it significantly improves the mechanical properties of rubber, making it indispensable for many industrial applications. However, the environmental impact and petroleum dependence of carbon black have prompted researchers to seek alternative "white" reinforcing agents (Sukumar and Menon 2008; El-Nashar et al. 2012; Ondrušová 2021).

One promising alternative is bentonite, a type of clay rock primarily composed of the clay mineral montmorillonite. Montmorillonite is a layered silicate clay mineral with the chemical formula $Al_{2-x}Mg_x(Si_4O_{10})(OH)_2{\cdot}(Na_x{\cdot}nH_2O)$. Its structure features an octahedral metal–oxygen sheet sandwiched between two tetrahedral $[SiO_4]$ sheets, forming a sandwich-like arrangement with a thickness of 0.96 nm. Sodium cations (Na^+) and water molecules are intercalated within these layers to balance the negative charge, and these intercalated substances can be replaced with other cations, such as magnesium (Mg^{2+}) and calcium (Ca^{2+}), or organic tetra-alkyl ammonium groups. This flexibility allows for extensive chemical modifications and diverse applications (Drobíková et al. 2020; Wan et al. 2020; Kurahatti 2010).

Bentonite's high cation exchange capacity and large surface area contribute to its valuable properties, including thixotropy, hydration, substantial swelling, bonding ability, impermeability, and plasticity (Mapossa et al. 2023; Şahin et al. 2015; Ollier et al. 2013). Its abundance, environmental friendliness, and low cost make it an attractive option for various industrial uses. However, integrating nanoclays like bentonite into polymer matrices poses challenges, primarily due to difficulties in achieving uniform dispersion. Nanoclays tend to aggregate, which can limit their effectiveness in enhancing the mechanical properties of polymers. To maximize performance, it is crucial to fully exfoliate and evenly distribute the silicate layers within the polymer matrix, ensuring that individual silicate layers are well-separated and dispersed throughout the polymer (Mapossa et al. 2023; Ollier et al. 2013).

The main difficulty in achieving uniform dispersion arises from the hydrophilic nature of silicate-based nanofillers and the typically hydrophobic nature of most polymers. Silicates, including those in bentonite, attract water molecules, while hydrophobic polymers repel them, leading to poor interaction and dispersion within the polymer matrix. To address this issue, pretreatment of the clay is essential (Mapossa et al. 2023; Hedley et al. 2007; Bala et al. 2000).

To address this issue, it is essential to develop a technology that can alter the structure of layered silicates at the beginning of the process, thereby providing them with new physical and chemical properties that enhance their processing efficiency (Gerasimov and Eremina 2021). The method of heating plays a crucial role in this modification, with various techniques available, including convective radiation-thermal heating, infrared radiation, hydrothermal heating, and microwave heating (Kotova 2016; Arsentyev et al. 2017; Singh et al. 2015). Microwave heating is particularly notable because it converts electromagnetic energy into heat that is evenly distributed throughout the entire volume of the material, allowing for more uniform and precise changes in the material's structure compared to traditional methods. This even heating can contribute to increased efficiency and effectiveness in processing (Gerasimov and Eremina 2021).

Layered silicates are particularly amenable to microwave heating because of their numerous "water centers" that interact with electromagnetic radiation. According to Buger's law, the electromagnetic field of microwaves primarily affects the water molecules within the silicates. Water, with a high dielectric loss factor, efficiently converts electromagnetic energy into heat, more so than the dry substance (Gerasimov and Eremina 2021; Binner et al. 2014; Leluk et al. 2010). At a frequency of 2.45 GHz, water dipoles rotate at 4.5×10^9 times per second, leading to rapid heating of both adsorbed and bound water in layered silicates (Gerasimov and Eremina 2021). Microwaves, with frequencies ranging from 300 MHz to 300 GHz and corresponding wavelengths between 1 and 1 mm, effectively convert electromagnetic energy into heat. This uniform heating is beneficial for even and deep moisture removal from clay materials at lower temperatures, with less time and energy compared to traditional heating systems. In contrast, convection heating transfers heat from the hot boundary layer to the material's surface, resulting in a higher surface temperature and requiring more time and energy for gradual moisture removal. The uniform heating mechanism of microwaves is described by Maxwell's equations, where the magnetic and electric field vectors are perpendicular to each other and to the direction of wave motion (Průša et al. 2023; Procházka et al. 2016).

In this study, we explored the application of microwave heating as a novel method for the physical modification of bentonite clay. The primary objective was to assess the feasibility of integrating this modified bentonite as a partial substitute for carbon black in polymer matrices. To achieve this, we conducted a series of comprehensive analyses to evaluate both the properties of the modified bentonite and the resulting polymeric materials.

7.2 Experimental

7.2.1 Materials

In this study, we used montmorillonite phase bentonite from Jelšovy Potok, Slovakia, featuring an average particle size of 4.5 μm, as the initial material for the modified bentonite. The materials used for polymer blends in this study included the natural rubber elastomer SMR 10, carbon black N339, cure activators zinc oxide (ZnO) and stearic acid, a plasticizer (residual aromatic extract), and a cure system comprising N-tert-butyl-2-benzothiazyl sulfonamide and sulfur. These materials were obtained from CMR Púchov Ltd., Slovak Republic.

7.2.2 Preparation of Modified Bentonites

The modified bentonite (MW/bentonite) was prepared using a straightforward method involving microwave radiation. This process consisted of microwave-assisted heating for 3 min at a power of 900 W and a frequency of 2450 MHz.

7.2.3 Characterization of the Modified Bentonite

An energy dispersive X-ray (EDX) analyzer was used to analyze the chemical composition of both the bentonite microwave modification samples and the unmodified bentonite samples. Thermal analysis of both the bentonite microwave modification and unmodified bentonite samples was carried out under nitrogen using a TGA/DSC 2 STARe System from METTLER TOLEDO. The measurements were conducted over a temperature range of 25–1000 °C with a heating rate of 10 °C/min for both types of samples. Samples of both bentonite microwave modification and unmodified bentonite were prepared in KBr tablets and analyzed in transmittance mode using a Nicolet spectrometer, with FTIR spectra obtained in the range of 4000–400 cm^{-1} at a resolution of 4 cm^{-1} for both sample types.

7.2.4 Preparation of Polymer Blends

Polymer blends were prepared using a laboratory Brabender Plastograph at an initial temperature of 55 °C and a speed of 50 rpm. The additives were incorporated in the following sequence: natural rubber SMR 20, activators (ZnO and stearic acid), filler (carbon black type N339), plasticizer (residual aromatic extract), and finally, the vulcanization system (sulfur and TBBS). This method was used to create the standard polymer blend. For the other rubber compounds, the same procedure was followed, with the only variation being the partial replacement of carbon black with bentonite and bentonite microwave modification. A total of seven polymer blends were prepared. The polymer blend (S) contained the standard filler, carbon black type N339. In three mixtures, carbon black was partially replaced with bentonite in amounts of 10%, 20%, and 30% by weight, labeled as B10, B20, and B30, respectively. In the remaining three compounds, carbon black was partially substituted with bentonite microwave modification (MW/bentonite) in amounts of 10%, 20%, and 30% by weight, designated as MW/B10, MW/B20, and MW/B30.

7.2.5 Characterization of Polymer Blends and Vulcanizates

The rheological and curing properties of rubber blends were characterized using the PRPA 2000 rubber process analyzer, which featured a lower die oscillating sinusoidally at a fixed angle and frequency. The parameters, including the optimum curing period ($T_{c(90)}$), pre-curing time ($t_{s(02)}$), minimum torque (M_L), and maximum torque (M_H), were calculated based on the rheograph of polymer blends.

Tensile tests were conducted to evaluate the mechanical properties, specifically tensile strength and elongation at break, in accordance with ASTM D412-87 standards for the polymer vulcanizates. Dumbbell-shaped samples underwent a stress–strain test at a crosshead speed of 100 mm/min using a Shimadzu Autograph AG–X plus universal testing machine with a 5 kN capacity. The hardness of the polymer vulcanizates was measured using a Shore A Durometer hardness tester in accordance with ASTM D2240-86 standards.

7.2.6 Accelerated UV Exposure and Condensation

Polymer vulcanizate samples underwent UV exposure and condensation humidity testing in a QUV accelerated weathering test chamber fitted with UVA-340 + fluorescent lamps. The UVA radiation intensity was set at 0.76 W/m^2 with a wavelength of 340 nm at 60 °C for 8 h. Condensation was set at 50 °C for 4 h. This UV accelerated weathering process was conducted over a period of 48 h.

Table 7.1 Chemical composition of the bentonite and MW/bentonite

Sample	SiO_2	Al_2O_3	CaO	MgO	Fe_2O_3	SO_3	K_2O
Bentonite	64.88	20.98	5.32	3.12	3.07	1.05	0.96
MW/Bentonite	64.82	21.06	5.28	3.48	3.03	0.74	0.99

7.3 Results

7.3.1 EDX Analysis

The EDX analysis of both MW/bentonite and unmodified bentonite samples, presented in Table 7.1, showed that the main components are SiO_2, Al_2O_3, CaO, MgO, and Fe_2O_3. Based on these results, the bentonite can be classified as calcareous bentonite.

7.3.2 FTIR Analysis

Figure 7.1 shows the FTIR spectrum of both untreated bentonite and MW/bentonite. FTIR analysis was conducted on both bentonite samples (bentonite and MW/bentonite) to investigate changes in the smectite structure, chemical composition, and surface properties resulting from microwave modification.

Stretching vibrations of structural OH groups within smectite are indicated by absorption bands in the range of 3700–3200 cm^{-1}. In the spectrum of the bentonite,

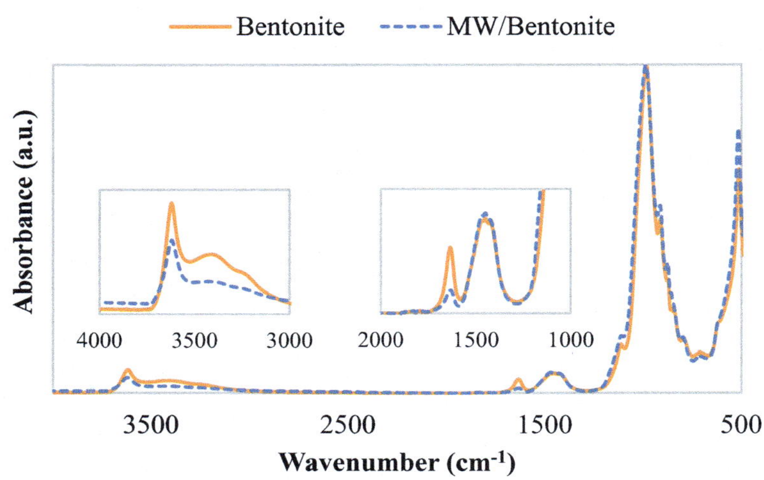

Fig. 7.1 FTIR spectrum of samples bentonite and MW/bentonite

absorption bands appear at 3413 cm^{-1} and 3621 cm^{-1}. The 3413 cm^{-1} band corresponds to the stretching vibrations of OH groups from water molecules adsorbed on the clay surface, while the 3621 cm^{-1} band indicates the stretching vibration of hydroxyl groups coordinated to octahedral Al^{3+} cations (Surendra et al. 2017; Motlagh et al. 2011). Absorption bands within the range of 1300–440 cm^{-1} indicate the presence of Si–O stretching vibrations and OH bending vibrations. The band at 1638 cm^{-1} corresponds to the bending vibration of the OH group in bound water, while the strong band at 1037 cm^{-1} signifies the stretching vibration of Si–O groups in the tetrahedral sheets (Pentrák et al. 2018; Caglar et al. 2016). The bands at 918 and 847 cm^{-1} correspond to the bending vibrations of the Al-Al–OH and Al–Mg-OH groups, respectively. Moreover, the asymmetric band at 798 cm^{-1} indicates the presence of silica admixtures (Pentrák et al. 2018; Mičicová, et al. 2023; Andrunik and Bajda 2019).

The reduced intensity of the bands related to structural valence and OH deformation vibrations in the FTIR spectra indicate that the modified bentonites have less water trapped between their layers compared to unmodified bentonite. These observations suggest that microwave treatment can improve the hydrophobic properties of bentonite (Cheng et al. 2024).

7.3.3 TG Analysis

The TG curves for untreated bentonite and bentonite microwave modification (MW/bentonite) are shown in Fig. 7.2.

The analysis identified two phases of weight loss in both samples. For untreated bentonite, the first phase occurred between 30–300 °C, resulting in a weight loss of 6.44%, attributed to the desorption of physically adsorbed water (Bousba 2018). In contrast, MW/bentonite exhibited a lower weight loss of 3.40% within the same temperature range. This reduction in mass is likely due to the removal of

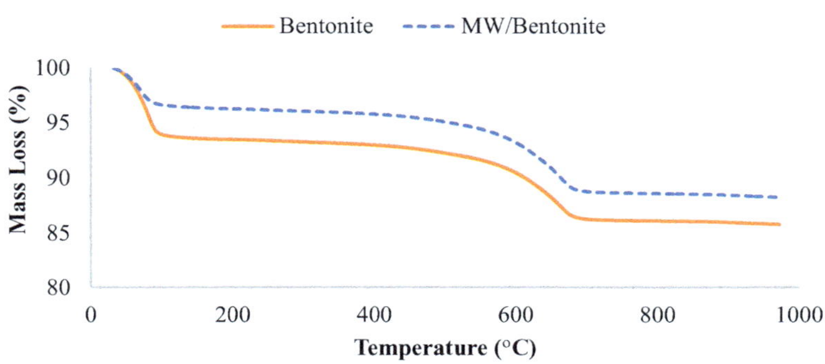

Fig. 7.2 Thermal decomposition of samples bentonite and MW/bentonite

adsorbed water molecules from the bentonite surface because of microwave treatment, suggesting that the modification process enhanced the bentonite's hydrophobic properties (Houhoune et al. 2016). The second phase of weight loss, occurring between 350–750 °C, is linked to the dehydroxylation of OH groups on the surface of natural bentonite (Erdem et al. 2010). In this phase, unmodified bentonite exhibited a weight loss of 7.25%, while MW/bentonite showed a slightly lower weight loss of 7.03%. The TG analysis shows that the MW/Bentonite sample modified by microwave irradiation showed the lowest overall weight loss of 10.43%.

7.3.4 Rheological and Curing Characteristics

The rheographs of polymer blends with partial replacement of unmodified bentonite (B10, B20, B30), MW/bentonite (MW/B10, MW/B20, MW/B30) and the polymer standard blend, obtained from PRPA 2000 are shown in Fig. 7.3. The curing characteristics derived from the rheographs, including parameters such as minimum torque (M_L), maximum torque (M_H), scorch time ($t_{s(02)}$), and optimum cure time ($t_{s(90)}$), are comprehensively detailed in Table 7.2.

The minimum torque (M_L), which reflects the viscosity of the compound during the initial stage of curing, was found to be lower in polymer blends with partial replacement of unmodified bentonite (B10, B20, B30), MW/bentonite (MW/B10, MW/B20, MW/B30) compared to the S standard. This reduction in M_L suggests that these blends containing unmodified bentonite and MW/bentonite exhibit lower viscosity at the onset of the curing process compared to the standard S blend (Ondrušová et al. 2020). In polymer blends containing unmodified bentonite and MW/bentonite, the maximum torque values were observed to be lower compared to those of the standard blend. The results show that polymer blends with increasing

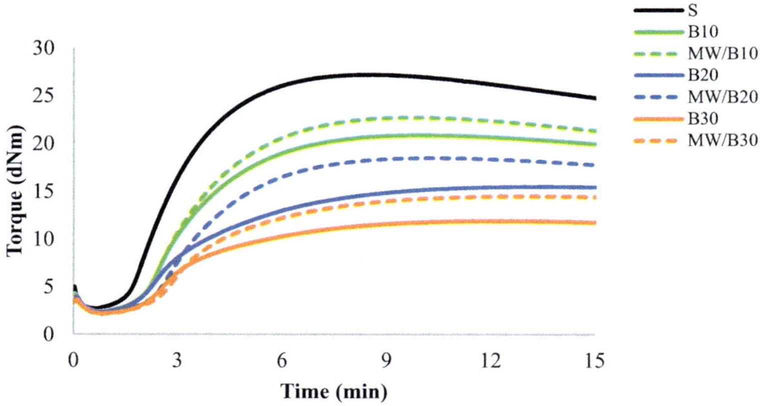

Fig. 7.3 Rheographs of polymer blends

Table 7.2 Rheological and curing characteristics of polymer blends

Polymer blends	M_L (dNm)	M_H (dNm)	$t_{s(02)}$ (min)	$t_{c(90)}$ (min)
S	2.72	27.16	1.66	5.13
B10	2.40	20.91	2.06	6.02
B20	2.40	15.48	2.09	7.51
B30	2.29	11.93	2.39	7.10
MW/B10	2.35	22.76	2.05	6.05
MW/B20	2.23	18.51	2.37	6.28
MW/B30	2.24	14.51	2.54	7.36

bentonite filler content exhibit an exponential decrease in maximum torque (M_H). This decrease is due to the weak interaction between the matrix and the filler, since bentonite tends to clump and form agglomerates at higher concentrations (Mat et al. 2016). On the other hand, polymer blends containing MW/bentonite (MW/B10, MW/B20, MW/B30) demonstrate higher maximum torque values compared to those polymer blends containing with unmodified bentonite (B10, B20, B30). This increase in maximum torque (M_H) reflects an enhancement in material stiffness, which is due to the improved distribution and interaction of the modified fillers within the polymer matrix (Božeková et al. 2023). Polymer blends that include unmodified bentonite and MW/bentonite show extended optimum curing times compared to standard blend. This phenomenon is caused by the adsorption of curative agents, which arises from the presence of hydroxyl groups on the bentonite's silicate layer. Hydroxyl groups can interfere with the vulcanization process, thereby lengthening the time required for the polymer to fully cure, but also an increase in the material' scorch time (Mat et al. 2016).

7.3.5 *Physical and Mechanical Properties Before and After QUV Exposure*

The results of tensile tests conducted on polymer vulcanizates with partial replacement of unmodified bentonite (B10, B20, B30), MW/bentonite (MW/B10, MW/B20, MW/B30) and the polymer standard vulcanizate, using the Shimadzu Autograph AG–X plus universal testing machine, are summarized in Table 7.3. The Table 7.3 includes data on tensile strength, elongation, and hardness before QUV exposure for each type of polymer vulcanizate.

The results showed that the tensile strength was lower for all polymeric vulcanizates with partial replacement of unmodified bentonite (B10, B20, B30) and MW/bentonite (MW/B10, MW/B20, MW/B30) compared to the standard vulcanizate. However, polymeric vulcanizates with MW/bentonite exhibited higher tensile strength than those with unmodified bentonite across all three replacement amounts.

Table 7.3 Physical and mechanical properties before QUV exposure

Vulcanizates	Tensile strength (MPa)	Elongation at break (%)	Shore A Hardness
S	28.65 ± 1.79	469 ± 4.44	64.86 ± 0.90
B10	21.49 ± 1.25	508 ± 19.24	57.86 ± 0.40
B20	15.10 ± 2.11	500 ± 45.35	48.26 ± 0.70
B30	12.52 ± 1.61	525 ± 43.29	39.72 ± 1.10
MW/B10	25.97 ± 1.68	548 ± 24.67	56.06 ± 0.50
MW/B20	21.36 ± 2.25	559 ± 37.39	47.06 ± 3.10
MW/B30	15.51 ± 1.61	549 ± 29.01	44.36 ± 1.40

The tensile strength of the MW/B20 vulcanizate was increased by 41% compared to the B20 vulcanizate. The interaction between the matrix and unmodified bentonite is weaker, which leads to an inability to distribute stress uniformly throughout the matrix under load. This results in reduced strength of polymer blends containing unmodified bentonite (Mičicová, et al. 2023; Mat et al. 2016). For elongation, all polymeric vulcanizates with partial replacement of unmodified bentonite (B10, B20, B30) and MW/bentonite (MW/B10, MW/B20, MW/B30) higher values compared to the standard vulcanizate. The polymer vulcanizates containing bentonite microwave modification, specifically MW/B10, MW/B20, and MW/B30, showed the highest elongation values. The hardness of a vulcanizate is closely related to its degree of crosslinking. As the crosslink density increases, the material becomes harder and less flexible due to the formation of more chemical bonds between polymer chains (Ramesan 2005). The hardness measurements of all polymeric vulcanizates with partial replacement of unmodified and bentonite microwave modification indicate that all samples show lower hardness. The results confirm that lower maximum torque values lead to a reduction in vulcanizate hardness. These findings are attributed to the material's stiffness, which results from the agglomeration of the filler within the elastomeric matrix and the crosslinking density, both of which directly affect the M_H value (Mičicová, et al. 2023).

The comparison of tensile strength, elongation and the hardness of the test samples before and after exposure to QUV test chamber are shown in Figs. 7.4, 7.5, 7.6, respectively.

After exposure to QUV test chamber, the tensile strength (Fig. 7.4) of the standard vulcanizate decreased significantly by 20%. In contrast, vulcanizates with partial replacement of unmodified bentonite and MW/bentonite showed a less pronounced reduction in strength compared to the standard. Specifically, the B20 vulcanizate showed 6% strength increase after exposure to QUV test chamber. This enhancement in mechanical strength is primarily attributed to the photostabilizing properties of bentonite, as the filler particles absorb UV radiation, thereby slowing down the photochemical degradation processes that occur during irradiation (Sit et al. 2023; Yousif and Haddad 2013). In the case of the tensile strength comparison before and after exposure to QUV test chamber, the tensile strength value of the MW/B10 vulcanizate was comparable to that of the S vulcanizate.

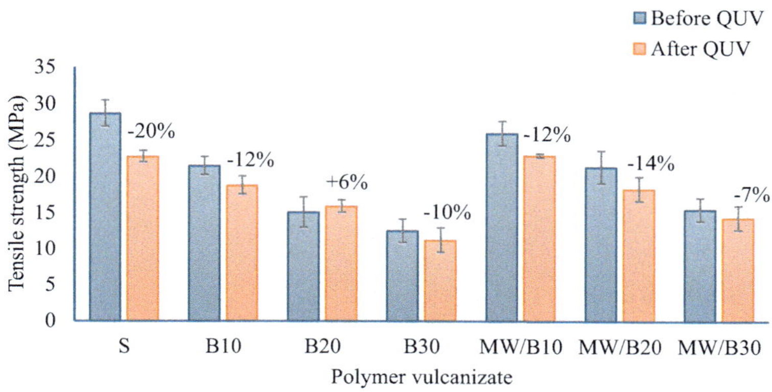

Fig. 7.4 Tensile strength of polymer vulcanizate before and after QUV exposure

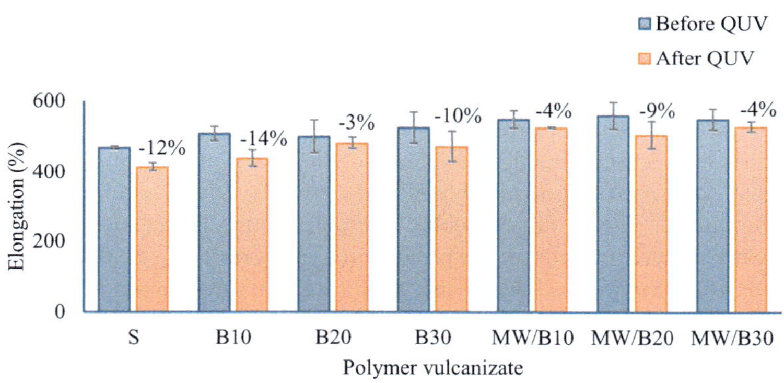

Fig. 7.5 Elongation of polymer vulcanizate before and after QUV exposure

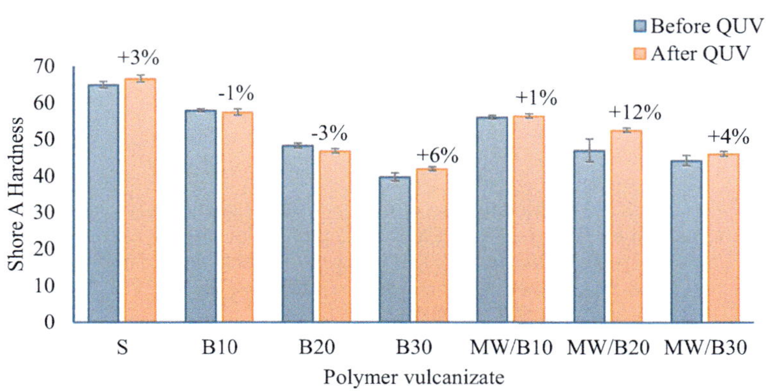

Fig. 7.6 Shore A Hardness of polymer vulcanizate before and after QUV exposure

A similar trend was observed for the elongation (Fig. 7.5) of polymer vulcanizates (Fig. 7.5).

After exposure to QUV test chamber, the Shorre A hardness (Fig. 7.6) of the polymeric vulcanizates changed minimally. The slight increase observed may be attributed to an increase in the crystallinity of the polymeric vulcanizates (Sit et al. 2023).

7.4 Conclusion

This study examined the rheological and mechanical properties of polymer blends filled with bentonite microwave modification. Characterization of the bentonite microwave modification was conducted using Fourier transform infrared spectroscopy and thermal analysis. The reduced intensity of the bands related to structural valence and OH deformation vibrations in the FTIR spectra suggests that microwave treatment can improve the hydrophobic properties of bentonite by reducing the amount of water trapped between its layers. In the case of bentonite microwave modification, however, improved dispersion and interaction of the fillers within the polymer matrix can influence the material's behavior during vulcanization. The mechanical properties of polymer blends vary significantly depending on the type of bentonite used. The results indicated that substituting conventional fillers with bentonite microwave modification in polymer materials resulted in enhanced strength and ductility when compared to polymer materials containing unmodified bentonite.

Acknowledgements This research work has been supported by the Operational Program Integrated Infrastructure, co-financed by the European Regional Development Fund by the project: Advancement and support of R&D for "Centre for diagnostics and quality testing of materials" in the domains of the RIS3 SK specialization, Acronym: CEDITEK II., ITMS2014+ code 313011W442 and the work was supported by the project KEGA 001TnUAD-4/2022 and VEGA 1/0265/24.

References

Andrunik M, Bajda T (2019) Modification of bentonite with cationic and Nonionic surfactants: structural and textural features. Mater 12(22):3772. https://doi.org/10.3390/ma12223772

Arawande JO, Orodepo GO, Olatide M (2023) Comparative study on thermodynamic properties of clay and carbon-filled natural rubber composites. Chem Eng Sci 5(12):24–29 (2023). https://doi.org/10.36686/Ariviyal.CSER.2023.05.12.071

Arsentyev VA, Gerasimov AM, Mezenin AO (2017) Kaolines beneficiation technology study with application of hydrothermal modification. Obogashchenie Rud 2:3–9. https://doi.org/10.17580/or.2017.02.01

Bala P, Samantaraya BK, Srivastava SK (2000) Synthesis and characterization of Na-montmorillonite-alkylammonium intercalation compounds. Mater Res Bull 35:1717–1724. https://doi.org/10.1016/S0025-5408(00)00368-8

Binner E et al (2014) A review of microwave coal processing. J Microw Power Electromagn Energy 48(1):35–60. https://doi.org/10.1080/08327823.2014.11689870
Bokobza L (2019) Natural rubber nanocomposites: a review. Nanomater 9(1):12. https://doi.org/10.3390/nano9010012
Bousba S, et al (2018) Adsorption removal of humic acid from water using a modified Algerian bentonite. Phys Chem Res 6(3):613–625. https://doi.org/10.22036/PCR.2018.129154.1482
Božeková S et al (2023) Rheological properties and curing features of polymer materials filled with biowaste Rheologische Eigenschaften und Härtungseigenschaften von mit Bioabfällen gefüllten Polymerwerkstoffen. Materialwiss Werkst 54(4):415–422. https://doi.org/10.1002/mawe.202200297
Caglar B et al (2016) Structural, thermal, morphological and surface charge properties of dodecyltrimethylammonium-smectite composites. J Mol Struct 1105:70–79. https://doi.org/10.1016/j.molstruc.2015.10.017
Chen G et al (2022) Effect of Nonisoprene degradation and naturally occurring network during maturation on the properties of natural rubber. Polym 14(11):2180. https://doi.org/10.3390/polym14112180
Cheng J, et al (2024) Preparation and characterization analysis of modified bentonite-based powder for improving explosion suppression effects. Powder Technol 440:119758. https://doi.org/10.1016/j.powtec.2024.119758
Drobíková K et al (2020) Magnetically modified bentonite: characterization and stability. Mater Today Proc 37:53–57. https://doi.org/10.1016/j.matpr.2020.10.375
El-Nashar DE, Ahmed NM, Yehia AA (2012) The role of ion-exchange bentonites in changing the properties of styrene–butadiene rubber composites. Mater des 34:137–142. https://doi.org/10.1016/j.matdes.2011.07.072
Erdem B, Özcan AS, Özcan A (2010) Preparation of HDTMA-bentonite: characterization studies and its adsorption behavior toward dibenzofuran. Surf Interface Anal 42(6–7):1351–1356. https://doi.org/10.1002/sia.3230
Gerasimov AM, Eremina OV (2021) Application microwave radiation for directional changes of layered silicates properties. Eurasian Min 1:55–60. https://doi.org/10.17580/em.2021.01.11
Hakimi NMF et al (2021) Surface modified Nanocellulose and its reinforcement in natural rubber matrix nanocomposites: a review. Polym 13(19):3241. https://doi.org/10.3390/polym13193241
Hedley CB, Yuan G, Theng B (2007) Thermal analysis of montmorillonites modified with quaternary phosphonium and ammonium surfactants. Appl Clay Sci 35:180–188. https://doi.org/10.1016/j.clay.2006.09.005
Houhoune F et al (2016) Behaviour of modified hexadecyltrimethyl-ammonium bromide bentonite toward uranium species. J Environ Chem Eng 4:3459–3467. https://doi.org/10.1016/j.jece.2016.07.018
Kotova OB, et al (2016) Radiation-thermal modification of ferruginous bauxite in the processes of their processing. Obogashchenie Rud 4:16–23. https://doi.org/10.17580/or.2016.04.03
Kurahatti RV, et al (2010) Defence applications of polymer nanocomposites. Def Sci J 60(5):551–563. https://doi.org/10.14429/dsj.60.578
Leluk K et al (2010) Dielectric permittivity of kaolinite heated to high temperatures. J Phys Chem Solids 71(5):827–831. https://doi.org/10.1016/j.jpcs.2010.02.008
Mapossa AB et al (2023) Thermal, morphological and mechanical properties of multifunctional composites based on biodegradable polymers/bentonite clay: a review. Polym 15(16):3443. https://doi.org/10.3390/polym15163443
Mat NS, Ismail H, Othman N (2016) Curing characteristics and tear properties of bentonite filled ethylene propylene diene (EPDM) rubber composites. Procedia Chem 19:394–400. https://doi.org/10.1016/j.proche.2016.03.029
Medupin RO et al (2019) Carbon nanotube reinforced natural rubber nanocomposite for anthropomorphic prosthetic foot purpose. Sci Rep 9:20146. https://doi.org/10.1038/s41598-019-56778-0

Mičicová Z, et al (2023) Rheological and mechanical properties of rubber blends filled with modified bentonite. In: Öchsner A, Altenbach H (eds) Engineering design applications V. advanced structured materials. vol 171. pp 21–34. Springer, Cham (2023)
Motlagh MM, Youzbashi AA, Rigi ZA (2011) Effect of acid activation on structural and bleaching properties of a bentonite. Iran. J Mater Sci Eng 8(4):50–56. http://ijmse.iust.ac.ir/article-1-383-en.html
Ollier R, Rodriguez E, Alvarez V (2013) Unsaturated polyester/bentonite nanocomposites: Influence of clay modification on final performance. Compos A: Appl Sci 48:137–143. https://doi.org/10.1016/j.compositesa.2013.01.005
Ondrušová D, et al (2021) Targeted modification of the composition of polymer systems for industrial applications. Bull Pol Acad Sci Tech Sci 69(2):e136721. https://doi.org/10.24425/bpasts.2021.136721
Ondrušová D, et al (2020) Application of alternative additives in the polymer composite systems used in automotive industry. IOP Conf Ser: Mater Sci Eng 776:012101. https://doi.org/10.1088/1757-899X/776/1/012101
Pentrák M et al (2018) Alteration of fine fraction of bentonite from Kopernica (Slovakia) under acid treatment: a combined XRD, FTIR, MAS NMR and AES study. Appl Clay Sci 163:204–213. https://doi.org/10.1016/j.clay.2018.07.028
Procházka M et al (2016) Microwave radiation and its application on construction materials. Eng Struct Technol 8(4):150–156. https://doi.org/10.3846/2029882X.2016.1257372
Průša D et al (2023) Effect of microwave radiation on the compressive strength of solid ceramic brick. Buildings 13:1018. https://doi.org/10.3390/buildings13041018
Ramesan MT (2005) The effects of filler content on cure and mechanical properties of dichlorocarbene modified styrene butadiene rubber/carbon black composites. J Polym Res 11:333–340. https://doi.org/10.1007/s10965-005-6571-y
Şahin Ö, Kaya M, Saka C (2015) Plasma-surface modification on bentonite clay to improve the performance of adsorption of methylene blue. Appl Clay Sci 116:46–53. https://doi.org/10.1016/j.clay.2015.08.015
Singh S et al (2015) Microwave processing of materials and applications in manufacturing industries: a review. Mater. Manuf. Process. 30:1–29. https://doi.org/10.1080/10426914.2014.952028
Sit M et al (2023) Inorganic fillers and their effects on the properties of flax/PLA composites after UV degradation. Polym 15:3221. https://doi.org/10.3390/polym15153221
Sukumar R, Menon ARR (2008) Organomodified kaolin as a reinforcing filler for natural rubber. J Appl Polym Sci 107(6):3476–3483. https://doi.org/10.1002/app.27469
Surendra BS et al (2017) Microwave assisted physico-chemical modification of Bentonite clay: characterization and photocatalytic activity. Mater Today: Proc 4(11):11727–11736. https://doi.org/10.1016/j.matpr.2017.09.089
Wan H et al (2020) Montmorillonite: A structural evolution from bulk through unilaminar nanolayers to nanotubes. Appl Clay Sci 194:105695. https://doi.org/10.1016/j.clay.2020.105695
Yousif E, Haddad R (2013) Photodegradation and photostabilization of polymers, especially polystyrene: review. Springerplus 2:398. https://doi.org/10.1186/2193-1801-2-398

Chapter 8
Influence of Carbon Black on Mechanical Behaviour of Polymer Blends Before and After Thermo-Oxidation Aging

Juliána Vršková, Ivan Labaj, Ivan Kopal, and Darina Ondrušová

Abstract The most widely used reinforcing nanomaterial for the production of tires and other rubber products is carbon black. Their production processes as well as the aggregate dimension determine their unique properties which can be used like a UV stabilizer, conductive or insulating agent in polymer composites. The degree of their dispersion, surface activity, structure, type, and their amount in the matrix also play an important role in their properties. The most commonly used property is their stiffening effect in the polymer matrix, which is influenced by the interaction between polymer molecules and carbon black particles (Spahr and Rothon 2016; Song et al. 2017). This study deals with the preparation of polymer blends with different types of industrially used carbon blacks of the type N121, N339, N550, and N660. The influence of the type and amount of carbon black (45; 50; 55; 60 phr) on the properties of polymer blends based on natural rubber was investigated. The prepared blends were studied for hardness, tensile strength, and elongation at break before and after thermo-oxidation aging. The amount of filler in the polymer blends mainly affected the hardness, which increased with the increasing filler amount, but hardness values decreased with increasing filler size in blends. The amount and type of filler also affected tensile strength and elongation at break of polymer blends before and after thermo-oxidation aging.

Keywords Natural rubber · Carbon black · Properties · Thermo-oxidation aging

J. Vršková (✉) · I. Labaj · I. Kopal · D. Ondrušová
Faculty of Industrial Technologies, Alexander Dubček University of Trenčín, Púchov, Slovakia
e-mail: juliana.vrskova@tnuni.sk

I. Labaj
e-mail: ivan.labaj@tnuni.sk

I. Kopal
e-mail: ivan.kopal@tnuni.sk

D. Ondrušová
e-mail: darina.ondrusova@tnuni.sk

A. Öchsner and H. Altenbach (eds.), *Engineering Design Applications VII*, Advanced Structured Materials 230, https://doi.org/10.1007/978-3-031-84346-4_8

8.1 Introduction

From a chemical standpoint, carbon black is a synthetic material composed of carbon particles tightly bound by covalent bonds. Its structure is three-dimensional and arranged in parallel layers that are not ordered (turbostratic structure). Typical properties include a density of 1.7–1.9 g/cm^3, primary particle sizes ranging from 14 to 250 nm, and a specific surface area of 7–560 m^2/g. Carbon black can be obtained by burning carbon gases as well as viscous residues of aromatic hydrocarbons. In thermal furnaces, carbon black particles can reach sizes of several hundred nanometres, but in certain stages of the furnace, particles with a diameter of less than 20 nm can be produced. The properties of carbon black are significantly influenced by the manufacturing process and the type of furnace used. Generally, furnace carbon black has a higher specific surface area and smaller particle size compared to the thermal type of carbon black. The surface of the particles contains many functional groups, such as carbon–oxygen, carbon–hydrogen, and carbon–nitrogen groups. These functional groups on the particle surface mainly affect wettability, dispersion, and moisture adsorption. The industrial applications of carbon black lie in its ability to mechanically reinforce, introduce colouring pigments, provide electrical conductivity to polymer mixtures, and act as a protective filter against UV radiation, thereby improving the polymer's resistance to degradation (Martín-Martínez 2002; Mohammad et al. 2006; Takamura and Garche 2009; Rodgers et al. 2013; Farida et al. 2019). In demanding rubber applications, such as automotive tyres and mechanical rubber goods, carbon black is the most widely used type of filler. Global production of carbon black amounts to up to 15 million metric tonnes annually, with more than 70% utilised in the manufacturing of tyres in the automotive industry. The production of non-tyre goods, as well as the manufacturing of paints, coatings, inks, and plastic composites, accounts for only 27%. Due to its unique properties and wide range of applications, carbon black holds significant industrial importance (Robertson and Hardman 2021; Fan et al. 2019). In this study, we focus on the preparation of polymer blends of natural rubber with industrially utilised types of carbon black (N121, N330, N550, N660) in contents ranging from 45 to 60 phr. The work provides valuable information from the perspective of the mechanical properties of polymer materials. To ensure the performance and longevity of finished products in industrial applications, it is crucial to study the mechanical response even after exposure to thermo-oxidation aging. Studying the influence of the type and amount of carbon black in the polymer blend enables researchers to design new polymer materials tailored to the specific needs of various industrial applications. Choosing the correct type and adequate quantity of carbon black in the polymer matrix is an effective way to optimise production costs, improve efficiency, and enhance production quality while maintaining the desired properties.

Table 8.1 Composition of studied polymer blends

Material	Contents (phr)	Producer	Function
Natural rubber SMR 10	100	Panorama Global Trade Sdn Bhd, Kuala Lumpur, Malaysia	Matrix
Carbon black N121, N339, N550, N660	45; 50; 55; 60	Makrochem Sp. z o.o., Lublin, Poland	Filler
Zinc Oxide	2.1	Bayer, Kuala Lumpur, Malaysia	Activator of vulcanization
Stearic acid	1.5	Setuza a.s., Ústí nad Labem, Czech Republic	Activator of vulcanization
Sulphur	2.7	Bayer, Kuala Lumpur, Malaysia	Agent of vulcanization
TBBS	1.2	Duslo a.s., Šaľa, Slovakia	Accelerator of vulcanization

8.2 Materials and Methods

8.2.1 Materials

The composition of the investigated polymer blends, the function of individual ingredients in them, and their manufacturers are provided in Table 8.1.

8.2.2 Preparation of Polymer Blends

Polymer blends were prepared at 90 °C in one-step by using laboratory mixer Plastograph Brabender type with chamber volume 80 cm^3 and 50 rpm. The blend was subsequently homogenised using a laboratory twin roll VOGT LaboWalz W 80 T 1 (VOGT, Berlin, Germany) into plates of 2 ± 0.04 mm thickness and then allowed to stand for 24 h at a laboratory temperature (25 °C ± 2 °C).

8.2.3 Determination of Hardness Shore A

By using a hand hardness tester SHORE A scale (Wallace LTD, Croydon, UK) were measured all prepared blends. According to standard ASTM D2240-15, a measurement was performed 5 times. All results were statistically evaluated.

8.2.4 Measurement of Mechanical Properties

Tensile strength and elongation at break were measured on a tensile tester device, Shimadzu AG–X Plus (Shimadzu, Tokyo, Japan), with two pneumatic clamping jaws. By using Instron cutter machine (Instron, Norwood, MA, USA) in the shape of dumb-bell type II according to ASTM D 412 were prepared samples. The speed of tensile tests was determined at 500 mm.min^{-1}.

8.2.5 Determination of Accelerated Thermo-Oxidation Aging in Air (TOA)

Pre-measured and marked samples were placed using metal hooks in a thermostat heated to 100 °C so that the distance between the given test bodies was at least 10 mm. A distance of at least 50 mm was complying with the conditions between the test samples and the walls. The test samples did not fill 10% of the volume of the thermostat. The test lasted 72 h and was performed according to the ISO 188:2023 standard. After the accelerated thermal aging in air, the test samples were removed from the thermostat, cooled, and left to rest in a desiccator. After aging, the hardness was measured again and the mechanical properties of the test samples were determined. The obtained results after thermo-oxidation aging were statistically evaluated and compared with the results of tensile strength, elongation at break, and hardness values of double-sided blades before thermo-oxidation aging.

8.3 Results and Discussion

In this results and discussion section, we summarize the values of hardness and mechanical properties before and after thermo-oxidation aging of polymer blends.

Fig. 8.1 shows the values of hardness for polymer blends with different types and content of carbon black. With the increasing filler content, the hardness values of all polymer blends increased linearly, which can be attributed to the reinforcing effect of the given types of filler (Boukfessa and Bezzazi 2021; Chuayjuljit et al. 2002). From the perspective of the impact of filler type, the highest hardness values were recorded for the N121 filler type. The increase in hardness values for blends with this filler type can be attributed to the higher surface activity of its smaller particles, as well as their better integration into the rubber matrix. This leads to improved chemical bonding between the filler and the matrix, resulting in increased stiffness of the polymer blend (Omnès et al. 2008).

Comparable hardness values of polymer blends with the N121 filler were also observed in polymer blends with the N339 filler, which is characterised by high surface activity similar to the N121 carbon black type. The lowest hardness values

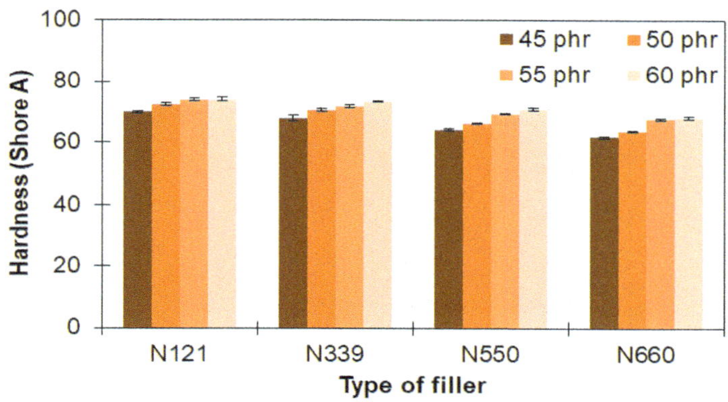

Fig. 8.1 Effect of filler type on hardness of NR blends before thermo-oxidation aging

were measured in blends with the N660 filler type. The particle size, surface activity, and quantity of this carbon black type affect the filler-matrix interaction, leading to weaker incorporation of the filler into the rubber matrix. The decrease in hardness values can be attributed mainly to the lower surface activity of the larger N660 filler particles (34–36 m^2/g) and their quantity. Similar hardness values to the N660 filler were also recorded for the N550 filler type, with particle surface activity ranging from 39 to 41 m^2/g (Bera et al. 2019).

In Fig. 8.2, it can be seen the values of hardness for polymer blends with different types and content of carbon black after thermo-oxidation aging. The results of hardness values after thermo-oxidation aging (TOA) clearly show that with increasing filler content, the hardness values increased for all polymer blends. Compared to the polymer blends before TOA, there was an increase in hardness values for all types and content of filler. In addition to the matrix and filler, TOA also affects the formed vulcanisation network, where the decomposition and recombination of sulphur bridges occur due to higher temperatures (100 °C) and prolonged exposure (72 h). The breakdown and recombination of sulphur bridges into mono- and di-bonds resulted in higher stiffness of the polymer blends (Li et al. 2023). The most significant increase in hardness values after thermo-oxidation aging was observed for the filler types N121 and N339. The highest value was recorded for the blend containing 55 phr of the N121 filler type after thermo-oxidation aging (77.26 Shore A).

The tensile strength values were examined for all polymer blends and are shown in Fig. 8.3. The results demonstrated that both the type and amount of filler used had a significant impact on the tensile strength values of the blends. The highest tensile strength value was recorded even with a minimal amount of filler for the filler type N121 (28.67 MPa). With the increasing content of the filler N121 in blend, there was a recorded decrease in the tensile strength values. Comparable tensile strength values were measured for the filler N339 while maintaining the same content of filler. A linear decreasing trend in tensile strength values with increasing

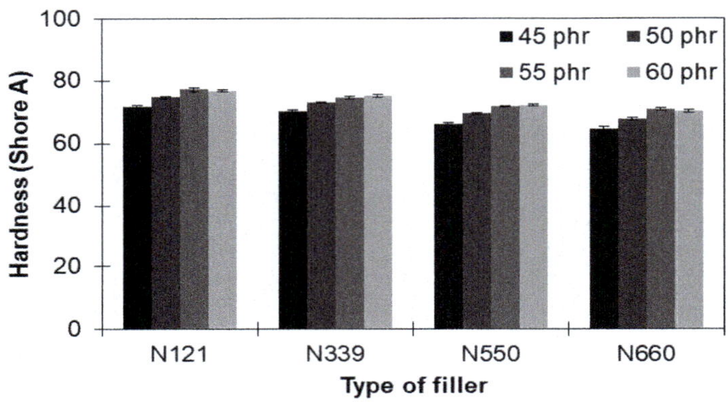

Fig. 8.2 Effect of filler type on hardness of NR blends after thermo-oxidation aging

filler content can be observed in all polymer blends. Polymer blends filled with filler types N550 and N660, which have larger particle sizes and lower active surfaces, achieved significantly lower tensile strength values compared to filler types N121 and N339. The lowest tensile strength value (22.01 MPa) was recorded for the blend with a filler content of 60 phr and the filler type N660. It can be assumed that the larger particles are also present in the interlayer spaces of the matrix, thereby restricting the mobility of individual chains and acting as stress concentrators (Labaj et al. 2023; Rogovina et al. 2013). The measurement results indicate that for applications requiring a tensile strength of the final product of at least 25 MPa, it is advisable not to exceed a filling limit of 50 phr regardless of the particle size and surface activity of the studied filler types (N121, N339, N550, N660).

Figure 8.4 shows the values of tensile strength of polymer blends after thermo-oxidation aging (TOA). The lowest tensile strength value after TOA was recorded for the polymer blend with a minimal content of filler type N660. The most significant

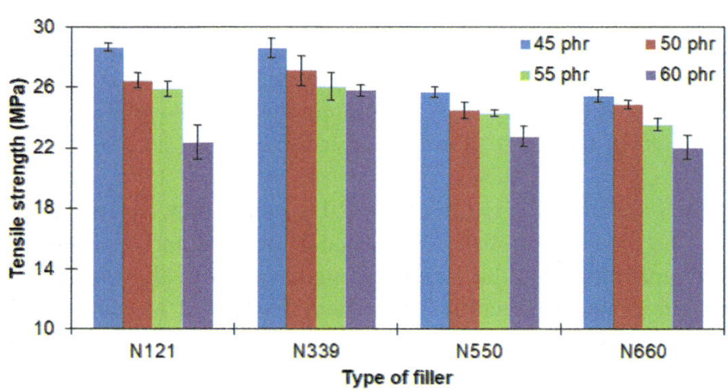

Fig. 8.3 Effect of filler type on tensile strength of NR blends before thermo-oxidation aging

decreases in values occurred in blends with a filling content of up to 50 phr, regardless of the type of filler used, compared to the same blends before TOA. These decreases in values can be attributed to the low content of filler in the polymer blend, which resulted in the formation of a small number of bonds between the filler and the polymer matrix, as well as the effect of the TOA itself (Labaj et al. 2024). The exceptions were blends with filler types N339 and N660, which recorded the highest tensile strength values compared to other polymer blends with filling up to 50 phr after TOA. After exceeding the filler amount of 45 phr in polymer blends, there was a non-linear increase in values after TOA. Adding the maximum filler content of 60 phr caused decreases in tensile strength values for all polymer blends, regardless of the filler type. These decreases in values can be attributed mainly to the effects of temperature and time of thermo-oxidation aging, as well as the influence of the high content of filler in the interlayer spaces of the polymer blend. Thermo-oxidation aging also had a significantly negative impact on the formed vulcanization network, with degradation processes of sulphur bridges predominating, which caused increased stiffness and thus reduced their ability to withstand tensile stress (Grasland et al. 2019). In polymer blends after TOA, there was an average decrease in values of up to 63% compared to the same blends before TOA. The highest tensile strength value after TOA was achieved by the blend with a filler content of 55 phr with filler N660 (17.60 MPa). The results of tensile strength measurements after TOA clearly show that a filler content of 60 phr significantly reduces the strength characteristics after TOA for all the studied fillers (N121, N339, N550, N660).

As the second mechanical property, the elongation at break of all prepared polymer blends was studied, which is graphically processed in Fig. 8.5. At the lowest filler content in polymer blends, regardless of the type used, the polymer matrix plays the main role on elongation at break values. Since the blends contain the same matrix content, their elongation at break does not differ significantly. By increasing the filler content above 45 phr, the reinforcing effect of the fillers begins to play a role depending on the particle size of the given filler type in terms of elongation at break.

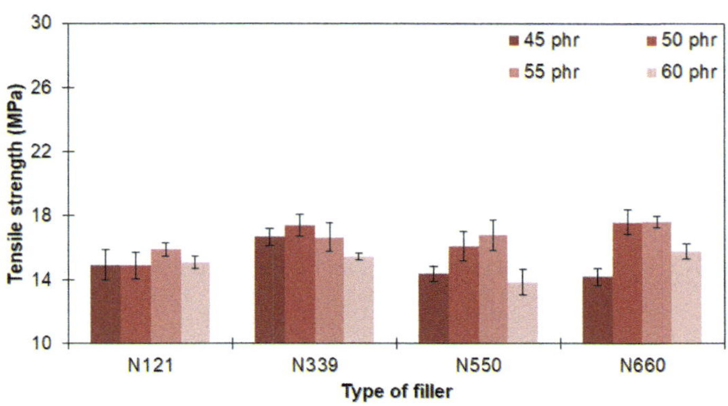

Fig. 8.4 Effect of filler type on tensile strength of NR blends after thermo-oxidation aging

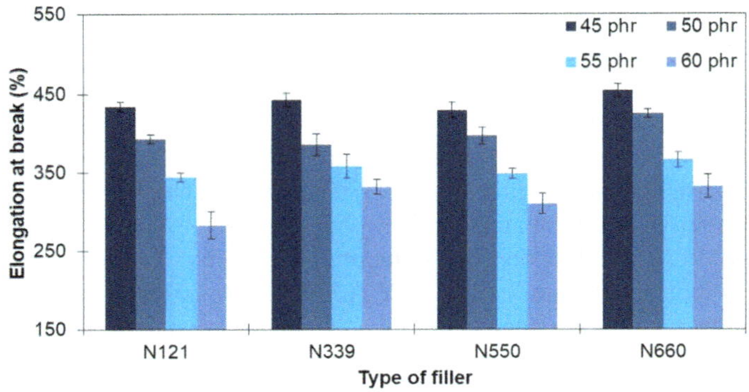

Fig. 8.5 Effect of filler type on elongation at break of NR blends before thermo-oxidation aging

With increasing filler content in the blend, the elongation at break values decreased for all blends. The most significant decreases in elongation at break values occurred in all polymer blends with a 60 phr content. The blend with a filler content of 60 phr and carbon black type N121 achieved the lowest value (283.044%) of elongation at break.

In Fig. 8.6, the elongation at break values of all polymer blends after TO are presented. In blends with minimal filler content, thermo-oxidation aging has a significant impact on elongation at break values, especially affecting the rubber matrix. In blends with minimal filler content, no significant changes in values were recorded regardless of the type of filler used. With increasing filler content in the blend, a decreasing trend in elongation at break values was observed for all types of fillers. An exception was the blend with filler type N660 with a content of 50 phr, which achieved the highest elongation at break value (253.60%) after TOA. The highest measured value can be attributed to the particle size of the filler. The amount of 50 phr for the given filler type can be considered optimal in the polymer blend because the effect of TOA does not result in such significant degradation of the rubber matrix as in other blends. The lowest elongation at break value was recorded for the blend with the maximum filler content (60 phr) and carbon black type N550. Thermo-oxidation aging has the most significant impact on the polymer matrix and the sulphur bridges formed during vulcanization, which degrade due to thermo-oxidation, as reflected in the elongation at break values (Grasland et al. 2019; Boženková et al. 2023; Yin et al. 2022). The most significant decreases were recorded in blends containing fillers with the most pronounced reinforcing effects (N121, N339). Conversely, in the case of fillers type N550 and N660, the decreases in values were the smallest.

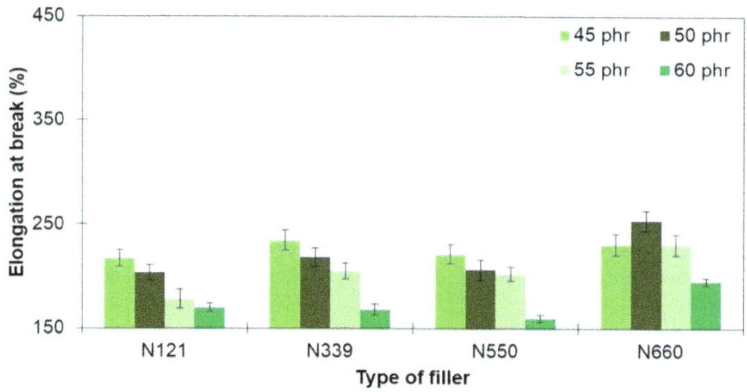

Fig. 8.6 Effect of filler type on elongation at break of NR blends after thermo-oxidation aging

8.4 Conclusion

From the results of the study, it is clear that the increasing filler content in the blend reduces the tensile strength values in the case of all types of fillers used. With filler type N121 after thermo-oxidative aging (TO), there are no significant changes in tensile strength, even in the case of an increase in the content of filler in the blend. By increasing the filler content above 45 phr, the stiffening effect of the fillers starts to play a role depending on the particle size of the given type of filler in case of elongation at break. From the results of the hardness values, it can be seen that with increasing filler content in the blend, the values increased due to the stiffening effect of all types of fillers. In addition to the matrix, TO has an effect on the created vulcanization network, where due to the influence of higher temperature (100 °C) and time (72 h), changes (decomposition and recombination) of sulphur bridges occur, most often to the disintegration of polysulfide bonds into mono- and di-bonds.

Acknowledgements This research work has been supported by the Operational Program Integrated Infrastructure, co-financed by the European Regional Development Fund by the project: Advancement and support of R&D for "Centre for diagnostics and quality testing of materials" in the domains of the RIS3 SK specialisation, Acronym: CEDITEK II., ITMS2014+ code 313011W442 and Scientific Grant Agency VEGA 1/0265/24.

References

Bera M, Gupta P, Maji PK (2019) Structural/load-bearing characteristics of polymer-carbon composites. In: Rahaman M, Khastgir D, Aldalbahi A (eds) Carbon-Containing Polymer Composites. Springer, Singapore, pp 457–502

Boukfessa H, Bezzazi B (2021) The effect of carbon black on the curing and mechanical properties of natural rubber/ acrylonitrile- butadiene rubber composites. JART. 19(3). https://doi.org/10.22201/icat.24486736e.2021.19.3.1686

Božeková S, Ondrušová D, Pajtášová M, et al (2023) Effect of alternative carbon-based filler on rubber compounds properties. Polimery 68:585–592. https://doi.org/10.14314/polimery.2023.11.2

Chuayjuljit S, Imvittaya A, Na-ranong N, Potiyaraj P (2002) Effects of particle size and amount of carbon black and calcium carbonate on curing characteristics and dynamic mechanical properties of natural rubber. JMMM 12(1):51–57

Fan Y, Fowler GD, Zhao M (2019) The past, present and future of carbon black as a rubber reinforcing filler—a review. J Clean Prod 247:119115. https://doi.org/10.1016/j.jclepro.2019.119115

Farida E, Bukit N, Marlina GE, Fisikanta BB (2019) The effect of carbon black composition in natural rubber compound. Case Stud Therm Eng 16:100566. https://doi.org/10.1016/j.csite.2019.100566

Grasland F, Chazeau L, Chenal JM, Schach R (2019) About thermo-oxidative ageing at moderate temperature of conventionally vulcanized natural rubber. Polym Degrad Stab 161:74–84. https://doi.org/10.1016/j.polymdegradstab.2018.12.029

Labaj I, Skalková P, Kopal I, et al (2023) Study of the influence of Chitosan on selected properties ofelastomeric blends. Materialwiss. Werkstofftech 54(4):429–440

Labaj I, Skalková P, Ondušová D, et al (2024) Effect of amount of soy protein on properties of elastomeric blends. Continuum Mech Therm 36:245–259. https://doi.org/10.1007/s00161-022-01079-7

Li ZX, Kong YR, Chen XF, Huang YJ, Lv YD, Li GX (2023) High-temperature thermo-oxidative aging of vulcanized natural rubber nanocomposites: evolution of microstructure and mechanical properties. Chinese J Polym Sci 41(8):1287–1297

Martín-Martínez JM (2002) Rubber base adhesives. In: Dillard DA, Pocius AV, Chaudhury M (eds) Adhesion science and engineering, pp. 573–675. Elsevier Science BV

Mohammad A, Simon GP (2006) Rubber-clay nanocomposites. In: Mai Y-W, Yu Z-Z (eds) Polymer Nanocomposites. Woodhead Publishing Series in Composites Science and Engineering, Woodhead Publishing, pp 297–325

Omnès B, Thuillier S, Pilvin P, Grohens Y, Gillet S (2008) Effective properties of carbon black filled natural rubber: Experiments and modeling. Compos—A: Appl Sci Manuf 39. https://doi.org/10.1016/j.compositesa.2008.04.003

Robertson CG, Hardman NJ (2021) Nature of carbon black reinforcement of rubber: perspective on the original polymer nanocomposite. Polymers 13(4):538. https://doi.org/10.3390/polym13040538

Rodgers B, Waddell W (2013) The science of rubber compounding. In: Mark JE, Erman B, Roland CM (eds) The Science and Technology of Rubber. Elsevier, New York, pp 417–471

Rogovina S, Aleksanyan K, Prut E, Gorenberg A (2013) Biodegradable blends of cellulose with synthetic polymers and some other polysaccharides. Eur Polym J 49(1):194–202

Song K (2017) Micro- and nano-fillers used in the rubber industry. In: Thomas S, Maria HJ (eds) Progress in rubber nanocomposites. Woodhead Publishing Series in Composites Science and Engineering, Sawston, United Kingdom, pp 41–80

Spahr ME, Rothon R (2016) Carbon black as polymer filler. In: Palsule S (eds) Polymers and polymeric composites: a reference series, pp 1–31. Springer, Berlin, Heidelberg

Takamura T (2009) Chemistry, electrochemistry, and electrochemical applications | Carbon. In: Garche J (ed) Encyclopedia of Electrochemical Power Sources. Elsevier, New York, pp 709–743

Yin B, Wen H, Luo W (2022) Effect of thermo-oxidative aging on the Payne effect and hysteresis loss of carbon-black filled rubber vulcanizates. npj Mater Degrad 6:94. https://doi.org/10.1038/s41529-022-00306-5

Chapter 9
The Influence of Chemicals Order in Mixing Process on the Thermal History and Properties of the Elastomeric Blends

Ivan Labaj, Petra Skalková, Juliána Vršková, Ivan Kopal, and Darina Ondrušová

Abstract Many scientific publications and articles deal with the properties of elastomeric blends (rheological properties, vulcanization characteristics, mechanical properties, etc.) regardless of the so-called thermal history of the blend depending on the mixing process (Sivaselvi et al. 2021; Jong 2019). Each elastomeric blend is created by a mixing process that merges the individual materials into a whole and partially forms the future properties of the blend itself. Mixing conditions (mixing speed, mixing temperature, etc.) of elastomeric blends can be adjusted, which affects its thermal history. In addition to the mentioned conditions, the blend itself is made up of several components (their effect is known and described), which, by their content, composition, and type, also affect the thermal history of the prepared elastomeric blend (Hasan et al. 2018). The presented work deals with the influence of the order of adding individual chemicals to the mixing process, their influence on the measured indicators of the mixing process (mixing torque, real blend temperature), and subsequently on the rheological properties, vulcanization characteristics, and mechanical properties. The influence of the order of adding chemicals to the mixing process can be an alternative to changing the recipe or adjusting the mixing parameters with an economically neutral footprint (since the time, speed, temperature of mixing, and the

I. Labaj (✉) · P. Skalková · J. Vršková · I. Kopal · D. Ondrušová
Faculty of Industrial Technologies in Púchov, Alexander Dubček University of Trenčín, Púchov, Slovak Republic
e-mail: ivan.labaj@tnuni.sk

P. Skalková
e-mail: petra.skalkova@tnuni.sk

J. Vršková
e-mail: juliana.vrskova@tnuni.sk

I. Kopal
e-mail: ivan.kopal@tnuni.sk

D. Ondrušová
e-mail: darina.ondrusova@tnuni.sk

A. Öchsner and H. Altenbach (eds.), *Engineering Design Applications VII*, Advanced Structured Materials 230, https://doi.org/10.1007/978-3-031-84346-4_9

content of individual chemicals do not change). The obtained results show a significant influence especially in rheological properties and vulcanization characteristics, on the contrary, the tensile properties obtained minimal values change due to the change in the order of the chemicals in the mixing process.

9.1 Introduction

Elastomeric blends and their development are well described in countless scientific studies and books. The standard for scientific articles containing the study of elastomeric blends is the reporting of their recipes used for the preparation, for the possibility of verification or repeating of the results. Often, however, the order or mixing time of each chemical is not indicated, which can have a significant impact through the thermal history of elastomeric blend as well as on the resulting vulcanizate properties or the final product itself (Sivaselvi et al. 2021; Jong 2019; Hasan et al. 2017, 2018; Narongthong et al. 2018). As shown in this study, the order of chemicals itself affects the thermal history of elastomeric blend, and their influence is subsequently reflected in the properties of the blend and vulcanizate. If it is possible to achieve significant changes just by changing the order of chemicals and thus influencing the thermal history with the commonly used silica filler, the results can be more significant in the case of other fillers. In particular, new alternative fillers require a detailed study not only of the properties of the prepared elastomeric blends but also of the mixing process itself (Pajtášová et al. 2018; Mičicová et al. 2023). It is therefore possible to correct the resulting properties of blends not only by changing the recipe but also by changing the order of chemicals, which gives a new possibility of using the recipes of blends where some of the properties were not satisfactory for the author.

9.2 Materials and Methods

Elastomeric blends for this study were mixed using a laboratory universal torque mixer Brabender Plastograph EC Plus (Brabender, Duisburg, Germany) with a chamber volume of 80 cm^3 at a speed of 50 ± 1 rpm and a set mixing temperature of 90 ± 1 °C according to the recipe in Table 9.1.

The purpose of the study was to demonstrate and describe the influence of the order of each chemical during the mixing process on the properties of elastomeric blends, the individual order of each chemical for each prepared blend is shown in Table 9.2.

Using the WinMix software, the parameters of the mixing process were recorded, namely the mixing torque and the real mixing temperature. To eliminate the effect of multi-step mixing, a single-step process was chosen.

Table 9.1 Recipe of elastomeric blend

Chemicals	Content (phr)
SMR10	100
Zinc oxide (ZnO)	1.2
Stearic acid (SA)	0.8
Silica	40
TBBS	1.5
Sulphur	2

Table 9.2 The order of chemicals for each prepared blend

	Blend					
	A	B	C	D	E	F
← Chemicals order	SMR10	SMR10	SMR10	SMR10	SMR10	SMR10
	ZnO	SA	Silica	TBBS	Sulphur	TBBS
	SA	ZnO	ZnO	ZnO	ZnO	Sulphur
	Silica	Silica	SA	SA	SA	ZnO
	TBBS	TBBS	TBBS	Silica	Silica	SA
	Sulphur	Sulphur	Sulphur	Sulphur	TBBS	Silica

After mixing, each blend was subjected to additional homogenization on a laboratory two-roll mill LaboWalz (Vogt, Berlin, Germany) and then left to rest for 24 h at laboratory temperature.

Rheological properties and vulcanization characteristics were determined at a vulcanization temperature of 160 °C using a PRPA 2000 rheometer (Alfa Technologies, Akron, OH, USA).

For determination of tensile properties, samples were prepared in the form of plates with a thickness of 2 ± 0.02 mm and vulcanized for an optimal cure time (t_{C90}), for samples for abrasion loss 1.5 times t_{C90} was applied in a Fontijne LabEcon 600 vulcanization press machine (Fontijne Presses, Vlaardingen, Netherlands).

Tensile properties of prepared blends were determined by a universal tensile tester Shimadzu AG–X Plus (Shimadzu, Tokyo, Japan) by a test speed of 500 ± 5 mm·min^{-1} (according to ISO 37) on type A2 samples which were cut using an Instron cutter machine (Instron, Norwood, MA, USA).

Abrasion loss of each blend was determined by an Abrasion tester machine Standard Fe05000 (Bareiss, ON, Canada) with sample thickness 8 and 16 mm diameter according to norm ISO 4649.

9.3 Results and Discussion

In all the mixing torque (MT) recording curves (Fig. 9.1) during the mixing of elastomeric blends, an initial increase in value can be seen, caused by the entry of a matrix with a high viscosity (at room temperature) into the mixing process. After reaching the maximum of the MT curve, the plasticization phase of the matrix (reduction of its viscosity and break of rubber chain) begins due to the temperature of the chamber and the rotation of the mixing blades (Majesté and Vincent 2015). A time of 180 s was specified for the plasticization of the matrix of all blends after dosing the entire content of the matrix. The trend of the MT curves after reaching the plastic state of the matrix changes the trend from decreasing to an oscillating plateau (oscillation around the mean value). The matrix has reduced its viscosity (due to chain shortening) and is ready for the entry of other chemicals and their dispersion and incorporation into its volume (Akindele et al. 2019). The filler has the most significant influence on the MT value not only for its stiffening character, but also for the content added to the blend. The earlier the filler was in the order of chemical adding, the sharp increase shifts towards to lower mixing times. The addition of filler increases the viscosity of the mixed blend and thus causes an increase in torque. The increase in viscosity and thus also the internal resistance of the blend to the mixing blades rotation has an additional effect on the matrix (from a mechanical point of view), where additional breaking of rubber chain can occur and therefore shorten them (Klie et al. 2014; Kraibut et al. 2024). The order in which the filler is added also affects the dispersion of chemicals that follow in the mixing process after it and can have a negative effect, especially for increasing the viscosity of the blend. When looking in more detail at the filler addition area, one can see a rapid linear increase in the MT curve. The size of this linear increase also depends on the order of filler addition to the blend. The earlier the filler is added to the blend, the longer this part is, since the filler enters the blend, which does not contain any other chemicals and the matrix take in a significantly higher content of filler at once, on the contrary, if the filler is added later, or as the last chemical, this area is shorter. The V-shaped effect appearing on the MT curves is either due to the addition of stearic acid (it temporarily reduces the shear forces in the mixing chamber by changing its state due to the temperature) or the release of pressure in the chamber after it is opened before the dosing of the chemical (occurring mainly in the last stages of the mixing process, when the blend in the mixing chamber contains majority of chemicals, which is related to its viscosity) (Patti et al. 2021). An interesting finding is that the V shape does not appear on the curve of blend F in the last phase of mixing, due to the dosing of the filler as the last chemical and thus the absence of increased pressure in the chamber due to it. Chemicals such as zinc oxide and sulphur, and vulcanization accelerator due to their content in the blend and character, their addition to the blend does not have a significant effect on the value of the MT curves (neutral chemicals).

The real mixing temperature (RMT) (Fig. 9.2) is a direct expression of the temperature of the mixed blend in the chamber of the mixing device, and its value is partly influenced by the values of the mixing torque (Fig. 9.1) and forms the thermal history

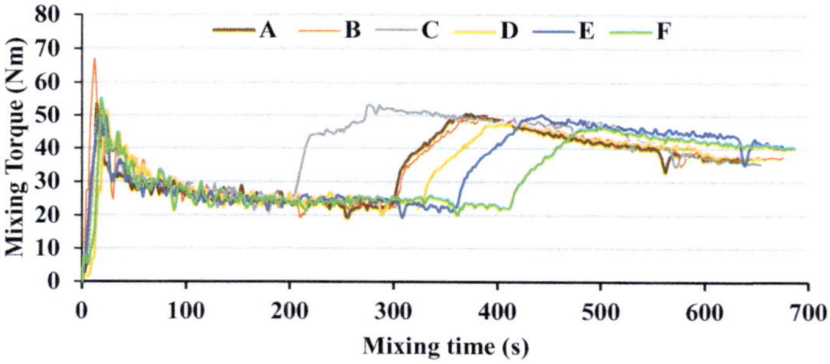

Fig. 9.1 Measured parameters of the mixing process—mixing torque

of rubber blends. The RMT is often confused with the value of the mixing temperature, which, however, indicates the set temperature of the chamber, and in the case of the mixing process, it only has the function of heating the matrix in the initial phase of mixing the blend and accelerating its plasticization and tempering of the mixed blend (for blends containing plasticizers or non-hardening types of fillers) (Sarkawi et al. 2014). From the view of Fig. 9.2 the RMT values are much higher than the mixing temperature (90 °C) in almost the entire mixing process. The initial decrease in the value of RMT is related, as in the case of mixing torque, to the entry of the matrix with laboratory temperature into the mixing process, and heating from the surfaces of the mixing chamber. In this section (until the mixing temperature is exceeded), the mechanical degradation of the matrix chains due to the rotation of the mixing blades prevails. With the subsequent decrease in viscosity, the RMT begins to rise mainly due to the shear stress in the blend in the chamber of the mixing device (Bruno et al. 2016). After the matrix plasticization process, the RMT value slightly stabilized. As with MT values, the filler has the most significant effect on RMT. Since the filler increases the viscosity as well as the shear forces in the blend, after its addition, the RMT value starts to rise until it reaches a maximum. The earlier the filler is added to the blend, the higher is the maximum value and the longer the blend is exposed to a higher temperature (more expressive thermal history of rubber blend) (Zhu et al. 2020). Exposure to higher temperature for a long time can have positive as well as negative consequences on other properties of rubber blends. The positive is the reduced viscosity of the matrix and thus the possibility of better intercorporation and dispersion of the dosed chemicals. The negative is directly affecting the matrix, which may be subject to degradation due to the temperature and speed of the mixing blades (increased efficiency of rubber chain breaks). In this study and with the given recipe (Table 9.1), the main factor affecting the thermal history of rubber blends is the filler. Before adding the filler, a slight decrease can always be seen on the curves, which is caused by the filler entering the chamber of the mixing device with a lower temperature than the mixed blend (this effect is not visible on the MT curves). After the addition, there is a continuous increase in the RMT value until reaching a maximum and the subsequent stabilization of the value in the case of blends A to C, i.e. blends where the filler was in the order of maximum 4th place.

Stabilization of the value indicates the filler dispersion, as the filler was dispersed in the volume of the matrix, the MT value (Fig. 9.1) decreases, and the filler no longer causes an increase in shear forces during the mixing process (Sökmen et al. 2021). In the case of blends D to F (filler order fifth to sixth place), on the other hand, a continuous increase in the RMT value can be seen without stabilization. In the last stages of mixing, a slight drop in the curve is visible (except for blend F), which causes the opening of the chamber during the dosing of the last chemicals (release of pressure as well as cooling through the dosing slot). From the point of view of RMT, chemicals such as zinc oxide, sulphur, curing accelerator and, in addition, stearic acid are neutral and do not significantly affect its value. Changing the status of stearic acid at RMT (effect on the MT curve), the mixed blend is resistant to short-term effects that will cause a decrease in MT. The heated blend can accumulate temperature for a certain time (Fig. 9.3).

The rheological properties of elastomeric blends express the stiffness at the beginning and end of the vulcanization process. The minimum torque (M_L) corresponding

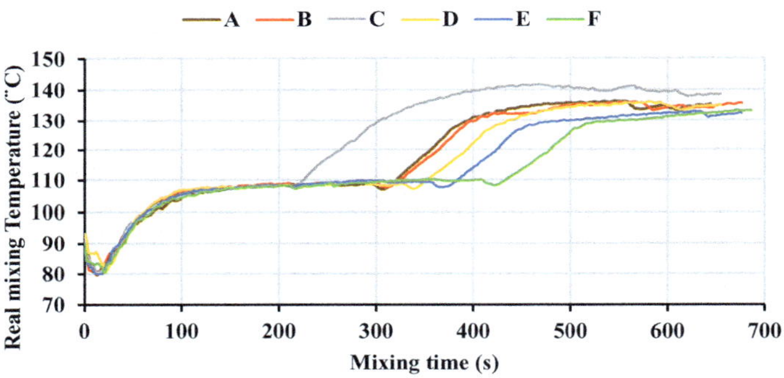

Fig. 9.2 Measured parameters of the mixing process – real mixing temperature

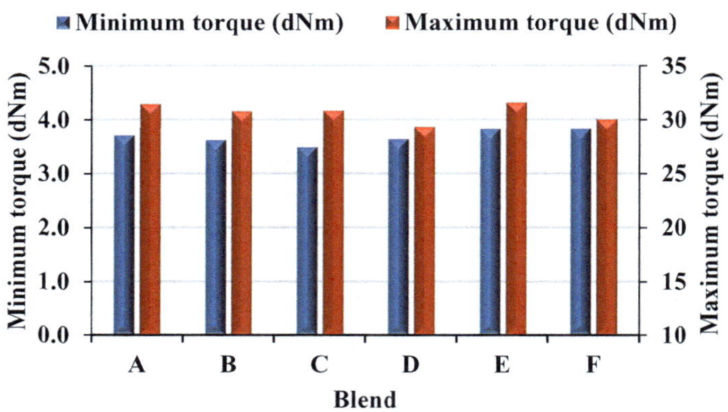

Fig. 9.3 Rheological properties

to the stiffness of the blend at the beginning of the vulcanization process is not significantly affected by the order of chemicals during the mixing process (Labaj et al. 2020). The lowest M_L value was achieved by blend C, which during the mixing process was exposed to increased RMT (filler added after the matrix) for a longer time and thus had the more expressive thermal history. The stiffness of the raw blend and thus the M_L value depends on the thermal history during the mixing process and specifically on the RMT values. A higher value of RMT caused a more intense break of rubber chain and thus their shortening due to the mixing of the blend. A shorter length of rubber chains will cause a lower resistance to the oscillations of the rheometer disc and thus a lower M_L value. On the contrary, blends that had a less expressive thermal history (blends E and F) caused a higher M_L value due to a lower intensity of rubber chain break. The maximum torque (M_H) expresses the stiffness of the vulcanizate but also the effectiveness of the vulcanization process, as the stiffness is mainly formed by the crosslinking network (Al-Nesrawy and Al-Nesrawy 1973). The order of the chemicals has a significant influence (the thermal history of the blend is associated with it), especially with the accelerator TBBS and silica. Blends in which TBBS was added as the second chemical right after rubber and the last was silica filler among (blends E and F) showed the lowest M_H values among the prepared blends. TBBS is an accelerator with a delayed action, and adding it to the blend as a second chemical caused the delay phase, which normally takes place during the curing test, to start already during the mixing process (possible decrease in the effectiveness of the accelerator), in addition, the filler was added to the blend between the last ones, which did not allow a significant increase RMT which is necessary to start the vulcanization (Ghosh et al. 2018). Due to the lower course of the RMT, the rubber chains were not subjected to such intense breaking and thus their length was longer compared to the other blends, and due to the reduction of the TBBS efficiency, the blends showed a low M_H value. The highest M_H value was achieved by blend E, in which the main chemicals affecting this value (Silica, TBBS) were dosed last. There were no significant differences in the M_H value in the interval of blend A to C (Fig. 9.4).

The scorch time of the blend is an important indicator of the vulcanization process, as it determines the beginning of the process and therefore affects its length and the quality of the final product (especially in the case of products with complex shapes). As with the M_H values, it can be stated that in the interval of blends A to C and blend E, there is no significant influence of the order of chemicals and thus also thermal history. However, the statement was confirmed that if the accelerator is added as the second chemical in the mixing process, right after the rubber, its delay action phase starts already during the mixing process (blends D and F) (Lee et al. 2017). By eliminating the delay action phase, TBBS enters the curing process as a fast accelerator and the blends show the lowest values of scorch time (possibility of reducing the value by up to 31%). A similar trend can also be observed in the case of the optimum cure time (t_{C90}), which indicates the time required to create a 90% crosslinking network. Since the scorch time of the blend was shortened for blends D and F, the t_{C90} itself also moved to lower values (possibility of reducing the value by up to 35%). Here, one can see what a complex effect thermal history and order

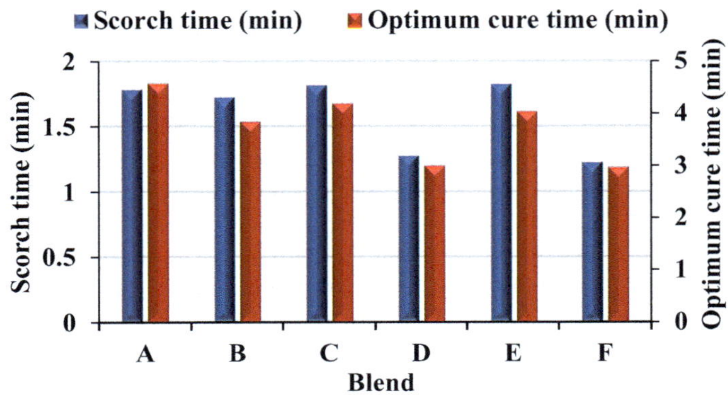

Fig. 9.4 Curing characteristics

of chemical can have on the properties of elastomeric blends. In the case of blend B, the t_{C90} value decreased by 16% compared to blend A (only the reverse order of chemicals ZnO and SA). SA and ZnO form zinc stearate in the vulcanization process, which acts as an activator of sulphur vulcanization. The change in order could cause a better dispersion of SA in the matrix and thereby increase the reactivity with ZnO. From the point of view of the length of the curing process, the use of the sequence of chemicals for blend A is therefore the most disadvantageous (Torani et al. 2019) (Fig. 9.5).

The tensile properties express the strength but also the flexibility of the resulting vulcanizate and thus also limit its applicability (Yao et al. 2023). The results of the study also demonstrate that the tensile properties are the most stable from the point of view of thermal history and order of chemicals, and in their case, only minimal changes occur. For tensile strength, the maximum change was at the level of approx.

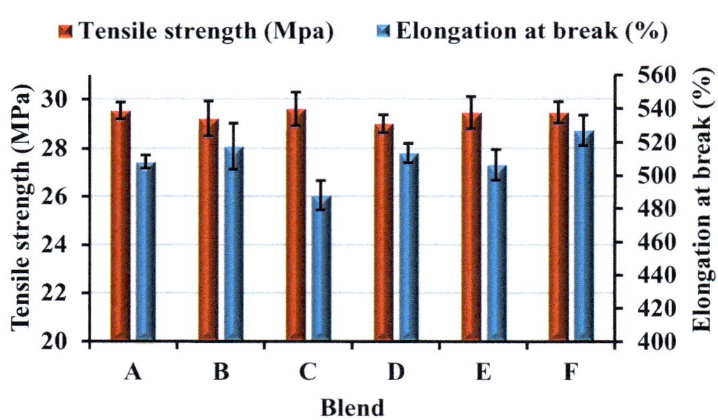

Fig. 9.5 Tensile properties of prepared elastomeric blends

1.8%. The highest value was achieved by blend C with the more expressive thermal history (compared to the lowest value for blend D, an increase of 0.6 MPa). This slight increase could be caused by several processes, namely: prolonged mixing time of the filler and therefore also the time of dispersion itself, intensity of mixing and thermal history, which caused more significant breaks of rubber chain (also visible on the value of M_L Fig. 9.3), shorter rubber chain supported the formation denser crosslink network (Kim et al. 2020). On the contrary, blend D achieved the lowest value at 29.01 MPa. This blend also reached the lowest M_H value (Fig. 9.3). Since the blend did not have a significant thermal history and the filler was added as one of the last chemicals, the rubber chains break was not significant, which also had an impact on the crosslinking process. Long rubber chains limit the density of the resulting crosslinking network. Elongation at break values showed the highest change at the level of approx. 8%, which is more pronounced than in the case of tensile strength, but insignificant compared to other properties. The lowest value was achieved by blend C with the more expressive thermal history, which was mainly caused by the short rubber chain, which is advantageous for tensile strength, but in the case of elongation at break causes a decrease, as their flexibility is limited (due to crosslinking bridges and the density of the network) and the possibility of movement due to impact stress. The best chemicals order from the point of view of elongation at break is with blend F, which reached the highest value. In this blend, the filler was dosed as the last chemical and thus the blend had the less expressive thermal history. In this case, the rubber matrix retains long chains, which are flexible under tensile stress, and the resulting crosslinking network does not affect their motion.

In order to achieve good elongation at break, it is advantageous to have a less expressive thermal history, a combination of long rubber chain of matrix and optimum crosslinking density (Fig. 9.6).

The last of the properties studied is resistance to abrasion, namely volume loss of each blend. Changes in volume loss were a maximum of 12% (from the comparison

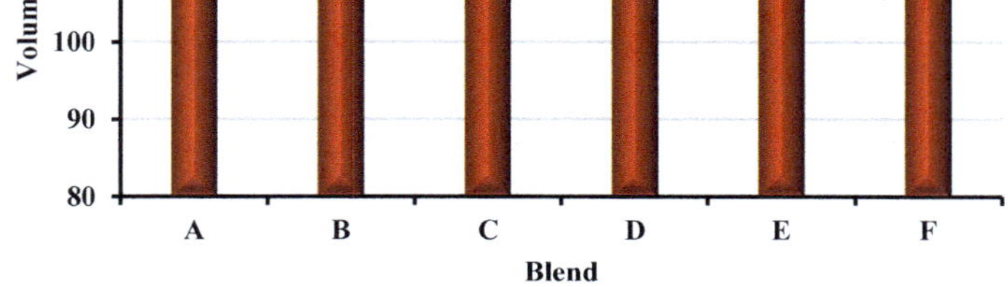

Fig. 9.6 Abrasion resistance of prepared elastomeric blends

of the minimum and maximum values). From the results of the study, the blends that achieved higher elongation at break values had a higher volume loss value. This fact causes the thermal history and order of chemicals. As mentioned above, the matrix of the blend with a lower thermal history (blends D and F) contained longer rubber chains, which ensured flexibility, but also abrasion of larger pieces of the blend when in contact with an abrasive surface (Chae et al. 2024). The only blend that escapes the fact is blend B, which had a similar thermal history as blend A. In this case, changing the order of the chemicals ZnO and SA caused an increased reactivity as well as affecting the crosslinking network density. Blends with a significant thermal history (blends A and C) achieved lower volume loss values, as their matrix has a short rubber chain and a denser crosslinking network. Again, there is a blend that, despite being assigned to the group of blends with an insignificant thermal history, reached a volume loss value corresponding to blends with a significant thermal history, namely blend E. It is possible to assume that with this blend, optimal dispersion of sulphur occurred due to its dosage as the second chemicals immediately after the matrix, which caused an equalization of the influence of longer rubber chain in matrix and crosslinking network density. To confirm this statement, it is necessary to carry out another study with the content of expanding analyses and tests of the order of chemicals in the E blend.

9.4 Conclusion

The results of the study showed a significant change in the case of some properties of elastomeric blends caused only by changing the order of each chemical and thus also their thermal history. The most significant influence was achieved in curing characteristics and tensile properties, the values of which are often the minimum requirements for the applicability of the blend in industry. By changing the order of each chemical, it is possible to achieve, for example, a shortening of the vulcanization process by a maximum of 35% (reduce the economic side of the given blend as well as to increase the efficiency in industrial application), or tensile properties, namely elongation at break by 8%. In addition, the volume loss of each blend was affected by approx. 12%. In the future, this study can serve as an aid in the process of preparing elastomeric blends, when the author wants to significantly change the selected property, without changing recipes.

Acknowledgements This research work has been supported by the Operational Program Integrated Infrastructure, co-financed by the European Regional Development Fund by the project: Advancement and support of R&D for "Centre for diagnostics and quality testing of materials" in the domains of the RIS3 SK specialization, Acronym: CEDITEK II., ITMS2014+ code 313011W442 and Scientific Grant Agency VEGA 1/0265/24.

References

Akindele OE, Kurian T, Gopi U (2019) Effect of sequence of mixing and internal mixer parameters on the viscosity of compounds for tire components. Int J Interdisp Res Innov 7(1):497–503

Al-Nesrawy EF, Al-Nesrawy SHH (2021) Rheological properties of (NR/SBR/CMC/C.B) Nanocomposites. J Phys Conf Ser 1973. https://doi.org/10.1088/1742-6596/1973/1/012161

Bruno F, Herblot M, Deme F (2016) Behavior of masticated natural Rubber highlighted by LAOS Rheology. Int. Tech. J. Polym. Mater. 69(10):44–50

Chae E, Yang SR, Choi SS (2024) Abrasion behaviors of silica-reinforced solution styrene-butadiene rubber compounds using different abrasion testers. Polymers 16:2038. https://doi.org/10.3390/polym16142038

Ghosh J, Ghorai S, Jalan AK, Roy M, De D (2018) Manifestation of accelerator type and vulcanization system on the properties of silica-reinforced SBR/devulcanize SBR blend vulcanizates. Adv Polym Technol 37:2636–2650. https://doi.org/10.1002/adv.21939

Hasan A, Rochmadi R, Sulistyo H, Honggokusumo S (2017) Rubber mixing process and its relationship with bound rubber and crosslink density. IOP Conf Ser: Mater Sci Eng 213:012048. https://doi.org/10.1088/1757-899X/213/1/012048

Hasan A, Rochmadi R, Sulistyo H, Honggokusumo S (2018) The effect of ingredients mixing sequence in rubber compounding upon vulcanization kinetics of natural rubber: an autocatalytic model study. Indones J Chem 18:709–717. https://doi.org/10.22146/ijc.25707

Jong L (2019) Improved mechanical properties of silica reinforced rubber with natural polymer. Polym Test 79. https://doi.org/10.1016/j.polymertesting.2019.106009

Kim DY, Park JW, Lee DY, Seo KH (2020) Correlation between the crosslink characteristics and mechanical properties of natural rubber compound via accelerators and reinforcement. Polymers 12:2020. https://doi.org/10.3390/polym12092020

Klie B, Teich S, Haberstroh E, Giese U (2014) New method for evaluating rubber mixing quality by means of alternative representation of the fingerprint chart. KGK-KAUT GUMMI KUNST 10(68):31–38

Kraibut A, Kaewsakul W, Sahakaro K, Saiwari S, Noordermeer JWM, Dierkes WK (2024) Degradation during mixing of silica-reinforced natural rubber compounds. Materials 17(2). https://doi.org/10.3390/ma17020341

Labaj I, Ondrušová D, Vršková J, Kohutiar M (2020) The effect of various alternative filler granularity on the properties of elastomeric vulcanizate. IOP Conf Ser: Mater Sci Eng 776:012098. https://doi.org/10.1088/1757-899X/776/1/012098

Lee KC, Yusoff NAMd, Othman N, Mohamad Aini NA (2017) Effect of vulcanization temperature on curing characteristic, physical and mechanical properties of natural rubber/palygorskite composites. IOP Conf Ser: Mater Sci Eng 223:012017. https://doi.org/10.1088/1757-899X/223/1/012017

Majesté JCH, Vincent F (2015) A kinetic model for silica-filled rubber reinforcement. J Rheol 59:405–427. https://doi.org/10.1122/1.4906621

Mičicová Z, Božeková S, Pajtášová M, Skalková P (2023) Rheological and mechanical properties of rubber blends filled with modified bentonite. In: Öchsner A, Altenbach H (eds) Engineering design applications V. advanced structured materials, vol 171, pp 21–33. Springer, Cham

Narongthong J, Sae-Oui P, Sirisinha CH (2018) Effects of mixing parameters and their interactions on carbon black filled styrene-butadiene rubber. Rubber Chem Technol 91:521–536. https://doi.org/10.5254/rct.18.82581

Pajtášová M, Mičicová Z, Ondrušová D, Božeková S, Janík R, Pecušová B, Raník L (2018) Use of waste materials in rubber matrix. MATEC Web Conf. https://doi.org/10.1051/matecconf/201815707009

Patti A, Lecocq H, Serghei A, Acierno D, Cassagnau P (2021) The universal usefulness of stearic acid as surface modifier: applications to the polymer formulations and composite processing. J Ind Eng Chem 96:1–33. https://doi.org/10.1016/j.jiec.2021.01.024

Sarkawi SS, Dierkes WK, Noordermeer JWM (2014) Reinforcement of natural rubber by precipitated silica: the influence of processing temperature. Rubber Chem Technol 87:103–119. https://doi.org/10.5254/rct.13.87925

Sivaselvi K, Varma VS, Harikumar A, Jayaprakash A, Sankar S, Krishna CY, Gopal K (2021) Improving the mechanical properties of natural rubber composite with carbon black (N220) as filler. Mater Today Proc 42:921–925. https://doi.org/10.1016/j.matpr.2020.11.836

Sökmen S, Oßwald K, Reincke K, Ilisch S (2021) Influence of treated distillate aromatic extract (TDAE) content and addition time on rubber-filler interactions in silica filled SBR/BR blends. Polymers 13:698. https://doi.org/10.3390/polym13050698

Torani D, Crespo JS, Brandalise RN (2019) Influence of ZnO on the properties of elastomeric compositions and their leached extract. Polímeros: Ciência e Tecnologia 29:e2019041. https://doi.org/10.1590/0104-1428.07218

Yao Q, Dong P, Zhao Z, Li Z, Wei T, Wu J, Qiu J, Li W (2023) Temperature dependent tensile fracture strength model of rubber materials based on Mooney-Rivlin model. Eng Fract Mech 292:109646. https://doi.org/10.1016/j.engfracmech.2023.109646

Zhu L, Tian X, Pan Y, Chang T, Wang K, Niu G, Zhang L, Wang C, Han W (2020) Optimization of serial modular continuous mixing process parameters for natural rubber composites reinforced by silica/carbon black. Polymers 12. https://doi.org/10.3390/polym12020416

Chapter 10
A System for the Recycling of Gray Water Through the Use of Mucilage as a Flocculant

Mauricio Aarón Pérez Romero, Luis Héctor Hernández Gómez, Luis Fernando Franco Guzmán, Alejandra Armenta Molina, Armando Josué Piña Díaz, and Felipe Escalona Cambray

Abstract A semi-automatic system for gray water recycling in a residential house located in a metropolitan area is reported in this paper. The purpose is to recover the water used in household cleaning activities (washing clothes, floors, etc.) that are normally sent to the local sewage system, without a treatment. Gray water recycling involves a physical–chemical process (flocculation, coagulation). Nopal mucilage can generate molecular networks and strongly retain large amounts of water. It is an alternative coagulant. The proposed prototype has double filtration and ultraviolet light disinfection. In addition, it has an ON–OFF control. The components of the prototype communicate with the controllers through the MQTT protocol. A star connection topology is used. The system is controlled with an interface within the Node-RED program. Under these conditions, the user can monitor and perform actions on the system. This is done by measuring the pH and turbidity variables. The

M. A. Pérez Romero · L. F. Franco Guzmán · A. J. Piña Díaz
Instituto Politécnico Nacional, Escuela Superior de Ingeniería Mecánica y Eléctrica, Ingeniería en Control y Automatización, Unidad Profesional Adolfo López Mateos "Zacatenco", Colonia Lindavista. Alcaldía Gustavo A. Madero, Ciudad de México, México
e-mail: maperezro@ipn.mx

L. F. Franco Guzmán
e-mail: lfrancog@ipn.mx

A. J. Piña Díaz
e-mail: apinad@ipn.mx

L. H. Hernández Gómez · A. Armenta Molina (✉) · F. Escalona Cambray
Instituto Politécnico Nacional, Escuela Superior de Ingeniería Mecánica y Eléctrica. Sección de Estudios de Posgrado E Investigación, Unidad Profesional Adolfo López Mateos "Zacatenco", Colonia Lindavista. Alcaldía Gustavo A. Madero, Ciudad de México, México
e-mail: aarmentamo@ipn.mx

L. H. Hernández Gómez
e-mail: lhhernandez@ipn.mx

F. Escalona Cambray
e-mail: fescalonac2200@alumno.ipn.mx

A. Öchsner and H. Altenbach (eds.), *Engineering Design Applications VII*, Advanced Structured Materials 230, https://doi.org/10.1007/978-3-031-84346-4_10

minimum permissible limits of the Mexican standard NOM-127-SSA1 are considered. Water is obtained, with a minimum of suspended solids for daily activities. However, it is not potable.

Keywords Gray water · Flocculation · Mucilage · pH and turbidity

10.1 Introduction

Water makes up 70% of our planet. It can be found in oceans, lakes, rivers, and even in the air and soil. It is essential for life. It is considered a renewable resource. However, it is finite and can run out if not managed properly. Wastewater comes from discharges from homes and industrial activities (SEMARNAT y CONAGUA 2023). The sea and rivers can be contaminated if it is treated improperly. As a result, flora, fauna, and human health are affected (Nguyen et al. 2021).

The amount of wastewater has increased steadily. According to SEMARNAT, 91.4% of the Mexican population uses the drainage system network or employs a septic tank (CONAGUA: National Water Information System 2023). Municipal wastewaters are generated in residential, public, and commercial areas. This type of water, when it has been generated recently, tends to be grayish and with a very peculiar smell of kerosene. As the time goes on, it becomes septic and pestiferous. Its smell transforms into one very similar to hydrogen sulfide. It has a characteristic black color (Zhu et al. 2021).

The World Health Organization (WHO) has recommended that 100 L per inhabitant in one day is sufficient to cover the basic needs of food and hygiene. In Mexico, the average water consumption per inhabitant is 380 L per day. Besides, the population is increasing steadily. As a result, the water demand is unsatisfied (CONAGUA: National Water Information System 2023). Water is intended only for household activities. However, it is used in different services. The drainage of gray water is demanding. It is sent to the sewage without any recycling. In this way, the demand for water for human use is challenging. In some cases, people buy drinking water.

Focusing on water consumption at home, 45% is required for potable consumption. The other 55% is intended for other purposes (We Are Water Foundation: The circles must be closed for everyone 2018). It is estimated that only 4% of water is recycled worldwide. Implementing a gray water recycling system at home would allow a partial solution.

10.2 Statement of the Problem

The purpose is the implementation of a domestic gray water recycling system. An average characterization of the physical, chemical, and bacteriological parameters of wastewater is not feasible. Each municipality has different factors that affect the water quality such as temperature, population, socioeconomic development, habits of the people, and type of sanitary devices, among others (Hutton and Chase 2016).

Domestic wastewater is used for hygienic purposes, such as cleaning bathrooms, kitchens, washing machines, etc. This type of water is discharged into sewage networks (Agarwal et al. 2022). It can be considered that 99.9% (by weight) is water and only 0.1% is suspended dissolved solids.

Gray water flows from washing machines, sinks, tubs, and showers, mainly. There are some quantities of soap, dirt, and bacteria. As far as possible, such water has a low bacterial index. According to Khan (2022), washing clothes is carried out with five different types of chemicals: concentrated detergent, fabric softener, alkaline degreaser, sanitizing bactericide, and bleach. The characterization of the water, at the outlet of the first load of a washing machine, is shown in Table 10.1. Figure 10.1 illustrates a conceptual scheme of a wastewater treatment system. It can be treated with chemical, physical, and biological processes (Chen et al. 2020).

Table 10.1 Results of the physicochemical and bacteriological characterization of typical outlet water of a washing machine

Parameters		Range	Average
pH	–	9.28–11.60	11.02 ± 0.84
Turbidity	UNT	103–172	141.18 ± 28.64
Conductivity	μS/cm	547–1985	1219 ± 581.5
Temperature	°C	24.53–28.50	26.59 ± 1.65
Sulfates	mg/L SO_4	14–50	35 ± 13.80
Total coliforms	UFC/100 mL	10–2420	700 ± 1156
Fecal coliforms	UFC/100 mL	0–299.7	75.73 ± 104.46

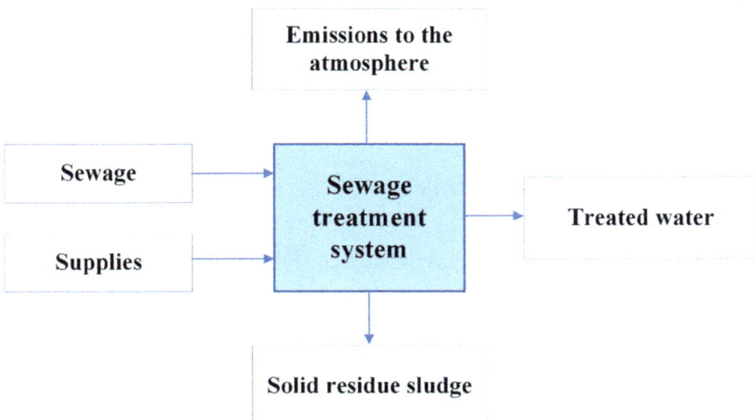

Fig. 10.1 Conceptual diagram of a wastewater treatment system

The quantity and quality of waste are not always constant (Cañón et al. 2046). It depends on the quality that is sought to be achieved, and the type of treatment to be carried out (Rodríguez Miranda et al. 2015).

Gray water is useful for irrigation in green areas or flushing toilets. On the other hand, the water from toilets and diaper washing should not be considered gray water. The water, that has been used to wash containers or materials containing sodium-based chemicals, is in a similar situation (Shaikh and Ahammed 2020).

A chemical wastewater treatment depends on the reagent and chemical properties of the substance. Iron and Oxygen can be eliminated, as well as phosphates and nitrates. It is necessary to carry out processes such as oxidation, ion exchange, coagulation, and electrochemical, among others (Sengupta et al. 2021).

Nopal mucilage is analogous to gums and is important for the food industry due to its viscosity properties (Cañón et al. 2046). It can generate molecular networks and strongly retain large amounts of water. Its importance lies in using nopal mucilage as an alternative coagulant to aluminum sulfate (Vargas-Solano et al. 2022). This technique uses low-cost equipment and materials. It is proven that the properties of the cactus can reduce and even eliminate contamination in water.

Solids are separated in a physical treatment. It depends on the physical properties of the water contaminant, such as size, buoyancy, viscosity, hardness, ductility, and malleability, among others. The main processes involved in this type of treatment are sieving, filtration of solids, precipitation, separation of dissolved solids, adsorption, flocculation, and clarification (Nguyen et al. 2021).

In a biological treatment, microorganisms and bacteria are eliminated (Hussain et al. 2021). Depending on the type of compound to be eradicated, the biological treatment is selected (Vargas-Solano et al. 2022). Treatments that eliminate carbon or nitrogen compounds can be used (Moghaddam et al. 2023). For this purpose, the aerobic and the anaerobic reactors can be used.

An aerobic system takes advantage of the microorganisms that are dissolved in water. It promotes oxidation (Li et al. 2022). On the other hand, an anaerobic system requires more energy to create dissolved sludge. Its selection depends on the substance to be treated.

10.3 Materials and Methods

Water recycling has been normalized in Mexico. For this purpose, Official Mexican Standards 127 was followed. The main aspects, which were considered are the following:

In a first proposal, the coarse solids in the water are eliminated. The filter selection depends on the characteristics of the water to be treated. An active sand filter is appropriate to remove most of the particles suspended in domestic gray water. The required crystallization level can be obtained.

Once the water has been filtered, it is recommended to add calcium carbonate. It reduces the hydrogen potential of the water. Afterward, the water stands for a few minutes and the carbonate is settled. Finally, the water is removed with the help of gravity. The outlet of the water is located above the height of the settling.

Reverse osmosis is another alternative considered by the standard mentioned above. A physical–chemical treatment eliminates the impurities and pathogens with the help of semi-permeable membranes. They allow the flow of water, and the bacteria are trapped. The solids are discarded down the drain. In the next step, the water goes through a post-carbon filter. In this case, the water meets the permissible limits. Unlike the previous methods, this one has a higher cost. It is inadvisable for residential application.

An alternative highly recommended by the same standard is the flocculation coagulation method. It has to be done before the water passes through the filter, where crystallization is obtained. The coagulants, which act as a catalyst, are added to the gray water. It separates the existing sludge. In addition, the sedimentation process takes place. It makes the filtration efficient, and the results are obtained on the same day. In a later stage, the pH of the water is improved. For these reasons, this procedure was followed.

For the problem at hand, the gray water is obtained from the first cycle discharge of a washing machine. It is expected that any softener has been used previously. Such water is stored in a collection container until the required volume is obtained. In the following step, the fluid flows through a washable disc filter. The larger solids are eliminated. Later, a physicochemical process will be carried out through a coagulation-flocculation batch. 15 L will be treated. The particles, that are not decanted in a second filter, are eliminated. The resultant mass of water is exposed to an ultraviolet light lamp. Finally, it is stored. Its pH and the number of suspended particles are monitored. The process is carried out correctly, if the pH is in the range between 6.5 and 8.5 and its turbidity is 5 units. Under these conditions, the water can be used in a secondary use, such as a toilet flush. If the pH and turbidity analysis are out of the acceptance range, the water is recirculated into the flocculation coagulation treatment vessel. The previous steps mentioned above are repeated.

10.4 Development of the Setup

Wastewater treatment requires effective solutions. It is required water, which is suitable for human use (Fig. 10.2). The two flocculants, used in this project, were cactus mucilage (organic) and aluminum sulfate. They allow greater density in the flocs. The sedimentation speed and the volume of the recovered water were increased.

Low-cost materials and open architectures for programming and configuration were considered, mainly. The components and communication architecture are illustrated below (Fig. 10.3).

Figure 10.4 illustrates the prototype described above. It has a valve at the outlet of the gray water storage vessel. It allows adequate maintenance of the filter. The next joint is a 90° elbow. It directs the flow of water to the next storage tank. It also

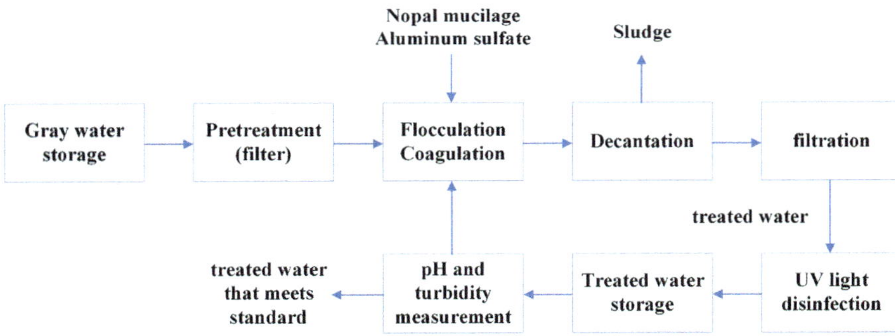

Fig. 10.2 Stages of the gray water recycling process

has a 3-way motorized valve and a solenoid valve. They transfer the gray water to the flocculation coagulation tank, both valves are operated in a coordinated manner.

The flocculation coagulation vessel has a two-level switch. A gear motor is attached to this vessel. It stirs and accelerates the flocculation and coagulation process. The sludge is decanted at the end of the process. The fluid flows through a 45° elbow. There is also a solenoid valve. Its main function is to purge the flocculation coagulation vessel and direct it to the second filtering stage.

The filtered water is sent to the ultraviolet light lamp. A flow sensor has been added. It pumps the water to the last vessel, where the pH and turbidity are determined. The pH sensor requires continuous contact with the water. Thus, an adaptation attached to a pipe, with three holes, is installed inside the vessel. Such holes keep flooded. In this way, the probe will never be left without water. Regarding the turbidity sensor, a hole is made in the container at 10 cm above the ground. The suspended particles can be evacuated.

In accordance with the results, the process can be finished, or the water is recirculated. For this purpose, there is a valve at the outlet of the treated water storage tank for maintenance purpose. Also, there is a solenoid valve. It allows or restricts the output flow.

A 3-way motorized valve has been added to the discharge of the pump. In one of the directions, the treated water is provided. In another direction and in conjunction with the motorized valve, the water is recirculated, and the process is repeated.

10.5 Programming of the Gray Water Recycling System

The stages of the recycling system are considered to determine the interface. Each of them is present through actuators and indicators. The process is operated consecutively to obtain the best possible results.

Filling stage. Gray water is conducted through primary filtration. It prevents the flow of large particles.

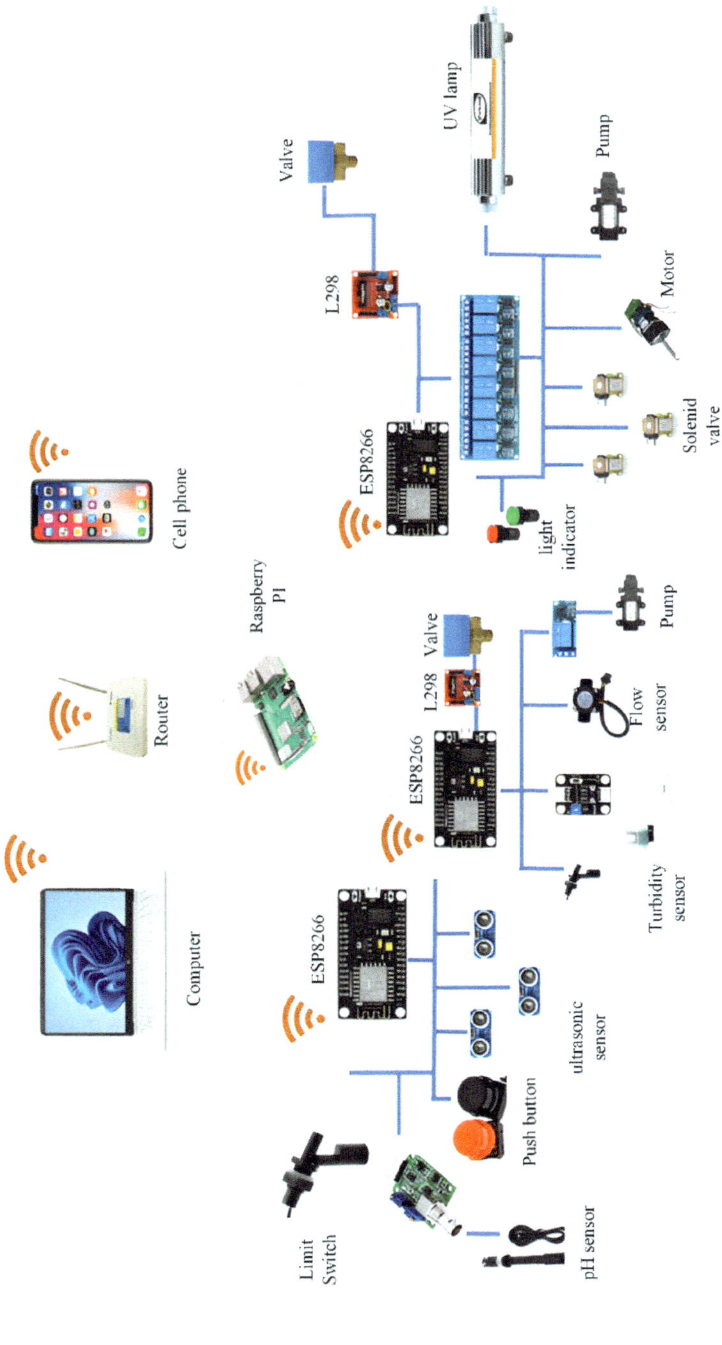

Fig. 10.3 The communication architecture of the proposed setup

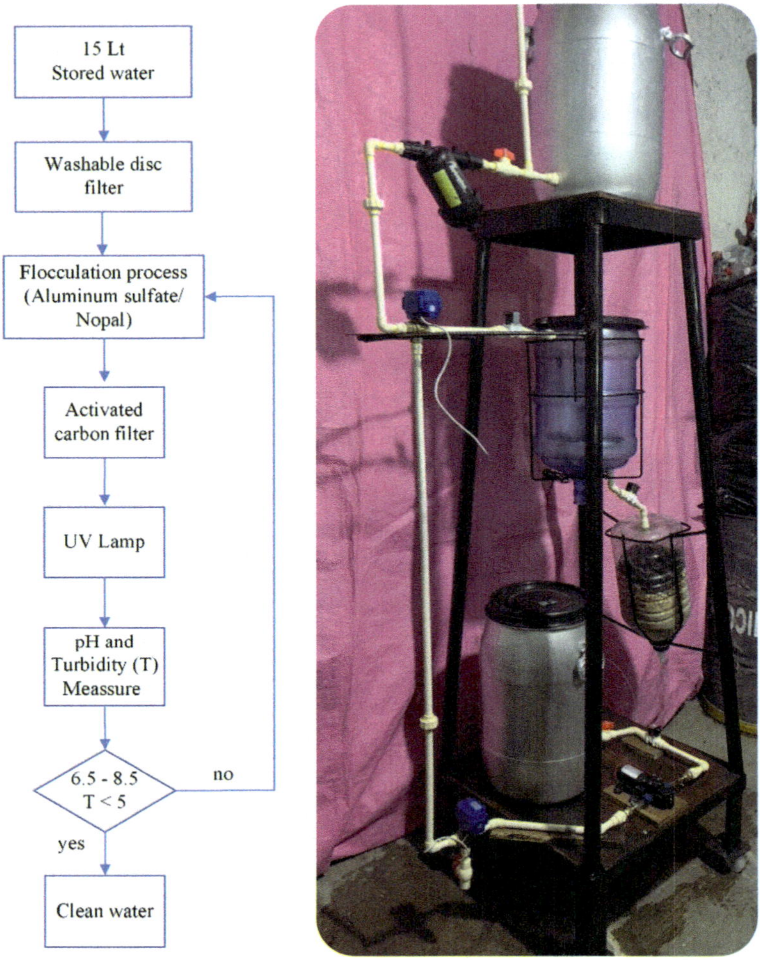

Fig. 10.4 Prototype for recycling gray water

Stir stage. The level of the flocculant is monitored. The stirring system is activated at the time established. The solids are decanted.

Secondary filtration stage. The pH is reduced, and the remaining solids are extracted. Besides, the disinfection is carried on with an ultraviolet lamp. At the end, the water is pumped to the last vessel of the process.

Analysis stage. pH and turbidity analyses are carried out. It is determined if these parameters are within the range established by the current NOM-127-SSA1. In this way, it is determined if the water has to be processed again or if it can be used.

The start and stop of the prototype are available virtually or physically. It will coordinate the valves and the liquid filter, generating the signal for the next procedure.

Once the filling of the stir container is activated, it will operate following the process requirements. The filtration button empties the agitator. The water is filtered through a sand and gravel filter and is disinfected by the UV lamp. Afterward, a pump is activated. It is operated automatically using a flow sensor. The user can stop this operation with the cancel pumping button.

Once the gray water has been processed, the flow direction of the valves can be decided using a Ros slider. Depending on the arrangement of the valves, the fluid is extracted or recirculated again. Once the direction of the fluid flow has been determined, pump 2 is activated. This operation can be canceled at any time. Finally, the user can determine the volume of water available, its pH, and turbidity. Figures 10.5 and 10.6 show the flowchart for the operation algorithm that will be executed.

10.6 Results

After several tests, the amount of flocculant required for 15 L of gray water was determined. Relevant results are summarized in Table 10.2 and are evaluated with the NOM-127-SSA1. A mixture of 70% cactus mucilage and 30% aluminum sulfate was used. Its NTU and pH were 6 and 8, respectively. The first parameter is within the acceptance range. However, the suspended particles are above the maximum level (5 NTU).

An adjustment of the flocculant mixture was made. It was 60% of nopal mucilage and 40% of the aluminum sulfate. The conditions of the first test were considered. The results were improved. The pH and turbidity obtained were 7 and 2, respectively. The minimum requirements of NOM-127-SSA1 were satisfied. The aluminum sulfate minimizes the sedimentation time.

10.7 Conclusions

The semi-automation of a gray water recycling system was carried out. The communication operated correctly on all devices with the MQTT protocol. However, there were some difficulties. This is the case of the ESP82266 slave controller. It did not have its GPIO layout. Since not all pins could be configured, a greater number of slave controllers was generated due to the demand for the inputs and outputs. Besides, some modifications were made to the final control elements. In the final scheme, the solenoid valves operated with the atmospheric pressure of the circulating flow.

The design of the pyramid-shaped prototype was appropriate for the condition of operation of the system. It operates with atmospheric pressure. It can be easily adapted and recycled water will be used in flush toilets and wash cars, among other activities.

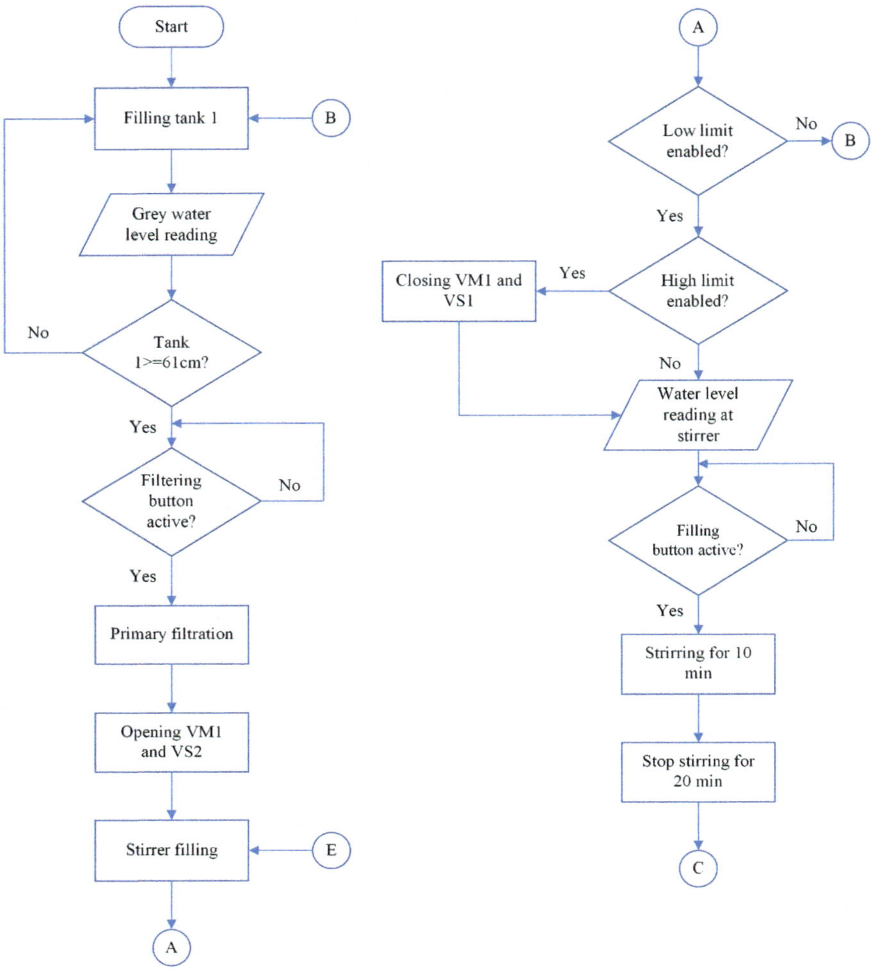

Fig. 10.5 Operation logic flowchart part 1

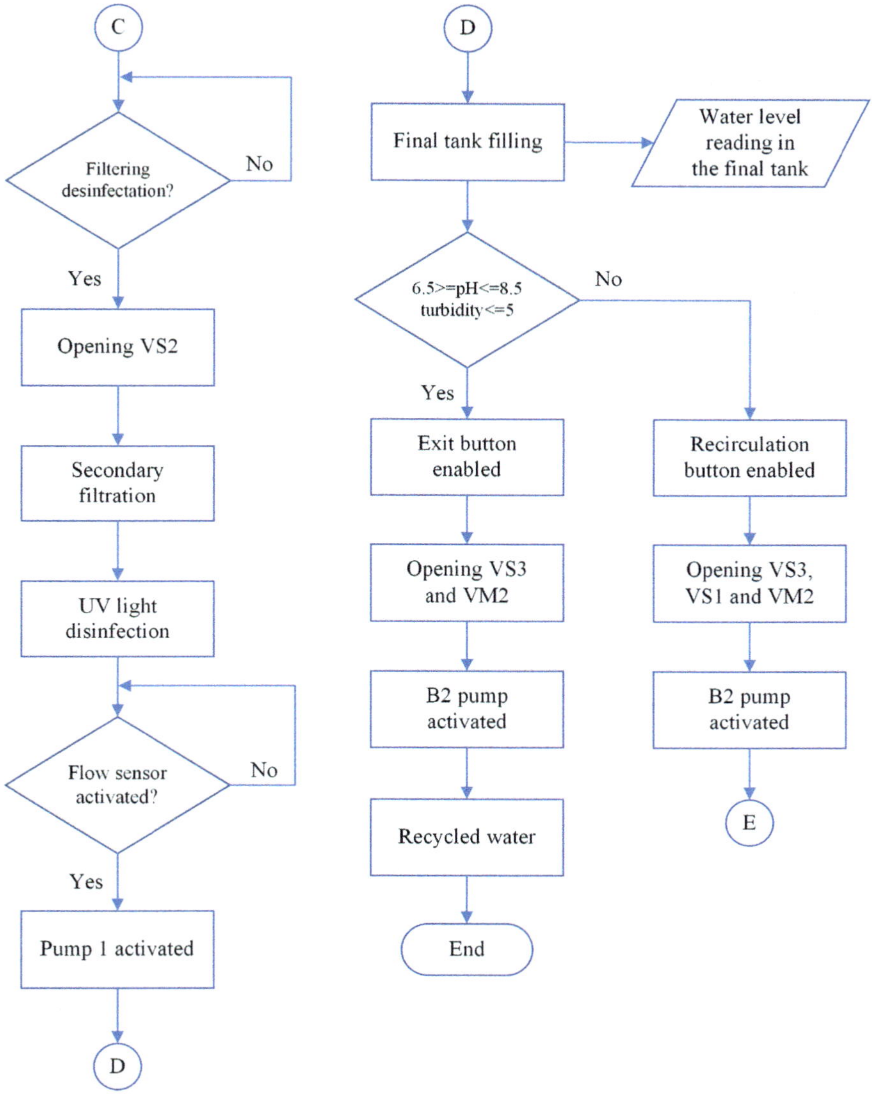

Fig. 10.6 Operation logic flowchart part 2

Table 10.2 Relevant results

Initial state	Quantity of flocculant	Final state	Result
	Nopal mucilage 70% Aluminum sulfate 30%		pH = 8 Turbidity = 6 NTU
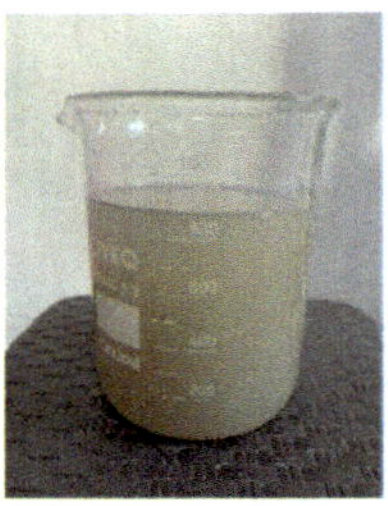	Nopal mucilage 60% Aluminum sulfate 40%		pH = 7 Turbidity = 2 NTU

Acknowledgements The authors kindly acknowledge the grant awarded by the Instituto Politécnico Nacional and COFAA for the development of this research.

References

Agarwal S, Darbar S, Saha S (2022) Challenges in the management of domestic wastewater for sustainable development. Curr Dir Water Scarcity Res 6:531–552. https://doi.org/10.1016/B978-0-323-91838-1.00019-1

Cañón AR, Carrillo CH, Cuaspud JG (2021) Evaluation of physicochemical properties of Nopal (Opuntia Ficus-Indica) as bio coagulant-flocculant for water treatment. J Phys Conf Ser 2046(1): 1–6. https://doi.org/10.1088/1742-6596/2046/1/012057

Chen GH, van Loosdrecht MC, Ekama GA, Brdjanovic D (eds) (2020) Biological wastewater treatment: principles, modeling and design. IWA publishing, London

CONAGUA: National water information system. https://sinav30.conagua.gob.mx:8080/ (2023). Accessed 28 Sept 2023

Hussain A, Kumari R, Sachan SG, Sachan A (2021) Biological wastewater treatment technology: advancement and drawbacks. Microb Ecol Wastewater Treat Plants 175–192. https://doi.org/10.1016/B978-0-12-822503-5.00002-3

Hutton G, Chase C (2016) The knowledge base for achieving the sustainable development goal targets on water supply, sanitation and hygiene. Int J Environ Res Public Health 13(6):536. https://doi.org/10.3390/ijerph13060536

Khan RA (2022) Detergents. Environ Micropollutants 117–130. https://doi.org/10.1016/B978-0-323-90555-8.00014-3

Li J, Ran X, Zhou M, Wang K, Wang H, Wang Y (2022) Oxidative stress and antioxidant mechanisms of obligate anaerobes involved in biological waste treatment processes: a review. Sci Total Environ 838:1–15. https://doi.org/10.1016/j.scitotenv.2022.156454

Moghaddam A, Khayatan D, Esmaeili Fard Barzegar P, Ranjbar R, Yazdanian M, Tahmasebi E, Tebyaniyan H (2023) Biodegradation of pharmaceutical compounds in industrial wastewater using biological treatment: a comprehensive overview. Int J Environ Sci Technol 20:5659–5696. https://doi.org/10.1007/s13762-023-04880-2

Nguyen PY, Carvalho G, Reis MA, Oehmen A (2021) A review of the biotransformations of priority pharmaceuticals in biological wastewater treatment processes. Water Res 188:116446. https://doi.org/10.1016/j.watres.2020.116446

Rodríguez Miranda JP, García Ubaque CA, Pardo Pinzón J (2015) Selection of technologies for the treatment of municipal wastewater. Tecnura 19(46):149–164. https://doi.org/10.14483/udistrital.jour.tecnura.2015.4.a03

SEMARNAT y CONAGUA: Official Mexican Standards. https://www.gob.mx/conagua/acciones-y-programas/normas-oficiales-mexicanas-66122 (2023). Accessed 28 Sept 2023

Sengupta S, Pal CK (2021) Chemistry in wastewater treatment. In: Sreedevi Upadhyayula S, Chaudhary A (eds) Advanced Materials and Technologies for Wastewater Treatment. CRC Press, Boca Raton, pp 129–152

Shaikh IN, Ahammed MM (2020) Quantity and quality characteristics of greywater: a review. J Environ Manage 261:110266. https://doi.org/10.1016/j.jenvman.2020.110266

Vargas-Solano SV, Rodríguez-González F, Martínez-Velarde R, Morales-García SS, Jonathan MP (2022) Removal of heavy metals present in water from the Yautepec River Morelos México, using Opuntia ficus-indica mucilage. Environ Adv 7:1–10. https://doi.org/10.1016/j.envadv.2021.100160

We Are Water Foundation: The circles must be closed for everyone (2018). https://www.wearewater.org/es/los-circulos-deben-cerrarse-para-todos_300561. Accessed 23 Oct 2023

Zhu TT, Su ZX, Lai WX, Zhang YB, Liu YW (2021) Insights into the fate and removal of antibiotics and antibiotic resistance genes using biological wastewater treatment technology. Sci Total Environ 776:145906. https://doi.org/10.1016/j.scitotenv.2021.145906

Chapter 11
Sensitivity Analysis of Acoustic Performance in Double Walls with Porous Layers Using the Finite Transfer Matrix Method and Kriging Metamodel

Soraya Bakhouche, Walid Larbi, Philippe Macquart, and Jean-François Deü

Abstract This work explores the application of the finite transfer matrix method (FTMM) for evaluating the sound transmission loss, of a double wall incorporating a porous layer. In order to optimize the acoustic performance of the system, a global sensitivity analysis using Sobol indices incorporating 11 parameters is performed, encompassing both material and geometric factors. Recognizing the computational demands of this sensitivity study with the FTMM model, the use of metamodels is proposed. A kriging metamodel is implemented to efficiently reduce the computation time while preserving accuracy. The sensitivity analysis results obtained from the original FTMM model are compared to those derived from the kriging metamodel. The study demonstrates that the acoustic performance of the double wall depends on the thickness of each layer, varying in importance across frequencies, and Young's modulus of the porous material at specific frequencies.

Keywords Double-wall · Porous material · Transmission loss · Finite transfer matrix method · Global sensitivity analysis · Metamodels

---

S. Bakhouche · W. Larbi (✉) · J.-F. Deü
Conservatoire National Des Arts Et Métiers (Cnam), Laboratoire de Mécanique Des Structures Et Des Systèmes Couplés (LMSSC), Paris, France
e-mail: walid.larbi@lecnam.net

S. Bakhouche
e-mail: soraya.bakhouche2@lecnam.net

J.-F. Deü
e-mail: jean-francois.deu@lecnam.net

P. Macquart
Union Des Fabricants de Menuiseries (UFMA), Courbevoie, France
e-mail: p.macquart@ufme.fr

A. Öchsner and H. Altenbach (eds.), *Engineering Design Applications VII*, Advanced Structured Materials 230, https://doi.org/10.1007/978-3-031-84346-4_11

11.1 Introduction

Double wall structures with a porous layer are widely utilized across diverse industries, due to their exceptional acoustic and thermal insulation capabilities. Characterizing the sound transmission loss of such multilayer systems necessitates various approaches, based on the frequency range under consideration. The finite element method is a common choice for analyzing low-frequency responses (Larbi et al. 2016; Soussi et al. 2021; Bakhouche et al. 2023a) while more efficient methods such as the transfer matrix method (TMM) (Allard and Atalla 2009) and statistical energy method (SEA) (Cherif et al. 2021) come into play for medium frequencies and high frequencies. The SEA conceptualizes the structure as interconnected subsystems with energy transfer pathways, while the TMM examines sound wave transmission through multilayer systems using transfer matrices. An extension of TMM, known as the finite transfer matrix method (FTMM), offers enhanced performance at low frequencies (Rhazi and Atalla 2010). Although, FTMM is slightly more resource-intensive compared to the classical transfer matrix method but less time-consuming than the finite element method.

Sensitivity analysis is important for assessing model robustness and optimizing performance. It comes in two common types: local and global. Local sensitivity analysis scrutinizes small perturbations around a nominal value, while global sensitivity analysis explores the impact of each input parameter across its entire range of variability. Qualitative methods, such as the Morris method (Morris 1991), and quantitative methods, like Sobol indices (Sobol′ 2001), contribute to global sensitivity analysis by ranking input parameters based on their relevance. While sensitivity analysis has found application in various models, Ouisse et al. (2012) employ quantitative methods, including Sobol indices and FAST, to quantify the sensitivity of porous models to parameter variations. Bakhouche et al. (2023b) conduct local and global sensitivity analysis using TMM to identify key parameters for a double wall structure (plate-air-porous-plate).

Calculating Sobol indices through Monte Carlo simulations can be computationally intensive, especially in the context of an important number of parameters. In such scenarios, the use of metamodels becomes essential. In this research, the Kriging metamodel is chosen. Kriging is a metamodeling technique that uses Gaussian processes to relate an output to its input parameters. Its origins can be traced back to the 1950s when Krige pioneered its use as a spatial interpolation tool in geostatistics (Krige 1951). Subsequently, Matheron formalized its principles in the 1960s (Matheron 1963). The application of Kriging expanded when Sacks et al. (Sacks et al. 1989) introduced it in the domain of metamodeling and computer experiments, where it was employed to represent the input/output mapping of computationally intensive models.

The study investigates the sensitivity of the sound transmission loss to uncertainties in geometric, mechanical, and acoustic parameters. The FTMM is used to model the system's vibroacoustic response to diffuse field excitation. Global sensitivity analysis techniques, including the Morris method and Sobol indices, are applied to identify influential parameters. A case study, starting with the Morris method to eliminate non-influential parameters (not presented here for brevity), subsequently utilizes Sobol indices to rank uncertain parameters by importance. All geometric variables follow a uniform distribution, and material variables follow a Gaussian distribution. A Kriging metamodel is implemented for the sensitivity study. Finally, the Sobol indices given by the original FTMM model are compared to those obtained with the Kriging metamodel.

11.2 FTMM for Double Wall with a Porous Layer

11.2.1 General Formulation

The transfer matrix method (TMM) stands as an efficient and simple technique employed for predicting the acoustic characteristics of laterally infinite multilayered systems. It utilizes transfer matrices to represent the propagation of plane waves across various layers and employs interface matrices to describe boundary conditions. In this investigation, we apply the TMM to predict the sound transmission loss in a double wall structure, encompassing plate, poroelastic, and plate layers (Fig. 11.1). The system is excited by an incident plane acoustic wave at an angle θ and is surrounded by two external fluids (fluids 1 and 2). The considered problem is two-dimensional, with the incidence plane identified as (x_1, x_3).

A transfer matrix noted $[T]$ characterizes the sound wave propagation within each layer of the system (Allard and Atalla 2009):

$$\vec{V}(M_n) = [T]\vec{V}(M_{n+1}) \tag{11.1}$$

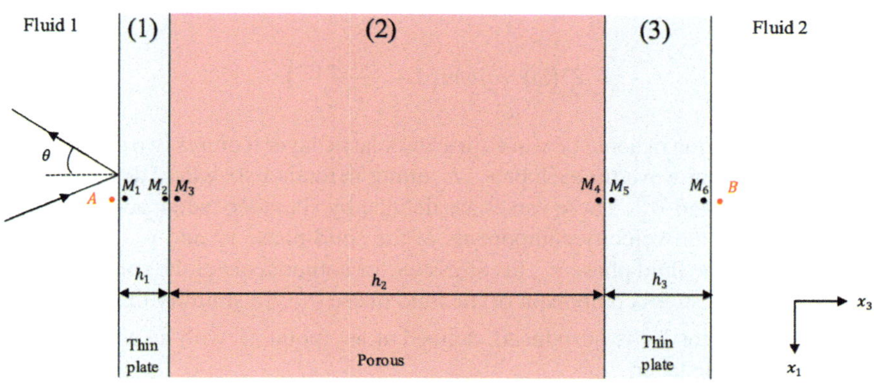

Fig. 11.1 A double wall with a porous layer subjected to an incident plane wave

The transfer matrix [T] is based on both the thickness h_i of each layer i, with $i = 1, 2, 3$, and the physical properties of the respective layer. It relates the propagation of sound waves between two points, M_n and M_{n+1}, with $n = 1, 2, 3, 4, 5$ positioned on the front and back faces of the layer, respectively. The vector $\overrightarrow{V}$ includes the variables delineating the acoustic field at a specific point M within the layer.

11.2.2 Construction of the Transfer Matrix for Each Layer

For a thin elastic plate, the consideration is limited to bending waves for wave propagation. The harmonic equation (Eq. (11.2)) of motion establishes the connection between the normal stress σ_{33}^s and normal velocity v_3^s of the plate at two points, M_1 and M_2. Equation (11.3) describes the continuity of velocities.

$$Z^s(\omega)v_3^s(M_2) = \sigma_{33}^s(M_2) - \sigma_{33}^s(M_1) \tag{11.2}$$

$$v_3^s(M_2) = v_3^s(M_1) \tag{11.3}$$

Equations (11.2) and (11.3) can be expressed in matrix form using the transfer matrix $[T_s]$ and the acoustic field vector $\overrightarrow{V^s}$.

$$\begin{pmatrix} \sigma_{33}^s(M_1) \\ v_3^s(M_1) \end{pmatrix} = \begin{bmatrix} 1 & -Z^s(\omega) \\ 0 & 1 \end{bmatrix} \begin{pmatrix} \sigma_{33}^s(M_2) \\ v_3^s(M_2) \end{pmatrix} = [T_s]\overrightarrow{V^s}(M_2) \tag{11.4}$$

The mechanical impedance $Z^s(\omega)$ (Eq. (11.5)) expresses the characteristics of the plate, including bending stiffness D, mass density per unit area m, wave number k, and incidence angle θ. Damping can be introduced through a complex Young's modulus.

$$Z^s(\omega) = j\omega m\left(1 - \frac{D(k\sin\theta)^4}{\omega^2 m}\right) \tag{11.5}$$

The propagation of acoustic waves in a poroelastic layer requires two compression waves and a shear wave for each phase resulting in total of six waves denoted as ϕ_1^S, ϕ_2^S, ψ_2^S ϕ_1^F, ϕ_2^F and ψ_2^F. These waves are defined by six independent acoustic quantities including the velocity components of the solid phase v_1^S and v_3^S, the velocity component of the fluid phase v_3^F, the stress tensor components of the solid phase σ_{33}^S and σ_{13}^S, and the stress component of the fluid phase σ_{33}^F. To simplify the representation, a state vector $\overrightarrow{V^p}$ is introduced, defined at any point M with an abscissa of x_3 in the poroelastic layer.

$$\overrightarrow{V^p}(x_3) = \left[v_1^S v_3^S v_3^F \sigma_{33}^S \sigma_{13}^S \sigma_{33}^F\right]^T \tag{11.6}$$

By analyzing the material, a matrix $[\Gamma]$ emerges as a representation its characteristics. This matrix builds a relation between the values of velocities and stresses at $x_3 = 0$ (one end) and $x_3 = h_2$ (opposite end) of the poroelastic layer, where h_2 is the thickness of the layer.

$$\overrightarrow{V^p}(0) = [\Gamma(0)][\Gamma(h_2)]^{-1}\overrightarrow{V^p}(h_2) \tag{11.7}$$

$$\left[T_p\right] = [\Gamma(0)][\Gamma(h_2)]^{-1} \tag{11.8}$$

The transfer matrix $\left[T_p\right]$ identifies the poroelastic layer and its components are given in reference (Allard and Atalla 2009). In practical applications, to simplify calculations and avoid matrix inversion, the origin of the x_3 axis can be shifted, allowing the use of the transfer matrix $\left[T_p\right] = [\Gamma(-h_2)][\Gamma(0)]^{-1}$.

11.2.3 Coupling Conditions

For a thin plate in contact with a porous layer, the continuity equations at the interface can be expressed as follows:

$$v_3^S(M_4) = v_3^s(M_5) \tag{11.9}$$

$$v_3^F(M_4) = v_3^s(M_5) \tag{11.10}$$

$$\sigma_{33}^S(M_4) + \sigma_{33}^F(M_4) = \sigma_{33}^s(M_5) \tag{11.11}$$

$$\sigma_{13}^S(M_4) = 0 \tag{11.12}$$

Equation (11.9) and Eq. (11.10) signify the continuity of the normal velocity at the porous-plate interface. Equation (11.11) denotes the continuity of the stresses. Finally, Eq. (11.12) reflects absence of shear at the interface.

These equations can be expressed in matrix form as follows:

$$\left[I_{ps}\right]\overrightarrow{V^P}(M_4) + \left[J_{ps}\right]\overrightarrow{V^s}(M_5) = \overrightarrow{0} \tag{11.13}$$

$$\left[I_{ps}\right] = \begin{bmatrix} 0\,-1 & 0\,0 & 0\,0 \\ 0\,0 & -1\,0 & 0\,0 \\ 0\,0 & 0\,-1 & 0\,-1 \\ 0\,0 & 0\,0 & -1\,0 \end{bmatrix} and \left[J_{ps}\right] = \begin{bmatrix} 0\,1 \\ 0\,1 \\ 1\,0 \\ 0\,0 \end{bmatrix} \tag{11.14}$$

11.2.4 Computation of the Transmission Loss of the Multilayer with Finite Size Correction

The implementation of the transfer matrix method results in the following matrix problem that needs to be solved:

$$[D]\overrightarrow{V} = \overrightarrow{0} \tag{11.15}$$

where $[D]$ is the assembled global transfer matrix given by:
***

$$[D] = \begin{bmatrix} [T0]\,[0] & [0]\,[0] & [-(1+R)0] \\ [I_{fs}]\,[J_{fs}][T_s] & [0]\,[0] & [0] \\ [0]\,[I_{sp}] & [J_{sp}][T_p]\,[0] & [0] \\ [0]\,[0] & [I_{ps}]\,[J_{ps}][T_s] & [0] \\ [0]\,[0] & [0]\,[I_{sf}] & [J_{sf}] \end{bmatrix} \tag{11.16}$$

and
***

$$\overrightarrow{V} = \left[\overrightarrow{v^f}(A)\,\overrightarrow{v^s}(M_2)\,\overrightarrow{v^p}(M_4)\,\overrightarrow{v^s}(M_6)\,\overrightarrow{v^f}(B)\right]^T \tag{11.17}$$

and where T is the transmission coefficient, R the reflection coefficient.

The sound transmission loss (TL) is calculated using the following equation employing the FTMM:

$$TL = -10\log\left[\frac{\int_{\theta_{min}}^{\theta_{max}}\int_0^{2\pi}\tau_f(\theta,\varphi)\sin\theta\cos\theta\, d\varphi\, d\theta}{\int_{\theta_{min}}^{\theta_{max}}\int_0^{2\pi}\sin\theta\cos\theta\, d\varphi\, d\theta}\right] \tag{11.18}$$

with $\tau_f = \tau_\infty(\sigma_R\cos\theta)$, where τ_∞ is the transmission coefficient calculated using the TMM for a given angle of incidence θ and σ_R is a correction factor and is homogeneous to radiation efficiency given by Allard and Atalla (2009) and (Atalla et al. 2006). In our case, $\theta_{min} = 0°$ and $\theta_{max} = 90°$.

11.3 Global Sensitivity Analysis Using Sobol Indices

11.3.1 Decomposition of Sobol

The decomposition of Sobol, introduced by Sobol (2001), serves as a widely utilized approach for analyzing the influence of input variables on an output of a system. This method quantitatively evaluates the relative importance of each input variable by measuring its contribution to the overall output variance. By fixing a variable, X_i, the decrease in the output variance is observed. The method determines if X_i is a significant contributor to the variance, indicating its importance. The decomposition of Sobol dissolves a function f into its constituent parts based on the contributions of each input variable to the output variance:

$$\begin{aligned} Y &= f\left(\vec{X}\right) = f(X_1, \ldots, X_M) \\ &= f_0 + \sum_{i=1}^{M} f_i(X_i) + \sum_{1 \le i \le j \le M} f_{ij}(X_i, X_j) + \cdots + f_{1,2,\ldots,M}(X_1, \ldots, X_M) \end{aligned} \tag{11.19}$$

where the constant f_0 and conditional expectations $f_i, f_{ij}, \ldots, f_{1,2,\ldots,M}$ are expressed as follows:

$$f_0 = E(Y) \tag{11.20}$$

$$f_i(X_i) = E(Y|X_i) - f_0 \tag{11.21}$$

$$f_{ij}\left(X_i, X_j\right) = E\left(Y|X_i, X_j\right) - f_0 - f_i - f_j \tag{11.22}$$

The total output variance $V(Y)$ can be expressed as:

$$\begin{aligned} V(\boldsymbol{Y}) = V[\mathbb{E}(\boldsymbol{Y}|\boldsymbol{X}_i)] + V\left[\mathbb{E}\left(\boldsymbol{Y}|\boldsymbol{X}_i, \boldsymbol{X}_j\right)\right] - V[\mathbb{E}(\boldsymbol{Y}|\boldsymbol{X}_i)] - V\left[\mathbb{E}\left(\boldsymbol{Y}|\boldsymbol{X}_j\right)\right] + \\ \cdots + V\left[\boldsymbol{f}_{1,2,\ldots,M}(\boldsymbol{X}_1, \ldots, \boldsymbol{X}_M)\right] \end{aligned} \tag{11.23}$$

11.3.2 Sobol Indices

Sobol indices are categorized into several types, including first order, second order, up to M-order (where M is the number of input variables), and total sensitivity indices. The first-order sensitivity index of Y to the parameter X_i, denoted as S_i, exclusively considers the contribution due to the parameter X_i independently, excluding interactions. It is defined as follows (Saltelli et al. 2008):

$$S_i = \frac{V(\mathbb{E}(Y|X_i))}{V(Y)} \tag{11.24}$$

The total sensitivity index S_{Ti} evaluates the contribution due to the parameter X_i alone, which corresponds to the index S_i, and the contribution due to the interaction of X_i with the other parameters:

$$S_{Ti} = 1 - \frac{V(\mathbb{E}(Y|X_{\sim i}))}{V(Y)} \tag{11.25}$$

The term $V(\mathbb{E}(Y|X_{\sim i}))$ represents the variance of the conditional expectation knowing all parameters except X_i.

11.4 Kriging Metamodel

11.4.1 Generalities

Kriging is widely recognized as a valuable tool for approximating the response variable Y in deterministic computer models. The fundamental concept behind Kriging is that the N deterministic response $Y\left(\overrightarrow{X}\right) = \left\{\overrightarrow{X}^{(i)}, Y^{(i)}|i = 1, \ldots, N\right\}$ can be represented as follows (Sacks et al. 1989):

$$Y\left(\overrightarrow{X}\right) \equiv \mathcal{M}^K\left(\overrightarrow{X}\right) = G\left(\overrightarrow{X}\right) + Z\left(\overrightarrow{X}\right) \tag{11.26}$$

where $G\left(\overrightarrow{X}\right) = \sum_{j=1}^{P} \beta_j f_j\left(\overrightarrow{X}\right)$ is defined as the sum of P arbitrary functions f_j with coefficients β_j for the uncertain variables $\overrightarrow{X}$, and $Z\left(\overrightarrow{X}\right)$ is a stochastic process (random function). In this context, the ordinary kriging is employed. Specifically, $G\left(\overrightarrow{X}\right)$ can be represented as a constant term $\widehat{\beta}$ with $P = 1$ and $f_j\left(\overrightarrow{X}\right) = \overrightarrow{1}$ (Schobi et al. 2015). $G\left(\overrightarrow{X}\right)$ serves as a global approximation of the design space, while $Z\left(\overrightarrow{X}\right)$ introduces localized deviations, allowing the kriging model to interpolate the N sampled data points. $Z\left(\overrightarrow{X}\right)$ is assumed to have a mean of zero, a variance of σ^2, and a non-zero covariance, which is expressed as:

$$Cov\left[Z\left(\vec{X}\right), Z\left(\vec{X}'\right)\right] = \sigma^2[\mathrm{R}]$$
$$= \sigma^2 \begin{bmatrix} 1 & \cdots & R\left(\vec{X}^{(1)}, \vec{X}^{(N)}; \vec{\theta}\right) \\ \vdots & \ddots & \vdots \\ R\left(\vec{X}^{(N)}, \vec{X}^{(1)}; \vec{\theta}\right) & cdots & 1 \end{bmatrix} \quad (11.27)$$

The matrix [R] is an $N \times N$ correlation matrix, which is symmetric, positive definite, and has a unit diagonal. This matrix plays a crucial role in determining the smoothness of the resulting model. The function $R(\overrightarrow{X}, \overrightarrow{X}'; \overrightarrow{\theta})$ represents the correlation between two data points $\overrightarrow{X}$ and $\overrightarrow{X}'$, with $\overrightarrow{\theta} = [\theta_1, \ldots, \theta_M]$ denoting the vector of unknown parameters associated with the correlation function, where M represents the number of uncertain variables.

11.4.2 Correlation Function

In our specific case, the Matérn kernel was chosen as the correlation family, with a shape parameter of $\upsilon = 5/2$. Thus, the correlation function type is ellipsoidal $R\left(\overrightarrow{X}, \overrightarrow{X}'; \overrightarrow{\theta}\right)$ and is defined as follows (Lataniotis et al. 2022):

$$R\left(\overrightarrow{X}, \overrightarrow{X}'; \overrightarrow{\theta}, \upsilon = \tfrac{5}{2}\right) = R(h) = \left(1 + \sqrt{5}h + \tfrac{5}{3}h^2\right)\exp\left[-\sqrt{5}h\right] \quad (11.28)$$

$$h = \left[\sum_{k=1}^{M} \left(\frac{|X_k - X_{k'}|}{\theta_k}\right)^2\right]^{\frac{1}{2}} \quad (11.29)$$

11.4.3 Maximum-Likelihood Estimation

To derive a Kriging metamodel, it is typically necessary to estimate the vector of hyperparameters $\overrightarrow{\theta}$, which is initially unknown. One commonly used approach for estimating $\overrightarrow{\theta}$ is the maximum-likelihood (ML) method, achieved by maximizing the likelihood function, as outlined in previous works (Dubourg 2011).

$$L\left(\overrightarrow{Y} | \widehat{\beta}, \sigma^2, \overrightarrow{\theta}\right) = \frac{1}{\sqrt[N]{2\pi\sigma^2}\sqrt{\det\left[R\left(\overrightarrow{\theta}\right)\right]}} \exp\left[-\frac{\left(\overrightarrow{Y} - \overrightarrow{F}\widehat{\beta}\right)^T \left[R\left(\overrightarrow{\theta}\right)\right]^{-1} \left(\overrightarrow{Y} - \overrightarrow{F}\widehat{\beta}\right)}{2\sigma^2}\right] \quad (11.30)$$

where $\overrightarrow{\mathrm{F}}$ represents a N -vector of basic functions, which is filled with ones when $G\left(\overrightarrow{X}\right)$ is considered as a constant. The corresponding negative log-likelihood can be expressed as follows (Baklouti et al. 2022):

$$
\begin{aligned}
&-\ln\mathcal{L}\left(\overrightarrow{Y}\,|\widehat{\beta},\sigma^{2},\overrightarrow{\theta}\right)\\
&=\frac{\left(\overrightarrow{Y}-\overrightarrow{F}\widehat{\beta}\right)^{T}\left[R\left(\overrightarrow{\theta}\right)\right]^{-1}\left(\overrightarrow{Y}-\overrightarrow{\mathrm{F}}\widehat{\beta}\right)}{2\sigma^{2}}+\frac{N\ln(2\pi)}{2}+\frac{N\ln\left(\sigma^{2}\right)}{2}+\frac{\ln\left(\det\left[R\left(\overrightarrow{\theta}\right)\right]\right)}{2}
\end{aligned}
\tag{11.31}
$$

With a specified value $\overrightarrow{\theta}$, the calibration of the Kriging model parameters $\left\{\widehat{\beta}\left(\overrightarrow{\theta}\right),\sigma^{2}\left(\overrightarrow{\theta}\right)\right\}$ can be determined through an empirical best linear unbiased estimator (BLUE) and are expressed as follows (Schobi et al. 2015; Lataniotis et al. 2022; Baklouti et al. 2022):

$$
\widehat{\beta}\left(\overrightarrow{\theta}\right)=\left(\overrightarrow{\mathrm{F}}^{T}\left[R\left(\overrightarrow{\theta}\right)\right]^{-1}\overrightarrow{\mathrm{F}}\right)^{-1}\overrightarrow{\mathrm{F}}^{T}\left[R\left(\overrightarrow{\theta}\right)\right]^{-1}\overrightarrow{Y}
\tag{11.32}
$$

$$
\sigma^{2}\left(\overrightarrow{\theta}\right)=\frac{1}{N}\left(\overrightarrow{Y}-\overrightarrow{F}\widehat{\beta}\right)^{T}\left[R\left(\overrightarrow{\theta}\right)\right]^{-1}\left(\overrightarrow{Y}-\overrightarrow{\mathrm{F}}\widehat{\beta}\right)
\tag{11.33}
$$

We streamline the first term of Eq. (11.31) by incorporating Eq. (11.32) into Eq. (11.30), resulting in a novel expression that is contingent solely upon $\overrightarrow{\theta}$:

$$
-\ln\mathcal{L}\left(\overrightarrow{Y}\,|\widehat{\beta},\sigma^{2},\overrightarrow{\theta}\right)=\frac{N}{2}\ln\left(\sigma^{2}\left(\overrightarrow{\theta}\right)\left(\det\left[R\left(\overrightarrow{\theta}\right)\right]\right)^{\frac{1}{N}}\right)+\frac{N}{2}(\ln(2\pi)+1)
\tag{11.34}
$$

The maximum likelihood estimates of $\overrightarrow{\theta}$ is ultimately characterized as the following global minimizer (Schobi et al. 2015; Dubourg 2011):

$$
\overrightarrow{\theta}=\operatorname{argmin}\left[\sigma^{2}\left(\overrightarrow{\theta}\right)\left(\det\left[R\left(\overrightarrow{\theta}\right)\right]\right)^{\frac{1}{N}}\right]
\tag{11.35}
$$

11.4.4 Estimation of Kriging

The estimation of a Kriging model for a prediction point $\overrightarrow{X}$ is treated as a Gaussian random variable. In this context, the prediction mean $\mu_{Y}\left(\overrightarrow{X}\right)$ serves as a surrogate for the original model, while the variance $\sigma_{\hat{Y}}^{2}\left(\overrightarrow{X}\right)$ provides a local error indicator that conveys information about precision (Schobi et al. 2015; Mohammadi 2016):

$$\mu_Y\left(\overrightarrow{X}\right) = \mathcal{M}^K\left(\overrightarrow{X}\right) = \widehat{\beta} + \overrightarrow{r}[R]^{-1}\left(\overrightarrow{Y} - \overrightarrow{\mathrm{F}}\widehat{\beta}\right) \tag{11.36}$$

$$\sigma_Y^2\left(\overrightarrow{X}\right) = \sigma^2\left(1 - \overrightarrow{\mathrm{r}}^T[R]^{-1}\overrightarrow{r}\right) \tag{11.37}$$

where $\overrightarrow{r} = \left[R\left(\overrightarrow{X}, \overrightarrow{X}^{(1)}; \overrightarrow{\theta}\right), R\left(\overrightarrow{X}, \overrightarrow{X}^{(2)}; \overrightarrow{\theta}\right), \ldots, R\left(\overrightarrow{X}, \overrightarrow{X}^{(N)}; \overrightarrow{\theta}\right)\right]^T$ is the correlation vector of length N.

11.4.5 Error Measurement

To assess the validity of the surrogate model, we employ the leave-one-out (LOO) error calculation on the original experimental design $\overrightarrow{X}$, paired with their corresponding responses $\overrightarrow{Y} = \mathcal{M}\left(\overrightarrow{X}\right)$, and the procedure is as follows (Lataniotis et al. 2022):

$$\epsilon_{LOO} = \frac{1}{N}\left[\frac{\sum_{i=1}^{N}\left(\mathcal{M}\left(\overrightarrow{X}^{(i)}\right) - \mu_{Y,(-i)}\left(\overrightarrow{X}^{(i)}\right)\right)^2}{\mathrm{Var}\left[\overrightarrow{Y}\right]}\right] \tag{11.38}$$

where $\mu_{Y,(-i)}\left(\overrightarrow{X}^{(i)}\right)$ represents the Kriging metamodel derived using all points from the experimental design $\overrightarrow{X}$, except $\overrightarrow{X}^{(i)}$.

11.5 Case Study

11.5.1 Description of the Studied Double-Wall

The given finite double-wall system consists of two plates. The first plate is made of steel and the second plate is a laminated plate. To improve sound transmission reduction, a poroelastic layer (mineral wool) is introduced between and bonded to both plates. The double wall measures 1.3 × 1.3 m and is subjected to a diffuse excitation field with a power spectral density of 1 Pa^2/Hz. The double wall is surrounded by a fluid domain.

The properties of the steel plate are described: Young's modulus $E_1 = 2.1 \times 10^{11}$ Pa, Poisson's ratio $\nu_1 = 0.3$, density $\rho_1 = 7850$ kg/m^3, loss factor $\eta_1 = 0.03$, and the thickness $e_1 = 0.75$ mm. The properties of the porous material are given by 9 material parameters (using Biot theory) and 1 geometric parameter: porosity $\phi = 0.95$, flow resistivity $\sigma = 34000$ Ns/m^4, tortuosity $\alpha_\infty = 1$, viscous length

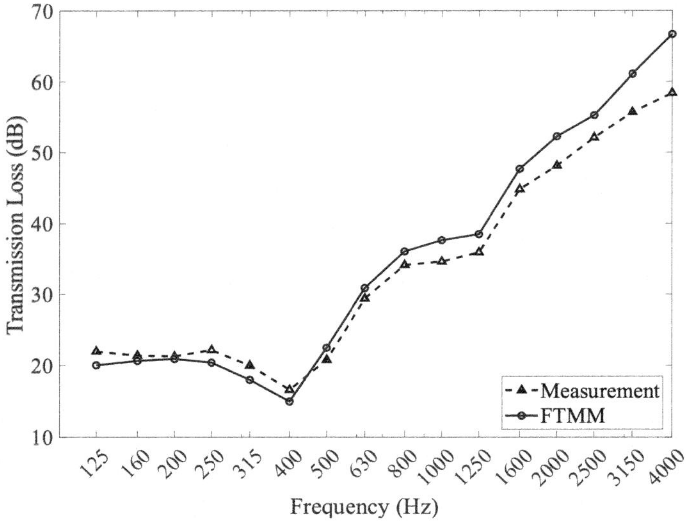

Fig. 11.2 Comparison between measurements (Villot et al. 2001) and FTMM result

$\Lambda = 40\mu$ m, thermal length $\Lambda\prime = 80\mu$ m, Young's modulus $E_2 = 40 \times 10^4$ Pa, Poisson's ratio $\nu_2 = 0$, density $\rho_2 = 90$ kg/m^3, loss factor $\eta_2 = 0.18$ and the thickness $e_2 = 30$ mm. The properties of the laminated plate are: Young's modulus $E_3 = 6 \times 10^8$ Pa, Poisson's ratio $\nu_3 = 0.15$, density $\rho_3 = 1360$ kg/m^3, loss factor $\eta_3 = 0.15$, and the thickness $e_3 = 3$ mm. In Fig. 11.2, a good agreement is observed between the experimental measurements (Allard and Atalla 2009; Villot et al. 2001) and the numerical result using FTMM of the sound transmission loss in the third band octave.

Considering the lateral dimensions fixed, we have conducted a sensitivity analysis using the Morris method, considering all material parameters and 3 thicknesses of each layer. In total 20 input parameters were considered. Using Morris, we removed 9 parameters that have no influence on sound transmission loss of the studied double wall. For the steel plate, we deducted 3 negligible parameters: Young's modulus, Poisson's ratio, and loss factor. For the poroelastic material, resistivity, thermal length, and Poisson's ratio were deduced as non-influential parameters. Finally, Young's modulus, Poisson's ratio, and loss factor of the laminated plate were found to be non-influential. This study is not shown for the sake of brevity.

Table 11.1 presents uncertain parameters and their corresponding bounds of the 11 parameters found to be potentially influential after Morris analysis, including 2 parameters for the steel plate (density and thickness), 7 for the mineral wool (porosity, tortuosity, viscous length, density, Young's modulus, loss factor, and thickness), and 2 parameters for the laminated plate (density and thickness). Material parameters are assumed to follow a Gaussian distribution, while geometric parameters follow a uniform distribution.

Table 11.1 Uncertain parameters for the double wall with a porous layer

Domain	X_i	Parameter	Variable	Nominal value	Unit	Lower bound	Upper bound
Steel plate	X_1	Density	ρ_1	7850	kg/m^3	7700	8000
	X_2	Thickness	e_1	0.00075	m	0.0006	0.001
Porous	X_3	Porosity	ϕ	0.95	–	0.90	0.99
	X_4	Tortuosity	α_∞	1.0	–	0.8	1.2
	X_5	Viscous length	Λ	40×10^{-6}	m	30×10^{-6}	50×10^{-6}
	X_6	Density	ρ_2	90	kg/m^3	80	100
	X_7	Young's modulus	E_2	400,000	Pa	300,000	500,000
	X_8	Loss factor	η_2	0.18	–	0.1	0.2
	X_9	Thickness	e_2	0.03	m	0.02	0.04
Laminated plate	X_{10}	Density	ρ_3	1360	kg/m^3	1000	1730
	X_{11}	Thickness	e_3	0.003	m	0.002	0.004

11.5.2 Metamodeling Using Kriging

Using Kriging method, our experimental design involved conducting 1000 simulations of the original model, utilizing the finite transfer matrix method. The sampling method chosen is Latin Hypercube sampling. Our initial focus was on comparing the leave-one-out error (ϵ_{LOO}) for every frequency within the third-octave band, ranging from 125 to 4000 Hz (Fig. 11.3). We can see a low error below 10^{-2}. This level of accuracy is acceptable, laying a solid foundation for the subsequent sensitivity analysis.

In Fig. 11.4, we employed the Kriging metamodel to plot the response of the double-wall system at nominal values. We compare in the same figure the results obtained from the Kriging metamodel with those generated by the original model using the FTMM. With the good agreement between the two models, the metamodel can be validated and used for a sensitivity study to explore the impact of variations in input parameters on the response of the system.

11.5.3 Global Sensitivity Analysis Using Sobol Indices

Using the Sobol method, a sensitivity analysis was conducted on the 11 parameters identified by the Morris method. Figure 11.5 displays the first order S_i and total S_{Ti} Sobol indices for these parameters across a frequency range of 125 to 4000 Hz per third band for the true model and the kriging metamodel. A sample size of $N = 10^4$ is used, resulting in a computational cost equivalent to 130,000 model simulations

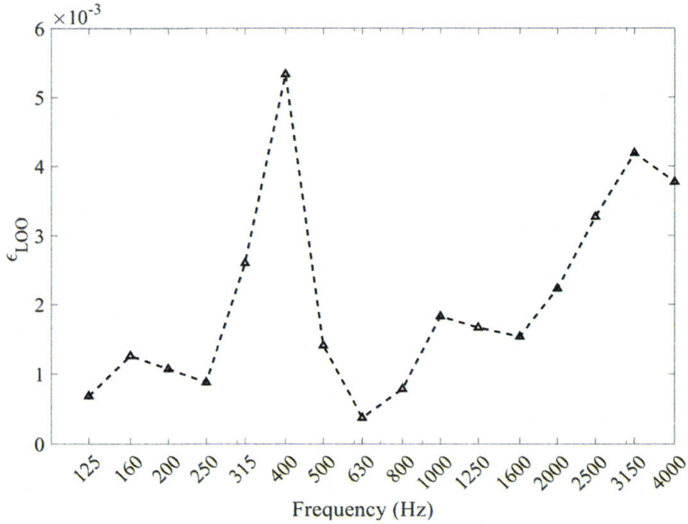

Fig. 11.3 The leave-one-out error of the Kriging metamodel between 125 and 4000 Hz

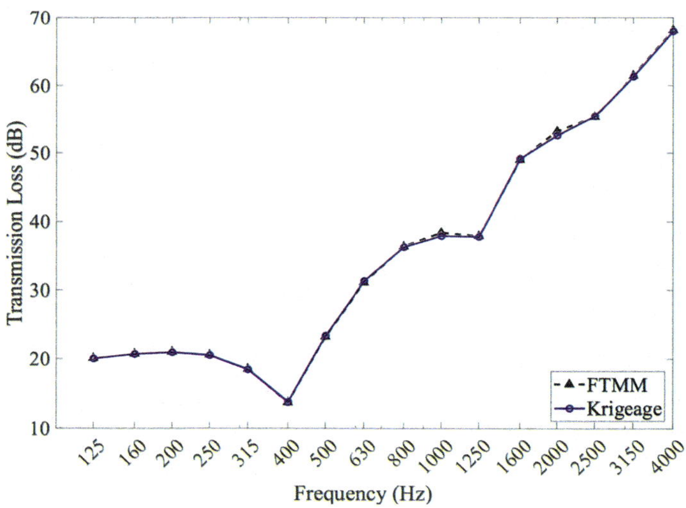

Fig. 11.4 Comparison between FTMM model and Kriging metamodel

$((M + 2) \times N)$ for each frequency for the true model. For each frequency, the computational cost of the sensitivity analysis is 24 h for the true model and 16 min for the Kriging model.

There exist five non-influential parameters, namely steel plate density ρ_1, porosityϕ, tortuosityα_∞, viscous length Λ, and porous material density ρ_2 (that is

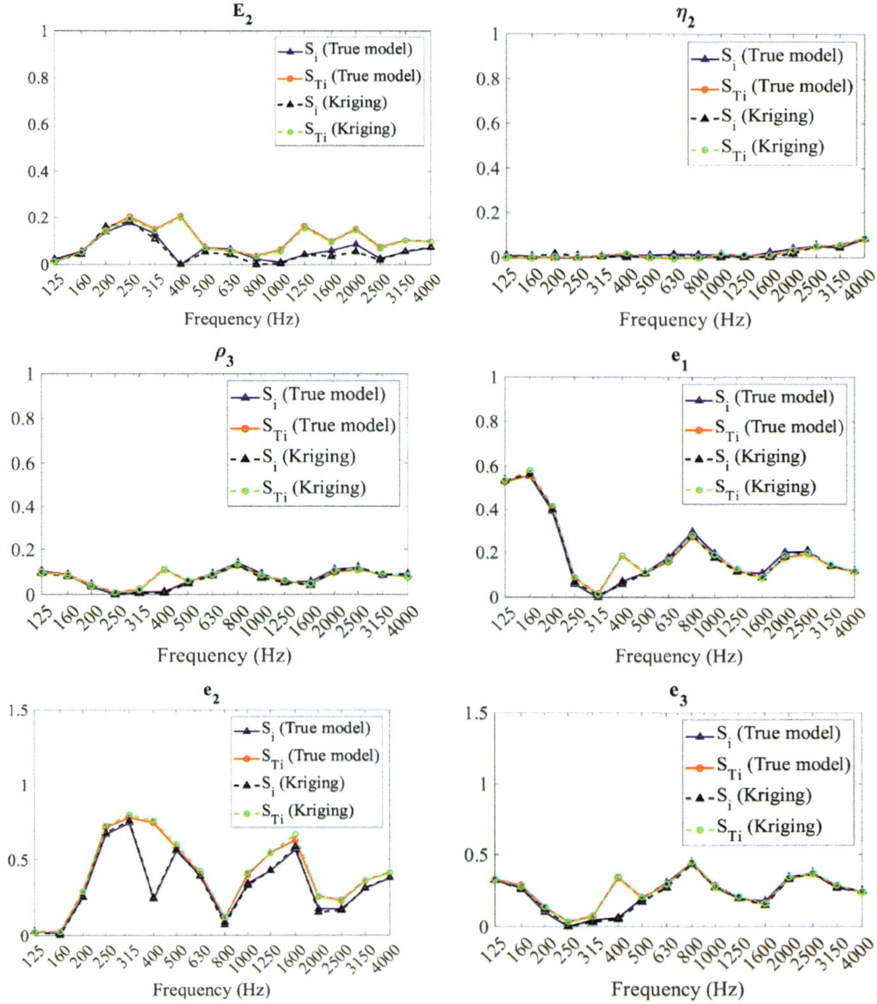

Fig. 11.5 Sobol indices of the eleven most influential parameters

not showed here). Conversely, the three most impactful parameters encompass the thicknesses of the porous materiale_2, laminated plate thickness $e_{3,}$ and steel platee_1. Then, there are three parameters with a slight influence: Young's modulus E_2 of the porous material, the density of the laminated plateρ_3, and the loss factor of the porous materialη_2.

Concerning the Young's modulus E_2 of the porous layer, its significance is paramount within four frequency peaks (250, 400, 1250, and 2000 Hz). For other frequencies, the parameter shows a slight influence. Beyond 315 Hz, this parameter exhibits sensitivity to interactions with other parameters, as the total index is higher than the first index. Examining the loss factor η_2 of the porous layer, it manifests no discernible influence up to 2000 Hz. Beyond 2000 Hz, a slight influence becomes evident, with minimal sensitivity to interactions with other parameters. Concerning the laminated plate density ρ_3, its impact is remarkable at three frequency peaks: 125, 400, and 800 Hz. Overall, this parameter has a minor impact. Its susceptibility to interactions is minimal except for the 400 Hz frequency.

The thickness of the steel plate e_1 has a remarkable influence at frequencies of 125, 160, 200, and 800 Hz. Its influence remains non-negligible at 800 Hz and above. The thickness of the steel plate remains insensitive to interactions, except for the 400 Hz frequency. The thickness of the porous material e_2 has an important influence at three frequency peaks: 315, 500, and 1600 Hz when we focus on the influence of the parameter alone. Notably, two peaks indicate sensitivity to interactions with other factors, with heightened total indices at 400 and 1600 Hz. A gradual decline in influence occurs beyond 1600 Hz, followed by stabilization between 2000 and 2500 Hz, marking the parameter's heightened significance beyond 2500 Hz. The laminated plate thickness e_3 has a peak influence at the following frequencies: 125, 400, and 800 Hz. This parameter remains influential as frequency rises above 800 Hz. This parameter exhibits insensitivity to interactions, except at 400 Hz.

For Young's modulus of the porous layer E_2, the laminated plate density ρ_3, and thicknesses of steel plate e_1 and laminated plate e_3, a remarkable peak surface in the total index at 400 Hz. At this frequency, the system demonstrates heightened sensitivity to interactions with the parameters previously mentioned.

11.6 Conclusion

This study explores the acoustic performance of a double-wall system with a poroelastic layer using the finite transfer matrix method. The Kriging method is employed as a surrogate model to mitigate computational demands in sensitivity analysis, proving effective for complex models. Global sensitivity analysis, using the Sobol method, identifies influential parameters among eleven considered in the study, revealing the significance of layer thicknesses in determining acoustic performance across frequencies.

The analysis identifies five parameters with negligible influence: steel plate density, porosity, tortuosity, viscous length, and porous material density. The influence of the loss factor of the porous material and the density of the laminated plate is subtle, but their roles intensify as frequency increases. Young's modulus of the porous material proves influential at specific frequencies, attributed to interactions with other parameters. Notably, the key determinants of acoustic performance are the thicknesses of each layer, with their hierarchical importance varying across frequencies.

The study underscores the importance of addressing uncertain parameters and employing sensitivity analysis to optimize the design of double-wall systems, offering potential for improved noise reduction in various applications.

Acknowledgements The authors would like to express gratitude to CODIFAB (Comité professionnel de Développement des Industries Française de l'Ameublement et du Bois) and UFME (Union des Fabricants de Menuiseries) for their cooperation.

References

Allard JF, Atalla N (2009) Propagation of sound in porous media: modelling sound absorbing materials (2nd ed), Wiley, Hoboken, New Jersey

Atalla N, Sgard F, Amedin CK (2006) On the modeling of sound radiation from poro-elastic materials. J Acoust Soc Am 120:1990–1995

Bakhouche S, Larbi W, Deü JF, Macquart P (2023a) Numerical and experimental investigation of sound transmission through roller shutter boxes. Appl Acoust 214:109679

Bakhouche S, Larbi W, Aloui R, Macquart P, Deü JF (2023b) Global sensitivity analysis of sound attenuation in double-wall system with porous layer. Mech Adv Mater Struct 1–13

Baklouti A, Dammak K, El Hami A (2022) Uncertainty analysis based on kriging meta-model for acoustic-structural problems. Appl Sci 12:1503

Cherif R, Wareing A, Atalla N (2021) Sound transmission paths through a statistical energy analysis model of mechanically linked aircraft double-walls. Noise Control Eng J 69:411–421

Dubourg V (2011) Adaptive surrogate models for reliability analysis and reliability-based design optimization. Dissertation, University of Blaise Pascal—Clermont II

Krige DG (1951) A statistical approach to some basic mine valuation problems on the Witwatersrand. J South Afr Inst Min Metall 6:119–139

Larbi W, Deü JF, Ohayon R (2016) Finite element reduced order model for noise and vibration reduction of double sandwich panels using shunted piezoelectric patches. Appl Acoust 108:40–49

Lataniotis C, Wicaksono D, Marelli S, Sudret B (2022) UQLab user manual—Kriging (Gaussian process modeling), Report UQLab-V2.0–105, Chair of Risk, Safety and Uncertainty Quantification, ETH Zurich, Switzerland

Matheron G (1963) Principles of geostatistics. Econ Geol 58:1246–1266

Mohammadi H (2016) Kriging-based black-box global optimization: analysis and new algorithms. Dissertation, Ecole Nationale Supérieure des Mines de Saint-Etienne

Morris MD (1991) Factorial sampling plans for preliminary computational experiments. Technometrics 33:161–174

Ouisse M, Ichchou M, Chedly S, Collet M (2012) On the sensitivity analysis of porous material models. J Sound Vib 331:5292–5308

Rhazi D, Atalla N (2010) A simple method to account for size effects in the transfer matrix method. J Acoust Soc Am 127:EL30–EL36

Sacks J, Schiller SB, Welch WJ (1989) Designs for computer experiments. Technometrics 1:41–47

Saltelli A, Ratto M, Andres T, Campolongo F, Cariboni J, Gatelli D, Saisana M, Tarantola S (2008) Global sensitivity analysis: the primer. John Wiley & Sons, Chichester, England

Schobi R, Sudret B, Wiart J (2015) Polynomial-Chaos-based Kriging. Int J Uncertain Quantif 5:171–193

Sobol′ IM (2001) Global sensitivity indices for nonlinear mathematical models and their Monte Carlo estimates. Math Comput Simul 55:271–280

Soussi C, Aucejo M, Larbi W, Deü JF (2021) Numerical analyses of the sound transmission at low frequencies of a calibrated Insulating Glazing Unit. Appl Acoust 179:108065

Villot M, Guigou C, Gagliardini L (2001) Predicting the acoustical radiation of finite size multilayered structures by applying spatial windowing on infinite structures. J Sound Vib 245:433–455

Chapter 12
Acoustic Interactions in a Rectangular Cavity with a Forced Vibrating Top Elastic Plate Filled with a Compressible Fluid

Manuel Gascón-Pérez

Abstract Acoustic cavities are found in many technological applications, such as in naval or aerospace engineering. In this paper, the vibroacoustic characteristics of a forced vibrating plate at top position of a rectangular cavity filled with an ideal fluid are studied. The velocity potential function is obtained by the method of separation of variables applied to the corresponding wave equation, and the pressure field is deduced from the momentum's linearized equation. After an integration over the surface of the plate of its dynamic motion equation considering the orthogonality of the functions involved expressed in series expansion of the associated modes, analytical expressions are obtained for the calculus of the plate's natural or free frequencies of vibration, and the system frequency response for the case of forced vibration. The frequency response or spectrum of the plate deformation, as well as the sound pressure level at different points of the cavity, are obtained, and the influence of various parameters such as container height, fluid density, and load position on these spectrums are analyzed. Also, the mean quadratic sound pressure level over the plate and the mean quadratic deformation level of the plate in vacuum and with fluid are obtained, and finally, the variation of the mean quadratic sensibility with the frequency is calculated. Validation of the method is made by comparing the results with those obtained by other authors and theories.

Keywords Top plate of rectangular cavity · Fluid–structure interaction · Natural frequencies · Frequency response forced vibrating plate · Compressibility effects · Fluid velocity potential

M. Gascón-Pérez (✉)
Escuela Técnica Superior de Ingeniería Aeronáutica y del Espacio, Universidad Politécnica de Madrid, Madrid, Spain
e-mail: manuel.gascon@upm.es

A. Öchsner and H. Altenbach (eds.), *Engineering Design Applications VII*, Advanced Structured Materials 230, https://doi.org/10.1007/978-3-031-84346-4_12

12.1 Introduction

Fluid–structure interaction regarding an elastic vibrating plate with an acoustic cavity is encountered in many practical aerospace, naval, and civil engineering applications such as, for instance, aircraft and ship cabins, bedrooms, sonar cavities, etc.

The noise generated by these vibrating structures (structure-borne noise) in cavities must be considered due to the influence on the comfort of operator and passengers in the cabins of surface vehicles, aircrafts, or ships, that are simplified and represented by a cavity structure with a vibrating side wall. So, a good comprehension of the fluid–structure interaction regarding a structure and a sound cavity is of critical significance in order to control the sound behavior on this compartments or cabins and has been analyzed since many years ago. High values of the sound pressure level SPL can affect seriously the physiological and mental conditions of the passengers or pilots. Hence, care must be taken regarding the vibroacoustic performance of these structures in order to achieve a noise reduction characteristic.

Also there are devices used in micro electromechanical systems with similar characteristics of the acoustic cavities, in which is interesting to know the natural frequencies in order to control the performance efficiency and power consumption considering the resonant conditions, such as micropumps, see Laser et al. (2004), Pan et al. (2003) or Shabani et al. (2013), and pressure sensors, see Defay et al. (2002).

A vibrating plate of a rectangular container or cavity is a typical modellization of many engineering problems and has been studied since many years ago. Early in 1963, Dowell and Voss (1963) studied the effect of a rectangular cavity on panel vibration, considering a compressible and inviscid fluid, expanding in a Fourier series the analytic solutions for the structure deformation and fluid pressure. Similarly, Pretlove (1965, 1966) analyzed the free and forced vibrations of a rectangular plate backed by a closed rectangular cavity.

Recently, researching is more focused on the active control of the coupled system in relation to room acoustics, see Pan et al. (1990), active structural acoustic control of sound radiation, see Qiu et al. (1995) and active sound transmission control, see Kim et al. (2000). Among these studies, a characteristic approach is the modal coupling theory where the structural vacuum modes and the acoustic rigid cavity modes are determined previously, and then combined to find the performance of the coupled system, but this approach has limitations and is only suitable for weak coupling, and not the case of strong coupling for example for a thin plate, or small cavity height or heavy fluid, and also the continuity on the interface boundary can not be fulfilled.

For the case of complex geometries of engineering structures, the solution is obtained employing finite element methods as well as boundary element methods. Zienkiewicz and Newton (1969), analyzed fluid–structure interaction problems employing a finite element technique for the modellization for the structure and the compressible fluid. Sestieri et al. (1984) analyzed a structural–acoustic coupling in cavities with complex geometries, making use of the Helmholtz integral equation formulation. Xie et al. (2016) analyzed a coupled system of an irregular cavity with an elastic plate and general wall impedances and boundary edge conditions, taking into account a variational based method.

Gascón-Pérez (2020) analyzed the hydro-elastic oscillations of the bottom membrane of a rectangular container by considering the fluid incompressible and inviscid; in Gascón-Pérez (2023), the same author analyzed the compressibility effects on the vibration of a plate at an off-center position of a rectangular container filled with fluid, and finally in Gascón-Pérez (2024) the acoustic radiation characteristics of a forced vibrating elastic panel are investigated.

The acoustic characteristics of a harmonically exited vibrating top plate of a rectangular cavity filled with a compressible and non-viscous fluid is analyzed in this paper. Considering the wave equation for the fluid velocity potential and the pressure field with the momentum's linearized equation, the dynamic equation of the plate is expressed in a series expansion of the associated modes, and after an integration procedure considering the orthogonality condition of the functions involved, an analytical expression is obtained in an elegant manner for the calculus of the plate natural frequencies of vibration for the case of free vibrations, and the system frequency response for the case of forced vibration. Validation of the method is made by comparing the results with those obtained by other authors and theories.

12.2 Problem Formulation

The dynamic response of the plate of a rectangular cavity subjected to a forcing load q_0 located at x_0, y_0 position, is governed by the equation, see Wang (2014):

$$D\nabla^4 w(x, y, t) + \rho_m t_h \frac{\partial^2 w}{\partial t^2} = p(x, y, t) + \delta q_0(x, y, t) \qquad (12.1)$$

where $D = \frac{E \cdot t_h^3}{12(1-\nu^2)}$ is the flexural rigidity, E is the elasticity modulus of the material, t_h is the plate thickness, ν is the Poisson‘s ratio, ρ_m is the material density of the plate, and w is the plate deflection in normal direction.

The elastic plate is in contact with a fluid and p is the fluid pressure acting on the plate.

And $\delta q_0(x, y, t)$ is the harmonic excitation force per unit area expressed as a Dirac-delta located at x_0, y_0 position associated to a load q_0 of 1N so that

$$\delta q_0(x, y, t) = \delta \tilde{q}_0(x, y) \cdot e^{i\omega t} \qquad (12.2)$$

and

$$q_0 = \iint_{S_p} \delta \tilde{q}_0(x - x_0, y - y_0)\, dx\, dy \qquad (12.3)$$

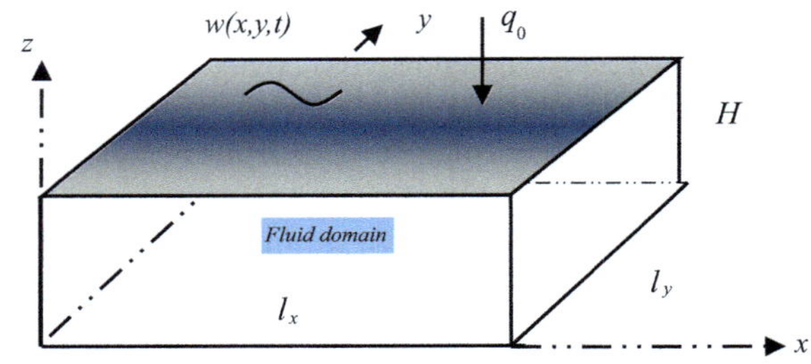

Fig. 12.1 Scheme of the rectangular cavity with a top plate showing the different geometrical parameters

For a plate placed at a position $z = H$, and with l_x, l_y the side lengths of the plate, being H the cavity total height, see Fig. 12.1.

Considering harmonic motion for both the fluid and the plate, the solutions for the pressure and deformation are expressed as:

$$w(x, y, t) = \tilde{w}(x, y) \cdot e^{i\omega t} \text{ and } p(x, y, z, t) = \tilde{p}(x, y, z) \cdot e^{i\omega t} \tag{12.4}$$

The deflection of the plate which is supposed to be simply supported, when coupled with the fluid, will be expressed as a linear combination of the normal modes of the plate in vacuum:

$$w(x, y, t) = \sum_{m,n} w_m^n(x, y) \cdot \overline{d}_m^n \cdot e^{i\omega_m^n t} \tag{12.5}$$

where $\overline{d}_m^n$ are the weight coefficients associated to each mode that contribute to the total deformation of the rectangular plate in contact with fluid.

The functions $w_m^n(x, y)$ for a rectangular plate simply supported along its edges, have the general expression, see Wang (2014):

$$w_m^n(x, y) = sin\left(\frac{m\pi x}{l_x}\right) \cdot sin\left(\frac{n\pi y}{l_y}\right) \tag{12.6}$$

The indexes $m-1$, $n-1$, represent the number of nodal lines in the y, x directions respectively.

Expanding in series of the deformation modes in vacuum the concentrated load:

$$\delta\tilde{q}_0(x, y) = \sum_m \sum_n \overline{q}_m^n \cdot w_m^n(x, y) \tag{12.7}$$

where the coefficients have the expression:

$$\overline{q}_m^n = \frac{4}{l_x l_y} \iint_{S_p} \delta\tilde{q}_0(x - x_0, y - y_0) \cdot w_m^n(x, y)dxdy = \frac{4}{l_x l_y} q_0 \cdot w_m^n(x_0, y_0) \tag{12.8}$$

In Fig. 12.1 the geometrical characteristics of the rectangular cavity are represented, in order to be defined and clarified.

12.2.1 Calculus of the Fluid Velocity Potential and Pressure Over the Plate

Assuming inviscid fluid and in absence of cavitation, the fluid motion induced by the vibration of the rectangular plate is described using the velocity potential function (Munson (2013). This velocity potential $\varphi(x, y, z, t)$ satisfies the wave equation for the fluid region:

$$\Delta\varphi - \frac{1}{c^2}\frac{\partial^2\varphi}{\partial t^2} = 0 \tag{12.9}$$

And considering harmonic motion $\varphi(x, y, z, t) = \tilde{\varphi}(x, y, z) \cdot e^{i\omega t}$, the Helmholtz's equation is deduced:

$$\Delta\tilde{\varphi} + k^2\tilde{\varphi} = 0 \tag{12.10}$$

Being k the wave number, $k = \frac{\omega}{c}$, Δ the Laplace operator, and c the sound speed that can be expressed as $c = \frac{B}{\rho_f}$ with B the fluid bulk modulus and ρ_f the fluid density, together with the boundary conditions:

- For the rigid wall boundaries, i.e., the lateral and bottom surfaces of the rectangular cavity

$$\frac{\partial\varphi}{\partial n} = 0, \tag{12.11}$$

being n the normal to the wall surfaces:

- For the plate boundary:

$$\frac{\partial\varphi}{\partial z} = \frac{\partial w}{\partial t} \text{ for } z = H \;\; 0 \le x \le l_x \; 0 \le y \le l_y \tag{12.12}$$

Applying a method of separation of variables in space–time and expanding in series, the velocity potential function for the fluid domain and the plate deformation are expressed:

$$\varphi(x, y, z, t) = \sum_{m,n} \varphi_m^n(x, y, z) \cdot f_m^n(t) \tag{12.13}$$

$$w(x, y, t) = \sum_{m,n} w_m^n(x, y) \cdot \; d_m^n(t) \tag{12.14}$$

where for harmonic motion

$$f_m^n(t) = \overline{f}_m^n \cdot e^{i\omega_m^n t} \; and \; d_m^n(t) = \overline{d}_m^n \cdot e^{i\omega_m^n t} \tag{12.15}$$

Considering Eq. (12.9) for the velocity potential and the boundary conditions (12.11), and applying again separation of variables in space coordinates, the solution takes this form:

$$\varphi(x, y, z, t) = \sum_{m,n} \varphi_m^n(x, y) \cdot \cosh\left(\lambda_m^n z\right) \cdot f_m^n(t) \tag{12.16}$$

where the function

$$\varphi_m^n(x, y) = \cos\left(\frac{m\pi x}{l_x}\right) \cdot \cos\left(\frac{n\pi y}{l_y}\right) \tag{12.17}$$

and the parameter

$$\lambda_m^n = \sqrt{\left(\frac{m\,\pi}{l_x}\right)^2 + \left(\frac{n\,\pi}{l_y}\right)^2 - k^2} \tag{12.18}$$

Taking into account the boundary condition (12.12) of the fluid-plate interface, the expression for the fluid velocity potential (12.16) and the plate deformation, Eq. (12.14), and after integration over the plate surface of the squared series expressions involved, the following expressions are obtained:

$$\begin{aligned} &\iint_{S_p} \left(\sum_{m,n} \left(f_m^n \cdot \varphi_m^n(x, y) \cdot \lambda_m^n \cdot \sinh\left(\lambda_m^n H\right)\right)\right)^2 dxdy \\ &= \iint_{S_p} \left(\sum_{m,n} \left(\dot{d}_m^n \cdot w_m^n(x, y)\right)\right)^2 dxdy \end{aligned} \tag{12.19}$$

from where it is deduced:

$$f_m^n(t) = \frac{1}{\lambda_m^n \sinh\left(\lambda_m^n H\right)} \dot{d}_m^n(t) \tag{12.20}$$

To evaluate the fluid pressure field in the fluid domain, neglecting gravity effects and for small displacements, the momentum's linearized equation is deduced, Munson (2013):

$$\frac{\partial \varphi}{\partial t} + \frac{p}{\rho_f} = 0 \tag{12.21}$$

Considering the Eq. (12.16) for the velocity potential and Eq. (12.21), the fluid pressure field is deduced for the cavity domain with the fluid density ρ_f :

$$p(x, y, z, t) = -\rho_f \sum_{m,n} \varphi_m^n(x, y) \cdot \cosh\left(\lambda_m^n z\right) \cdot \dot{f}_m^n(t) \tag{12.22}$$

By substituting (12.20) in (12.22), the fluid pressure over the plate $z = H$ is:

$$p(x, y, t) = -\sum_{m,n} \frac{\rho_f \cot h\left(\lambda_m^n H\right)}{\lambda_m^n} \varphi_m^n(x, y)\, \ddot{d}_m^n(t) \tag{12.23}$$

And considering (12.15) for harmonic motion:

$$p(x, y, t) = \sum_{m,n} P_m^n(x, y)\, \overline{d}_m^n e^{i\omega_m^n t} \tag{12.24}$$

Being P_m^n, the pressure mode:

$$P_m^n(x, y) = \frac{\rho_f \coth\left(\lambda_m^n H\right)}{\lambda_m^n} \varphi_m^n(x, y) \omega_m^{n2} \tag{12.25}$$

12.2.2 Calculus of the Plate Free Frequencies

First, the free frequencies of vibration of the plate are calculated solving Eq. (12.1) without the forcing load. Considering the expressions in the Eqs. (12.5) and (12.24) for the plate deformation and pressure over the plate, this equation is expressed in a series modal expansion as:

$$\sum_{m,n} D\nabla^4 w_m^n(x, y) d_m^n(t) - \sum_{m,n} \rho_m t_h \omega_m^{n2} w_m^n(x, y) d_m^n(t) = \sum_{m,n} P_m^n(x, y) d_m^n(t) \tag{12.26}$$

Considering Eq. (12.25) for the pressure and (12.4) for the plate deflection mode and taking into account the expression (12.15) for harmonic motion, and after integration over the plate surface of the squared series expressions of the plate equation, it is deduced:

$$\iint\limits_{S_p} \left\{ \sum_{m,n} \left(D\left[\left(\frac{m\pi}{a}\right)^2 + \left(\frac{n\pi}{b}\right)^2 \right]^2 - \rho_m t_h \omega_m^{n2} \right) \cdot w_m^n(x, y) \right\}^2 dxdy$$

$$= \iint\limits_{S_p} \left\{ \sum_{m,n} \left(\frac{\rho_f \coth\left(\lambda_m^n H\right)}{\lambda_m^n} \omega_m^{n2} \right) \cdot \varphi_m^n(x, y) \right\}^2 dxdy \tag{12.27}$$

And with the orthogonality properties of the w_m^n and φ_m^n functions in mind, an analytic simple quadratic formula is deduced to obtain the modal natural frequencies:

$$A_m^n \cdot \omega_m^{n4} + B_m^n \cdot \omega_m^{n2} + C_m^n = 0 \tag{12.28}$$

where:

$$A_m^n = (\rho_m t_h)^2 - \left(\frac{\rho_f \cot h\left(\lambda_m^n H\right)}{\lambda_m^n} \right)^2 \tag{12.29a}$$

$$B_m^n = -\left(2\rho_m\, t_h D\, \Lambda_m^{n4}\right) \tag{12.29b}$$

$$C_m^n = \left(D\, \Lambda_m^{n4}\right)^2 \tag{12.29c}$$

With

$$\Lambda_m^n = \sqrt{\left(\frac{m\pi}{a}\right)^2 + \left(\frac{n\pi}{b}\right)^2}$$

An easy iteration procedure due to the fast convergence is employed for finding the roots or solution, because of the dependence of the parameter λ_m^n with the frequency (wave number k), Eq. (12.18).

12.2.3 Calculus of the Forced Frequency Response of the Plate

Similarly, the frequency response of the plate is calculated solving Eq. (12.1) but now with the forcing load of frequency ω. Considering the expressions in the Eqs. (12.5), (12.7) and (12.24) for the plate deformation, concentrated forcing load and pressure over the plate, for harmonic motion, this equation is expressed in a series modal expansion as:

$$\sum_{m,n} D\nabla^4 w_m^n(x, y)\bar{d}_m^n - \sum_{m,n} \rho_m t_h \omega^2 w_m^n(x, y)\bar{d}_m^n$$
$$= \sum_{m,n} P_m^n(x, y)\bar{d}_m^n + \sum_{m,n} w_m^n(x, y) \cdot \bar{q}_m^n \tag{12.30}$$

Considering Eq. (12.25) for the pressure and (12.6) for the plate deflection mode, and after integration over the plate surface of the squared series expressions of the plate equation, it is deduced the following equation:

$$\iint\limits_{S_p} \left\{ \sum_{m,n} \left(\left(D\left[\left(\frac{m\pi}{a}\right)^2 + \left(\frac{n\pi}{b}\right)^2 \right]^2 - \rho_m t_h \omega^2 \right) \cdot \overline{d}_m^n - \overline{q}_m^n \right) \cdot w_m^n(x, y) \right\}^2 dxdy$$
$$= \iint\limits_{S_p} \left\{ \sum_{m,n} \frac{\rho_f \coth\left(\lambda_m^n H\right)}{\lambda_m^n} \omega^2 \cdot \varphi_m^n(x, y) \cdot \overline{d}_m^n \right\}^2 dxdy \tag{12.31}$$

With the orthogonality properties of the w_m^n and φ_m^n functions in mind, an analytic quadratics expression is obtained relating the deformation and forcing coefficients $\overline{d}_m^n$ and $\overline{q}_m^n$:

$$Y_m^n \cdot \left(\frac{\overline{d}_m^n}{\overline{q}_m^n} \right)^2 + Z_m^n \cdot \left(\frac{\overline{d}_m^n}{\overline{q}_m^n} \right) + 1 = 0 \tag{12.32}$$

And the deformation coefficient is obtained with the frequency transfer function that relates it with the forcing coefficient:

$$\overline{d}_m^n = TF(\omega) \cdot \overline{q}_m^n \tag{12.33}$$

where the transfer function is expressed:

$$TF(\omega) = \frac{-Z_m^n \pm \sqrt{\left(Z_m^n\right)^2 - 4 \cdot Y_m^n}}{2 \cdot Y_m^n} \tag{12.34}$$

And the coefficients have the expressions:

$$Y_m^n = A_m^n \cdot \omega^4 + B_m^n \cdot \omega^2 + C_m^n \tag{12.35}$$

$$Z_m^n = \frac{1}{\rho_m t_h} \left(B_m^n + 2 \cdot \rho_m^2 \cdot t_h^2 \cdot \omega^2 \right) \tag{12.36}$$

12.3 Results

For the verification of the method, in Table 12.1, the frequencies f_{lm}^n (in Hz) of a simply supported rectangular plate at the top of a rectangular cavity filled with an incompressible fluid, are compared with the results of Gascón-Pérez (2023) that analyzes the frequencies of the top rectangular plate of a rectangular container filled with fluid, and are the same, and also calculating the change in the natural frequencies with respect to the vacuum case f_{vm}^{n}. So the validity of this method can be concluded.

Also for the verification of the method, in Table 12.2, the non-dimensional frequency parameters Ω_m^n defined as $\Omega = \sqrt{\frac{\rho_m t_h}{D}} l_x^2 \cdot \omega$ of a simply supported rectangular plate at the top of a rectangular container filled with an incompressible fluid, are compared with the results of Zhou et al. (2001) that analyze the frequencies of the rectangular plate of a rectangular container filled with a fluid, and the results are approximately the same with slight discrepancy. The plate is made of aluminum and the relation of fluid-plate material densities is $\rho_f / \rho_m = 0.125$.

The next results are obtained for the case of an aluminum rectangular plate of a rectangular cavity filled with a compressible fluid of density $\rho_f = 1.225 \frac{\text{kg}}{\text{m}^3}$, and the following reference characteristics are considered:

Side lengths $l_x = 2$ m, $l_y = 1.5$ m, plate thickness $t_h = 1$ mm, container height $H = 1$ m, and material density $\rho_m = 2700 \frac{\text{kg}}{\text{m}^3}$.

Table 12.3 presents different natural frequencies of the plate regarding the vacuum case and in contact with fluid.

The sound pressure level at an interior point of the rectangular cavity (the reference point is located at $x = 0.25\, l_x$ $y = 0.25\, l_y$ $z = 0.5\, H$) and the deflexion level of a point of the plate (the reference point is located the center position) are expressed in the following Figs. 12.2, 12.3, 12.4, 12.5, 12.6, 12.7, 12.8, 12.9 and 12.10 in logarithmic scale as SPL (dB) and DL(dB), considering an excitation load q_0 of 1N

Table 12.1 Natural frequencies (Hz) for the bottom simply supported steel plate (in vacuum and water) for the present method and the work of Gascón-Pérez (2023), of side lengths $l_x = 4$ m, $l_y = 5$ m, container height $H = 3$ m, and thickness $t_h = 1$ cm

m, n	f_v	f_l (Gascón-Pérez (2023))	f_l (present method)
1, 2	5.40	1.74	1.74
2, 1	7.04	2.41	2.41
2, 2	9.96	3.68	3.68

Table 12.2 Natural frequency parameters for the simply supported aluminum plate (in vacuum and dense fluid) for the present method and the work of Zhou et al. (2001), of side lengths $l_x = 1$ m, $l_y = 1$ m, container height $H = 1$ m and thickness $t_h = 5$ mm

Modal number	Ω_v	Ω_l(Zhou)	Ω_l(present method)	Relative discrepancy (%)
1	49.35	39.2	40.4	3.06
2	49.35	39.2	40.4	3.06
3	78.96	66.6	67.7	1.65
4	98.70	85.6	86.1	0.56
5	128.30	112.7	114.0	1.15

Table 12.3 Free modal frequencies (Hz) of the reference plate in vacuum and with air for the different modes

Modal number	f_v	f_a
1	1.669	1.539
2	3.471	3.279
3	4.872	4.643
4	6.475	6.207
5	6.675	6.403
6	9.679	9.348
7	10.213	9.873
8	10.680	10.332
9	12.015	11.645
10	13.884	13.485

located at point $x_0 = 0.25\,l_x$, $y_0 = 0.25\,l_y$ and have the expressions:

$$SPL\,(dB) = 20\log\left(\frac{P}{P_{ref}}\right) \text{ with } P_{ref} = 10^{-5}\text{Pa}$$

$$DL\,(dB) = 20\log\left(\frac{w}{w_{ref}}\right) \text{ with } w_{ref} = 10^{-12}\text{m}$$

Figure 12.2 presents for the reference values of the panel, the variation of the sound pressure level and the deflexion level with the frequency. The different peaks correspond to the resonant frequencies and some of them, but not all, correspond

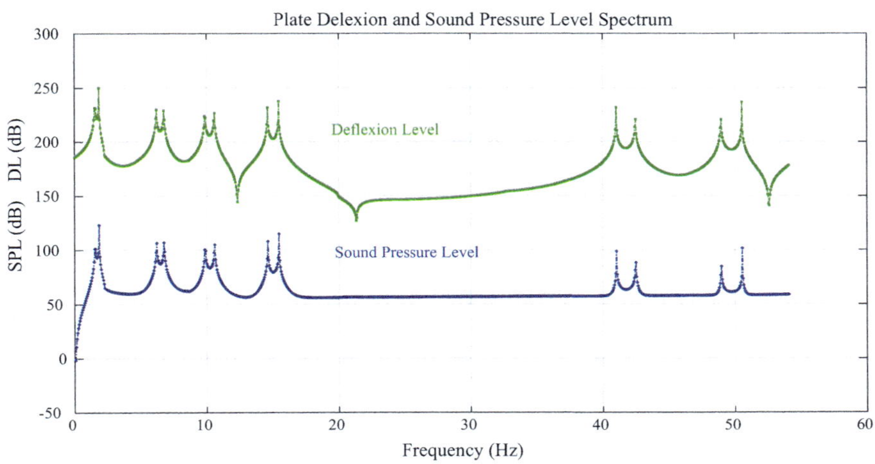

Fig. 12.2 Variation of the sound pressure Level SPL and plate deflexion level DL with the frequency for the reference conditions

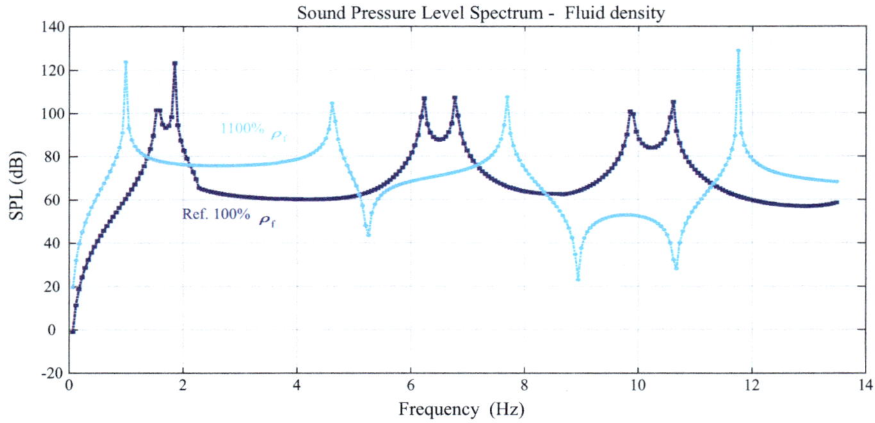

Fig. 12.3 Variation of the sound pressure level SPL with the frequency for different values of the fluid density

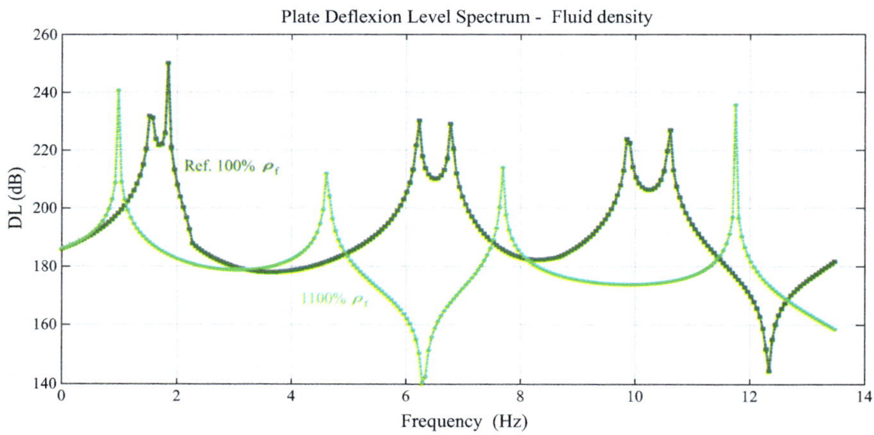

Fig. 12.4 Variation of the plate deflexion level DL with the frequency for different values of the fluid density

to the natural frequencies of vibration of the plate in contact with the fluid, because there are higher value of natural frequencies of the coupled system than resonant peaks in the showed spectrum range. A peak wavy variation of the plate deflexion level with the frequency is observed in the frequency spectrum (0–15 Hz), while for the interval (15–40 Hz) there is an anti-resonant peak, and for higher values of the frequency presents again different peaks. The sound pressure level initially presents a peak wavy variation in the range (0–15 Hz) and almost remains constant in the interval (15–40 Hz), and for higher values of the frequency appear again different

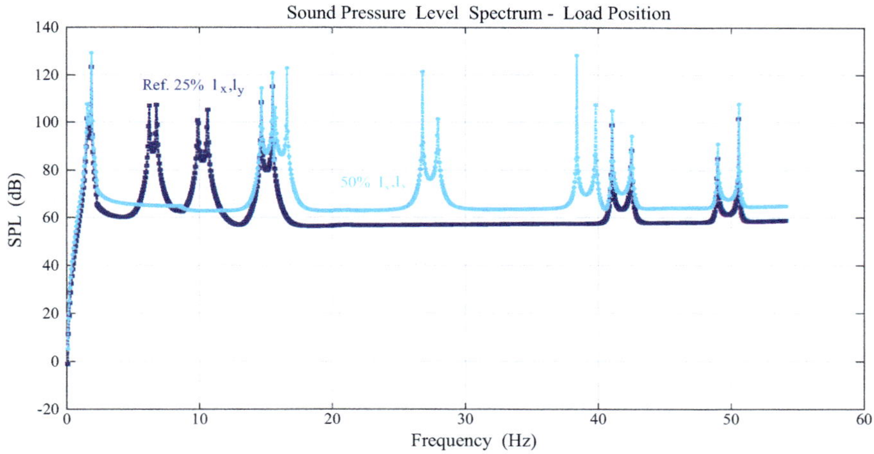

Fig. 12.5 Variation of the sound pressure level SPL with the frequency for different values of the load position

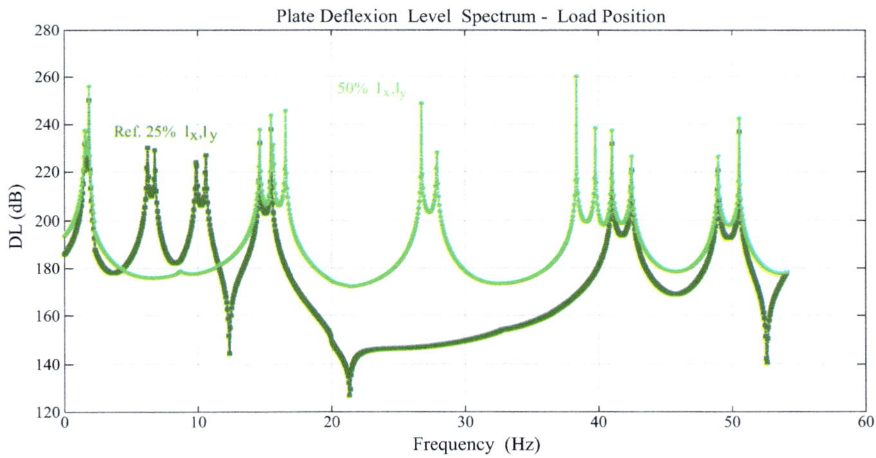

Fig. 12.6 Variation of the plate deflexion level DL with the frequency for different values of the load position

peaks. These peaks appear in the form of a doublet peak. All the resonant frequencies coincide for the sound pressure level and the plate deflexion level.

Figure 12.3 shows the variation of the sound pressure level SPL with the frequency of the excitation load (located at $x_0 = 0.25\, l_x$ and $y_0 = 0.25\, l_y$) or the sound pressure level spectrum at a point located at position $x = 0.25\, l_x$ $y = 0.25\, l_y$ $z = 0.5\, H$ of the rectangular cavity, for different values of the fluid density. For the case of heavier fluid, the resonant peaks appear at lower values of the frequency and in the form of a

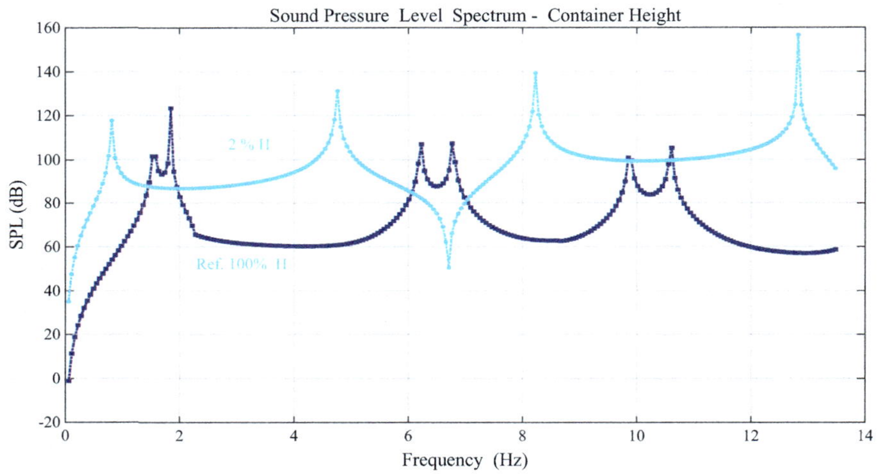

Fig. 12.7 Variation of the sound pressure level SPL with the frequency for different values of the cavity height

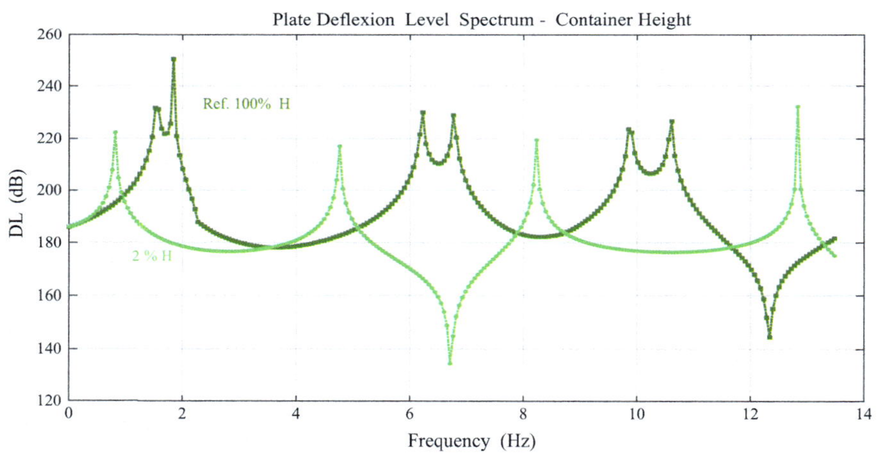

Fig. 12.8 Variation of the plate deflexion level DL with the frequency for different values of the cavity height

single peak (not doublet). Also for heavier fluid appear anti-resonant peaks and not as in the reference case (lighter fluid).

Figure 12.4 shows the variation of the plate deflexion level DL with the frequency of the excitation load (located at $x_0 = 0.25\, l_x$ and $y_0 = 0.25\, l_y$) or the plate deflexion level spectrum of the center point of the plate, for different values of the fluid density. For the case of heavier fluid, the resonant peaks appear at lower values of the frequency and in the form of a single peak (not doublet). In both cases appear

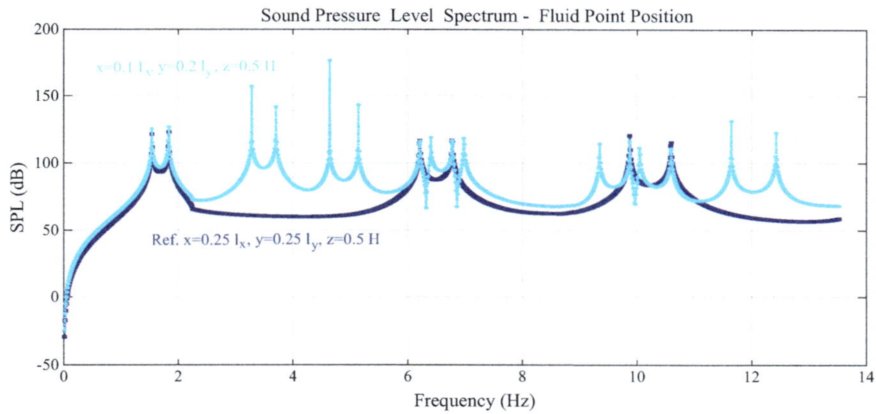

Fig. 12.9 Variation of the sound pressure level SPL with the frequency for different values of the fluid point position

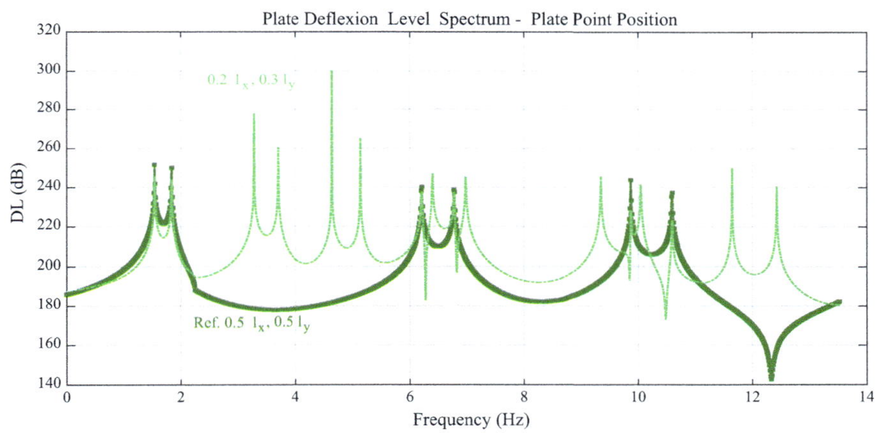

Fig. 12.10 Variation of the plate deflexion level DL with the frequency for different values of the plate point position

anti-resonant peaks. All the resonant frequencies coincide with the corresponding for the sound pressure level of Fig. 12.3.

Figure 12.5 shows the variation of the sound pressure level SPL with the frequency of the excitation load or the sound pressure level spectrum at a point located at position $x = 0.25\, l_x\; y = 0.25\, l_y\; z = 0.5\, H$ of the rectangular cavity, for different values of the load position x_0,y_0. Some of the resonant peaks for the not-reference position appear at the same frequency that in the reference case, but there are peaks with different frequencies for the two cases. Also, there are triplet and quadruplets peaks for the not-reference position. As commented in Fig. 12.2, some but not all of the peaks have the same frequency than the corresponding natural frequencies of the system. The first resonant peak coincides in frequency for the two load positions.

Figure 12.6 shows the variation of the plate deflexion level DL with the frequency of the excitation load with the reference position (located at $x_0 = 0.25\,l_x$ and $y_0 = 0.25\,l_y$) or the plate deflexion level spectrum of the center point of the plate, for different values of the load position x_0,y_0. The first resonant peak coincides in frequency for the two load positions. For the reference load point position appear some anti-resonant peaks, but none in the case of the other load point position. The resonant peaks for the two load positions occur at the same frequency as for the sound pressure level of Fig. 12.5.

Figure 12.7 shows the variation of the sound pressure level SPL with the frequency of the excitation load (located at $x_0 = 0.25\,l_x$ and $y_0 = 0.25\,l_y$) or the sound pressure level spectrum at a point located at position $x = 0.25\,l_x$ $y = 0.25\,l_y$ $z = 0.5\,H$ of the rectangular cavity, for different values of the cavity height. For the case of lower height the resonant peaks appear at lower values of the frequency and in the form of a single peak (not doublet). Also for lower height appear an anti-resonant peak and not as in the reference case (higher height).

Figure 12.8 shows the variation of the plate deflexion level DL with the frequency of the excitation load (located at $x_0 = 0.25\,l_x$ and $y_0 = 0.25\,l_y$) or the plate deflexion level spectrum of the center point of the plate, for different values of the cavity height. The behavior is analogous the previous case of Fig. 12.7. The frequencies of the resonant peaks coincide in the two figures.

The frequency spectrum or variation of the sound pressure level SPL and plate deflexion level DL with the frequency, presents no variation with the fluid sound speed c.

Figure 12.9 shows the variation of the sound pressure level SPL with the frequency of the excitation load (located at $x_0 = 0.25\,l_x$ and $y_0 = 0.25\,l_y$) or the sound pressure level spectrum at two different points of the rectangular cavity, the reference point located at position $x = 0.25\,l_x$ $y = 0.25\,l_y$ $z = 0.5\,H$ and the point located at position $x = 0.1\,l_x$ $y = 0.2\,l_y$ $z = 0.5\,H$ of the rectangular cavity. This sound pressure level spectrum depends on the location of the fluid point considered inside the cavity. For the new fluid point, the position appears more resonant peaks, and some of them coincide with the corresponding to the reference case. Many of them, but not all, are coincident with the natural frequencies of the system, so the number of resonant peaks are higher than the natural frequencies of the coupled system.

Figure 12.10 shows the variation of the sound plate deflexion level DL with the frequency of the excitation load (located at $x_0 = 0.25\,l_x$ and $y_0 = 0.25\,l_y$) or the plate deflexion level spectrum at two different points of the rectangular plate, the reference point located at center position $x = 0.5\,l_x$ $y = 0.5\,l_y$ and the point located at position $x = 0.2\,l_x$ $y = 0.3\,l_y$ of the rectangular plate. Similarly as in Fig. 12.9, the frequency response of the plate deflexion depends on the location of the point of the plate considered and the number of resonant peaks is higher for the new plate point position.

Figure 12.11 shows the variation of the mean quadratic plate sound pressure level MQSPL with the frequency of the excitation load (located at $x_0 = 0.25\,l_x$

and $y_0 = 0.25\, l_y$) or the mean quadratic plate sound pressure level spectrum. There are a lot of resonant peaks, and many of them coincide in frequency with the natural frequencies of the coupled system, but there are more resonant peaks than the number of the natural frequencies of the coupled system. This mean quadratic plate sound pressure level is calculated as:

$$MQSPL\,(dB) = 20\log\left(\frac{MQP}{P_{ref}}\right) \text{ with } P_{ref} = 10^{-5}\text{Pa}$$

where *MQP* is the mean quadratic sound pressure over the plate

$$MQP = \sqrt{\frac{1}{l_x l_y}\iint_{S_p}\left[\sum_{m,n} P_m^n(x,y)\,\overline{d}_m^n\right]^2 dxdy} \tag{12.37}$$

Figure 12.12 shows the variation of the mean quadratic plate deflexion level MQDL with the frequency of the excitation load (located at $x_0 = 0.25\, l_x$ and $y_0 = 0.25\, l_y$) or the mean quadratic plate deflexion level spectrum. The resonant peaks appear at the same frequency as in the case of Fig. 12.11 for the mean quadratic plate sound pressure. This mean quadratic plate deflexion level is calculated as:

$$MQDL(dB) = 20\log\left(\frac{MQD}{w_{ref}}\right) \text{ with } w_{ref} = 10^{-12}\text{m}$$

where *MQD* is the Mean Quadratic Deflexion of the plate.

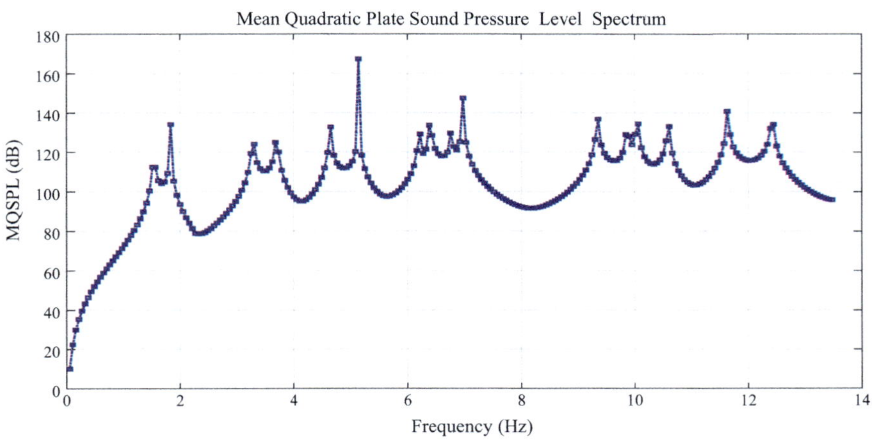

Fig. 12.11 Variation of the mean quadratic plate sound pressure level MQSPL with the frequency

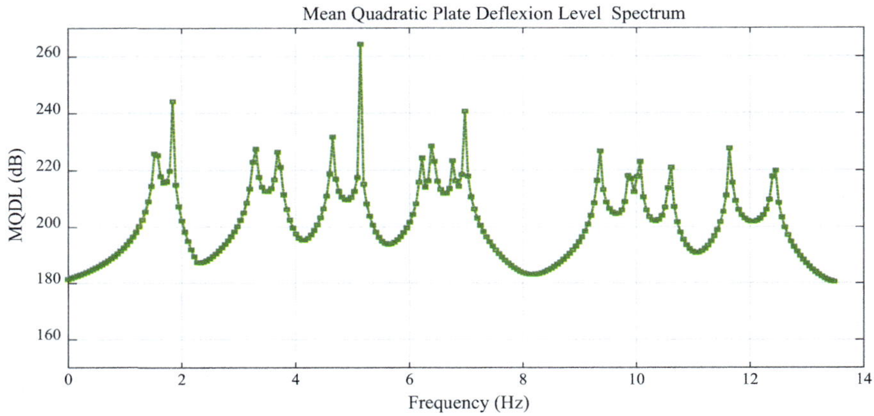

Fig. 12.12 Variation of the mean quadratic plate deflexion level MQDL with the frequency

$$MQD = \sqrt{\frac{1}{l_x l_y} \iint_{S_p} \left[\sum_{m,n} w_m^n(x, y)\, \overline{d}_m^n \right]^2 dxdy} \tag{12.38}$$

Many of the peak frequencies correspond to the natural frequencies of the plate in contact with fluid inside the rectangular cavity, but there are more peak frequencies than the natural frequencies of the coupled system.

Figure 12.13 shows the variation of the mean quadratic plate deflexion level MQDL in vacuum with the frequency of the excitation load (located at $x_0 = 0.25\, l_x$ and $y_0 = 0.25\, l_y$) or the mean quadratic plate deflexion level spectrum. In this vacuum case, the frequencies of the peaks correspond to the natural frequencies of vibration of the plate in vacuum.

Figures 12.14 and 12.15 show the variation of the sensibility level SL with the frequency of the excitation load (located at $x_0 = 0.25\, l_x$ and $y_0 = 0.25\, l_y$) or the sensibility level spectrum for different range of the frequency spectrum. This sensibility level is calculated as:

$$SL(dB) = 20 \log\left(\frac{MQS}{S_{ref}}\right) \text{ with } S_{ref} = 10^{-7} \frac{\text{m}}{\text{Pa}}$$

where MQS is the mean quadratic sensibility of the plate defined as $MQS = \frac{MQD}{MQP}$.

The sensibility decreases with the frequency in the range of excitation in the range (0–14 Hz) and the higher values correspond to the low frequency spectrum. In Fig. 12.15 for a wider range spectrum it can be seen a wavy variation in the range (15–35 Hz) and a decreasing tendency for higher values of the frequency. This is an interesting result for the design of pressure sensors based on the vibration of a plate or diaphragm in an air cavity because the higher values of the sensitivity occur at lower frequencies of excitation of these micro electromechanical systems MEMS.

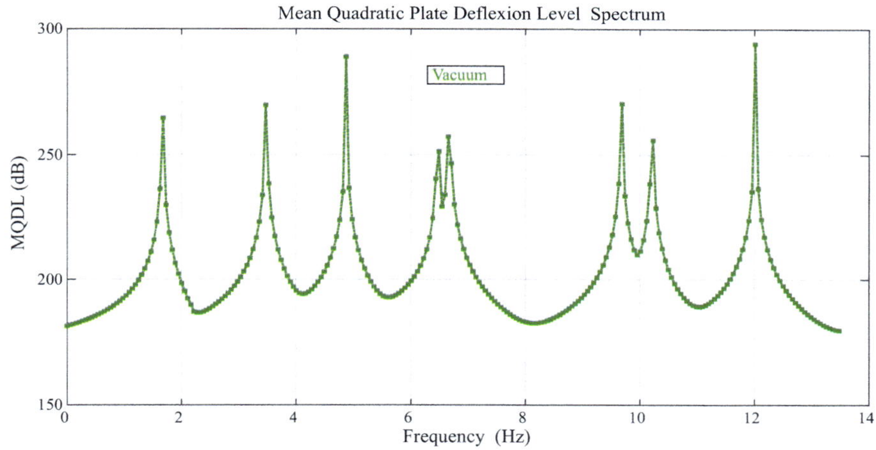

Fig. 12.13 Variation of the mean quadratic plate deflexion level MQDL in vacuum with the frequency

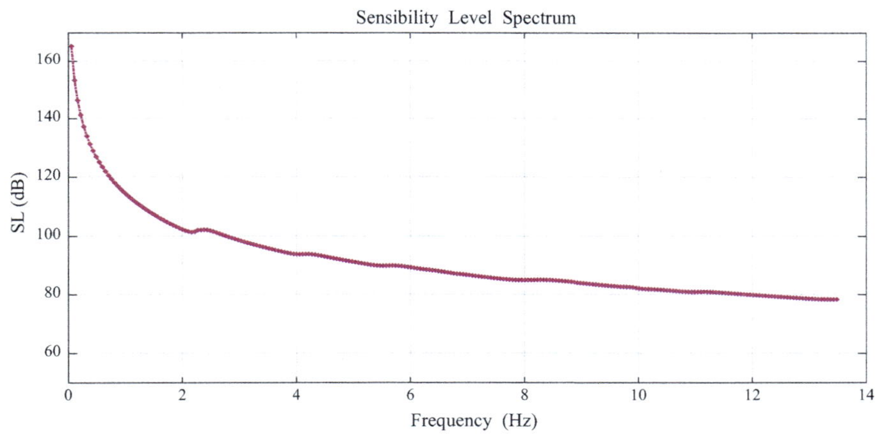

Fig. 12.14 Variation of the sensibility sevel SL with the frequency (0–14 Hz)

12.4 Conclusions

Acoustic interaction of a vibrating structure with fluid in an acoustic cavity has been studied for the case of a forced, excited plate at the top position of a rectangular container filled with compressible and non-viscous fluid. In this regard, the wave equation for the fluid velocity potential is solved by a method of separation of variables, and the acoustic pressure is deduced from the momentum's linearized equation, so that considering the equation of motion of the forced vibrating and after an integration procedure over its surface considering the orthogonality of the modes involved

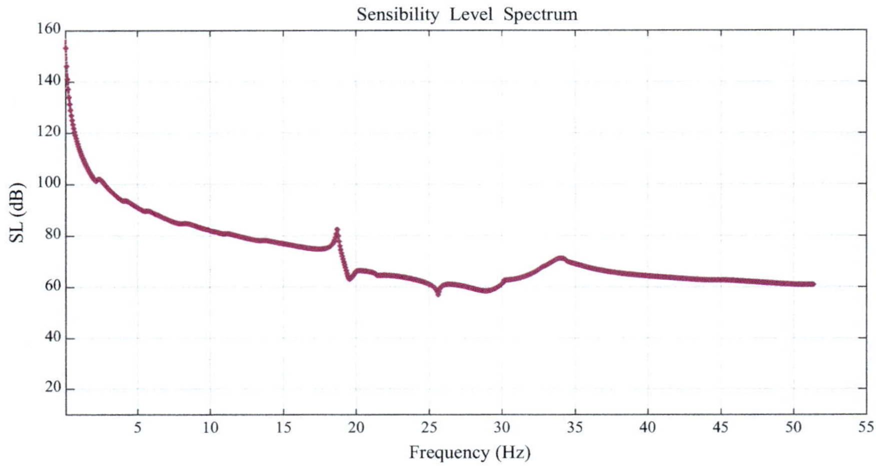

Fig. 12.15 Variation of the sensibility level SL with the frequency (0, 50 Hz)

expressed in series expansion of the associated velocity potential and plate deformation, a simple analytical expressions in an elegant manner is obtained for the calculus of the free or natural vibration frequencies, and the system frequency response for the case of forced vibrating plate of the coupled system. The frequency response or spectrum of the plate deformation of the middle point of the plate expressed in logarithmic scale (dB) as a deformation level and well as the sound pressure level at a point inside the cavity is obtained considering its variation with different parameters such as the container height, fluid density and forced excitation load position. Also is studied the SPL spectrum for different fluid point of the acoustic cavity and the plate deformation spectrum DL for a different plate point position. Furthermore, the variation of the mean quadratic plate sound pressure level MQSPL with the frequency is analyzed as well as the mean quadratic plate deflexion level MQDL spectrum for the case of vacuum and in contact with fluid. Finally, the variation of the mean quadratic sensibility *MQS* with the frequency is obtained. Validation of the method is made by comparing the results with those obtained by other authors and theories.

References

Defay E, Millon C, Malhaire C et al (2002) PZT Thin Films integration for the realisation of a high sensitivity pressure Microsensor based on a vibrating membrane. Sensors and Actuators 99:64–67

Dowell EH, Voss HM (1963) The effect of a cavity on panel vibration. AIAA J 1(2):476–477

Gascón-Pérez M (2020) Hydro-elastic oscillations of the bottom membrane of a rectangular container filled with fluid. Defect and Diffusion Forum. 399. Trans Tech Publications Ltd, Switzerland, pp 76–86

Gascón-Pérez M (2023) Compressibility effects on the hydroelastic vibration of a plate at an off-center position of a rectangular container filled with fluid. Engineering Design Applicatons V. Advanced Structured Materials. Springer Nature, Cham

Gascón-Pérez M (2024) Acoustic radiation characteristics of a forced vibrating elastic panel under thermal environments. J Theor Comput Acoust 32(4)

Kim SM, Brennan MJ (2000) Active control of harmonic sound transmission into an acoustic enclosure using both structural and acoustic actuators. J Acoust Soc Am 107:2523–2534

Laser DJ, Santiago JG (2004) A review of micropumps. J Micromech Microeng 14(6):35–64

Munson BR, Okiishi TH, Huebsch WW, Rothmayer AP (2013) Fundamentals of fluid mechanics. John Wiley & Sons, Hoboken, NJ

Pan J, Bies DA (1990) The effect of fluid-structural coupling on sound waves in an enclosure Theoretical part. J Acoust Soc Am 87(2):691–707

Pan LS, Ng TY, Wu XH et al (2003) Analysis of valveless micropumps with inertial effects. J Micromech Microeng 13:390–399

Pretlove AJ (1965) Free vibrations of a rectangular panel backed by a closed rectangular cavity. J Sound Vib 2(3):197–209

Pretlove AJ (1966) Forced vibrations of a rectangular panel backed by a closed rectangular cavity. J Sound Vib 3(3):252–261

Qiu XJ, Sha JZ, Yang J (1995) Mechanisms of active control of noise transmission through a panel into a cavity using a point force actuator on the panel. J Sound Vib 182:167–170

Sestieri E, Del Vescovo D, Lucibello P (1984) Structural-acoustic coupling in complex shaped cavities. J Sound Vib 96(2):219–233

Shabani R, Golzar FG, Tariverdilo S, Taraghi H et al (2013) Hydroelastic vibration of a circular diaphragm in the fluid chamber of a reciprocating micro pump. Int J Eng 27:643–650

Wang CY, Wang CM (2014) Structural vibration: exact solutions for strings, membranes, beams and plates. CRC Press, Boca Raton FL

Xie X, Zheng H, Qu Y (2016) A variational formulation for vibro-acoustic analysis of a panel backed by an irregularly-bounded cavity. J Sound Vib 373:147–216

Zhou D, Cheung YK (2001) (2001) Vibrations of elastic bottom plate of a sealed rectangular container fully filled with liquid. Int J Struct Stab Dyn 1(1):145–162

Zienkiewicz OC, Newton RE (1969) Coupled vibrations of a structure submerged in a compressible fluid. Stuttgart In: Proceeding international symposium on finite element techniques, pp 1–9

Chapter 13
Machine Learning-Based Biomimetic Optimization Models for Engineered Materials

Shishir Barai, Joe Sgarrella, and Christian Peco

Abstract The slime mold *Physarum polycephalum* develops networks that can solve high combinatorial problems and optimize material properties. This emergent behavior is heavily based on the complex microstructural tuning of its cytoplasm. In this study, we part from a phase-field model of the slime mold and propose a machine-learning strategy to optimize the microstructural response, hence capturing the optimization of engineered materials with embedded networks. By representing transport networks as composite mesh structures—analogous to biological organisms with distinct regions mimicking functionalities like nuclei and cellular matrices—we bridge the gap between biological phenomena and computational modeling of advanced materials. Using continuum equations, we simulate protoplasmic flux through these meshes, generating data that captures the essence of transport dynamics. A machine learning model is trained to predict the resulting pressure and flow fields given a composite mesh structure. The trained model rapidly approximates the physics and enables iterative optimization of the mesh topology to maximize material efficiency. This framework synergistically combines composite mesh modeling, biophysics simulations, and machine learning to discover new bio-inspired strategies for optimizing engineered material designs with embedded networks.

Keywords *Physarum policephalum* · Phase-field · Neural networks · Finite elements

---

S. Barai · J. Sgarrella · C. Peco (✉)
Department of Engineering Science and Mechanics, The Pennsylvania State University, State College, PA, USA
e-mail: christian.peco@psu.edu

S. Barai
e-mail: svb6406@psu.edu

J. Sgarrella
e-mail: jfs5687@psu.edu

C. Peco
Institute for Computational and Data Sciences, The Pennsylvania State University, State College, PA, USA

A. Öchsner and H. Altenbach (eds.), *Engineering Design Applications VII*, Advanced Structured Materials 230, https://doi.org/10.1007/978-3-031-84346-4_13

13.1 Introduction and Background

Biological networks exhibit remarkable capabilities to dynamically self-organize and adapt to environmental stimuli. Organisms ranging from slime molds to vascular networks develop internal morphologies that maximize efficiency and resilience, solving highly complex optimization challenges with emergent decentralized intelligence (Clauset et al. 2008; Nakagaki et al. 2000). These networks are often embedded within the material structure of the organism, playing a crucial role in determining its overall properties and performance. Mathematical models encapsulating the underlying biophysical mechanisms have provided fundamental insights into natural network adaptations (Jones 2010; Tero et al. 2007). However, computational expense often limits iterative analysis and discovery over large parameter spaces. Recent advances in machine learning now offer pathways to develop surrogate models that rapidly approximate solutions to computationally intensive physics-based simulations (Karniadakis et al., 2021; Raissi et al., 2019).

The plasmodium stage of the slime mold *Physarum polycephalum* (*P. polycephalum*) demonstrates notable abilities to adapt its tubular network morphology to optimize transport efficiency. When presented with spatially distributed nutrients, *P. polycephalum* will reconfigure its vein network to connect the nutrients through shortest possible pathways (Adamatzky 2012; Nakagaki et al. 2000). Some examples of *P. polycephalum* forming optimized pathways can be observed in Fig. 13.1. Key mechanisms deduced from biological experiments include tube diameter expansion in response to higher fluxes of protoplasm and atrophy of tubes with lower fluxes (Alim et al. 2017; Kamiya 1950; Tero et al. 2007). This intelligent adaptation relies on simple decentralized rules enacted locally by the organism, resulting in a material with embedded networks that are optimized for its specific needs. The concept of materials with embedded networks has gained significant attention in recent years, with researchers exploring how the principles observed in biological systems can be applied to the design of engineered materials (Studart 2016; Wegst et al. 2015; Wijerathne et al. 2022). For example, 3D printing techniques have been used to create materials with embedded vascular networks that mimic the transport efficiency and self-healing properties of biological tissues (Gao et al. 2019; Kolesky et al. 2016). The incorporation of embedded networks in engineered materials has the potential to revolutionize fields such as tissue engineering, soft robotics, and advanced manufacturing (Wegst et al. 2015). Similarly, bio-inspired microvascular networks have been integrated into structural composites to improve their multifunctional capabilities, including self-healing and thermal management (Patrick et al. 2014). Recent computational models integrating the dynamics of biological networks through systems of partial differential equations have shown good agreement with experimental observations (Ghanbari et al. 2020, 2023a, b). These models combine a novel reference map-based phase-field approach (Ghanbari and Peco 2023) with a large-scale amenable description from fluid to solid phases (Shimada et al. 2021, 2022). This

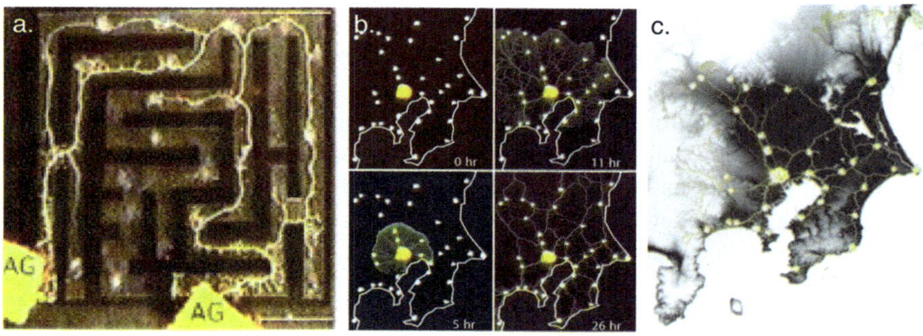

Fig. 13.1 **a** *Physarum polycephalum* solving a maze (Tero et al. 2007). This is an intermediate state of the maze solving process. **b** Time-lapse images of *P. polycephalum* network formation of Tokyo railway system (Tero et al. 2010). The network begins with a small plasmodium surrounded by food sources that mark major cities within the region. As time progresses, the plasmodium grows out from the initial food source to the additional food sources. An interconnected network of the food sources is left behind the growth front. **c** Reimagined Tokyo railway system.

approach eliminates the need for boundary tracking and greatly facilitates the representation of evolving geometries. Phase-field methodologies have proven effective for tackling problems characterized by intricate geometries. They have been instrumental in advancing the modeling of various systems in soft matter, including biomembranes (Peco et al. 2013; Rosolen et al. 2013) and particle raft fracture mechanics (Guével et al. 2023; Liu et al. 2019; Peco et al. 2017, 2019; Spencer et al. 2016). To numerically solve the associated partial differential equations, we adopt a variational approach and leverage MOOSE, a finite element method (FEM) implementation (Gaston et al. 2009; Sgarrella et al. 2023). However, challenges to fully characterize *P. polycephalum*'s adaptation space remains, including the complexity of the microstructural response and its associated computational cost, which makes direct simulation impractical.

In response to these challenges, recent works have begun leveraging machine learning to approximate solutions to the complex biophysics models of slime mold network evolution for accelerated simulation and optimization (Jones 2015; Whiting et al. 2016). These engineered materials with embedded networks are represented as weighted graphs, with edge weights encoding the conductivity of tubes. Continuum equations govern the flux of protoplasm through the network for a given conductivity state. Machine learning models are then trained on simulation data to predict network evolution dynamics orders of magnitude faster than direct numerical simulation. These learned models enable extensive iterative improvement of network parameters to optimize morphology for internal efficiency. In this work, we extend the principles observed in slime molds to a broader context of biological organisms, represented as composite mesh structures. These structures, analogous to biological organisms, consist of distinct regions mimicking functionalities like nuclei and cellular matrices. By simulating protoplasmic flux through these meshes using continuum equations, we generate data that mirrors the transport dynamics in such materials. A machine learning model, trained on this data, is then capable of predicting how changes in the material's structure might affect its internal pathways. The model's predictions, based on the mesh structure, can swiftly approximate complex biophysical behaviors, facilitating iterative optimization of mesh designs for enhanced transport efficiency. This integrated approach not only showcases the potential of machine learning in biophysical simulations but also highlights the incredible adaptive capabilities of organisms like slime molds and their bio-inspired counterparts in the context of advanced materials design.

13.2 Continuum Modeling of the Fungal Mycelium

Continuum modeling provides a comprehensive framework to describe the behavior of systems with evolving geometries and complex interactions. In the context of engineered materials with embedded networks inspired by slime mold, continuum modeling offers a mathematical representation of the material's adaptive behavior in response to external stimuli. The growth dynamics of the embedded networks within the material are captured using a multispecies formulation (Ghanbari et al. 2023b). This approach allows for the modeling of the complex interplay between the material matrix and the embedded network, enabling the study of how the network's evolution influences the overall properties and performance of the engineered material. By drawing insights from the adaptive behavior of slime mold transport networks, we can develop continuum models that accurately describe the behavior of materials with embedded networks, paving the way for the design and optimization of novel, bio-inspired materials with enhanced functionality and efficiency. This approach considers the availability and distribution of nutrients in the medium and the specific metabolism of each species. Within the domain of study, Ω, we consider the existence of $i = 1, 2, \ldots, N$ slime mold species. The free energy functional for the ith species is defined as

$$\mathcal{G}_i(\phi, \nabla\phi, c_s, c_c) = \int_\Omega F_i(\phi, \nabla\phi, c_s, c_c)\, \mathrm{d}\Omega. \tag{13.1}$$

In this formulation, ϕ is a phase-field variable that represents the existence or absence of biomass. It ranges in values from $\phi = 1$ (existence of biomass) to $\phi = 0$ (absence of biomass). Additionally, the variables c_c and c_s represent the concentrated nutrient source and the substrate nutrient source respectively. The free energy density, F_i, that corresponds to the free energy functional is defined as

$$F_i(\phi, \nabla\phi, c_s, c_c) = I_\mathcal{G}(\nabla\phi) + W_\mathcal{G}(\phi) + B_{\mathcal{G}i}(\phi, c_s, c_c). \tag{13.2}$$

The free energy density is composed of three terms that represent the interfacial energy ($I_\mathcal{G}$) and the contribution of the bulk phase energy ($W_\mathcal{G} + B_{\mathcal{G}i}$). The two terms that make up the bulk phase energy contribution are a symmetric double-well potential ($W_\mathcal{G}$) and the nutritional bias ($B_{\mathcal{G}i}$). These three terms are defined as

$$I_{\mathcal{G}} = \frac{1}{2}\epsilon_{\mathcal{G}}^2 |\nabla\phi|^2, \tag{13.3}$$

$$W_{\mathcal{G}} = \phi^2(1-\phi)^2, \tag{13.4}$$

$$B_{\mathcal{G}i} = n_i(c_s, c_c)s(\phi). \tag{13.5}$$

In the interfacial energy equation, $\epsilon_{\mathcal{G}}$ is a parameter that is a length scale proportional to the width of the diffuse interface. In the nutritional bias equation, $n_i(c_s, c_c)$ is a tilting function that accounts for the growth or decay of the slime mold biomass due to the presence or absence of nutrients within the domain Ω. The function $s(\phi)$ in the nutritional bias equation is a monotonically increasing function with lower and upper bounds. The shape of the overall bulk free energy landscape ($W_{\mathcal{G}}^* = W_{\mathcal{G}} + B_{\mathcal{G}}$) double-well is determined by the presence or absence of the total nutrient.

The model also considers the metabolism and nutrient dynamics for the slime mold species. It is assumed that the concentrated nutrient source, c_c, is the primary nutrient source utilized by the slime mold for growth during foraging. The nutrient source present in the substrate, n_s, is responsible for the front morphologies and branching behavior exhibited by the slime mold's network. The evolution of the substrate nutrient source is dependent on the substrate nutrient diffusivity (D_{c_s}), the substrate nutrient (c_s), and a sink term that defines how the slime mold metabolizes the substrate nutrient ($\Xi(\phi, c_s)$). Similarly, the evolution of the concentrated nutrient source is dependent on the concentrated nutrient diffusivity (D_{c_c}), the concentrated nutrient (c_c), and a source term ($\varpi(\phi, c_c)$) that includes that replenishment and consumption of the concentrated nutrient source.

With the free energy potential and the metabolism and nutrient dynamics, we formulate the governing equations for the multispecies growth dynamics to be

$$\frac{\partial\phi_i}{\partial t} = \lambda_{\mathcal{G}i}\left[\epsilon_{\mathcal{G}}^2\Delta\phi - \phi(4\phi^2 - 6\phi + 2) + 6(\phi - \phi^2)n_i(c_s, c_c)\right] \quad \text{in } \Omega \times [0, t_f], \tag{13.6}$$

$$\frac{\partial c_{si}}{\partial t} = \nabla \cdot (D_{c_s i}\nabla c_s) - \Xi_i(\phi, c_s) \quad \text{in } \Omega \times [0, t_f], \tag{13.7}$$

$$\frac{\partial c_{c_i}}{\partial t} = \nabla \cdot (D_{c_c i}\nabla c_c) + \varpi_i(\phi, c_c) \quad \text{in } \Omega \times [0, t_f]. \tag{13.8}$$

These governing equations are subjected to the following initial and boundary conditions:

$$\phi = \phi_D \quad \text{on } \partial\Omega_D \times [0, t_f], \tag{13.9}$$

$$c_s = c_{s_D} \quad \text{on } \partial\Omega_D \times [0, t_f], \tag{13.10}$$

$$c_c = c_{c_D} \quad \text{on } \partial\Omega_D \times [0, t_f], \tag{13.11}$$

$$\nabla c_s \cdot \mathbf{n} = q_{c_s} \quad \text{on } \partial\Omega_N \times [0, t_f], \tag{13.12}$$

$$\nabla c_c \cdot \mathbf{n} = q_{c_c} \quad \text{on } \partial\Omega_N \times [0, t_f], \tag{13.13}$$

$$\phi = \phi_0 \quad \text{in } \Omega(t = 0), \tag{13.14}$$

$$c_s = c_{s_0} \quad \text{in } \Omega(t = 0), \tag{13.15}$$

$$c_c = c_{c_0} \quad \text{in } \Omega(t = 0). \tag{13.16}$$

The Dirichlet and Neumann boundary conditions are denoted by $\partial\Omega_D$ and $\partial\Omega_N$, respectively. The boundaries representing the sidewalls of the Petri dish are considered to be impenetrable, therefore $\phi_D = 0$ for the simulations presented in this work. Additionally, to represent the slime mold growth being constrained to the Petri dish, $c_{s_D} = c_{c_D} = q_{c_s} = q_{c_s} = 0$ due to no-nutrient generating boundaries and no nutrient exchange with the surroundings. Table 13.1 summarizes the main parameters of the model and their corresponding values used in the simulations.

Within this framework, the slime mold's network dynamics are described using a Poisson equation that relates pressure differences across the network to the conductivities and lengths of the tubes (Tero et al. 2007). This equation captures how the tubes' conductivities change over time-based on the flux (flow rate) through them, affecting the pressure distribution across the network. For our project, the specific equations from the continuum modeling of slime mold dynamics will be adapted and utilized to simulate the behavior of the mold in our designed maze-like structure.

Table 13.1 Geometrical data and material properties

Property	Description	Value	Unit
ϕ	Fungal marker	0–1	–
c_s	Substrate nutrient	0–1	None
c_c	Concentrated nutrient	0–1	–
c^*	Nutrient threshold	0.4	–
l	Domain size (diameter or length), 2D	100	mm
L	Domain size (diameter × height), 3D	100 × 20	mm
$\kappa_1^{c_s}$	Consumption rate, substrate nutrient	0.5, 1.0	h^{-1}
$\kappa_2^{c_s}$	Mass creation coefficient	0.7, 0.8, 2.0	–
κ^{c_c}	Consumption coefficient, concentrated nutrient	0.05	none
$\lambda_{\mathcal{G}}$	Growth mobility	3.33×10^3	mm h
D_{c_s}	Substrate nutrient diffusivity	1.0	$mm^2\ h^{-1}$
$\epsilon_{\mathcal{G}}$	Growth regularization parameter	0.01	mm

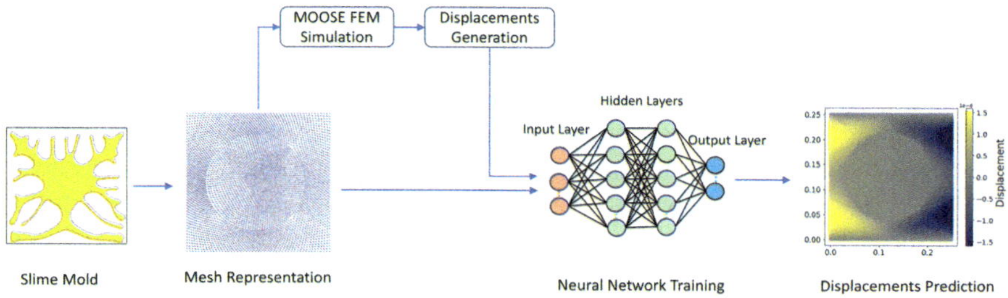

Fig. 13.2 Schematic of feedforward neural network and MOOSE FEM simulation

To generate data for training and evaluating machine learning models, we represent the embedded network within the engineered material as a 2D composite mesh. This representation is achieved using the open-source Gmsh package (Geuzaine and Remacle 2009). Within this mesh, distinct regions are demarcated to exhibit high and low conductivity, symbolizing the tubular veins and the surrounding plasmodium, respectively, shown in Fig. 13.2. The protoplasmic flux through these composite meshes is then simulated using the MOOSE Frameworks, An open-source, parallel finite element framework (Lindsay et al. 2022). By discretizing the poroelastic flow equations over the mesh geometry, we can solve for the resulting pressure and velocity fields. The data generated from these simulations serves as the foundation for training a deep neural network model using TensorFlow in Python (Abadi et al. 2016). This data becomes the input to the neural network, while the outputs are the simulated pressure and velocity fields. As the network undergoes training on a plethora of simulations, it becomes adept at swiftly predicting the physics of protoplasmic flux based on the given mesh topology. This approach allows for the rapid exploration and optimization of engineered materials with embedded networks, inspired by the adaptive behavior of slime mold transport networks. By leveraging the power of machine learning and continuum modeling, we can efficiently design and develop novel materials with enhanced properties and performance, opening up new possibilities in fields such as bioengineering, soft robotics, and advanced manufacturing.

13.3 Numerical Examples: Model Testing and Validation

In this work, we generated composite mesh structures representing embedded network morphologies inspired by the slime mold. An example structure is shown in Fig. 13.1, containing an inclusion region with distinct material properties mimicking the vein-like tubes.

The matrix and inclusion regions were assigned Young's moduli of 9 and 210 GPa and Poisson's ratios of 0.33 and 0.22, respectively. This composite mesh was discretized using 4,449 nodes for simulation. Uniaxial tensile displacement of 10^{-6} was applied in the y-direction. Protoplasmic flux through these composite meshes was simulated using the finite element module in the MOOSE framework. The poroelastic flow equations were solved to obtain pressure and velocity fields reflecting the composite transport physics. The inclusion region with higher conductivity exhibited greater pressure gradients and flux compared to the matrix, as expected.

The simulation data was used to train a fully connected deep neural network model in TensorFlow to predict pressure and velocity fields within engineered materials with embedded networks. The remarkable adaptation abilities of slime molds like *P. polycephalum* emerge from the intricate microstructural changes in their cytoplasm and tubular networks, serving as inspiration for the design of adaptive materials. Capturing this complex mechanical response is key to elucidating the biophysical mechanisms behind their emergent behavior and translating these principles to engineered systems. However, directly simulating the dynamics at the microstructural level over large iterative spaces is computationally prohibitive. To address this challenge, we leverage machine learning to develop surrogate models that accurately approximate the material's complex behavior under various conditions. We represent the microstructure as a composite mesh containing distinct domains, as inspired by biological cells. Physics-based simulations on these meshes generate datasets reflecting the rich nonlinear dynamics of internal protoplasmic flux and network remodeling within the material. A neural network model trained on this simulation data learns to rapidly predict the mechanical response and evolution of the encoded microstructures. This machine learning framework bypasses expensive simulations to enable extensive topology optimization driven by data-science approximations of the intricate multiscale phenomena. The integration of biophysics, composite material modeling, and machine learning provides both novel biological insights and tools to engineer smart, resilient systems.

For this work, the neural network was configured with the mesh node coordinates and material properties as input. Figure 13.2 shows the schematic of the neural network which consists of one input layer, two hidden layers, and an output layer, all composed of fully connected layers. We used ReLU as activation function, Adam as optimizer and mean squared error as the loss function for the training. To optimize the model's efficiency and accuracy, we conducted hyperparameter tuning, focusing on learning rate and the number of hidden layers. We explored learning rates ranging from 0.0001 to 0.01 and 1 to 5 hidden layers. Our results, as shown in Fig. 13.4a, indicate that a learning rate of 0.001 and 2 hidden layers yield the best balance between accuracy and computational cost, minimizing the mean squared error loss. Each hidden layer contains 16 neurons. The model takes the mesh structure nodes as input data for training and displacements as target data in order to predict the displacements of each node. We split the data into 80% for training and 20% for testing, allowing the model to learn from unseen data. The model was trained for 100 epochs with a batch size of 32. After training on numerous mesh simulations, the model achieved high accuracy in rapidly inferring the composite materials physics.

Figure 13.3 illustrates the strong agreement between the simulated and predicted pressure fields for a sample test case. To validate its predictive capability, the trained model was evaluated on a distinct composite mesh containing an inclusion of diameter 0.1. As shown in Fig. 13.3, the neural network accurately predicted the nodal displacements compared to those obtained from direct finite element simulation. The model's performance is further demonstrated by the loss curve in Fig. 13.3a, which shows a smooth decrease without signs of underfitting or overfitting. Also, our model achieved average mean absolute and mean squared errors of 7.78×10^{-8} and 3.95×10^{-14}, respectively, demonstrating reliable generalization, shown in Fig. 13.3b. R^2 scores for x and y displacement predictions, illustrated in Fig. 13.4b, c, are 0.989 and 0.991 respectively, indicating the model's high accuracy in predicting displacements within the engineered materials.

This machine learning framework establishes a foundation to leverage data-driven approximations of complex biomimetic material physics for accelerated discovery of optimal embedded network designs. Ongoing work is focused on encoding the meshes into a compact latent space to enable differentiable topology optimization through the trained neural network model. As an example and benchmark of the results that can be achieved, we present here the triangle of Steiner, a 3-point network optimization problem applied to materials with embedded networks. In this example, three nutrient sources are placed at the corners of an equilateral triangle. A slime mold culture is then placed at the center of the triangle. The results of this simulation can be seen in Fig. 13.5. In the simulation, the slime mold grows to the nutrient sources during the foraging phase but begins to reconfigure to optimize the transport network between the three nutrient sources. The final results of this simulation agree nicely with the solution to the Steiner tree problem.

13.4 Conclusions

This work demonstrates a machine learning framework to rapidly optimize biomimetic designs for engineered materials with embedded networks. By representing these materials as composite meshes and generating simulation data of internal flux, we train models to approximate complex biophysics. The trained networks then enable iterative topology improvements purely through fast inference. Encoding the meshes accelerates this optimization loop. The integrated methodology synergistically leverages composite mesh modeling, physics simulations, and machine learning to illuminate bio-inspired pathways for intelligent material adaptation. Our approach provides a powerful tool for designing advanced materials with optimized internal structures, inspired by the efficient networks observed in organisms like *P. polycephalum*. The high accuracy of our model in predicting nodal displacements and pressure fields demonstrates its potential for rapid prototyping and optimization of material designs without the need for extensive physical experimentation. Ongoing efforts involve applying the framework to additional transport problems and exploring generative models to directly propose optimized designs. Overall, this

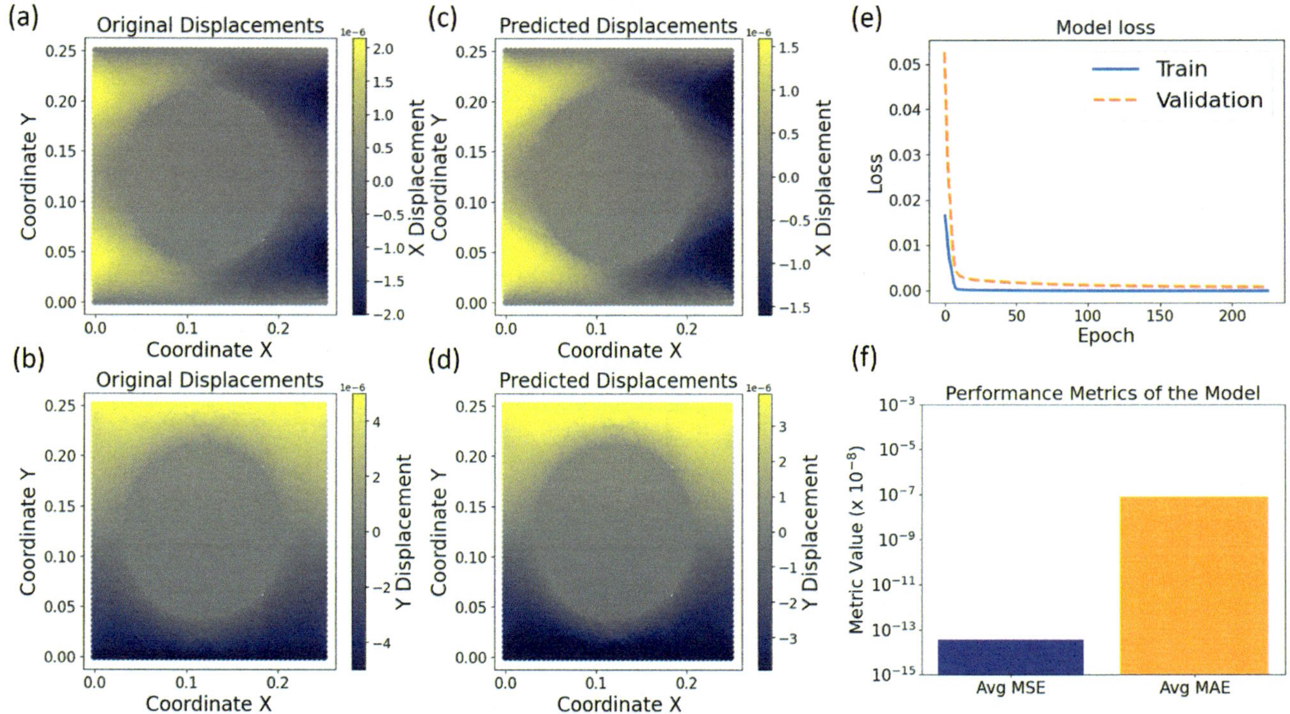

Fig. 13.3 Results of machine learning prediction model. **a–d** Original and predicted displacements of nodes. **e** Train versus validation model loss with epochs. **f** Average mean squared errors and average mean absolute errors of the prediction model

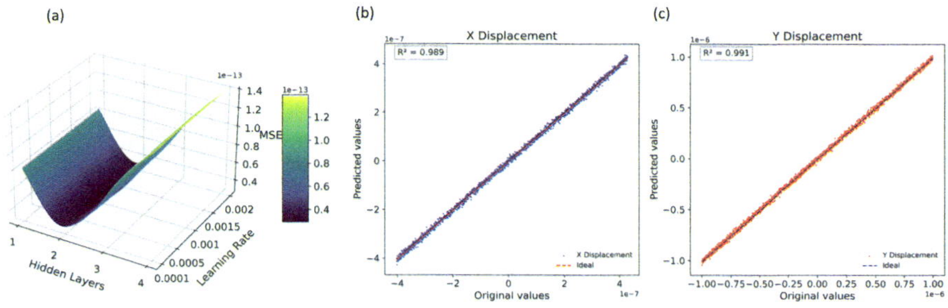

Fig. 13.4 **a** Hyperparamaters tuning of the surrogate model, **b** R^2 scores for predicted x displacement, **c** R^2 scores for predicted y displacement

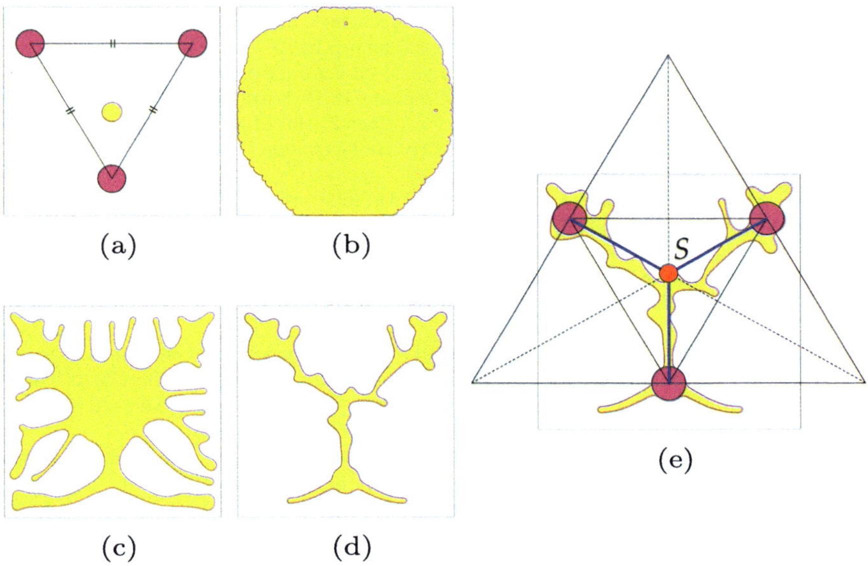

Fig. 13.5 Numerical results for a three-point optimization problem. **a** The nutrient sources, indicated by pink circles, are placed on the vertices of an equilateral triangle. The initial slime mold culture is placed at the center of the equilateral triangle. **b–d** Display the growth of the slime mold and the process of optimizing the branching. **e** Numerical results overlapped with the geometry of the Steiner point, marked by S and solid blue lines (Ghanbari et al. 2023b)

fusion of computational modeling and data-driven techniques provides both novel biophysical insights and practical tools to engineer smart, resilient materials with embedded networks. As we continue to refine this methodology, we expect it to play a crucial role in the design of next-generation materials for various applications in engineering and technology.

Acknowledgements This work has been funded by NSF grant 2339373. Any opinions, findings, and conclusions or recommendations expressed in this material are those of the authors and do not necessarily reflect the views of the National Science Foundation.

References

Abadi M, Barham P, Chen J, Chen Z, Davis A, Dean J, Devin M, Ghemawat S, Irving G, Isard M et al (2016) {TensorFlow}: a system for {Large-Scale} machine learning. In: 12th USENIX symposium on operating systems design and implementation (OSDI 16), pp 265–283

Adamatzky A (2012) Slime mold solves maze in one pass, assisted by gradient of chemo-attractants. TNB 11(2):131–134. https://doi.org/10.1109/TNB.2011.2181978

Alim K, Andrew N, Pringle A, Brenner MP (2017) Mechanism of signal propagation in Physarum polycephalum. PNAS 114(20):5136–5141. https://doi.org/10.1073/pnas.1618114114

Clauset A, Moore C, Newman ME (2008) Hierarchical structure and the prediction of missing links in networks. Nature 453(7191):98–101. https://doi.org/10.1038/nature06830

Gao Q, Niu X, Shao L, Zhou L, Lin Z, Sun A, Fu J, Chen Z, Hu J, Liu Y et al (2019) 3d printing of complex GelMA-based scaffolds with nanoclay. Biofabrication 11(3):035006. https://doi.org/10.1088/1758-5090/ab0cf6

Gaston D, Newman C, Hansen G, Lebrun-Grandié D (2009) MOOSE: a parallel computational framework for coupled systems of nonlinear equations. Nucl Eng Des 239(10):1768–1778. https://doi.org/10.1016/j.nucengdes.2009.05.021

Geuzaine C, Remacle J-F (2009) Gmsh: a 3-d finite element mesh generator with built-in pre-and post-processing facilities. Int J Numer Meth Eng 79(11):1309–1331. https://doi.org/10.1002/nme.2579

Ghanbari F, Costanzo F, Hughes DP, Peco C (2020) Phase-field modeling of constrained interactive fungal networks. J Mech Phys Solids 145:104160. https://doi.org/10.1016/j.jmps.2020.104160

Ghanbari F, Peco C (2023) Eulerian finite-strain elasticity with Phase-field and the reference map technique. Trans JSCES 27:8–13. https://doi.org/10.26207/w4p2-mw24

Ghanbari F, Sgarrella J, Peco C (2023a) Coding soft matter with bionetworks-inspired emergent principles. Bioinspir Biomim Biorepl XIII 12481:5–14. https://doi.org/10.1117/12.2657691

Ghanbari F, Sgarrella J, Peco C (2023b) Emergent dynamics in slime mold networks. J Mech Phys Solids 179:105387. https://doi.org/10.1016/j.jmps.2023.105387

Guével A, Meng Y, Peco C, Juanes R, Dolbow JE (2023) A Darcy-Cahn-Hilliard model of multiphase fluid-driven fracture. J Mech Phys Solids 181:105427. https://doi.org/10.1016/j.jmps.2023.105427

Jones J (2010) Characteristics of pattern formation and evolution in approximations of physarum transport networks. Artif Life 16(2):127–153. https://doi.org/10.1162/artl.2010.16.2.16202

Jones JD (2015) The emergence and dynamical evolution of complex transport networks from simple low-level behaviours. Int J Unconv Comput 6:125–144

Kamiya N (1950) The rate of the protoplasmic flow in the myxomycete plasmodium. I. Cytologia 15(3–4):183–193. https://doi.org/10.1508/cytologia.15.183

Karniadakis GE, Kevrekidis IG, Lu L, Perdikaris P, Wang S, Yang L (2021) Physics-informed machine learning. Nat Rev Phys 3(6):422–440. https://doi.org/10.1038/s42254-021-00314-5

Kolesky DB, Homan KA, Skylar-Scott MA, Lewis JA (2016) Three-dimensional bioprinting of thick vascularized tissues. PNAS 113(12):3179–3184. https://doi.org/10.1073/pnas.1521342113

Lindsay AD, Gaston DR, Permann CJ, Miller JM, Andrš D, Slaughter AE, Kong F, Hansel J, Carlsen RW, Icenhour C, Harbour L, Giudicelli GL, Stogner RH, German P, Badger J, Biswas S, Chapuis L, Green C, Hales J, Hu T, Jiang W, Jung YS, Matthews C, Miao Y, Novak A, Peterson JW, Prince ZM, Rovinelli A, Schunert S, Schwen D, Spencer BW, Veeraraghavan S, Recuero A, Yushu D,

Wang Y, Wilkins A, Wong C (2022) 2.0—MOOSE: enabling massively parallel multiphysics simulation. SoftwareX 20:101202. https://doi.org/10.1016/j.softx.2022.101202

Liu Y, Peco C, Dolbow J (2019) A fully coupled mixed finite element method for surfactants spreading on thin liquid films. Comput Methods Appl Mech Eng 345:429–453 (2019) https://doi.org/10.1016/j.cma.2018.10.045

Nakagaki T, Yamada H, Tóth Á (2000) Maze-solving by an amoeboid organism. Nature 407(6803):470–470. https://doi.org/10.1038/35035159

Patrick JF, Hart KR, Krull BP, Diesendruck CE, Moore JS, White SR, Sottos NR (2014) Continuous self-healing life cycle in vascularized structural composites. Adv Mater 26(25):4302–4308. https://doi.org/10.1002/adma.201400248

Peco C, Chen W, Liu Y, Bandi MM, Dolbow JE, Fried E (2017) Influence of surface tension in the surfactant-driven fracture of closely-packed particulate monolayers. Soft Matter 13(35):5832–5841. https://doi.org/10.1039/c7sm01245d; arXiv:1706.08729

Peco C, Liu Y, Rhea C, Dolbow JE (2019) Models and simulations of surfactant-driven fracture in particle rafts. IJSE 156–157:194–209. https://doi.org/10.1016/j.ijsolstr.2018.08.014

Peco C, Rosolen A, Arroyo M (2013) An adaptive meshfree method for phase-field models of biomembranes. Part II: A Lagrangian approach for membranes in viscous fluids. J Comput Phys 249:320–336. https://doi.org/10.1016/j.jcp.2013.04.038

Raissi M, Perdikaris P, Karniadakis GE (2019) Physics-informed neural networks: a deep learning framework for solving forward and inverse problems involving nonlinear partial differential equations. J Comput Phys 378:686–707. https://doi.org/10.1016/j.jcp.2018.10.045

Rosolen A, Peco C, Arroyo M (2013) An adaptive meshfree method for phase-field models of biomembranes. Part I: approximation with maximum-entropy basis functions. J Comput Phys 249:303–319. https://doi.org/10.1016/j.jcp.2013.04.046

Sgarrella J, Ghanbari F, Peco C (2023) I-STL2MOOSE: from STL data to integrated volumetrical meshes for MOOSE. SoftwareX 21:101273. https://doi.org/10.1016/j.softx.2022.101273

Shimada T, Nishiguchi K, Peco C, Okazawa S, Tsubokura M (2021) Eulerian formulation using Lagrangian marker particles with reference map technique for fluid-structure interaction problem. In: 9th Edition of the international conference on computational methods for coupled problems in science and engineering, pp 1–7

Shimada T, Nishiguchi K, Peco C, Okazawa S, Tsubokura M (2022) Eulerian unified formulation for fluid-structure interaction problems using marker particles with Reference map. https://doi.org/10.11421/jsces.2022.20220002

Spencer BW, Jiang W, Dolbow JE, Peco C (2016) Pellet cladding mechanical interaction modeling using the extended finite element method. In: Top fuel 2016: LWR fuels with enhanced safety and performance, pp 929–938

Studart AR (2016) Additive manufacturing of biologically-inspired materials. Chem Soc Rev 45(2):359–376. https://doi.org/10.1039/C5CS00836K

Tero A, Kobayashi R, Nakagaki T (2007) A mathematical model for adaptive transport network in path finding by true slime mold. J Theor Biol 244(4):553–564. https://doi.org/10.1016/j.jtbi.2006.07.015

Tero A, Takagi S, Saigusa T, Ito K, Bebber DP, Fricker MD, Yumiki K, Kobayashi R, Nakagaki T (2010) Rules for biologically inspired adaptive network design. Science 327(5964):439–442. https://doi.org/10.1126/science.1177894

Wegst UG, Bai H, Saiz E, Tomsia AP, Ritchie RO (2015) Bioinspired structural materials. Nat Mater 14(1):23–36. https://doi.org/10.1038/nmat4089

Whiting JG, Jones J, Bull L, Levin M, Adamatzky A (2016) Towards a physarum learning chip. Sci Rep 6(1):19948. https://doi.org/10.1038/srep19948(2016)

Wijerathne B, Liao T, Ostrikov K, Sun Z (2022) Bioinspired robust mechanical properties for advanced materials. Small Struct 3(9):2100228. https://doi.org/10.1002/sstr.202100228

Chapter 14
Experimental Investigation of the Overall Efficiency of a Centrifugal Blood Pump Using Glycerin Solution and Xanthan Gum Solutions as Blood Analog Fluid

Ahmet Onder, Omer Incebay, and Rafet Yapici

Abstract The most critical issues to consider in the design of the blood pumps are the hydraulic performance and hemocompatibility of the pump. It is important to perform hydraulic performance tests during the blood pump development process before clinical applications. Using substitute fluids instead of blood as the working fluid during the testing of blood pumps provides convenience for the tests. In this study, solid models of all components of a centrifugal blood pump were designed, and then the pump volute and impeller were manufactured. The hydraulic performance of the manufactured centrifugal blood pump is experimentally tested using an aqueous glycerin solution and xanthan gum solution as working fluids. The overall efficiency of the pump at four rotation speeds which are 2700, 3000, 3300, and 3600 rpm was determined for both fluids. Experiments were carried out in the flow rate range of 1–8 L/min. The experimental overall efficiencies of the two working fluids were compared at all rotational speeds and throughout the flow rate range. As a result of the study, there were approximately 2% differences in overall efficiency for the two working fluids at the same rotational speeds and flow rates.

Keywords Blood analog fluid · Centrifugal blood pump · Non-Newtonian fluid · Overall efficiency

---

A. Onder (✉)
Technical Sciences Vocational School, Mechanical and Metal Technologies Department, Konya Technical University, Konya, Turkey
e-mail: aonder@ktun.edu.tr

O. Incebay · R. Yapici
Faculty of Engineering and Natural Science, Mechanical Engineering Department, Konya Technical University, Konya, Turkey
e-mail: oincebay@ktun.edu.tr

R. Yapici
e-mail: ryapici@ktun.edu.tr

A. Öchsner and H. Altenbach (eds.), *Engineering Design Applications VII*, Advanced Structured Materials 230, https://doi.org/10.1007/978-3-031-84346-4_14

14.1 Introduction

At the present time, centrifugal pumps are widely used for fluid transfer in many areas such as heavy industry, agriculture, mining, chemical industry, energy plants, water supply plants, and biomedical devices (Thin et al. 2008; Qiu et al. 2019; Zhou et al. 2021). In the medical area, the use of centrifugal and axial type roto-dynamic blood pumps achieved widespread acceptance in the treatment of patients with heart failure (Reul and Akdis 2000; Antaki et al. 2008; Hijikata et al. 2008). Centrifugal type pumps are widely preferred in continuous-flow rotary blood pump applications because of higher hydraulic efficiency, flatter H-Q curves, and wider peak efficiency zones (Moazami et al. 2013; Princer et al. 2022; Onder et al. 2022). In the development process of centrifugal blood pumps, the main considerations are to provide good hematological compatibility as well as adequate hydraulic performance (Behbahani et al. 2009). The potential of centrifugal blood pumps to meet the intended performance at the pre-design should be tested in the laboratory before clinical applications.

Even though whole blood usage as the working fluid in hydraulic tests for centrifugal blood pumps is much more realistic, it is necessary to use a blood substitute fluid in laboratory tests due to the hassle associated with the use of in vitro blood (Pinho et al. 2019). Moreover, ethical concerns and animal welfare usage of blood substitute fluid reduce the high volume of blood consumption significantly (Brookshier and Tarbell 1993; Walker et al. 2014). In most of the studies, blood is assumed to be Newtonian and water or aqueous glycerin solution (GS) is used as a substitute fluid (Onder et al. 2023; Yu et al. 2021; Okamoto et al. 2023). However, many researchers have used aqueous xanthan gum solution (XGS) as a non-Newtonian substitute fluid, to reach a better match with the rheological properties of blood (Miklosovic 1999; Zhang et al. 2007).

In this study, a centrifugal blood pump is designed and manufactured, and then the overall efficiency of the pump is investigated. The overall efficiency of the pump was determined in detail using both Newtonian GS and non-Newtonian XGS over a wide range of pump flow rates at different rotation speeds.

14.2 Pump Configuration

The pump model is a centrifugal blood pump designed to meet the nominal blood flow needs of the human body which is a pressure head of 100-mmHg at a flow rate of 5 L/min (Behbahani et al. 2009). Conventional design techniques were used in determining the dimensions of the centrifugal blood pump (Gülich 2008). The most important component in a pump is the impeller, in which energy is transferred to the fluid (Djerroud et al. 2011). The outer diameter of the impeller, whose three-dimensional design is shown in Fig. 14.1, is 37 mm, the number of blades is 5 and the blade thickness is 0.58 mm. Due to the complex three-dimensional structure and

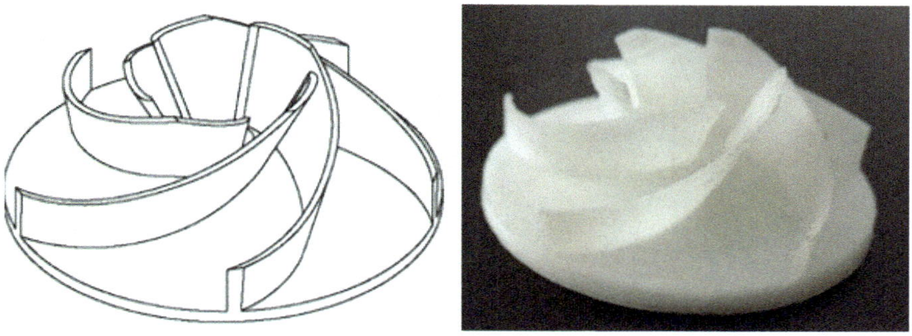

Fig. 14.1 Left: Three-dimensional view of the impeller Right: İmpeller produced with a 3D printer

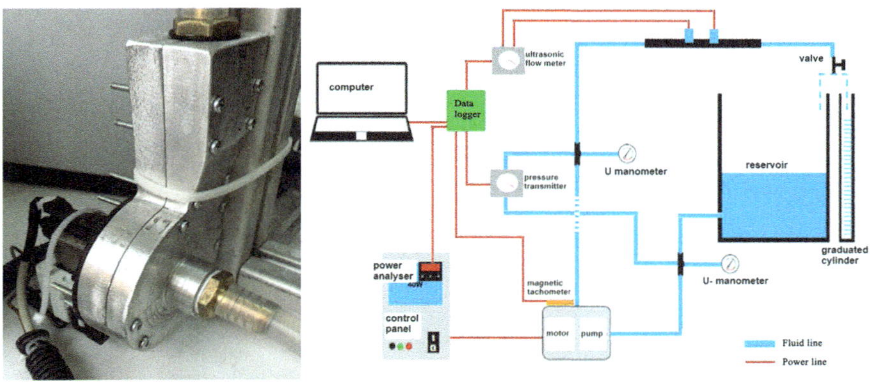

Fig. 14.2 Left: an assembled version of the pump. Right: experimental setup of the pump

narrow impeller side wall gaps, the geometry of the impeller of the designed model pump was produced with a 3D printer.

The volute geometry suitable for the manufactured impeller was designed in two parts and manufactured of aluminum alloy. An O-ring is used to ensure sealing between the lower and upper body. Then, the impeller was placed inside the volute and the volute was mounted on the motor as seen in Fig. 14.2.

14.3 Experimental Setup

The experimental setup created to determine the overall efficiency of the centrifugal blood pump is given in Fig. 14.2. Transparent silicone hoses 10 mm in diameter are used for piping of the experimental setup. A stainless-steel reservoir with a volume of 3.5 L was used to provide the required fluid to the pump.

Table 14.1 Specifications of the measuring devices used in the experimental setup

Measurement	Device company	Model	Accuracy	Validation
Pressure head	Mesens	MPS550	0.5%	Mercury U manometer
Flow rate	VFA	QT (ultrasonic flow meter)	1%	Graduated cylinder
Rotational speed	Melexis	US5881 (magnetic sensor)	0.5%	Digital photo tachometer (Lutron DT2236)
Power consumption	KAEL	Mutiser-XX-TFT	0.5%	–

In the experiments, pressure head, flow rate, rotational speed, and motor power consumption were measured to calculate the overall efficiency. The device company, model, accuracy specifications, and validation methods of the measuring instruments used in the experimental setup are given in Table 14.1.

As working fluids non-Newtonian XGS of 600 ppm and Newtonian 40% GS which has a density of 1050 kg/m^3 and a viscosity of 0.0035 Pa.s are used. The overall efficiency of the pump is defined as the ratio of the hydraulic power P_h to motor power P_m:

$$\eta = \frac{P_h}{P_m} \tag{14.1}$$

where $P_h = HQ$ and H is the pressure head, Q is the flow rate.

14.4 Results and Discussion

The pump's overall efficiency was determined by measuring pressure heads at various flow rates in the 1–8 L/min range at 2700, 3000, 3300, and 3600 rpm rotational speeds. The overall efficiency calculated from the measurements from the experiments carried out at four different rotational speeds using XGS as the working fluid is given in Fig. 14.3. As seen in this figure, the overall efficiency varies between 9–10% at different rotational speeds at our design flow rate of 5 L/min. The highest overall efficiency is approximately 11.2% at a rotational speed of 3600 rpm.

In Fig. 14.4, the overall efficiency calculated from the measurements from the experiments carried out at four different rotational speeds using GS as the working fluid is given. As can be clearly observed, the overall efficiency varies between 10–11% at different rotational speeds at our design flow rate of 5 L/min. The highest overall efficiency is approximately 11.6% at a rotational speed of 3600 rpm.

A comparison of data for the two working fluids shows that there were approximately 1–2% differences in overall efficiency for the two working fluids at the same rotational speeds and flow rates. GS shows approximately 0.4% higher maximum

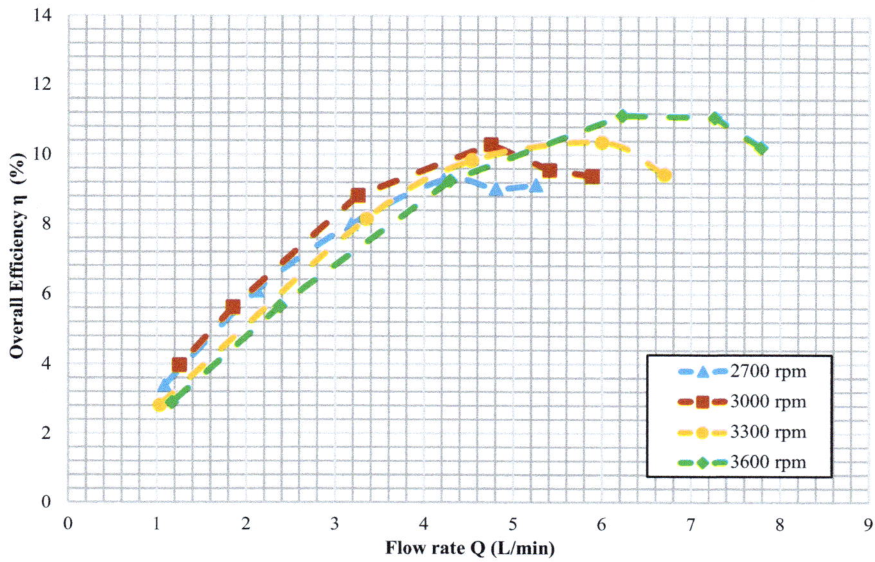

Fig. 14.3 The overall efficiency curves of XGS solutions at 2700, 3000, 3300, and 3600 rpm rotational speeds

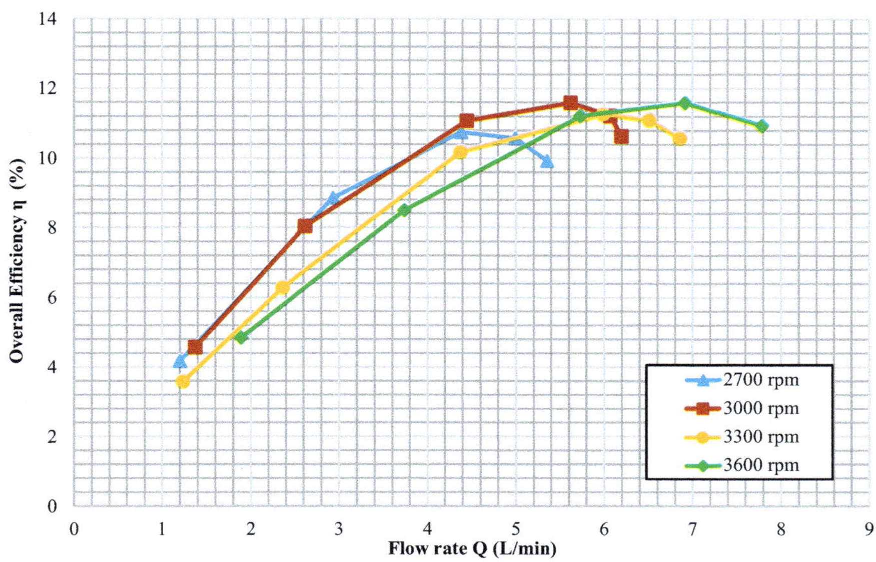

Fig. 14.4 The overall efficiency curves of GS solutions at 2700, 3000, 3300, and 3600 rpm rotational speeds

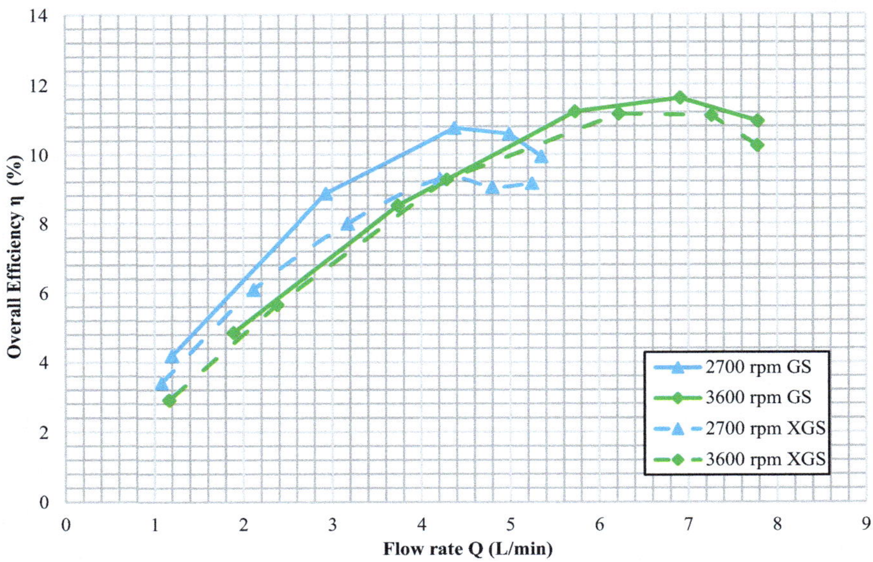

Fig. 14.5 The overall efficiency curves of GS and XGS solutions at 2700 and 3600 rpm rotational speeds

overall efficiency than XGS at 3600 rpm. This result can be attributed to XGS having a higher average apparent viscosity than GS (Onder et al. 2022).

Figure 14.5 shows that the difference in overall efficiency in experiments conducted with GS and XGS solutions at 2700 rpm rotational speed is quite high. On the contrary, the overall efficiencies in experiments conducted with GS and XGS at 3600 rpm rotational speed are close to each other. This can be attributed to the fact that the drag-reduction effect becomes more effective as the Reynolds number increases and that the drag-reduction effect plays an important role in the use of xanthan gum at high rotational speeds (Hu et al. 2012; Zhang et al. 2020).

In many applications, 80% to 90% pump efficiencies are achievable. However, such peak efficiency values are unachievable for any rotary blood pump (Matsuda 2000). Okamoto et al. (2019) used an axial pump with outflow of 5 L/min against a pump head of 100 mm Hg at a rotational speed of 9600 rpm in vitro tests using glycerin solution as the working fluid and stated that the hydrodynamic efficiency of the pump as 7–8%. Yamane et al. (2007) tested the performance of a hydrodynamic levitation centrifugal blood pump using glycerin solution as a working fluid. These researchers achieved an overall efficiency of about 14% at 3000 rpm rotational speed at 5 L/min flow rate. In this study, overall efficiency varies between 9–11% at different rotational speeds at our design parameter of 5 L/min flow rate for both working fluids. The reason for the low overall efficiency of the pump can be attributed to the use of a low-efficiency single-phase AC motor in the pump system. It can also be thought that another reason for the low efficiency of the system is that the total energy use is very low and the ratio of mechanical losses in this total energy is relatively high.

14.5 Conclusion

In this study, the use of Newtonian GS and non-Newtonian XGS as blood analog fluid in vitro performance tests concerning the overall efficiency of a centrifugal blood pump was compared. The experiments were carried out at four different rotational speeds in the flow rate range of 1–8 L/min and their overall efficiencies were calculated. The overall efficiency curves showed that the difference between the overall efficiencies of the two working fluids at the lowest rotational speed was high. On the contrary, the overall efficiencies for the two working fluids at the highest rotational speed were very close. This situation can be attributed to the drag-reducing effect of the use of xanthan gum in the solution. The maximum efficiency obtained from the experiments using GS as the working fluid was higher than the maximum efficiency obtained from the experiments using XGS as the working fluid. Data comparison shows that using only Newtonian fluid in vitro performance tests of blood pumps will not fully mimic the rheological properties of blood. It was observed that the effects of using non-Newtonian fluids should not be ignored.

Acknowledgements The authors would like to thank Konya Technical University Scientific Research Projects Coordination Unit for financial support under project Grant No. 18101019.

References

Antaki J, Diao C, Shu F, Wu J, Zhao R, Kameneva M (2008) Microhaemodynamics within the blade tip clearance of a centrifugal Turbodynamic blood pump. Proc Inst Mech Eng [h] 222:573–581

Behbahani M, Behr M, Hormes M, Steinseifer U, Arora D, Coronado O, Pasquali M (2009) A review of computational fluid dynamics analysis of blood pumps. Eur J Appl Math 20:363–397

Brookshier K, Tarbell J (1993) Evaluation of a transparent blood analog fluid: aqueous xanthan gum/glycerin. Biorheol 30:107–116

Djerroud M, Dituba Ngoma G, Ghie W (2011) Numerical identification of key design parameters enhancing the centrifugal pump performance: impeller, impeller-volute, and impeller-diffuser. Int Sch Res Notices 2011:794341

Gülich JF (2008) Operation of centrifugal pumps. Springer, Berlin

Hijikata W, Shinshi T, Shimokohbe A, Sobajima H, Takatani S (2008) Disposable MagLev centrifugal blood pump utilizing cone-shaped impeller. In IEEE/ASME international conference on advanced intelligent mechatronics. IEEE, pp 570–575

Hu QH, Li JY, Zhang MY, Zhu XR (2012) An experimental study of Newtonian and non-Newtonian flow dynamics in an axial blood pump model. Artif Organs 36:429–433

Matsuda H (ed) (2000) Rotary blood pumps. Springer, Tokya

Miklosovic DS (1999) An experimental evaluation of the non-Newtonian scaling effects in a rotodynamic left ventricular assist device. Dissertation, The Ohio State University

Moazami N, Fukamachi K, Kobayashi M, Smedira NG, Hoercher KJ, Massiello A, Lee S, Horvath DJ, Starling RC (2013) Axial and centrifugal continuous-flow rotary pumps: a translation from pump mechanics to clinical practice. J Heart Lung Transplant 32:1–11
Okamoto E, Yano T, Inoue Y, Shiraishi Y, Yambe T, Mitamura Y (2019) Development of rear-impeller axial flow blood pump for realization of axial flow blood pump installed at aortic valve position. ArtIf Organs 43:828–833
Okamoto E, Yano T, Sekine K, Inoue Y, Shiraishi Y, Yambe T, Mitamura Y (2023) Development and initial performance of a miniature axial flow blood pump using magnetic fluid shaft seal. J Artif Organs 26:12–16
Onder A, Yapici R, Incebay O (2022) An experimental performance comparison of Newtonian and non-Newtonian fluids on a centrifugal blood pump. Proc Inst Mech Eng [h] 236:399–405
Onder A, Guzel MH, Incebay O, Sen MA, Yapici R, Kalyoncu M (2023) Fuzzy logic-based modeling of a centrifugal blood pump performance via experimental data of Newtonian and non-Newtonian fluids. J Mech Med Biol 23:2350016
Pinho D, Muñoz-Sánchez BN, Anes CF, Vega E, Lima R (2019) Flexible PDMS microparticles to mimic RBCs in blood particulate analogue fluids. Mech Res Commun 100:103399
Princer K (2022) Pump design and mechanics: a primer for ventricular assist device (VAD) clinicians. In: Scott S, Blood P (eds) A guide to mechanical circulatory support. Springer, Switzerland, pp 13–28
Qiu G, Huang S, Gu Y (2019) Experimental investigation and multi-conditions identification method of centrifugal pump using Fisher discriminant ratio and support vector machine. Adv Mech Eng 11:1687814019878041
Reul HM, Akdis M (2000) Blood pumps for circulatory support. Perfusion 15:295–311
Thin KC, Khaing MM, Aye KM (2008) Design and performance analysis of centrifugal pump. World Acad Sci, Eng Technol 46:422–429
Walker AM, Johnston CR, Rival DE (2014) On the characterization of a non-Newtonian blood analog and its response to pulsatile flow downstream of a simplified stenosis. Ann Biomed Eng 42:97–109
Yamane T, Maruyama O, Nishida M, Kosaka R, Sugiyama D, Miyamoto Y, Kawamura H, Kato T, Sano T, Okubo T (2007) Hemocompatibility of a hydrodynamic levitation centrifugal blood pump. J Artif Organs 10:71–76
Yu Z, Tan J, Wang S, Guo B (2021) Multiple parameters and target optimization of splitter blades for axial spiral blade blood pump using computational fluid mechanics, neural networks, and particle image velocimetry experiment. Sci Prog 104:00368504211039363
Zhang G, Zhang M, Zhu X (2007) Experimental investigation of rheological properties of blood and Xanthan gum solutions as a blood analog fluid. J Exp Fluid Mechanics 21:41–45
Zhang Y, Zhou F, Li J, Kang J, Zhang Q (2020) Application and research of new energy-efficiency technology for liquid ring vacuum pump based on turbulent drag reduction theory. Vacuum 172:109076
Zhou M, Qi Z, Xia Z, Li Y, Ling W, Yang J, Yang Z, Pei J, Wu D, Huo W (2021) Miniaturized soft centrifugal pumps with magnetic levitation for fluid handling. Sci Adv 7:eabi7203

Chapter 15
Analysis of Hepatic Steatosis Using Neural Networks

Verónica Guzmán-Mercado, Juan Alfonso Beltrán-Fernández, Enrique Granados-Sandoval, Alejandro Tonatiu Velázquez-Sánchez, Dra. María Rocío Thompson-Bonilla, Maricela García-Rojas, Cinthya Patricia Camacho-Galván, Ulises Vega-Acevedo, Erik Omar Alvarado-Alcántara, Karen Pamela Vázquez-Thierry, and Kenia Madeleyn Ortiz-Morales

Abstract Clinical cases of hepatic steatosis have increased dramatically worldwide. Contributing factors include obesity, insulin resistance, poor diet, and sedentary lifestyle, among others. In patients with low or no alcohol consumption, hepatic steatosis can be fatal, as most cases are asymptomatic during the treatable phase of the disease, making effective treatment challenging once detected. This study provides

V. Guzmán-Mercado · J. A. Beltrán-Fernández (✉) · A. T. Velázquez-Sánchez · E. O. Alvarado-Alcántara · K. P. Vázquez-Thierry · K. M. Ortiz-Morales
Sección de Estudios de Posgrado e Investigación, Escuela Superior de Ingeniería Mecánica y Eléctrica, Instituto Politécnico Nacional, Unidad Zacatenco, "Unidad Profesional Adolfo López Mateos", Ciudad de México, Mexico
e-mail: jbeltran@ipn.mx

V. Guzmán-Mercado
e-mail: vero_guzmanm94@hotmail.com

A. T. Velázquez-Sánchez
e-mail: atvelazquezs@ipn.m

E. O. Alvarado-Alcántara
e-mail: eomar_1@yahoo.com.mx

K. P. Vázquez-Thierry
e-mail: kappa.thierry@gmail.com

K. M. Ortiz-Morales
e-mail: kmadeleeyn@gmail.com

E. Granados-Sandoval · Dra. M. R. Thompson-Bonilla · M. García-Rojas · C. P. Camacho-Galván · U. Vega-Acevedo
Hospital Regional "1o de Octubre" del ISSSTE, Av. Instituto Politécnico Nacional 1669, Revolución IMSS, Gustavo A. Madero, Ciudad de México, CDMX, Mexico
e-mail: henryluz18@yahoo.com.mx

Dra. M. R. Thompson-Bonilla
e-mail: rociothompson@yahoo.com.mx

A. Öchsner and H. Altenbach (eds.), *Engineering Design Applications VII*, Advanced Structured Materials 230, https://doi.org/10.1007/978-3-031-84346-4_15

a tool for physicians to detect liver fat and offer timely treatment. Additionally, it aims to establish a database to gather Mexican statistics on the condition.

Keywords Steatosis · Hepatic · Liver · Neural networks · Computed tomography

15.1 Introduction

Non-alcoholic fatty liver disease (NAFLD) is a metabolic disorder characterized by the accumulation of fat in the liver of individuals with low or no alcohol consumption. It is the most common liver disease worldwide (Sayiner et al. 2016). It is estimated that between 20 and 30% of the global population suffers from this disorder (Méndez Sánchez et al. 2010). In Mexico, over 30% of the population has fatty liver issues (Mexicana and de Gastroenterología. (n.d.). 2023), with 17.05% being asymptomatic (Méndez Sánchez et al. 2010). According to a recent study to determine the prevalence of fatty liver in the Mexican population, 585 volunteers were registered, showing a prevalence of 41.3% (Bernal-Reyes et al. 2023).

NAFLD can manifest due to various conditions, with acquired metabolic conditions being the primary cause, resulting from sedentary behavior and/or poor diet. Among these, obesity is particularly significant, as studies have shown a correlation between body mass index and hepatic steatosis.

The treatment of fatty liver varies depending on the disease stage, potentially including pharmacological or surgical options. The rising prevalence of NAFLD has led to numerous studies aimed at improving its treatment and achieving a more accurate diagnosis.

Over time, research aimed at diagnosing fatty liver without invasive methods has advanced. Currently, diagnostic and treatment guidelines have been developed based on analytical and clinical parameters (Teixidó 2023).

Artificial neural networks are a computational technology that simulates the neural behavior of human neurons (Barraza et al. n d). In medicine, this technology enables the analysis of medical images (such as magnetic resonance imaging, nuclear medicine, and radiology), the diagnosis of heart disease using sound recordings, and the diagnosis of eye disease. For the network to perform these functions, it must be trained to learn to classify patterns in medical images (Delgado-Cortés et al. 2018).

In Mexico, there is no national registry for liver diseases. This study aims to develop a system of artificial neural networks capable of identifying the level of

M. García-Rojas
e-mail: dra.maricelagaroj2510@gmail.com

C. P. Camacho-Galván
e-mail: cinthyacamgal05@gmail.com

U. Vega-Acevedo
e-mail: druvega92@gmail.com

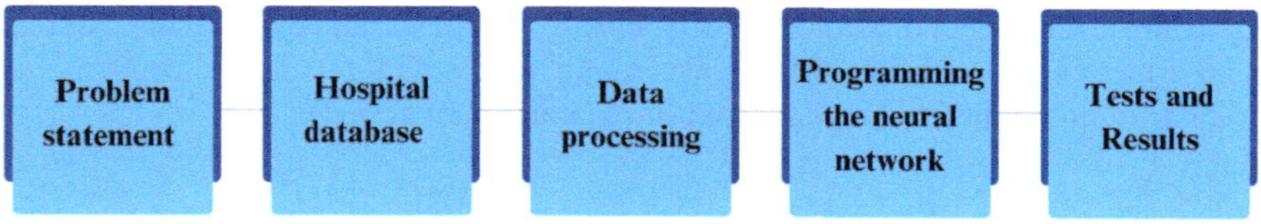

Fig. 15.1 Methodology

hepatic steatosis in patients using segmentation through grayscale images from MRI and ultrasound scans. The result will be a model that records the characteristics of each stage of NAFLD.

15.2 Materials and Methods

The development process of the artificial neural network system for detecting and classifying hepatic steatosis is outlined in this section. The process includes a description of the methodology for data acquisition, key concepts relevant to the study, and the procedures followed for data analysis. Ensuring the accuracy and reliability of the neural network in diagnosing hepatic steatosis from imaging data is a primary focus of this development phase.

15.3 Methodology

The methodology for developing and implementing the neural network for hepatic steatosis detection is detailed in this section, see Fig. 15.1. The preparation and preprocessing of imaging data, the design and training of the neural network, and the evaluation criteria for performance are described. Effective training of the neural network to recognize patterns indicative of hepatic steatosis and to provide accurate diagnostic support is ensured through this methodology.

15.4 Key Concepts

- **Live**: Defined as a crucial organ in the body's defense, the liver's parenchymal and non-parenchymal cells help combat infections. Additionally, it is the largest visceral organ (Otero and Sierra 2003).
- **Hepatic Steatosis**: Established as the accumulation or infiltration of fat in the liver, with a threshold of 5% fat required for diagnosis as fatty liver. Some patients diagnosed with this condition may not present any symptoms, potentially leading

to the development of steatohepatitis, fibrosis, cirrhosis, and even cancer. NAFLD is a metabolic syndrome that occurs in individuals with no alcohol consumption or with moderate intake (<20 g per day) (Barraza et al. n d).

15.5 Data Analysis

15.5.1 Database Acquisition

The Regional Hospital "1st of October" committed to providing studies conducted on various patients to implement the database for the neural network and ensure proper training. The studies provided were non-contrast computed tomography scans, as these are considered ideal by specialists for detecting hepatic steatosis at its initial stage.

To date, approximately 272 clinical cases have been received, with more cases in preparation to provide a solid foundation for the project.

15.5.2 Data Cleaning

Due to the hospital's collaboration with various research teams, some of the clinical cases received are not suitable for this research, necessitating proper data cleaning.

The initial step involved compiling the data, including the patient's name (which will remain confidential), the patient's age, the usefulness of the study for the project, the reason for exclusion if applicable, whether the patient suffers from hepatic steatosis, and the number of CT slices. An example of this can be seen in Fig. 15.2.

In the data cleaning stage, information from the columns Name, Useful, and Number of Scans was entered. A total of 100 cases have been cleaned from the received cases, representing 37% of the total cases (see Fig. 15.3).

In accordance to date, only three reasons for discarding cases have been identified, as illustrated in Fig. 15.4.

15.5.3 Data Processing

For the data processing stage, the programs RadiAnt® and 3DSlicer® were utilized. RadiAnt® was used to determine the Hounsfield units (HU), while 3DSlicer® was employed for liver segmentation. Three files were created for each case with the following purposes: the first file serves as a reference image to enable the neural network to identify the image it will analyze (Fig. 15.5); the second file is a segmented image of the liver, which allows the network to pinpoint the liver region in the

NAME	AGE	USEFUL	MOTIVE	NASH	CUTS
J.L.	62	NO	ANOTHER STUDY	NOT DISPLAYED	441
J.V.	65	YES	NA	NO	294
B.E.	49	YES	NA	YES	317
E.	66	YES	NA	NO	283
M.A.	64	YES	NA	NO	320
S.	73	YES	NA	NO	420
L.	57	YES	NA	NO	678
E.	12	YES	NA	NO	277
G.	53	YES	NA	NO	511
B.	38	YES	NA	YES	435
J.E.	70	YES	NA	NO	483
M.	58	YES	NA	NO	299
E.	63	YES	NA	NO	451
N.	52	YES	NA	NO	582
G.R.	79	YES	ANOTHER STUDY	NOT DISPLAYED	410
O.	54	YES	NA	YES	536
M.G.	69	YES	NA	NO	461
J.R.	70	YES	NA	NO	455
R.	60	YES	CONTRAST	NOT DISPLAYED	508
J.	10	YES	NA	NO	102

Fig. 15.2 Example of data consolidation

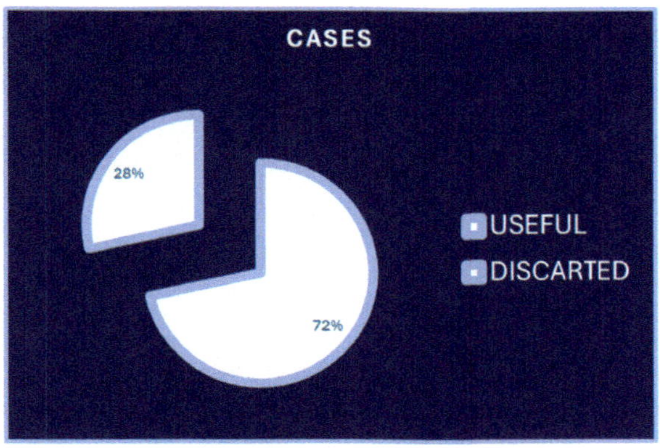

Fig. 15.3 Data records

reference image (Fig. 15.6); and the third file indicates, through HU values, whether the liver is healthy or has developed hepatic steatosis (Fig. 15.7).

15.6 Discussion of Results

The analysis of the collected data has revealed several key findings that contribute to the understanding of hepatic steatosis and its detection using artificial neural networks. It was observed that a substantial portion of the initial data set, specifically 37%, was excluded during the data cleansing phase. This exclusion rate highlights the

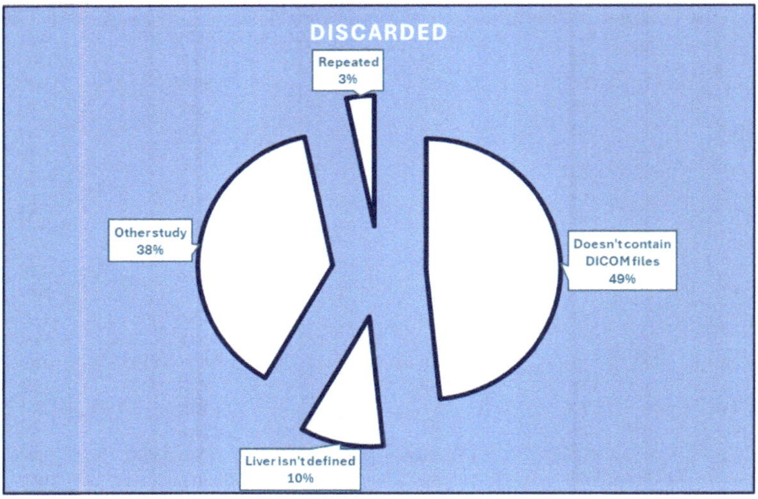

Fig. 15.4 Reasons for discarding cases

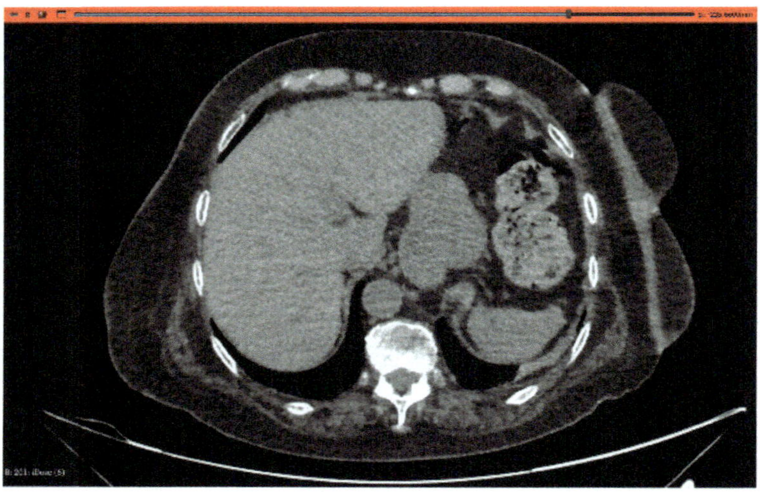

Fig. 15.5 Reference image

importance of stringent data verification processes to ensure the quality and relevance of the data used for training the neural network. The reasons for data exclusion, as shown in Fig. 15.4, were predominantly related to incomplete or non-diagnostic quality of the imaging studies.

In the data processing stage, the utilization of RadiAnt® and 3DSlicer® software proved effective in creating accurate reference images and segmenting liver tissues.

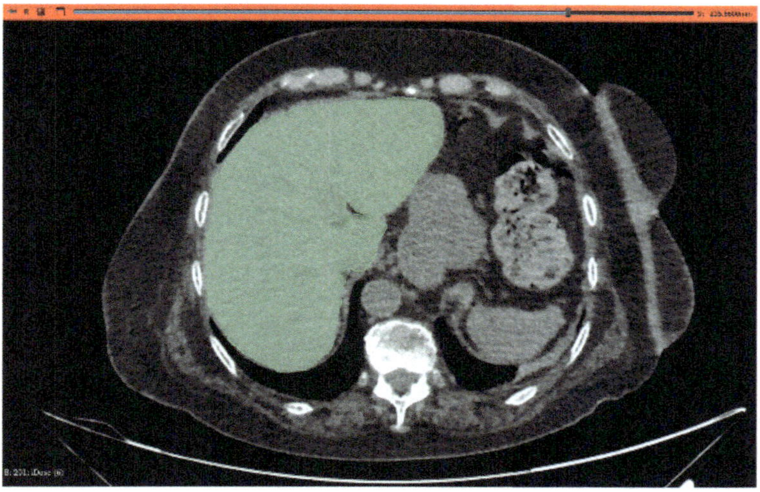

Fig. 15.6 Segmented liver

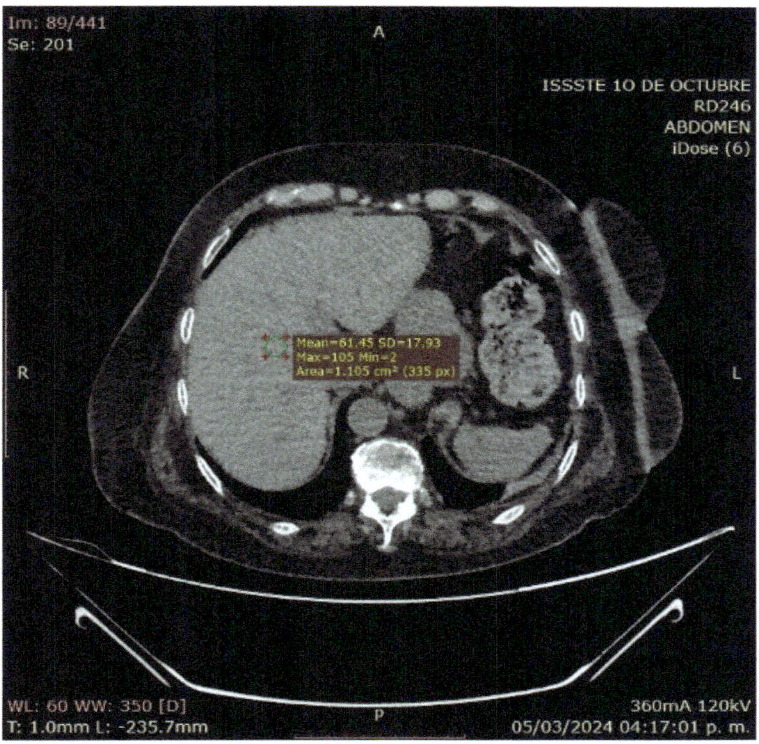

Fig. 15.7 Hounsfield units in the liver

The creation of three distinct files for each case—reference images, segmented liver images, and Hounsfield unit (HU) annotations—facilitated a comprehensive approach to training the neural network. The reference images provided a baseline for the neural network to identify liver regions, while the segmented images allowed for precise localization of hepatic tissue. The HU annotations were crucial in distinguishing between healthy liver tissue and areas affected by steatosis.

The segmentation process, illustrated in Figs. 15.5 and 15.6, was essential for delineating the liver from surrounding tissues and for training the neural network to focus on specific areas of interest. The ability to accurately segment the liver was directly correlated with the network's performance in identifying hepatic steatosis. The detailed HU analysis, depicted in Fig. 15.7, further supported the classification of liver conditions by providing quantitative measures of fat accumulation.

The successful implementation of these methodologies underscores the potential of artificial neural networks in enhancing the diagnostic capabilities for hepatic steatosis. It is anticipated that with further refinement and validation, the developed model will provide valuable insights into the prevalence and severity of hepatic steatosis in the Mexican population. The establishment of a robust and accurate database, as achieved in this study, is expected to contribute significantly to the ongoing research and development in the field of hepatic disease diagnosis.

Overall, the integration of advanced imaging techniques and neural network algorithms has shown promise in improving the detection and understanding of hepatic steatosis. Future work will focus on expanding the data set, refining the neural network model, and validating the results through clinical trials to ensure the reliability and applicability of the developed system in real-world scenarios.

15.7 Conclusions

The data cleaning stage is a critical component of the project, as it enables the identification and exclusion of data that may not be suitable for the study. The applied technique has successfully identified that 37% of the cases are not appropriate for the training phase.

Data processing is crucial for the next phase of the research, as creating a precise database with accurate definitions will ensure that the training is effective and accurate.

Image segmentation is essential because it allows for the delineation of the area of interest, enabling the neural network to identify the liver in any given image.

Acknowledgements Large projects require strong foundations to support and enable their completion. In the context of this research, appreciation is expressed to the key institutions that have been instrumental in its development. These institutions include:

- Instituto Politécnico Nacional (IPN)
- Comisión de Operación y Fomento de Actividades Académicas del IPN (COFAA)
- Consejo Nacional de Humanidades, Ciencia y Tecnologías (CONAHCyT)
- Programa Institucional de Formación de Investigadores (PIFI)
- Hospital Regional "1° de octubre"-ISSSTE.

References

Barraza EW, Stalnikowitz DK (n d) Esteatosis y esteatohepatitis no alcohólica. Manuscript in preparation

Bernal-Reyes R, Icaza-Chávez ME, Chi-Cervera LA, Remes-Troche JM, Amieva-Balmori M, Priego-Parra BA, Triana-Romero A (2023) Prevalencia y características clínico-epidemiológicas de una población mexicana con enfermedad del hígado graso asociada a disfunción metabólica: Un estudio en población abierta. Rev Gastroenterol Mex 88(3):199–207. https://doi.org/10.1016/j.rgmx.2022.08.001

Delgado-Cortés HM, García-Juárez FI, García-Juárez I (2018) La enfermedad por hígado graso no alcohólico y el trabajo del internista. Revista Del Hospital Juárez De México 85(2):86–93. https://doi.org/10.1016/j.rhjm.2018.03.002

Méndez Sánchez N, Gutiérrez Grobe Y, Chávez Tapia C, Kobashi Margain RA, Uribe M (2010) Hígado graso no alcohólico y esteatohepatitis no alcohólica: Conceptos actuales. Rev Gastroenterol Mex 75(2):143–148

Revista Mexicana de Gastroenterología. (n d) Enfermedad por hígado graso no alcohólico (EHGNA). ABC Salud. Retrieved 18 Jul 2023. https://www.abcsalud.mx/enfermedad-por-higado-graso-no-alcoholico-ehgna/

Otero W, Sierra F (2003) El hígado en cirugía. Rev Colomb Gastroenterol 18(4):230–238. https://doi.org/10.1016/S0120-9967(03)71145-3

Sayiner M, Koenig A, Henry L, Younossi ZM (2016) Epidemiology of nonalcoholic fatty liver disease and nonalcoholic steatohepatitis in the United States and the rest of the world. Clin Liver Dis 20(2):205–214. https://doi.org/10.1016/j.cld.2016.01.001

Teixidó CH (2023) Hígado graso no alcohólico: Clasificación e historia natural. Diabetes Práctica 4. https://doi.org/10.1016/j.dia.2023.04.002

Chapter 16
Biomechanical Evaluation of Upper Extremity Arthroplasty in Patients with Total Amputations and Its Effects Due to Phantom Limb Syndrome

Karen Pamela Vázquez-Thierry, Juan Alfonso Beltrán-Fernández, Alejandro Tonatiu Velázquez Sánchez, Erik Omar Alvarado-Alcántara, and Verónica Guzmán-Mercado

Abstract This study aims to investigate the phantom limb syndrome (PLS) by developing and implementing an experimental protocol involving a sample of patients with upper limb amputations. The objective is to systematically collect, compare, and identify individuals exhibiting symptoms of PLS, thereby enabling appropriate referral to specialized care and rehabilitation services. The focus of the analysis is on the assessment of myoelectric signals obtained from the study sample.

Keywords Phantom limb syndrome · Amputation · Myoelectric signals · Prosthetic device

K. P. Vázquez-Thierry (✉) · J. A. Beltrán-Fernández · A. T. V. Sánchez · E. O. Alvarado-Alcántara · V. Guzmán-Mercado
Sección de Estudios de Posgrado E Investigación, Instituto Politécnico Nacional—Escuela Superior de Ingeniería Mecánica y Eléctrica—Unidad Profesional Adolfo López Mateos "Zacatenco" Col. Lindavista, Ciudad de México, México
e-mail: kappa.thierry@gmail.com

J. A. Beltrán-Fernández
e-mail: jbeltran@ipn.mx

A. T. V. Sánchez
e-mail: atvelazquezs@ipn.m

E. O. Alvarado-Alcántara
e-mail: eomar_1@yahoo.com.mx

V. Guzmán-Mercado
e-mail: vero_guzmanm94@hotmail.com

A. Öchsner and H. Altenbach (eds.), *Engineering Design Applications VII*, Advanced Structured Materials 230, https://doi.org/10.1007/978-3-031-84346-4_16

16.1 Introduction

In the United States, approximately 1.7 million individuals are affected by limb amputation (Kaur and Guan 2018). Data from Mexico indicate that, according to the National Institute of Statistics and Geography (INEGI), 785,000 individuals were recorded with varying degrees of amputation in 2010, with this number increasing to 935,000 by 2014. Traumatic amputations, particularly those resulting from accidents, represent the predominant cause of permanent disability and are expected to rise in the coming years. Additionally, it is noted that only one in ten amputees undergoes rehabilitation, and among those rehabilitated, only 30% effectively utilize their prostheses (Armas Mestre et al. 2021).

The Amputee Coalition reports that approximately 80% of amputee patients globally experience symptoms of phantom limb syndrome (PLS) (Amputee Coalition: Controlando el dolor fantasma—Amputee Coalition. 2019). PLS is recognized as a condition following limb amputation that involves both physical and psychological changes, with substantial social and occupational repercussions. The management of PLS presents considerable challenges and imposes a significant financial burden on healthcare systems worldwide (Nikolajsen. n.d).

16.2 General Objective

To design and develop a myoelectric prosthetic device that enables the collection of data to evaluate the muscular condition of patients with complete upper limb amputations, and to facilitate the study of the phantom limb syndrome.

16.3 Methodology

The methodology for this study, Fig. 16.1, is designed to include a complementary device that measures the myoelectric biopotential of the residual limb in each patient. This device aims to provide data on the muscular status of patients and detect the presence of phantom limb syndrome, thereby facilitating their referral to appropriate rehabilitation services. The development process of the device is organized into four main phases, culminating in the production of the initial functional prototype.

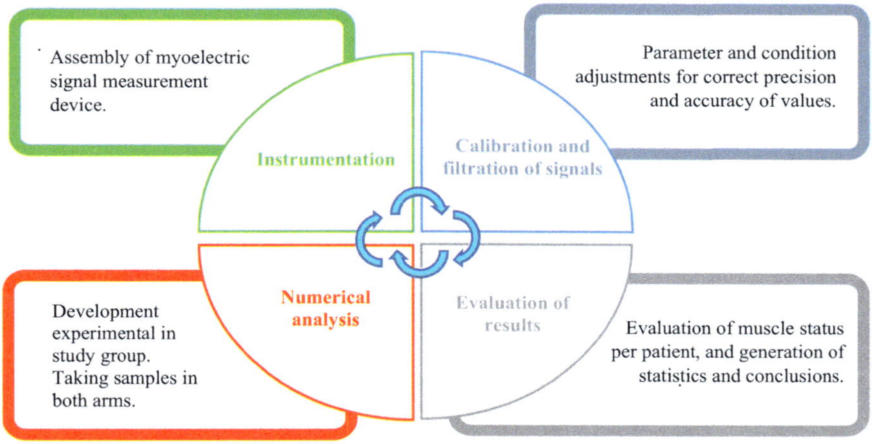

Fig. 16.1 Methodology

16.3.1 Objective of Data Collection

The hypothesis of the study intends to demonstrate in a group of study with total upper amputations, that their myoelectric signals are different between their healthy arm and their amputated arm.

It is intended to obtain a comparative myoelectric signal from both arms. The design of the experimental sampling test can recomplicate data in a ratio of 3 mV impulses per recorded second. The total duration of the test is 60 s, which can be expressed mathematically as follows:

$$x\text{x}\, t = y \tag{16.1}$$

where x represents the number of data points per second, t is the time in seconds, and y is the total amount of data collected per arm.

For this test, the calculation is

$$3 \times 60 = 180 \tag{16.2}$$

Since data is being collected from both arms, the total amount of data per patient is $180 \times 2 = 360$ per patient, Fig. 16.2.

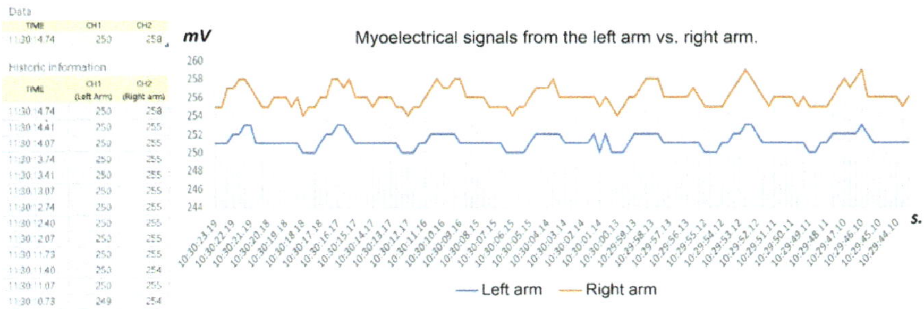

Fig. 16.2 Example of data collected and plotted from myoelectric signals

Fig. 16.3 Diagram of myoelectric signal processing and its four main stages

16.3.2 Instrumentation

The device is composed of several stages as illustrated in Fig. 16.3. Initially, nerve muscle fibers exhibiting the highest myoelectric activity must be identified. Subsequently, sensors and electrodes are positioned at these specific points. The data collection and processing are managed through the Arduino system, which converts the collected data into graphical representations (Fig. 16.4).

16.3.3 Calibration and Signal Filtering

The signal filtering process encompasses several stages aimed at ensuring that myoelectric signals are accurately captured as true representations of the muscle's biopotential, rather than being distorted by noise or interference that could lead to erroneous results.

To achieve precise data collection, several calibration steps are necessary:

- Precision Calibration: Ensures the instrument produces consistent results across multiple measurements.
- Accuracy Calibration: Ensures that the results closely align with known values of the test.

A comparative analysis of various calibration graphs was conducted to determine the optimal calibration point for effective data acquisition. Five calibration configurations were tested using the Arduino system, each under consistent exercise sequence

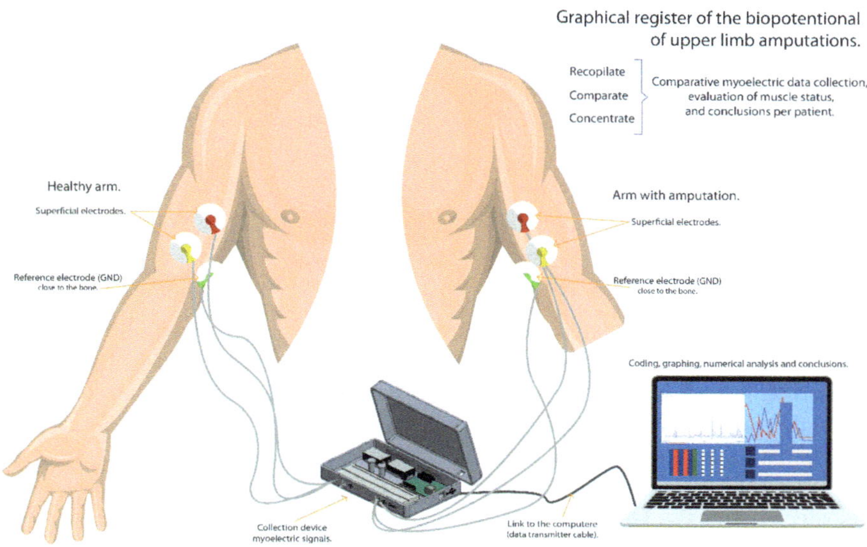

Fig. 16.4 Illustration of the operation of a device for acquiring myoelectric signals

parameters for a duration exceeding 5 s. This approach controlled for variable influences, enabling the identification of an ideal and consistent configuration for the exercises performed during the experiment.

The fifth calibration test, designated as B05, emerged as the optimal configuration for myoelectric signal acquisition. The graph in Fig. 16.5 (left) illustrates a congruence between the exercise sequence and the minimum and maximum points on the Y-axis. Satisfactory tuning was observed between the signals recorded from both arms, with appropriate displacement over time on the X-axis:

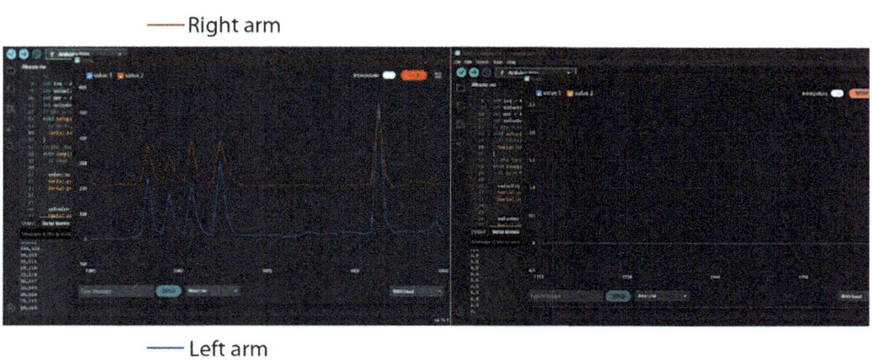

Fig. 16.5 (**Left**) Graphical result of the calibration configuration–B 05 (**Right**) Graphical result of the noise and interference test

Additionally, a noise test was conducted, as depicted in Fig. 16.5 (right). Calibration of the microcontroller board, in conjunction with the use of a rectifier diode, ensures that the acquired signals are as free from noise and static as possible. The absence of external noise recordings indicates effective filtering and signal amplification.

16.3.4 Numerical Analysis

The experimental study group was assembled during the free prosthesis donation campaign organized by the LN4 Hand Project, in collaboration with the Valle de México Rotary Club. The evaluation, Fig. 16.6, comprised 20 patients with ages ranging from 4 to 68 years, including both female and male participants, with a trend of amputation on the left arm.

Each of the 20 participants underwent a preliminary test to collect their specific myoelectric signals. The test was designed with the following classifications based on the patient's condition:

- A.1. Measurement of the patient's amputated arm without the use of a prosthesis.
- A.2. Measurement of the patient's healthy limb, with the amputated arm not fitted with a prosthesis.
- B.1. Measurement of the patient's amputated arm with the use of a prosthesis.
- B.2. Measurement of the patient's healthy limb, with the amputated arm fitted with a prosthesis.

The results were derived from readings of both arms, with 90% of the tests yielding significant data.

During the testing day, patients were evaluated both with and without prostheses, as illustrated in Fig. 16.7. This analysis provided valuable insights into how the muscular behavior of patients is altered with the application of a prosthesis, offering a comparative view of conditions before and after prosthesis fitting. The left side of Fig. 16.7 displays cases categorized under A.1 (measurement of the amputated arm without a prosthesis) and A.2 (measurement of the healthy limb without a prosthesis on the amputated arm). Conversely, the right side of Fig. 16.7 presents cases under

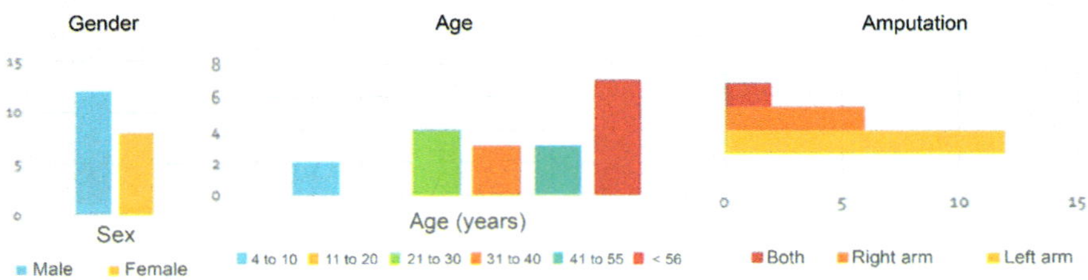

Fig. 16.6 Demographic graphs and causes of amputation in study group

B.1 (measurement of the amputated arm with a prosthesis) and B.2 (measurement of the healthy limb with a prosthesis on the amputated arm).

The electronic components depicted in Fig. 16.8 performed optimally and without issues throughout the study. The device's casing was only opened for the purpose of recharging the two 9 V batteries, a procedure conducted approximately every seven readings. This was done to ensure that the batteries were fully charged, and that recharging did not affect the results obtained.

Additionally, a table was utilized to assess muscle disturbance, muscle tension, and joint conditions. This table was designed in accordance with the classification of target muscles in the evaluation of myoelectric signals by Yoo et al. (2019) and was applied to provide a more comprehensive and complementary evaluation.

Based on this classification, five levels were established to assess muscle strength and tension parameters. Table 16.1 presents these parameters from left to right, in order of capability.

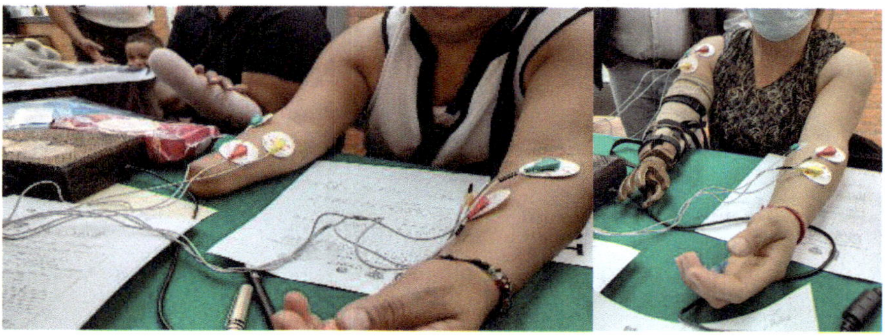

Fig. 16.7 (**Left**) Application of the experimental test in a patient with an amputation and without a prosthesis. (**Right**) Application of the test in a patient with an amputation and with a prosthesis in place

Fig. 16.8 Myoelectric device with open casing for battery charging

Table 16.1 Evaluation of disturbance level of the residual stump

Level of the muscle disturbance				
Domain	Active	Average	Passive	Deteriorated

- Domain: This phase is typically observed in patients with congenital amputations who have lived with the condition for their entire lives and have achieved a high level of adaptability, or in traumatic amputations where correct rehabilitation was undertaken immediately after the amputation, leading to routine strengthening exercises.
- Active: This stage reflects an excellent evaluation outcome. It is evident that the patient manages their stump effectively, demonstrating good force application and precision.
- Average: The patient is in a normal state but could benefit from additional rehabilitation and/or exercise sequences to further strengthen the residual stump.
- Passive: This level indicates that the patient is not in a normal state and requires attention due to inadequate control over force application and precision.
- Deteriorated: This final level shows severe muscle atrophy. The patient struggles to control strength voluntarily and is not a suitable candidate for a myoelectric prosthesis.

Numerical Analysis Per Patient

A review of 2 out of 20 records generated per patient is presented. Volunteer conditions:

- A.1. Measurement of the patient's arm with upper limb amputation without the use of a prosthesis.
- B.1. Measurement of the patient's arm with upper limb amputation without the use of a prosthesis.

A 58-year-old male patient with total right-hand amputation due to trauma was examined. Symptoms of phantom limb syndrome, such as tingling, squeezing, and the perception of cold, were detected (Table 16.2).

The signals depicted in Fig. 16.9 indicate a notable disparity in disturbance levels, with the amputated right arm exhibiting stronger signals compared to the left. This discrepancy may be attributed to the traumatic nature of the amputation, which resulted from an accident. The patient, having transitioned from being right-handed

Table 16.2 Evaluation of disturbance level of the residual stump of file 4

Level of the muscle disturbance				
Domain	~~Active~~	Average	Passive	Deteriorated

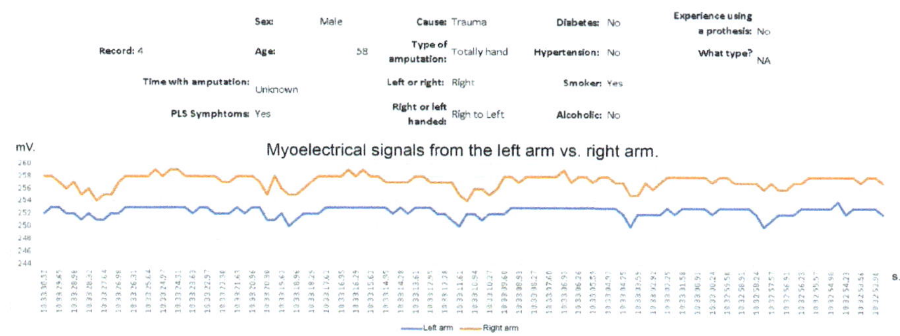

Fig. 16.9 Compilation of myoelectric signals from a patient with total amputation of the right hand

to left-handed, demonstrates symptoms consistent with phantom limb syndrome, including tingling, squeezing, and cold sensations. Given that this is the patient's first experience with a prosthesis, and considering that he underwent rehabilitation following the amputation, it is anticipated that he will require a relatively brief period to adapt to and effectively master the use of the prosthesis.

Protocol with donation of prosthesis placed.

Volunteer Conditioning: A.2. Measurement of the patient's healthy limb, without prosthesis on the amputated arm, and B.2. Measurement of the patient's healthy limb, with prosthesis on the amputated arm.

A 35-year-old male patient, who experienced right wrist disarticulation trauma 12 years ago, presents with phantom limb syndrome symptoms such as tingling, palpitations, prickling sensations, and varying perceptions of movement, heat, and cold approximately once a week. These symptoms are exacerbated by stress, poor circulation, prosthesis use, and cold weather. Notably, the patient has observed some relief through massage of the residual stump (Table 16.3).

The myoelectric test depicted in Fig. 16.10 illustrates a prominent behavior of the orange line corresponding to the right arm, which is fitted with the donated prosthesis. The immediate activation observed in the subject's perception upon fitting the prosthesis is noteworthy. This activation appears to enhance the subject's ability to focus on coordinating movements with greater precision compared to the healthy arm.

Table 16.3 Evaluation of the disturbance level of the residual stump of file 19

Level of the muscle disturbance				
Domain	~~Active~~	Average	Passive	Deteriorated

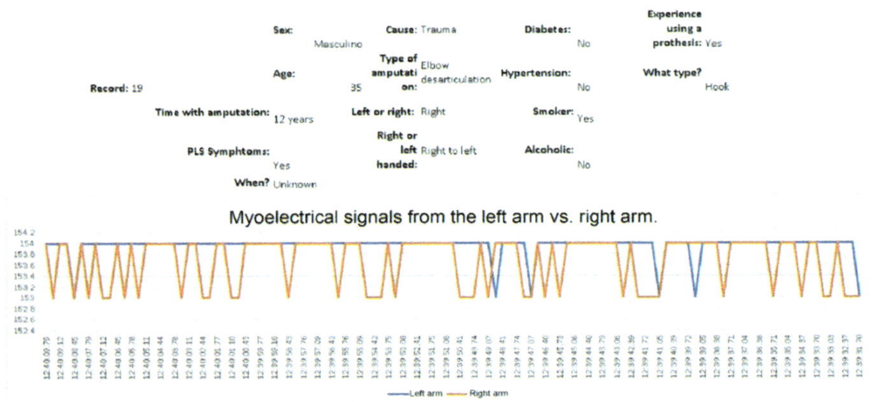

Fig. 16.10 Compilation of myoelectric signals from a patient with total amputation of the right hand

During the initial recovery phase, the patient underwent both psychological and physical therapy. The patient also had previous experience with a hook-type mechanical prosthesis, which presented significant challenges in performing precision activities.

16.3.5 Result Evaluation

Statistical analysis reveals that 12 out of 20 patients in this study exhibit symptoms and clinical behaviors indicative of phantom limb syndrome. Figure 16.11 demonstrates a significant increase in positive cases among patients with traumatic amputations, rising from 10 cases to 12. This suggests that congenital amputations are less frequently associated with PLS.

Fig. 16.11 Incidence of patients with phantom limb syndrome

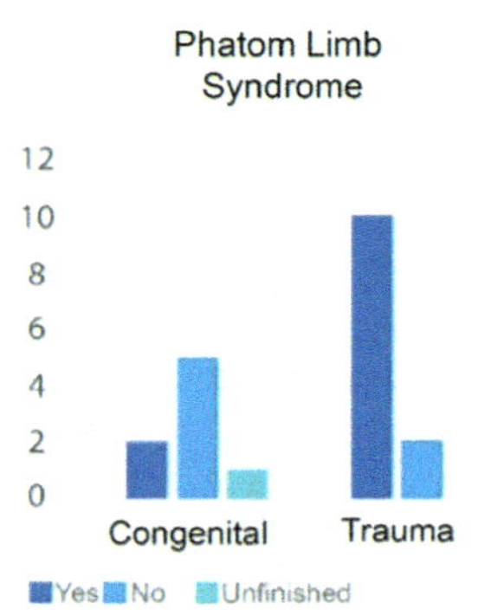

The symptoms detected in the affected cases include tingling, sensations of heat and/or cold, burning, sensations of movement, stinging, numbness, and burning.

The data indicate that most patients exhibit a satisfactory muscular condition. Notably, patients who have undergone progressive physical rehabilitation demonstrated significant improvements. Moreover, those who engaged in additional psychological rehabilitation achieved a more comprehensive recovery, as reflected in the improved control of their amputated limb.

Patients who participated in the study with a prosthesis in place exhibited intriguing results. The graphs revealed an immediate improvement effect during the exercise sequence. In some cases, the precision and application of force with the prosthetic limb were optimal, even surpassing the performance of the dominant arm. This suggests that the patient's motor function becomes more focused on the prosthetic limb, leading to enhanced manipulation capabilities.

Overall, the results suggest that most patients will undergo a favorable adaptation process to their new prosthesis, and many are well-suited for further evaluations for myoelectric prosthesis usage.

16.4 Conclusions

Electromyographic signals were found to differ significantly between the healthy and amputated arms. Additionally, phantom limb syndrome was observed to be rare in congenital amputations but prevalent in traumatic amputations.

The test successfully assessed whether patients were receiving appropriate rehabilitation of their residual stump, based on muscle disturbance and the precision of their exercise performance in terms of timing and force.

The study facilitated the referral of patients with phantom limb syndrome, and those lacking adequate rehabilitation, to appropriate support services for continued care and progression towards effective prosthetic use.

Acknowledgements • Instituto Politécnico Nacional for all the contributions and support in the development of the study.

• Higher research scholarships CONAHCYT for the economic support in the develop of the study.

• Club Rotario Zona Esmeralda for facilitating the database and study group.

References

Amputee coalition: controlando el dolor fantasma—amputee coalition. https://amputee-coalition.org/resources/spanish-managing-phantom-pain/

de Armas Mestre J, Soria Pérez R, Ibis González S, Bethencourt González M, Benavides Flores M, Rivera Somoano E (2021) Comportamiento del dolor fantasma en pacientes amputados en Cárdenas. Periodo 2015–2019. Rev Médica Electrónica 43(1):2759–2770

Kaur A, Guan Y (2018) Phantom limb pain: a literature review. Chin J Traumatol 21(6):366–368. https://doi.org/10.1016/j.cjtee.2018.04.006

Nikolajsen L (n.d) Postoperative pain, postamputation pain, treatment and prevention. In: Encyclopedia of Pain. Springer Berlin Heidelberg, Berlin, Heidelberg, pp 1925–1928. https://doi.org/10.1007/978-3-540-29805-2_3459

Yoo H-J, Park H-J, Lee B (2019) Myoelectric signal classification of targeted muscles using dictionary learning. Sensors 19(10):2370. https://doi.org/10.3390/s19102370

Chapter 17
Generative Design for a Case of Tumoral Resection, Reconstruction of the Mandibular Continuity

Carolina Alvarado-Moreno, Juan Alfonso Beltrán-Fernández, Mauricio González Rebattú, and Juan Carlos Hermida Ochoa

Abstract This study presents an innovative approach to the customized reconstruction of a lower jaw, with the primary objective of demonstrating an alternative process that integrates biocompatible materials, including PMMA, titanium, and stainless steel. The methodology includes the digital acquisition of the patient's morphological profile via computed tomography scans and introduces the use of generative design methods to optimize the implant's weight and rigidity. Using CAD/CAE tools, the missing bone segment was reconstructed, and fixation plates and screws for the surgical procedure were proposed. Additionally, the characteristics of the PVD coating on the plates were analyzed through a numerical study employing the finite element method.

Keywords Lower jaw · CAD · Generative design · Finite element method, DICOM · PMMA

Juan Carlos Hermida Ochoa has died before the publication of the book.

C. Alvarado-Moreno · J. A. Beltrán-Fernández (✉)
Sección de Estudios de Posgrado E Investigación, Instituto Politécnico Nacional—Escuela Superior de Ingeniería Mecánica y Eléctrica, Unidad Profesional Adolfo López Mateos "Zacatenco" Col. Lindavista, Ciudad de México, México
e-mail: jbeltranf@hotmail.com

C. Alvarado-Moreno
e-mail: jbeltran@ipn.mx

M. González Rebattú
Hospital Regional "1 de Octubre" del ISSSTE, Ciudad de México, CDMX, México
e-mail: mauriciorebattu@yahoo.com.mx

J. C. Hermida Ochoa
Hospital German Diaz Lombardo—Centro de Investigación y Laboratorio Biomecánico, Chimalistac San Ángel, CDMX, México
e-mail: hermidagroup@gmail.com

A. Öchsner and H. Altenbach (eds.), *Engineering Design Applications VII*, Advanced Structured Materials 230, https://doi.org/10.1007/978-3-031-84346-4_17

17.1 Introduction

Mandibular reconstruction primarily aims to maintain the height and geometry of the resected bone, ensuring a functional foundation for dental rehabilitation post-reconstruction. This process is crucial for the proper management of masticatory forces, the restoration of supporting muscles and surrounding tissues, and the ability to sustain repeated loading during mastication (Ali et al. 2019).

The surgical procedure typically involves the removal of cancer from the floor of the mouth, which often necessitates the extraction of any teeth attached to the affected areas. In initial mandibular reconstructions, commercial plates are frequently used before the ablation. However, when the tumor is extensive, a more effective method involves creating a custom template of the mandible using pre-operative computed tomography (CT) scans (Gullane et al. 2004).

This situation highlights the significance of applying computer-aided design (CAD) and computer-aided manufacturing (CAM) techniques, alongside 3D printing with biocompatible materials, for reproducing complex anatomical structures. These advancements offer the potential for personalized prostheses that replicate the original geometry and weight of the mandible, ensuring a more precise and functional reconstruction (Ali et al. 2019; Gullane et al. 2004).

17.2 Methodology

The implant generation process for this study was based on a pre-operative computed tomography scan of a female patient. The procedure is outlined in Figs. 17.1–17.3.

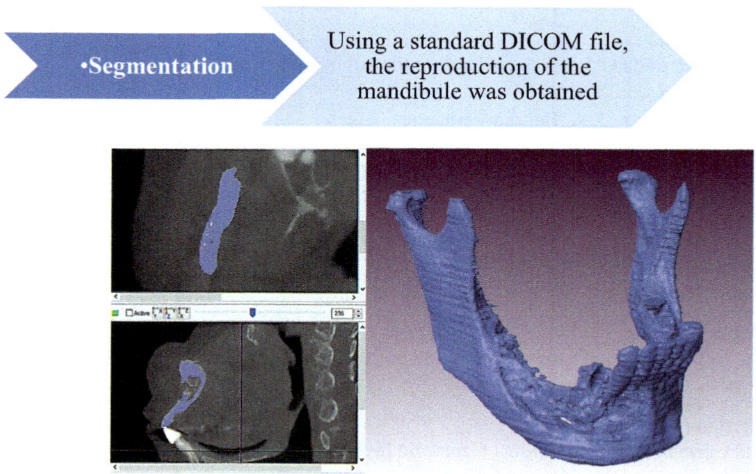

Fig. 17.1 Segmentation distribution, with the mandible generated

The 2D CT images, captured in the XYZ directions, were sequentially combined to generate a 3D model of the mandible. This initial bone model, defined by a triangular mesh, does not yet contain any physical properties. Upon review by the medical team, it was determined that the damage to the bone required the removal of the affected section, including the mandibular and alveolar regions, to ensure continuity of the mandible (Fig. 17.1).

Using ANSYS® SpaceClaim® (Swanson Analysis Systems 2024a) student 2024 R1, the triangular mesh was converted into a solid 3D object, enabling further modifications. This transformation allowed the user to manipulate the structure and design an implant that conformed to the patient's natural anatomical shape. In this case, the original design was adjusted to restore the mandibular continuity based on the surgical procedure, which involved removing parts of the sublingual fossa, alveolar process, mental foramen, and the mental protuberance (chin) of the mandible.

Within ANSYS® Mechanical® (Swanson Analysis Systems 2024b) student 2024 R1, the remaining bone, along with the implant, was prepared for finite element method (FEM) analysis. The parameters used in the analysis are outlined in Table 17.1 and include:

- Masticatory muscle forces to replicate the natural chewing process.
- Bone density specific to the patient, which was determined to be low.
- Material properties such as Young's modulus and Poisson's ratio for the implant, made of polymethyl methacrylate (PMMA), as well as for the surgical instruments, including Ti6Al4 V titanium alloy and 316L stainless steel.

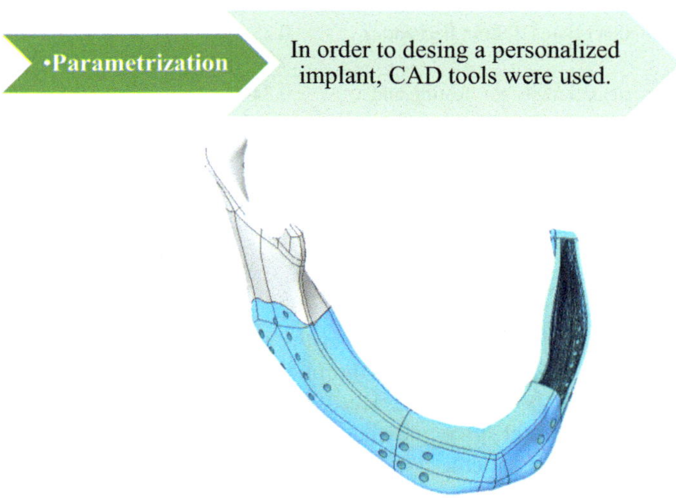

Fig. 17.2 Final design implant

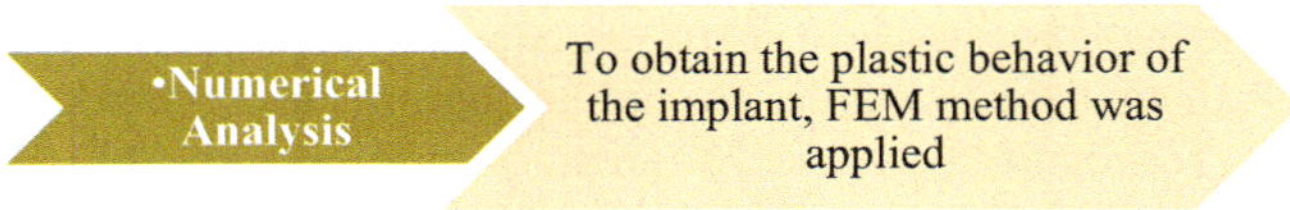

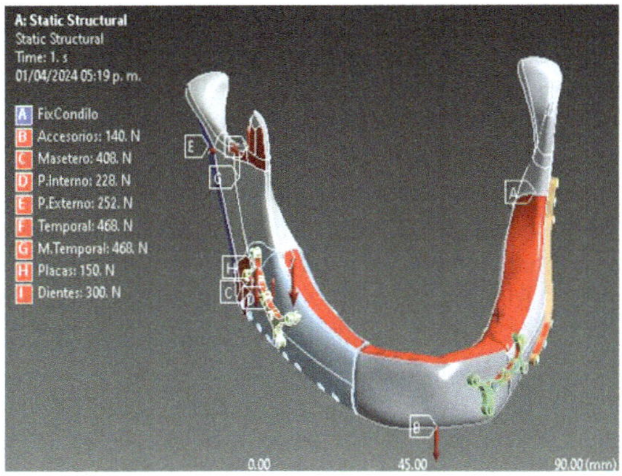

Fig. 17.3 Masticatory muscles placed in the mandible

Table 17.1 Mechanical properties of the materials

Material	Poisson's ratio	Young's modulus in MPa
Low cortical bone (Pan et al. 2019)	0.25	1000
Polymethyl-methacrylate (Beltrán Fernández 2019)	0.3	2,400
Ti6Al4V (American Society for Testing and Materials 2023)	0.34	114,000
316L stainless steel (American Society for Testing and Materials 2023)	0.28	190,000

17.3 Results

It is important to highlight the reaction at the forces on the different plates. This behavior in Table 17.2, indicates that the plate is absorbing the muscular force in order to maintain the bone with the necessary energy to promote the osteogenesis without applying the total force and cause an injury on the tissue and in the remain low-quality bone of the patient.

Table 17.2 Obtained values of equivalent (von Mises) stress

	Maximum value	Minimum value	Average value
	7,522.4 MPa	1.7943×10^{-6} MPa	23.326 MPa
Localization	Right ramus/plates	PMMA implant	Lower jaw

17.4 Discussion

The results of the static analysis (Fig. 17.4) indicate that the implant remained within the elastic limit of the PMMA material, and similarly, the surgical instrumentation retained its structural integrity. In contrast, the remaining bone of the patient, due to its low density, would likely suffer a critical fracture under normal chewing forces.

However, the average values obtained from the analysis suggest that failure could only occur in an extreme scenario where the patient applies the full force of their masticatory muscles. In typical chewing conditions, the entire reconstructed jaw would maintain its structural integrity.

It is also important to note that in such cases, patients often undergo muscle rehabilitation to strengthen the masticatory muscles. Nevertheless, despite rehabilitation efforts, the patient in this study did not fully recover the natural chewing force.

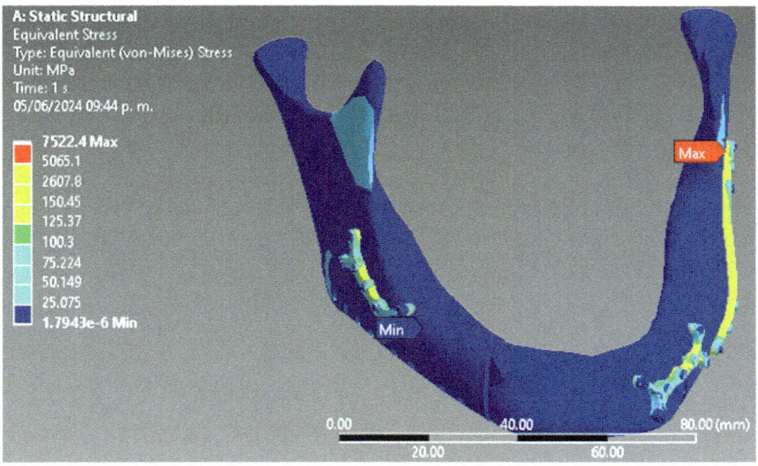

Fig. 17.4 Finite element method results

17.5 Conclusion

The present study demonstrates that the implant designed with polymethylmethacrylate provides favorable mechanical performance and successfully restores the continuity of the mandible without significant risk of failure. The static analysis showed that although the implant may undergo some plastic deformation, this deformation does not compromise the overall stability or functionality of the reconstructed mandible. The implant remains structurally sound under typical masticatory forces, and there is no indication that post-surgical removal or additional intervention would be required based on the current model.

Furthermore, the use of advanced CAD/CAE tools combined with patient-specific data from computed tomography scans enabled a precise and customized design that closely matches the patient's original anatomical structure. This level of customization ensures better integration with the surrounding bone and soft tissue, contributing to the patient's functional recovery. The generative design approach, focused on maintaining both appropriate weight and rigidity, played a crucial role in optimizing the implant for long-term use.

However, despite the success of the implant design, there are challenges related to the quality of the patient's bone. The low-density bone poses a risk of collapse, especially under extreme or repeated loading conditions. While the implant itself remains stable, the surrounding bone's capacity to sustain forces during chewing is significantly compromised. This underlines the importance of ongoing monitoring and possibly supplementary treatments to enhance bone density in future procedures.

This study highlights the potential of combining biocompatible materials such as PMMA with generative design methods and finite element analysis for mandibular reconstruction. The approach offers a viable alternative to traditional techniques, especially in complex cases where extensive bone loss is present. Future studies could explore the use of alternative materials or reinforcement techniques to address the limitations posed by low-density bone, further improving the long-term success rates of such reconstructive surgeries.

Acknowledgements The grants to the national institutions, Consejo *Nacional de Humanidades, Ciencias y Tecnologías de México (CONAHCYT),* the *Instituto Politécnico Nacional (IPN),* and the *Comisión de Operación y Fomento de Actividades Académicas del Instituto Politécnico Nacional (COFAA), which support this research.*

To the *Hospital 1ro de Octubre* for all the contributions made during the research and to the *Centro de Investigación y Laboratorio de Biomecánica,* which actively continue with health care research.

References

Ali MN et al (2019) Mandibular reconstruction: a review. Updat Dent Coll J 9(2):50–54

American Society for Testing and Materials (2023) F1472 standard specification for wrought Ti-6Al-4V alloy for surgical implant applications. ASTM F1472/UNS R56400, pp 1–6. https://doi.org/10.1520/F1472-20A

American Society for Testing and Materials (2023) Standard specification for chromium and chromium-nickel stainless steel plate, sheet, and strip for pressure vessels and for general applications: 316/316L stainless steel plates physical properties. ASTM A240, pp 1–12. https://doi.org/10.1520/A0240_A0240M-23A

Beltrán Fernández JA (2019) Design and characterization of a mandibular prosthesis prototype by hemimandibulectomy. Adv Eng Mater 92:313–343. https://doi.org/10.1007/978-3-319-79005-3_22

Gullane PJ et al (2004) Management of the mandible in cancer of the oral cavity. OTO J 15(4):256–263. https://doi.org/10.1016/j.otot.2004.07.001

Pan C et al (2019) Effects of cortical bone thickness and trabecular bone density on primary stability of orthodontic mini-implants. J Dent Sci 14(4):383–388

Swanson analysis systems (2024a) Software spaceClaim. PA. Ansys, Inc., Canonsburg

Swanson analysis systems (2024b) Software mechanical. PA. Ansys, Inc., Canonsburg

Chapter 18
Biomechanical Study of Orthopedic Implants for Distal Fibula in Type B Fractures

Milton Alfredo Perez-Reyes, Juan Alfonso Beltrán-Fernández, Belén Alejandra Contreras-Mendoza, Elías Humberto Hermida-Ochoa, and Alejandro González Rebattú y González

Abstract This research presents the results of a structural mechanical analysis performed on an implant designed for the rehabilitation of type B ankle fractures, aimed at its application for patients undergoing therapeutic rehabilitation. The study focuses on a 40-year-old male, weighing 90 kg and standing 1.80 m tall. The bone model, which features a distal fibular fracture, was created using a high-resolution, non-contrast computed tomography scan with a resolution of 0.625 mm. The DICOM images were meticulously segmented with ScanIP® software to define the region of interest. The modeling and assembly of the implant were carried out using SpaceClaim® software. The structural mechanical analysis of the implant was conducted using the finite element method, implemented in the ANSYS 2021 R1® Academic program. This analysis considered the effects of muscle forces, bone mechanical properties, and the load experienced during the mid-stance phase of walking, which is equivalent to three times the subject's body weight, or 2648.7 N. Additionally, various experimental load application tests were performed on the implant. For this purpose, 3D printing in PLA was utilized to create models of the

M. A. Perez-Reyes · J. A. Beltrán-Fernández (✉) · B. A. Contreras-Mendoza
Sección de Estudios de Posgrado E Investigación, Instituto Politécnico Nacional—Escuela Superior de Ingeniería Mecánica y EléctricaUnidad Profesional Adolfo López Mateos "Zacatenco" Col. Lindavista, Ciudad de México, México
e-mail: jbeltran@ipn.mx

M. A. Perez-Reyes
e-mail: mperezre@ipn.mx

B. A. Contreras-Mendoza
e-mail: bcontrerasm@ipn.mx

E. H. Hermida-Ochoa
Centro de Investigación y Laboratorio Biomecánico, Chimalistac San Ángel, CDMX, México
e-mail: hermidagroup@gmail.com

A. G. Rebattú y González
Hospital Regional "1 de Octubre" del ISSSTE, Ciudad de México, CDMX, México
e-mail: alexrebattu@hotmail.com

A. Öchsner and H. Altenbach (eds.), *Engineering Design Applications VII*, Advanced Structured Materials 230, https://doi.org/10.1007/978-3-031-84346-4_18

tibia, fibula, and talus, allowing for the rigid fixation of the implant. An incremental load was systematically applied to the models, ranging from 882.9 N to 2648.7 N. This methodology enabled the determination of the implant's strain using the GOM Correlate® software. The results indicated a significant reduction in stresses in the regions proximal to the fracture following implant insertion, thereby substantially aiding in the optimal rehabilitation of the patient.

Keywords Orthopedic implant · 3D printing · Finite element method · Biomechanical modeling · Type B fibula fracture

18.1 Introduction

Ankle fractures are a common clinical issue, with an incidence rate of 187 per 100,000 people annually. Type B fractures, which are primarily caused by foot supination and external rotation, have also been linked to obesity and diabetes. The primary goal of treating type B fractures is to reduce the displacement of the fractured bone fragments and maintain proper anatomical alignment (Moore and Dalley 2018).

For treating distal fibula fractures, locking compression plates (LCP) are recommended. These plates help prevent screw loosening by providing a more rigid, threaded construction (Murray et al. 2002). This analysis focuses on a clinical case of a distal fibula fracture that was treated with an LCP and subjected to a vertical compressive force equivalent to the patient's body weight of 90 kg. The analysis considered a load of 2648.7 N, which represents three times the patient's body weight during the mid-stance phase of walking (Rose et al. 2006).

18.2 Methodology

The biomechanical analysis was conducted using the finite element method (FEM) (Öchsner 2023), requiring the generation of a precise CAD model. First, a high-resolution computed tomography (CT) scan was performed on the patient, producing DICOM images. These images were segmented using ScanIP® software, allowing for the separation of relevant structures and the generation of an accurate 3D model of the region of interest (Fig. 18.1).

For FEM analysis, a refined mesh of the model was created to ensure the accuracy of the results. The biomechanical properties of the bone and implant materials were incorporated, and a load corresponding to the mid-stance phase of walking (three times the body weight) was applied to simulate real-world forces experienced during rehabilitation. The results were evaluated to assess the stress distribution and strain experienced by the bone and implant (Arnold et al. 2001; Hallmann et al. 2019).

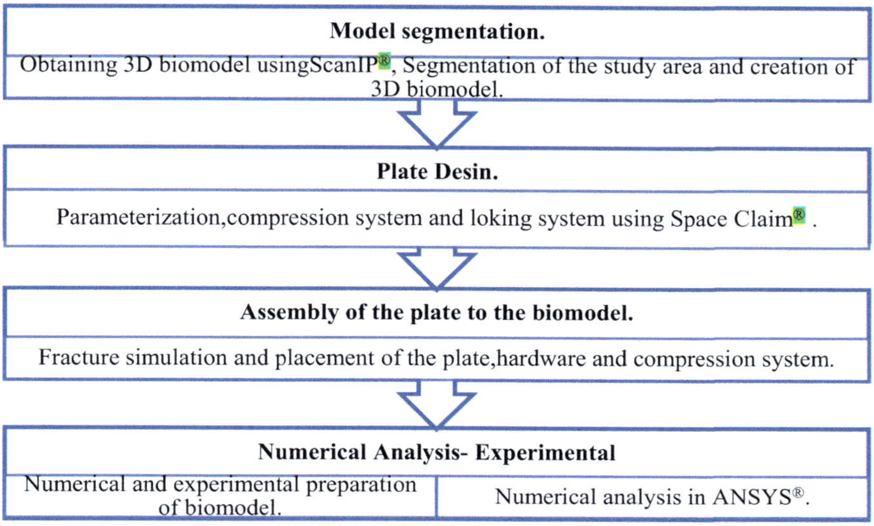

Fig. 18.1 Methodology

18.2.1 Segmentation

The acquisition of high-quality imaging data is crucial for developing the required skeletal model. In this case, the model is generated in a neutral, unloaded position (Fig. 18.2). The segmentation process begins by isolating the region of interest from the high-resolution DICOM images obtained through computed tomography. The ScanIP® software was used to perform the segmentation, allowing for precise delineation of bone structures and surrounding anatomical features, ensuring an accurate representation for subsequent biomechanical analysis.

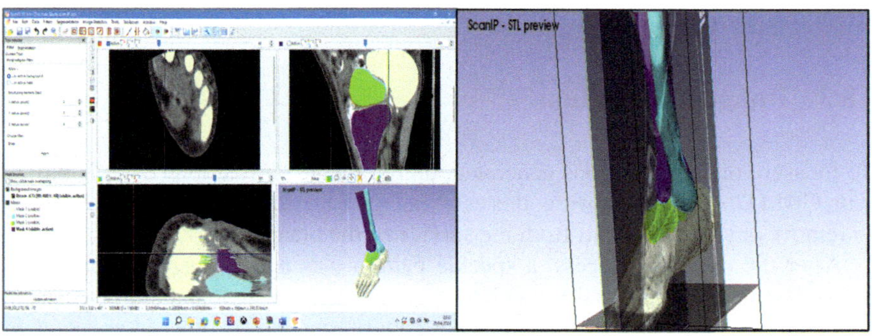

Fig. 18.2 Mask definition for each anatomical section to be segmented

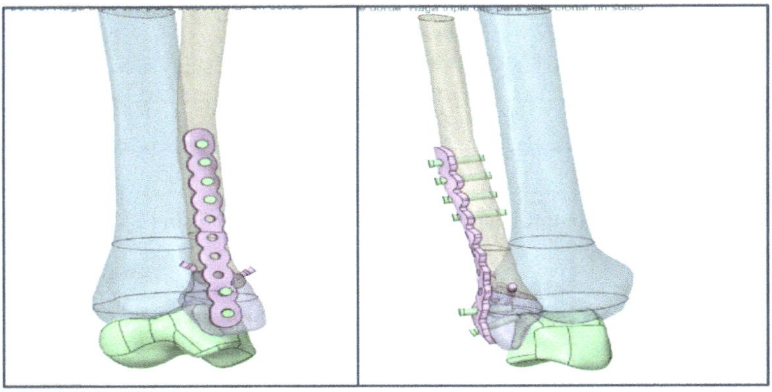

Fig. 18.3 Design facets

18.2.2 Plate Design

The design of the implant plate was conducted using the ANSYS® SpaceClaim® interface. The extrusion and combination tools within the software were applied to model the geometry of the plate. Subsequently, six cortical screws and two cancellous screws were integrated into the model (Fig. 18.3) (Ghosh 2021), ensuring the appropriate fixation for biomechanical analysis and surgical application.

18.2.3 Load Conditions

A compression force of 2648 N was applied to the contact surface, reflecting the mid-stance phase in the human gait cycle (Fig. 18.4) (Avila et al. 2007; Hertzberg 1968; Asla et al. 2006).

18.3 Image Correlation

To measure the deformations generated, digital image correlation (DIC) is utilized with GOM Correlate® software. It is essential to clean the implant with 70% ethanol to remove any grease or dirt from the surface of the plate (Fig. 18.5).

After the cleaning process, a speckle pattern was applied using black surface paint. The application was done with a pneumatic spray gun, maintaining a distance of 20 cm from the sample. The speckle pattern enables the measurement of the implant's deformation (Fig. 18.6) (Crammond et al. 2013; Su et al. 2016).

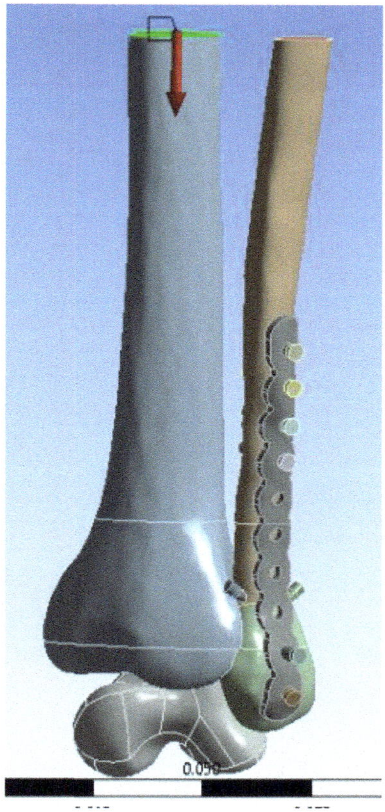

Fig. 18.4 Load conditions for type B fracture ANSYS® Mechanical

The implant is shown securely holding the specimen holder on the testing machine's platform to ensure no movement underload. Control limits are established for maximum and minimum loads and displacements to protect the sample from excessive loads (Fig. 18.7).

18.4 Results

The results indicate that the maximum stress is 149.39 MPa in the diaphysis of the plate (Fig. 18.8). In the area of the fracture, the stress is 12.917 MPa with a maximum lateral displacement of 0.36343 mm, attributed to the natural anatomical movement (Fig. 18.9).

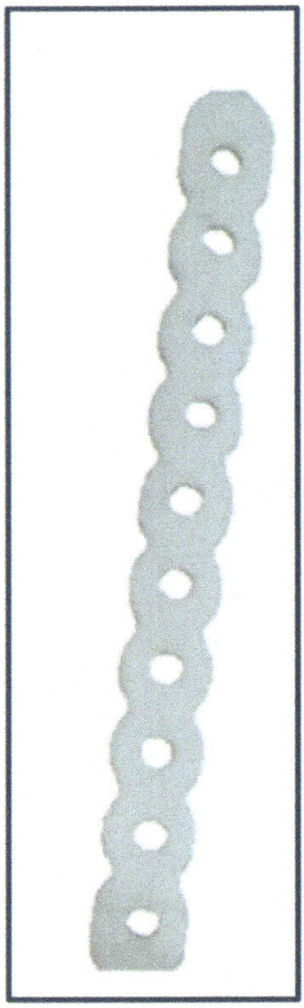

Fig. 18.5 Implant for type B ankle fracture before applying the speckle pattern

The test was conducted on a PLA-printed compact sample of the plate loaded on a hydraulic testing machine (Fig. 18.10). The analysis was performed using a single-camera DIC setup in the GOM Correlate® software, resulting in deformation values ranging from 0.00 to 0.127 in the color spectrum scale (Helgason et al. 2014).

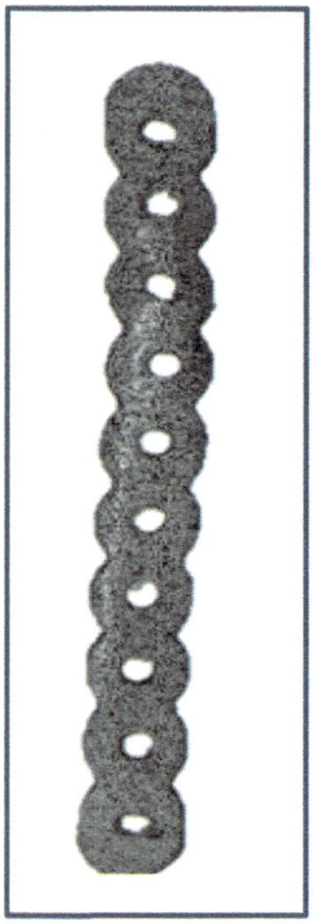

Fig. 18.6 Speckle pattern on the implant plate

Fig. 18.7 Fixation of the plate for the implant

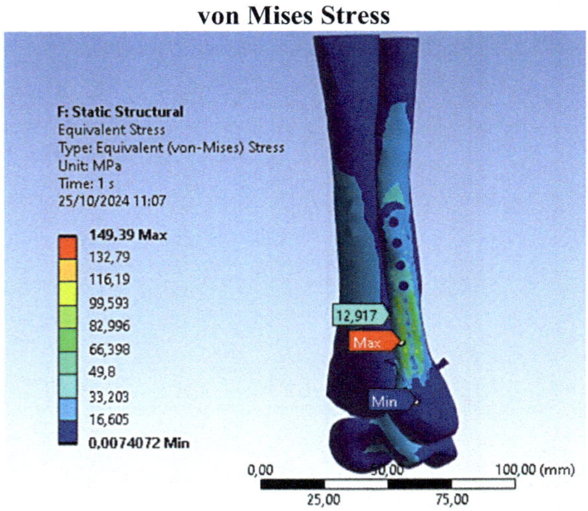

Fig. 18.8 Von Mises stress results in the plate for the implant

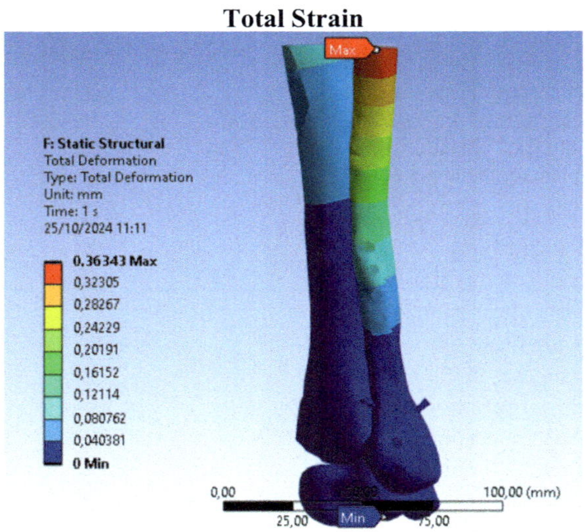

Fig. 18.9 Displacement results in the plate implant

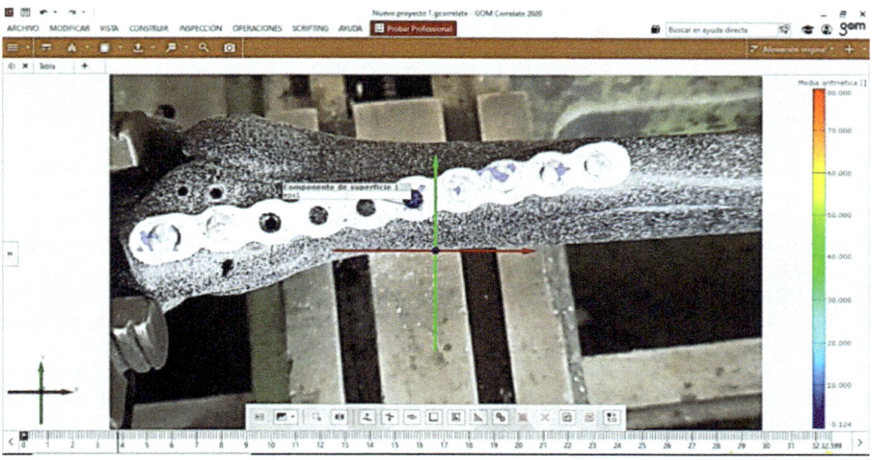

Fig. 18.10 Fixation of the plate for the implant

18.5 Conclusion

The insertion of the implant significantly reduces the stress at the location of the type B fracture. The findings demonstrate that the implant effectively supports the forces exerted on the ankle, contributing positively to the patient's rehabilitation process.

However, the study has limitations related to the loading conditions and the ankle position considered in the model for analyzing force distribution and stress. Additionally, the analysis does not account for muscular forces due to a lack of relevant literature.

In the digital image correlation tests, data acquisition, analysis, and presentation, the results were evaluated with biomechanical considerations. Comparisons with finite element models, specifically those for type B fracture implants, indicate that this type of implant can mitigate deformation and stress. Surgeons can use this information to assess the mechanical advantages of such implants, alongside understanding loading conditions and essential medical criteria for numerical and experimental evaluations.

Acknowledgements

- The authors grant to the Instituto Politécnico Nacional and the Hospital 1° de Octubre for their invaluable contributions to this research.
- A special thanks to the Biomechanical Research Center and Laboratory, located at Carmen #18, Chimalistac San Angel, 01070 Mexico, CDMX.

References

Arnold AS, Blemker SS, Delp SL (2001) Evaluation of a deformable musculoskeletal model for estimating muscle-tendon lengths during crouch gait. Ann Biomed Eng 29:263–274. https://doi.org/10.1114/1.1355277

Avila R, Prado L, Gonzalez E (2007) Anthropometric dimensions of the latin American population: Mexico, Cuba, Colombia, Chile (2nd ed). Universidad de Guadalajara

Crammond G, Boyd SW, Dulieu-Barton JM (2013) Speckle pattern quality assessment for digital image correlation. Opt Lasers Eng 51(12):1368–1378. https://doi.org/10.1016/j.optlaseng.2013.03.014

De Asla RJ, Wan L, Rubash HE, Li G (2006) Six DOF in vivo kinematics of the ankle joint complex: Application of a combined dual-orthogonal fluoroscopic and magnetic resonance imaging technique. J Orthop Res 24(5):1019–1027. https://doi.org/10.1002/jor.20142

Ghosh M (2021) Construction and simulation of a novel high altitude versatile armor of X-Aerogel and carbon fiber composite with ansys 2020 R2. Mater Today Proc 44:3045–3049. https://doi.org/10.1016/j.matpr.2021.02.441

Hallmann M, Goetz S, Schleich B (2019) Mapping of GD&T information and PMI between 3D product models in the STEP and STL format. Comput Aided des 115:293–306. https://doi.org/10.1016/j.cad.2019.06.006

Helgason B, Gilchrist S, Ariza O, Chak JD, Zheng G, Widmer RP, Ferguson SJ, Guy P, Cripton PA (2014) Development of a balanced experimental–computational approach to understanding the mechanics of proximal femur fractures. Med Eng Phys 36(6):793–799. https://doi.org/10.1016/j.medengphy.2014.02.019

Hertzberg HTE (1968) The conference on standardization of anthropometric techniques and terminology (a report). Am J Phys Anthropol 28(1):1–16. https://doi.org/10.1002/ajpa.1330280110

Moore KL, Dalley AF (2018) Clinically oriented anatomy. Wolters Kluwer India Pvt Ltd

Murray WM, Buchanan TS, Delp SL (2002) Scaling of peak moment arms of elbow muscles with upper extremity bone dimensions. J Biomech 35(1):19–26. https://doi.org/10.1016/S0021-9290(01)00122-8

Öchsner A (2023) Computational statics and dynamics: An Introduction Based on the Finite Element Method. Springer, Cham

Rose J, Gamble JG, Adams JM (2006) Human walking (3rd ed). Lippincott Williams & Wilkins

Su Y, Zhang Q, Xu X, Gao Z (2016) Quality assessment of speckle patterns by consideration of both systematic errors and random errors. Opt Lasers Eng 86:132–142. https://doi.org/10.1016/j.optlaseng.2016.05.019

Chapter 19
Kinematic Implications of a Wedging Vertebral Fracture in Spine Flexion Movements

Claudia F. Romero-Flores, Jose M. Amorocho, Marcos Moya Bencomo, Rogelio Bustamante-Bello, and Iñaki Zenteno Aguirrezabal

Abstract Compression fractures can be caused by high axial loading or sudden spine flexion, which could happen during sports training or playing. However, information about thoracic compression fractures in sports is scarce and almost nonexistent. The literature suggests a strong relationship between spine injuries and local mobility restriction. Computer models help address spine mobility because they allow for calculating the motion of specific vertebrae. This work investigates the biomechanical implications of a healed T8 wedging fracture on spine motion during flexion exercises. Four male competitive cheer athletes were recruited. One participant experienced a wedging fracture at the T8 level. The subjects were asked to perform four unrestricted and restricted spine flexion exercises. Motion was captured at 120 Hz using a Vicon motion capture system. The OpenSim Thoracolumbar Spine and Rib Cage were used to calculate the ranges of motion of the spine since this model has proven to determine kinematics differences in vertebral motion.

Keywords Spine biomechanics · Thoracolumbar spine · Vertebral kinematics

C. F. Romero-Flores · J. M. Amorocho · M. M. Bencomo (✉) · R. Bustamante-Bello · I. Z. Aguirrezabal
School of Engineering and Sciences, Tecnologico de Monterrey, Monterrey, Nuevo Leon, México
e-mail: marcosmoya@tec.mx

C. F. Romero-Flores
e-mail: claudia.rf@tec.mx

J. M. Amorocho
e-mail: A01760029@tec.mx

R. Bustamante-Bello
e-mail: rbustama@tec.mx

I. Z. Aguirrezabal
e-mail: inaki@tecsalud.mx

A. Öchsner and H. Altenbach (eds.), *Engineering Design Applications VII*, Advanced Structured Materials 230, https://doi.org/10.1007/978-3-031-84346-4_19

19.1 Introduction

Compression fractures are the most common type of vertebral bone fracture (Cigerci et al. 2024). The elderly population is the most affected by this condition (Al et al. 2024). Vertebral compression fractures often happen when osteoporosis is presented due to the loss of bone density. Without osteoporosis, compression fractures are caused by high axial loading or sudden spine flexion (Menzer et al. 2015). Contact and acrobatic sports involve high axial loads and rapid flexion movements. Indicating these activities could be a risk factor for developing compression fractures. However, information about thoracic compression fractures in sports is scarce and almost nonexistent (Huang et al. 2015). Nevertheless, high incidences have been reported. For instance, nearly 800 fractures in the thoracolumbar spine were reported between 1990 and 2002 in junior cheerleading squats (Shields and Smith 2006). Neck and thoracolumbar spine fractures have been reported to be 4% of total injuries in Canadian cheerleading (Hardy et al. 2015).

Computer musculoskeletal models have been used to address spine mobility because they allow for calculating the motion of specific bones. OpenSim is a specialised open-source simulation software for human and animal motion (AS., et al. 2018). It was developed to enable biomechanics researchers to share and compare musculoskeletal models. The software has several tools that allow researchers to create and manipulate biomechanical models and simulate dynamics and neuromuscular control (AS., et al. 2018). The models are made of bone geometry and muscle elements. Users can modify and adapt their muscle models, bone interaction, contact models, controllers and analyses (Delp et al. 2007). Models can be driven from joint kinematics or motion capture data (e.g. marker coordinates, inertial measurement unit orientation).

Some OpenSim models have been developed to study the movement of the human spine (Christophy et al. 2012; Raabe and Chaudhari 2016; Bruno et al. 2015). The Thoracolumbar Spine and Rib Cage model was created to correctly predict the thoracic and lumbar vertebral loading (Christophy et al. 2012). This model contemplated individual vertebrae in the thoracic and lumbar regions, ribs, and sternum and incorporated associated muscles. This means thoracic and lumbar vertebrae can be completely articulated between each other. Compared to more traditional models, this allows for more precise studies of the spine's biomechanical implications. It has been reported that an articulated spine in musculoskeletal models results in fewer vertebrae position errors when compared to a fixed spine (Glover et al. 2021). A multi-segmental spine approach allows researchers to identify motion relationships and patterns between different sections of the spine (Gou et al. 2024).

The literature suggests a strong relationship between injuries and local mobility alterations. Patients suffering from a spine injury have shown movement alteration patterns during several activities (Tsang et al. 2014; Hooker et al. 2021). Therefore, this work aims to investigate the thoracolumbar spine motion of a healed T8 compression fracture patient and compare it to similar healthy participants, looking for differences in movement patterns between the vertebral pairs during complete flexion exercises.

19.2 Materials and Experimental Details

Ethical clearance was obtained from Copamel in 2023. After informing the research and experiment, four male cheerleaders accepted to participate in the experiment. They all matched in age, height, and weight (Table 19.1). Participant coded P01 had suffered from a wedging fracture in T8 a year before the experiment was performed, but for the session, he was healed and resumed regular training. The other three participants did not have any history of spine injuries. The experiment consisted of a single motion capture session. At the beginning of the session, participants' anthropometric information and sports backgrounds were taken (Table 19.1).

A seven-camera Vicon motion capture system (Vicon Motion Systems, Oxford Metrics plc) recorded participants' motion at 120 Hz. After a guided warm-up, nine-millimetre spherical reflective markers were placed with double-sided tape on meaningful body landmarks, such as the spinous processes of C3, C5, and each vertebra from C7 to S1, lateral to lumbar markers, in the pelvis, scapula and sternum (Fig. 19.1). A static trial was captured for 20 s, with the participant in an anatomically neutral position. All participants performed four exercises in the same order: the first was an unrestricted torso flexion–extension in a standing position (unrest FlexExt), next was an unrestricted flexion in sitting (unrest Flex), followed by the restricted version of both exercises (rest FlexExt and rest Flex). Participants were allowed to move their torsos freely for the unrestricted exercises but were asked to move the rest of their bodies as little as possible. Participants were asked to keep their backs as neutral as possible in the restricted exercises. Entire flexion movements were performed in the standing exercises, while just full flexion was performed in the sitting exercises. The exercises were repeated five times at 30 bits per minute in each trial.

Table 19.1 Participant information

	Age [years]	Weight [kg]	Height [cm]	BMI [kg/cm^2]	Experience [moths]	Position
P01	18	72.75	179	22.71	4	Spotter, Gymnast
P02	20	72.2	178	22.79	12	Base, Gymnast
P03	21	71.3	180	22.01	18	Spotter
P04	21	72.4	187.8	20.53	60	Spotter, Gymnast

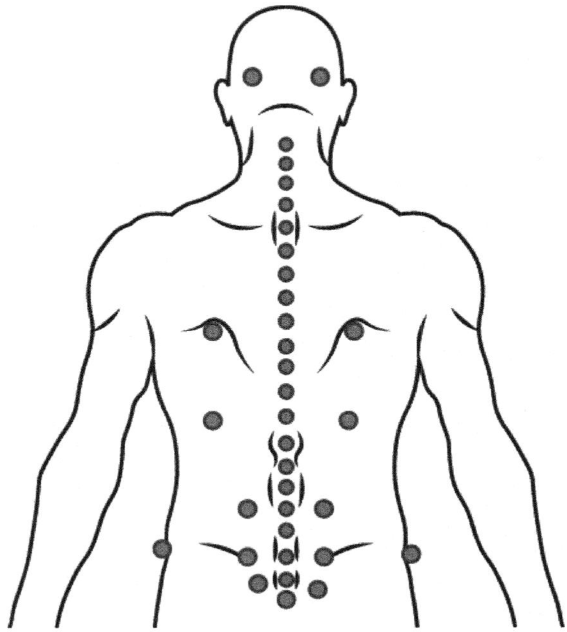

Fig. 19.1 Marker set used for the experiment. Markers were placed on the following body landmarks: right and left posterior superior iliac spine, right and left anterior superior iliac spine, the right and left tip of the iliac crest, spinous processes of C3, C5, and C7 to S1, laterally to L2 and L4, right and left 10th rib, inferior edge of right and left scapula, notch, xiphoid process, four at the head, head of the humerus, coracoid process, medial and lateral epicondyle of both arms, styloid and ulnar process of both wrists

Nexus (Vicon Motion Systems, Oxford Metrics plc) was used to label each trial, and MATLAB (R2023b, The MathWorks, Inc.) was used to prepare the data. A fourth-order Butterworth filter with a cut-off frequency of 5Hz was used in the marker coordinates. Gaps equal to or shorter than 1.5 s were interpolated with a cubic spline function. The axes were rotated, and units were converted to meters to meet the International Society of Biomechanics convention.

The OpenSim (AS., et al. 2018) software was used to simulate the motion of each participant. The Thoracolumbar Spine and Rib Cage Male model was utilised due to its capacity to allow unrestricted rotation of the thoracic and lumbar vertebrae in all three spatial directions (Bruno et al. 2015). For this study, the reduction of the system's degrees of freedom was accomplished by constraining the lateral bending and axial rotation of all spine elements and the movements of the wrist, ankle, and ribs in all directions. The model was scaled for each participant using the weight and one second of the static trial. Then, the Inverse Kinematics tool was used for each exercise to obtain the angles between each vertebral pair from T1 to S5.

Three cycles of the flexion movement were taken by visually looking for smoothness in the sagittal plane angles of C7, T1, and T2. The torso angle's minimum and maximum values were used to identify each cycle's start and end. The total range of movement in each cycle was calculated by setting the initial value to zero degrees.

A one-way ANOVA test performed in MATLAB was used to analyse the differences between the fractured participant and the healthy group at 25, 50, 75, and 100% of the cycle for each exercise.

19.3 Results

Table 19.2 shows the ANOVA p-values of the sagittal angles between vertebral pairs for all exercises. The mean sagittal angles for all exercises are shown in Figs. 19.2–19.4.

At the level of the injury, the pair T7–T8 presented significant differences at the beginning of the unrestricted Flex, the middle of unrestricted and restricted Flex exercise, and at the end of unrestricted FlexExt and restricted Flex. One significant value was presented at the T8–T9 pair level, at 25% of the unrestricted Flex exercise. No significant differences between groups were found in the restricted FlexExt or at 75% of any exercise.

The upper thoracic pairs (above T7) showed the most significant differences across all exercises. Differences can be seen throughout the cycle, especially in the pairs between T3 and T6. Thoracic pairs below the affected vertebrae show few significant differences, concentrating on unrest FlexExt and rest Flex exercise. No significant differences were found in the restricted FlexExt or at the beginning of the cycle (25%) in this region.

Lumbar pairs presented differences near the transition from the thoracic segment to the sacrum. The pair L1–L2 presented the most differences in the lumbar region: in restricted Flex exercise and the end of unrestricted FlexExt and restricted Flex. Only the L4–L5 pair did not show significant differences in any of the four exercises.

Transitions pairs (T12–L1 and L5–S1) showed significant differences in unrest FlexExt and unrest Flex, respectively.

19.4 Discussion of Results

Upper thoracic vertebral pairs concentrated most of the significant differences between groups. The pairs T3–T4, T4–T5, and T5–T6 presented significant differences throughout most of the cycle. When looking at the mean cycle values (Fig. 19.2), both groups separate from each other, with the fractured participant alternating more flexion (negative rotation) and extension (positive rotation), compared to the healthy group. At the same time, the variation between the healthy participants is minimal for these three pairs. Unexpectedly, differences up to 30° were found in the upper thoracic vertebrae. Unrestricted flex exercise had higher differences, which seem to be achieved at the middle of the cycle and maintained until the end.

Few significant differences were found in the lower thoracic (below T9) and the lumbar vertebrae. Overall, the unrestricted FlexExt exercise had more significant differences than the other exercise.

When looking at the mean values of all vertebrae (Figs. 19.2–19.4) the fractured participant showed more movement in the upper thoracic vertebrae than the lower thoracic and lumbar vertebrae. On the other hand, the healthy group showed little movement in the thoracic vertebra, tending to more movement and variation in the

Table 19.2 ANOVA p-values for the sagittal angles of all exercises. For each cell of percentage and vertebral pair, the values correspond as follows: upper left to unrest FlexExt, upper right to rest FlexExt, lower left to unrest Ext, and lower right to rest Flex. Significant differences are indicated in bold

Vertebral pairs	Cycle percentage							
	25%		50%		75%		100%	
T1–T2	0.22	0.09	0.58	0.18	0.50	0.18	0.21	0.11
	0.00	0.96	**0.00**	0.57	**0.00**	0.25	**0.00**	0.20
T2–T3	**0.02**	0.12	**0.00**	**0.01**	**0.00**	0.06	**0.00**	**0.02**
	0.36	0.80	0.67	**0.00**	0.53	**0.00**	0.52	**0.00**
T3–T4	**0.00**	0.46	**0.03**	0.43	0.59	**0.00**	**0.02**	**0.00**
	0.00	0.14	**0.00**	**0.00**	**0.00**	**0.00**	**0.00**	**0.00**
T4–T5	**0.00**	**0.00**	**0.00**	**0.00**	**0.00**	0.47	**0.00**	**0.01**
	0.00	**0.03**	**0.00**	**0.00**	**0.00**	**0.00**	**0.00**	**0.00**
T5–T6	**0.00**	**0.00**	**0.00**	**0.00**	**0.00**	0.26	**0.00**	**0.00**
	0.00	0.33	**0.00**	**0.00**	**0.00**	**0.00**	**0.00**	**0.00**
T6–T7	0.18	**0.03**	0.09	**0.02**	0.06	0.31	0.21	0.46
	0.02	0.12	0.08	0.06	**0.04**	0.05	**0.03**	0.07
T7–T8	0.85	0.86	0.52	0.89	0.83	0.93	**0.03**	0.96
	0.00	0.07	**0.00**	**0.02**	0.08	0.06	0.07	**0.03**
T8–T9	0.33	0.51	0.57	0.52	0.28	0.46	0.11	0.40
	0.01	0.26	0.06	0.18	0.08	0.14	0.07	0.11
T9–T10	0.86	0.30	0.08	0.12	**0.04**	0.52	0.50	0.39
	0.34	0.80	0.94	0.77	0.33	0.75	0.15	0.30
T10–T11	0.92	0.41	0.35	0.05	0.41	0.83	0.82	0.49
	0.36	0.73	0.22	0.07	0.42	**0.02**	0.28	0.10
T11–T12	0.56	0.38	**0.04**	0.37	**0.00**	0.63	**0.01**	0.94
	0.83	0.51	0.85	0.91	0.53	0.39	0.54	0.34
T12–L1	0.47	0.86	**0.03**	0.81	**0.00**	0.98	**0.00**	0.79
	0.89	0.23	0.10	0.87	0.09	0.48	0.10	0.30
L1–L2	0.85	**0.04**	0.06	**0.00**	**0.00**	**0.00**	**0.01**	**0.01**
	0.27	0.57	0.17	0.28	0.41	0.05	0.37	**0.04**
L2–L3	0.28	0.56	**0.01**	0.52	**0.03**	0.16	**0.04**	0.05
	0.18	0.66	0.68	0.96	0.52	0.66	0.45	0.78
L3–L4	0.75	0.07	0.27	0.14	0.15	0.07	0.18	**0.02**
	0.05	0.08	0.19	0.17	0.41	0.23	0.42	0.13
L4–L5	0.37	0.28	0.68	0.54	0.24	0.39	0.15	0.69
	0.15	0.17	0.42	0.29	0.95	0.55	0.69	0.61
L5–S1	0.24	0.69	0.47	0.93	0.45	0.45	0.93	0.46
	0.50	0.18	**0.00**	0.54	**0.00**	0.49	**0.01**	0.53

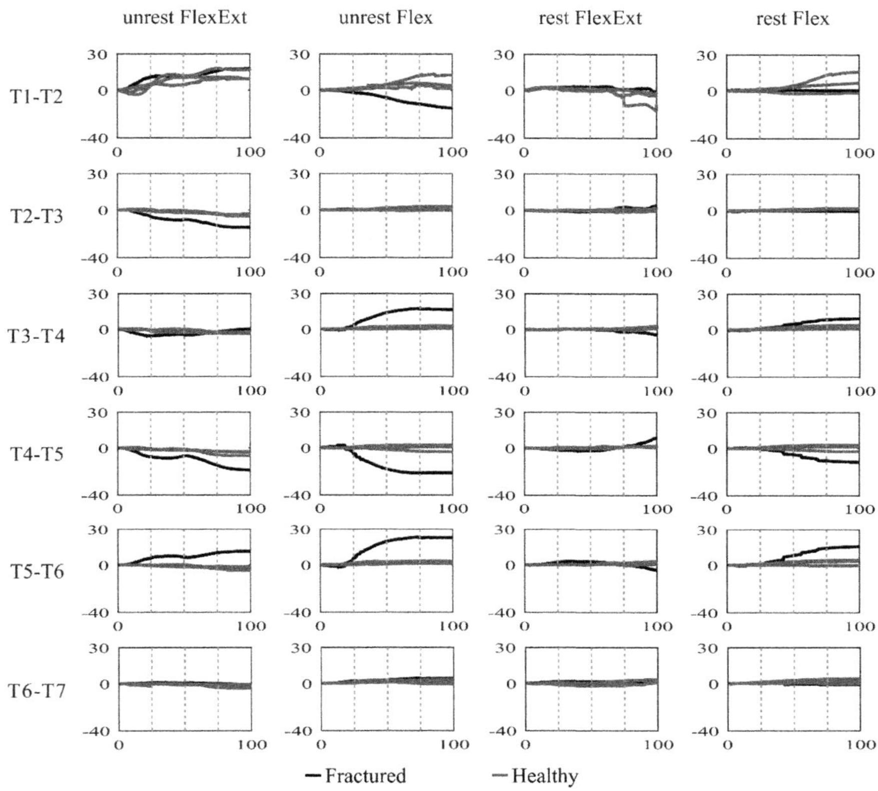

Fig. 19.2 Upper thoracic sagittal mean angles of the four participants. Vertical dotted lines indicate the 25, 50, and 75% of the cycle

lumbar vertebrae. Due to the articulation of the entire rib cage, thoracic vertebrae have less movement in between than the cervical and lumbar vertebrae. However, the results showed that the fractured participant had important movement between the upper thoracic vertebrae (Fig. 19.2). This indicates that the compression fracture could be related to more mobility in the thoracic vertebra. However, whether this was present before the incident is not known. Additionally, the healthy group showed more variability in the lumbar vertebral pairs than in the thoracic vertebral pairs. This variability could mitigate the differences between both groups, therefore not showing any statistical differences. Nevertheless, this motion pattern seems to be expected. In a similar study, where spine sagittal angles of healthy participants were analysed in flexion, more variability was found in the lumbar segments (between 17° and 11.2°) compared with the thoracic segments (between 7.7 and 1.6°) (Leardini et al. 2011).

It is worth mentioning that skin markers do not represent the exact location of the bony landmarks, resulting in positioning errors throughout the movement. The

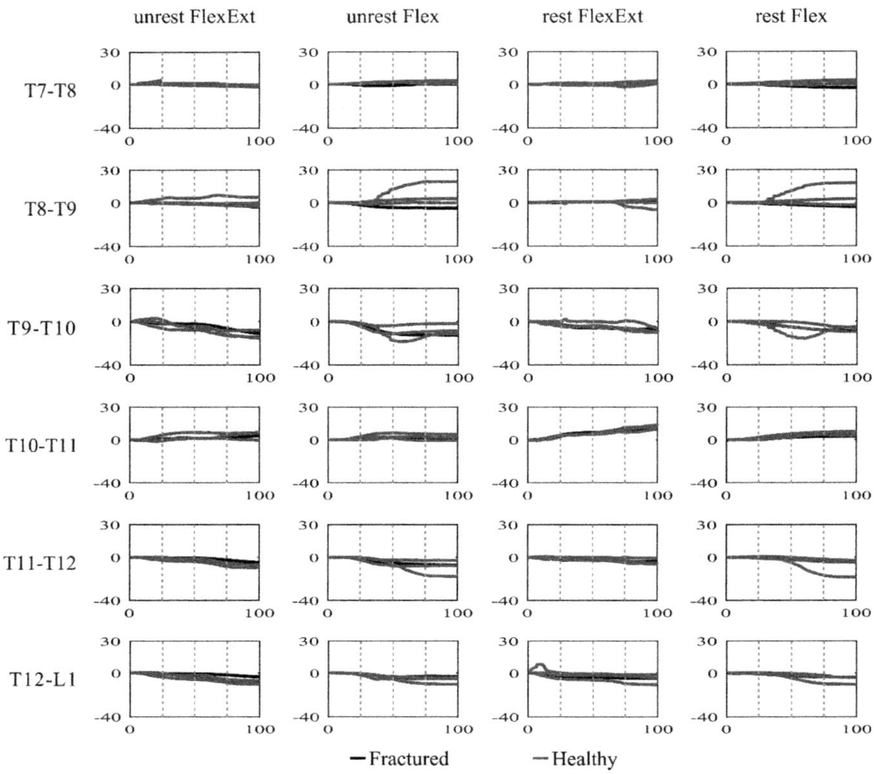

Fig. 19.3 Lower thoracic cycle sagittal mean angles of the four participants. Vertical dotted lines indicate the 25, 50, and 75% of the cycle

soft tissue artefact refers to the movement of the skin above the bones. This creates a discrepancy between markers and bony landmarks when the participant moves. It is expected that during maximum flexion, there will be some shift in the markers, which does not correctly represent the position of the spinous processes. However, in people with average body mass index, it has been reported that the error between the marker placement and the spinous process location is small (Zemp et al. 2014; Mörl and Blickhan 2006).

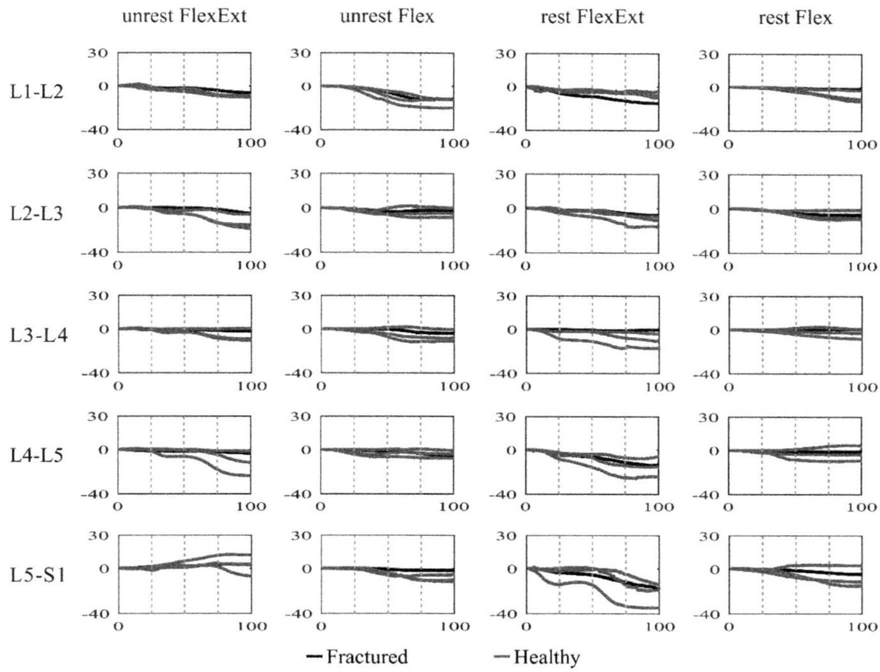

Fig. 19.4 Lumbar cycle sagittal mean angles of the four participants. Vertical dotted lines indicate the 25, 50, and 75% of the cycle

19.5 Conclusion

This case study showed that a thoracic compression fractured patient presented significant differences and more mobility in the upper thoracic vertebrae compared to healthy matching participants. The affected vertebra pairs showed fewer differences than those above, with no apparent pattern. Both unrestricted exercises presented more significant differences. These results conclude a relationship between high mobility in the thoracic vertebrae and thoracic compression fractures, with restricted mobility in the affected vertebra.

Acknowledgements The authors gratefully acknowledge the support of Tecnologico de Monterrey and Consejo Nacional de Humanidades, Ciencia y Tecnologia (CONAHCYT) Mexico, under scholarship number CVU 646998.

References

Al TK, Lauper N, Bauer DE et al (2024) Multidisciplinary and coordinated management of osteoporotic vertebral compression fractures: current state of the art. J Clin Med 13:930. https://doi.org/10.3390/JCM13040930

Bruno AG, Bouxsein ML, Anderson DE (2015) Development and validation of a musculoskeletal model of the fully articulated thoracolumbar spine and rib cage. J Biomech Eng 137(8):081003. https://doi.org/10.1115/1.4030408

Christophy M, Senan NAF, Lotz JC, O'Reilly OM (2012) A Musculoskeletal model for the lumbar spine. Biomech Model Mechanobiol 11:19–34. https://doi.org/10.1007/s10237-011-0290-6

Cigerci Ç, Şaşmaz Mİ, Cigerci NE (2024) Evaluation of vertebral fractures and associated injuries in multiple trauma patients. East J Med 29:186–192. https://doi.org/10.5505/EJM.2024.23500

Delp SL, Anderson FC, Arnold AS et al (2007) OpenSim: open-source software to create and analyze dynamic simulations of movement. IEEE Trans Biomed Eng 54(11):1940–1950. https://doi.org/10.1109/TBME.2007.901024

Glover NA, Kakar RS, Chaudhari AMW (2021) Effects of spinal coupling and marker set on tracking of spine models during running. J Biomech. 116:110217. https://doi.org/10.1016/j.jbiomech.2020.110217

Gou Y, Tao J, Lei H et al (2024) Trunk kinematic analysis of ascent and descent stairs in college students with idiopathic scoliosis: a case-control study. Spine J 24(9):1712–1722. https://doi.org/10.1016/J.SPINEE.2024.04.004

Hardy I, McFaull S, Saint-Vil D (2015) Neck and spine injuries in Canadian cheerleaders: an increasing trend. J Pediatr Surg 50(2):790–792. https://doi.org/10.1016/j.jpedsurg.2015.02.039

Hooker QL, Lanier VM, van Dillen LR (2021) Consistent differences in lumbar spine alignment between low back pain subgroups and genders during clinical and functional activity sitting tests. Musculoskelet Sci Pract 52:102336. https://doi.org/10.1016/J.MSKSP.2021.102336

Huang P, Anissipour A, McGee W, Lemak L (2015) Return-to-play recommendations after cervical, thoracic, and lumbar spine injuries: a comprehensive review. Sport Health 8(1):19–25. https://doi.org/10.1177/1941738115610753

Leardini A, Biagi F, Merlo A et al (2011) Multi-segment trunk kinematics during locomotion and elementary exercises. Clin Biomech 26(6):562–571. https://doi.org/10.1016/j.clinbiomech.2011.01.015

Menzer H, Gill GK, Paterson A (2015) Thoracic spine sports-related injuries. Curr Sports Med Rep 14(1):34–40. https://doi.org/10.1249/JSR.0000000000000117

Mörl F, Blickhan R (2006) Three-dimensional relation of skin markers to lumbar vertebrae of healthy subjects in different postures measured by open MRI. Eur Spine J 15:742–751. https://doi.org/10.1007/s00586-005-0960-0

Raabe ME, Chaudhari AMW (2016) An investigation of jogging biomechanics using the full-body lumbar spine model: model development and validation. J Biomech 49(7):1238–1243. https://doi.org/10.1016/j.jbiomech.2016.02.046

Sethu A, Hicks JL, Uchida TK et al (2018) OpenSim: simulating musculoskeletal dynamics and neuromuscular control to study human and animal movement. https://doi.org/10.1371/journal.pcbi.1006223

Shields BJ, Smith GA (2006) Cheerleading-related injuries to children 5 to 18 years of age: United States, 1990–2002. Pediatrics 117(1):122–129. https://doi.org/10.1542/peds.2005-1139

Tsang SMH, Szeto GPY, Lee RYW (2014) Altered spinal kinematics and muscle recruitment pattern of the cervical and thoracic spine in people with chronic neck pain during functional task. J Electromyogr Kinesiol 24,104–113. https://doi.org/10.1016/j.jelekin.2013.10.011

Zemp R, List R, Gulay T et al (2014) Soft tissue artefacts of the human back: Comparison of the sagittal curvature of the spine measured using skin markers and an open upright MRI. PLoS One 9(4):e95426. https://doi.org/10.1371/journal.pone.0095426

Chapter 20
Methodology for the Development of Numerical Modeling of Different Human Anatomical Structures

Martin Ivan Correa-Corona, Guillermo Urriolagoitia-Sosa, Beatriz Romero-Ángeles, Maria de la Luz Suarez-Hernandez, Yonatan Yael Rojas-Castrejon, Eduardo Enrique Carmona-Hernández, Alfonso Trejo-Enriquez, Rolando Manuel-Jose, and Guillermo Manuel Urriolagoitia-Calderón

Abstract Throughout history, the human body has been studied and analyzed from different perspectives, applying different techniques that only allowed obtaining basic knowledge about its functioning. Initially, dissections were developed on human corpses to visualize shapes and textures of organs, which later led to naming and classifying anatomical structures that make up the human being. Currently, various computational tools and medical equipment have been developed to facilitate the visualization of internal structures of the human body. This work proposes, in a general way, a methodology that exposes the process of developing three-dimensional models of various biological systems from computed tomography (*CT*) or magnetic resonance imaging (*MRI*), obtaining internal images of the human body, and making use of using *CAD*-type computational tools, the identification of images in the different planes (coronal, sagittal and transverse) is carried out to carry out the process of constructing a *3D* model of the desired biological elements as complex and precise as the user requires it. It represents a flexible process since the fundamental steps are mentioned to virtually replicate various biological tissues such as cortical bone,

M. I. Correa-Corona (✉) · G. Urriolagoitia-Sosa · B. Romero-Ángeles ·
M. de la L. Suarez-Hernandez · Y. Y. Rojas-Castrejon · E. E. Carmona-Hernández ·
A. Trejo-Enriquez · R. Manuel-Jose · G. M. Urriolagoitia-Calderón
Sección de Estudios de Posgrado E Investigación, Instituto Politécnico Nacional Escuela Superior de Ingeniería Mecánica y Eléctrica, UnidadProfesionalAdolfoLópezMateos, Zacatenco, Ciudad de México, México
e-mail: ingenieria.ivan.95@gmail.com

M. de la L. Suarez-Hernandez
e-mail: luzsuarez398@gmail.com

A. Trejo-Enriquez
e-mail: atrajoe@ipn.mx

R. Manuel-Jose
e-mail: rmanuel1800@alumno.ipn.mx

A. Öchsner and H. Altenbach (eds.), *Engineering Design Applications VII*, Advanced Structured Materials 230, https://doi.org/10.1007/978-3-031-84346-4_20

trabecular bone, cartilage, ligament, muscle, pulp, or dentin, presenting a model that is as real as possible in terms of its geometry and its location in the human body, regardless of the gender, age, height, weight of the individual, among other factors. The models that replicate the anatomical morphology are considered biomodels. Their development plays an essential and novel place in the area of medicine because, for their development, it is not necessary to perform invasive procedures that compromise the physical integrity of the patient, representing an alternative to the ethical and moral issues that arise when carrying out experimental research with living beings, in addition to saving time, economic and material resources. Biomodels improve the understanding of anatomical structures' functioning and study different pathologies that affect the human body. Likewise, biomodels are beneficial for studies on optimizing endoprostheses, endoprostheses, and various implants by proposing different materials that behave biocompatible with the human body. Additionally, with biomodels, surgical techniques can be proposed that optimize the treatment or replacement of any biological organ or tissue compromised by some pathology's action.

Keywords Numerical simulation · Methodology · Tomography · Bones · Finite element method · Soft tissue

20.1 Introduction

In different cultures and at distinct historical moments, the human body has been admired due to the complexity of the systems that are part of the human body, coming to be considered as a perfectly calibrated machine that allows the autonomous movement of the human in space. One of the most critical developments in the knowledge of the human body was the work done by philosophers and physicians such as *Hippocrates* (Panourias et al. 2005), who is considered the father of medicine, *Herophilus* and *Erasistratus* (Reverón 2014; Staden and Herophilus 1989), who performed vivisections that allowed them to describe the position, shape, color, or density of the internal structures, or Galen (Temkin and Straus 1946), whose observations and dissections on animals contributed to the understanding of human anatomy. However, knowledge about the human body was limited and often inaccurate due to the lack of technology and the prohibition of years of human dissection for religious reasons. Over time, the study and analysis of the human body have taken different approaches, including the development of biomodels, which are very convenient for understanding the behaviors of biological systems and helping generate hypotheses. Medicine and engineering work together to find solutions to problems affecting the structures of the human being, marking significant progress in the evolution of the study of the human body (Liang et al. 2018). Biomodels are characterized by virtually replicating the characteristics of the anatomical system to be studied, facilitating the numerical analysis of biological structures such as soft tissues or joints, and identifying the areas of discomfort or pain (Hernández-Vázquez et al. 2020; Li et al. 1999;

Table 20.1 Materials and values in the *Hounsfield* scale

Material	*Hounsfield* scale
Air	−1000 or less
Grease	−120
Water	0
Muscle	40
Contrast	130
Bone	400 or more

Sadeqi et al. 2023; Yang et al. 2023). The essential steps to develop a biomodel of the structures of the human body are explained below.

20.2 Biomodel Methodology

The methodology previously reported in several articles where biomodels of different biological tissues were developed (Urriolagoitia-Calderón et al. 2008; Hernández-Vázquez et al. 2018; Marquet-Rivera et al. 2021a; Maldonado et al. 1819; Antoniac et al. 1128; López 2012) was analyzed, highlighting the fundamental steps for their development. In the following, the fundamental steps to follow to develop a biomodel of the internal structures of the body are presented and explained.

20.2.1 Image Acquisition

It involves selecting the patient to be studied and generating the biological system to be examined using images obtained from a computerized tomography (*CT*) in *DICOM* format. The *CT* imaging process, renowned for its precision and efficiency, provides images in grayscale, and the *Hounsfield* units are used for their interpretation, determining the density of the tissues (Table 20.1) (Laguna-Canales et al. 2023) without the need for physical tests.

20.2.2 Processing the DICOM File in a CAD Type Software for Modeling

The biomodeling process of the biological structures begins with obtaining images of the tomography. Later, in a *cutting-edge CAD-type* program, the tomography is imported with the *DICOM* images, where a cloud of points is generated, locating the parts of the human body; in this specific case, the *Mimics Medical*® program, a

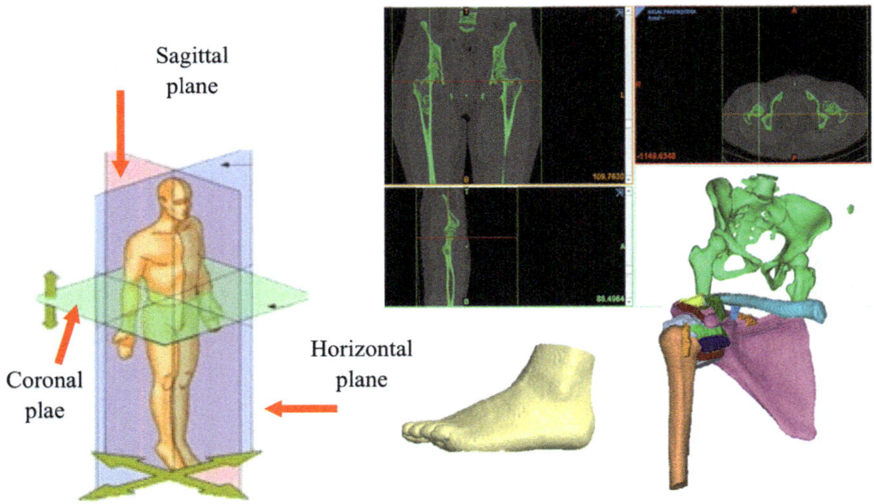

Fig. 20.1 View of the work areas and delimitation of the structures of the 3D model design

state-of-the-art tool, was used to develop the biomodel. After importing the *DICOM* file into the program, the program screen is divided into four windows, where the three views of each tomographic image (coronal, axial, and sagittal) are displayed, in addition to a fourth window where the simulation of the *3D* model is displayed.

Here is where the biological structures will be delimited and shaped, applying the threshold (using the *Hounsfield* scale) to define the different organic structures that will be part of the model. For each one of them, a new mask is created, where the area is drawn, filled or delimited for each one of the elements that will conform the model, generating a cloud of points that when being superimposed, a three-dimensional model is generated (Fig. 20.1). The project is saved with the version of the *Mimics Medical®* file format, to conserve the information provided by the *DICOM* file.

20.2.3 Design Optimization Software and Development of the Solid Model for Export

Then, the previously generated file is imported to another computer program that allows cleaning the raw data of a *CAD* design; in the development of this model, the *Materialise 3-matic®* program was used, where tools such as sorting, removing repeated triangular surfaces, repairing, smoothing, solid generation and uniform discretization according to the geometry were used for its evaluation, to be exported in a *STL* or *IGES* file.

20.2.4 Import of the Solid Model from a Finite Element Method Software

Once the biological structure biomodel is finished, the *STL or.IGES* file is imported into a program that performs numerical analysis using *FEM*, which divides the system into *n* elements of different geometries that are joined by nodes, where partial differential equations are established to reach an approximate final solution (Hernández et al. 2014). *Abaqus*® and *ANSYS*® are examples of computational programs that work with *FEM*, and the fundamental reason for using them is that they facilitate the resolution of problems with complex geometry, so their analytical evaluation would be difficult (Trejo-Enriquez et al. 2023). In this type of program, the analysis is carried out by defining the type of study and geometry, determining the mechanical properties of the material, applying boundary conditions, and applying loads, resulting in the visualization of the behavior of the biological system from the point of view of classical mechanics.

20.3 Importance of the Use of Biomodels Using FEM

With the incursion of *FEM* in medicine, significant contributions have been made to the analysis of models that characterize complex structures sufficiently accurately to offer results that are approximations to reality, predicting how the biological structure will behave. Biomodels are an alternative to the study of human organs and tissues without compromising the physical integrity of patients, saving time, economic resources, and materials (Fig. 20.2), making the process of biomodeling the appropriate solution to solve certain types of medical issues having a direct impact on the development of solutions such as surgical implants or prostheses customized for the patient, in addition to the diagnosis of various ailments (Hernández-Vázquez et al. 2018; Correa-Corona et al. 2023; Serrato-Pedrosa et al. 1650; Marquet-Rivera et al. 2021b).

20.4 Conclusion

The elaboration of the biomodel of biological systems is a complex activity to perform. However, with the methodology presented in this work, it is possible to develop biomodels of different biological tissues independently of their morphology in a shorter time since the fundamental steps for the realization of a complex model of its components are presented with a high level of precision, playing an important and novel role in the area of mechanics and medicine, where results of simulations close to reality are obtained with numerical analysis, facilitating the understanding

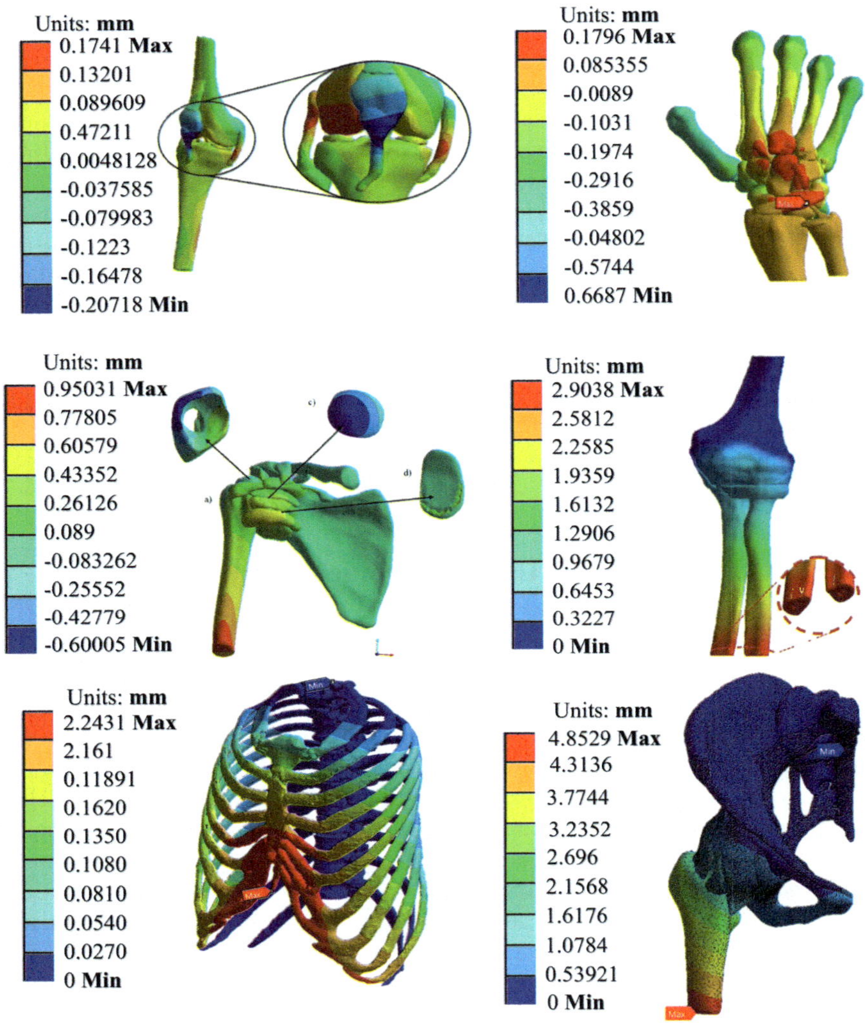

Fig. 20.2 Biomodels detailed made in last years with the technology help

specific pathologies, contributing to the development of new devices that improve the quality of life.

As seen in the elaboration of the 3D biomodel, despite the potential limitations, its development is an alternative to understanding biological processes with medical applications by applying non-invasive techniques to the patient. The models' complexity level is due to the evolution of medical equipment, computer programs, and applied techniques (Fig. 20.3). Considering that in the simulations performed, it

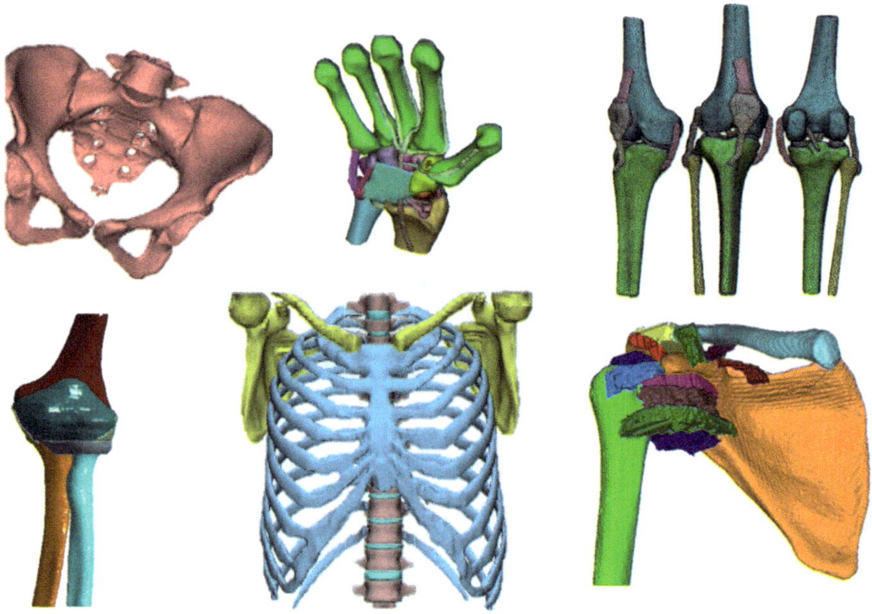

Fig. 20.3 Complex three-dimensional modeling examples for numerical analysis by finite element method

is impossible to include all of the influencing factors such as age, lifestyle, and diet, among other factors, they are approximate results.

References

Antoniac IV, Stoia DI, Ghiban B, Tecu C, Miculescu F, Vigaru C, Saceleanu V (2019) Failure analysis of a humeral shaft locking compression plate-surface investigation and simulation by finite element method. Materials 12(7):1128. https://doi.org/10.3390/ma12071128

Correa-Corona MI, Urriolagoitia-Sosa G, Romero-Ángeles B, Urriolagoitia-Luna A, Maya-Anaya D, Suarez-Hernández M de la L, Trejo-Enríquez A, Urriolagoitia-Calderón GM (2023) Application of the finite element model using 3D modeling of a human bone system with osteoporosis for biomedical and mechanical análisis. MOJ App Bio Biomech 7(1):14–15. https://doi.org/10.15406/mojabb.2023.07.00169

Hernández AF, Cañizo RGR, Huerta OS, Cruz EAM, Pineda JMS, Hernández EP (2014) Aplicación de metodología de modelado 3D utilizando Tomografías computarizadas para su análisis numérico. Revista Facultad de Ingeniería Universidad de Antioquía 72:116–126

Hernández-Vázquez RA, Romero-Ángeles B, Urriolagoitia-Sosa G, Vázquez-Feijoo JA, Vázquez-López AJ, Urriolagoitia-Calderón G (2018) Numerical analysis of masticatory forces on a lower first molar considering the contact between dental tissues. Appl Bionics Biomech 18:1–15. https://doi.org/10.1155/2018/4196343

Hernández-Vázquez RA, Urriolagoitia-Sosa G, Marquet-Rivera RA, Romero-Ángeles B, Mastache-Miranda OA, Vázquez-Feijoo JA, Urriolagoitia-Calderón G (2020) High-biofidelity biomodel generated from three-dimensional imaging (cone-beam computed tomography); a methodological proposal. Comput Math Methods Med. 2020(4):1–14. https://doi.org/10.1155/2020/4292501

Laguna-Canales AS, Urriolagoitia-Sosa G, Romero-Ángeles B, Martinez-Mondragon M, García-Laguna MA, Correa-Corona MI, Urriolagoitia-Calderón GM (2023) Mechanical design and numerical analysis of a new front wing for a formula one vehicle. Fluids 8(7):1–20. https://doi.org/10.3390/fluids8070210

Li G, Gil J, Kanamori A, Woo SY (1999) A validated three-dimensional computational model of a human knee joint. J Biomech Eng 121(6):657–662. https://doi.org/10.1115/1.2800871

Liang L, Liu M, Martin C, Sun W (2018) A deep learning approach to estimate stress distribution; a fast and accurate surrogate of finite-element analysis. J R Soc Interface 15(138):20170844. https://doi.org/10.1098/rsif.2017.0844

López CI (2012) Aplicación de un método no destructivo para la obtención propiedades físicas de tejido óseo basado técnica imagenológica y herramientas software CAD. Prospectiva 10(2):22–30

Maldonado JA, Puentes DA, Quintero ID, González-Estrada OA, Villegas DF (2023) Image-based numerical analysis for isolated type II SLAP lesions in shoulder abduction and external rotation. Diagnostics 13(10):1819. https://doi.org/10.3390/diagnostics13101819

Marquet-Rivera RA, Urriolagoitia-Sosa G, Hernández-Vázquez RA, Romero-Ángeles B, Mastache-Miranda OA, Urriolagoitia-Calderón G (2021) High biofidelity 3D biomodel reconstruction from soft and hard tissues (knee), FEM, and 3D printing; a three-dimensional methodological proposal. BioMed Res Int 2021:1–10. https://doi.org/10.1155/2021/6688164

Marquet-Rivera RA, Urriolagoitia-Sosa G, Romero-Ángeles B, Hernández-Vázquez RA, Mastache-Miranda OA, Cruz-López S, Torres-Yáñez A, Urriolagoitia-Calderón G (2021) Numerical analysis of the ACL, with sprains of different degrees after trauma. Comput Math Methods Med 2021(1). https://doi.org/10.1155/2021/2109348

Panourias IG, Skiadas PK, Sakas DE, Marketos SG (2005) Hippocrates a pioner in the treatment of head injuries. Neurosurgery 57(1):181–189. https://doi.org/10.1227/01.NEU.0000163416.66374.48

Reverón RR (2014) Herophilus and Erasistratus, pioneers of human anatomical dissection. Vesalius; Internationales Historiae Medicinae 20(1):55–58

Sadeqi S, Baumann AP, Goel VK, Lilling V, Sullivan SJ (2023) A validated open-source shoulder finite element model and investigation of the effect of analysis precision. Ann Biomed Eng 51(1):24–33. https://doi.org/10.1007/s10439-022-03018-8

Serrato-Pedrosa JA, Urriolagoitia-Sosa G, Romero-Ángeles B, Urriolagoitia-Calderón GM, Cruz-López S, Urriolagoitia-Luna A, Carbajal-López DE, Guereca-Ibarra JR, Murillo-Aleman G (2024) Biomechanical evaluation of plantar pressure distribution towards a customized 3D orthotic device; a methodological case study through a finite element analysis approach. Appl Sci 14(4):1650. https://doi.org/10.3390/app14041650

von Staden H (1989) The art of medicine in early Alexandria. 1 Edition, Ed. Cambridge University Press, pp 144–146

Temkin O, Straus WL (1946) Galen's dissection of the liver and of the muscles moving the forearm; translated from the anatomical procedures. Bull Hist Med 19(2):167–176

Trejo-Enriquez A, Urriolagoitia-Sosa G, Romero-Ángeles B, García-Laguna MÁ, Guzmán-Baeza M, Martínez-Reyes J, Rojas-Castrejon YY, Gallegos-Funes FJ, Patiño-Ortiz J, Urriolagoitia-Calderón GM (2023) Numerical evaluation using the finite element method on frontal craniocervical impact directed at intervertebral disc wear. Appl Sci 13(21):11989. https://doi.org/10.3390/app132111989

Urriolagoitia-Calderón G, Urriolagoitia-Sosa G, Hernández-Gómez LH, Merchán-Cruz EA, Vite-Torres M, Feria-Reyes CV, Beltrán-Fernández JA (2008) Análisis del desgaste de la articulación cabeza femoral-copa acetabular mediante simulación experimental con máquina perno-disco. Revista Colombiana de Biotecnología 10(1):94–110

Yang Z, Xu G, Yang J, Li Z (2023) Effect of different loads on the shoulder in abduction postures; a finite element analysis. Sci Rep 13(1):9490. https://doi.org/10.1038/s41598-023-36049-9

Chapter 21
Comparative Biomechanical Analysis by Finite Element Method of Healthy Hip Versus Hip with Osteomyelitis in the Initial Phase

Jonathan Rodolfo Guereca-Ibarra, Guadalupe Murillo-Aleman, Guillermo Urriolagoitia-Sosa, Beatriz Romero-Ángeles, Jorge Alberto Gomez-Niebla, Jesus Alejandro Serrato-Pedrosa, Alejandro Urriolagoitia-Luna, Gerardo Esteban Santiago-Lopez, and Iván Alejandro López-Zumarán

Abstract One of the most important structures of the human body is the bone structure, since it acts as a support and is the basis of the body's mobility. In order for the bone system to work optimally, it is necessary that it maintains a good quality, which can be affected due to injuries or any pathology. One of the pathologies that can most affect the bone system is osteomyelitis, because it is an infection that weakens the bones causing serious consequences that can even lead to bone destruction. In

J. R. Guereca-Ibarra · G. Murillo-Aleman (✉) · G. Urriolagoitia-Sosa · B. Romero-Ángeles · J. A. Gomez-Niebla · J. A. Serrato-Pedrosa · A. Urriolagoitia-Luna · G. E. Santiago-Lopez · I. A. López-Zumarán
Escuela Superior de Ingeniería Mecánica y Eléctrica, Sección de Estudios de Posgrado E Investigación, Instituto Politécnico Nacional, Unidad Profesional Adolfo López Mateos, Zacatenco, Ciudad de México, México
e-mail: Guadalupe.murillo.a95@gmail.com

J. R. Guereca-Ibarra
e-mail: jguerecai2200@alumno.ipn.mx

G. Urriolagoitia-Sosa
e-mail: guiurri@hotmail.com

B. Romero-Ángeles
e-mail: BROMEROA@ipn.mx

J. A. Gomez-Niebla
e-mail: george7gomez@gmail.com

J. A. Serrato-Pedrosa
e-mail: alejandroserrato@live.com.mx

A. Urriolagoitia-Luna
e-mail: Alex_ul56@hotmail.com

G. E. Santiago-Lopez
e-mail: gsantiagol1301@alumno.ipn.mx

A. Öchsner and H. Altenbach (eds.), *Engineering Design Applications VII*, Advanced Structured Materials 230, https://doi.org/10.1007/978-3-031-84346-4_21

the field of classical mechanics, the bone structure is studied to know its mechanical properties and biomechanical behavior in different circumstances, because there are several factors that can greatly modify such behavior such as bone size, age, gender, injuries, pathologies, etc. In the present work we intend to perform an analysis of the bone structure of the hip, formed by cortical and trabecular bone of the pelvis, cortical and trabecular bone of the femur, as well as the iliofemoral, ischiofemoral and pubofemoral ligaments, in which the comparison of a structure in clinically healthy conditions against one affected with osteomyelitis in the femur in an initial phase is made. Different analyses are performed with the help of the finite element method (FEM) to learn more about the biomechanical behavior of this structure with this pathology in order to identify the critical areas and to be able to implement a possible treatment for this infection before reaching a more advanced stage.

Keywords Biomechanical behavior · Bone · Finite element method · Osteomyelitis · Pathology

21.1 Introduction

The skeletal system represents a large part of the human body and it is essential to keep it in good condition because it fulfills vital functions for human beings. Among its main functions are that it serves as support for the body, as protection for the organs and is one of the main responsible for providing mobility (Lafita-Tejedor 2003). Bones can become damaged or fractured due to injuries or different pathologies such as osteoporosis, osteogenesis or osteomyelitis. Osteomyelitis is an infection that can affect people of all ages, it is important to detect it on time because it weakens the bones causing serious consequences that can even lead to bone destruction, but with timely treatment this can be avoided (Rao et al. 2011). In the field of classical mechanics, the bone structure is studied to know its mechanical properties and biomechanical behavior in different circumstances, since there are several factors that can greatly modify such behavior, such as bone size, age, gender, injuries, pathologies, etc. (Guede et al. 2013).

21.2 Methodology

For the realization of this work, it was used as support a computed tomography (CT), from where a series of images in *DICOM* format are obtained as seen in Fig. 21.1, which can be visualized by means of different computational software, through them

I. A. López-Zumarán
e-mail: ivan.zumaran@utd.edu.mx

it is possible to obtain a biomodel of any biological system by making masks of the desired structure, see Fig. 21.2 (López-Liévano et al. 2017; D'Amico et al. 2005).

There are several specialized software for the realization of biomodels, in this particular case it was used Materialise Mimics, which allows to generate masks of the desired structure (Serna-Serna et al. 2010). To obtain better results it is recommended to work with individual masks and at the end to work with the structure as an assembly so that the results are as accurate as possible.

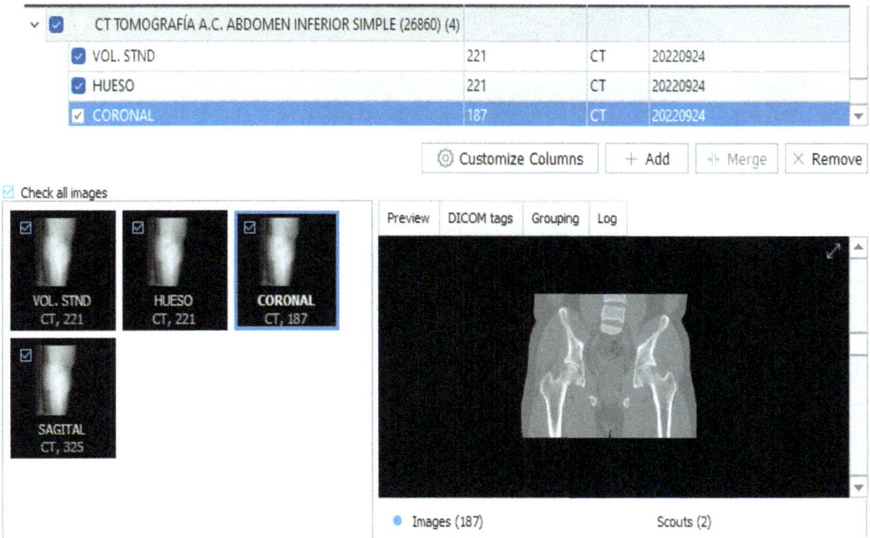

Fig. 21.1 Viewing *DICOM* images

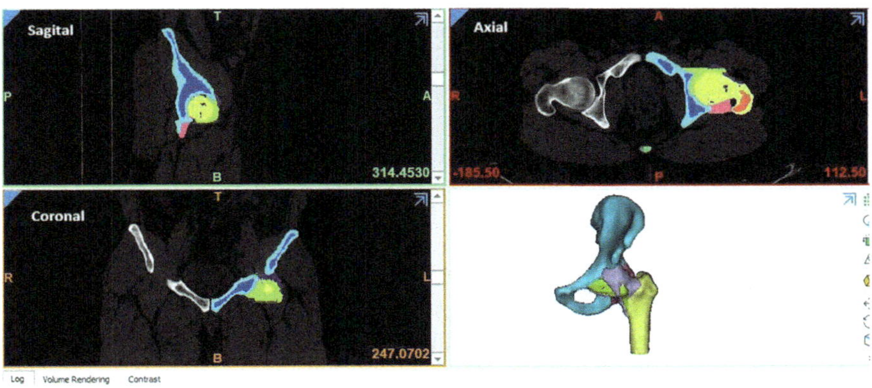

Fig. 21.2 Visualization of the generated hip structure masks

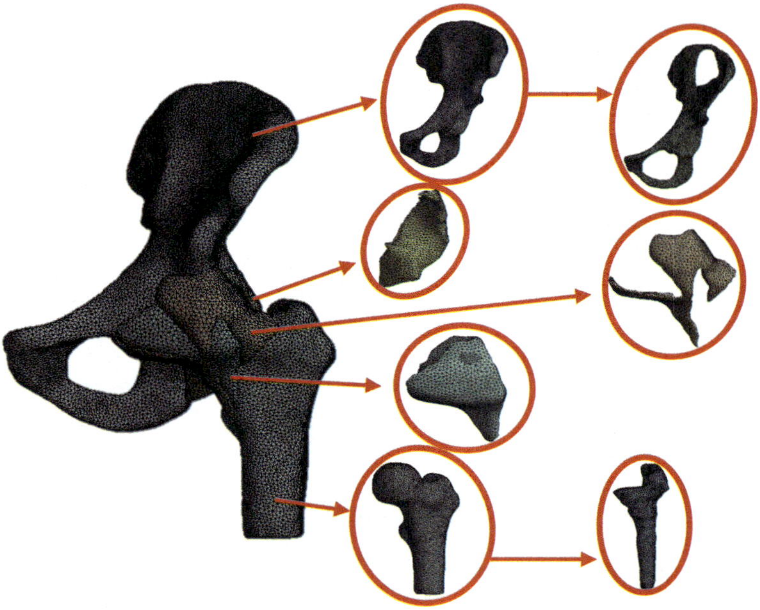

Fig. 21.3 Visualization of the discretized model in all its components

Once the masks are generated, they are exported to another software, in this case Materialise 3-Matic, in which possible errors in the construction of the model are corrected. It is also important to convert the model into a volume, because otherwise it is impossible to perform the corresponding analysis. After making sure that the model is free of errors, it is possible to pass it to another software, in this case ANSYS Workbench in which the analysis will be performed. For the analysis to run correctly, it is necessary to perform a uniform meshing of the model as seen in Fig. 21.3, establish the mechanical properties corresponding to each material, determine the boundary conditions and apply the desired load so that the results obtained are as close as possible to reality, see Fig. 21.4 (Ashman et al. 1984).

21.3 Results

The above process was to simulate a clinically healthy hip. To simulate phase 1 osteomyelitis, slight modifications were made to the trabecular bone of the femur, removing material from it in order to compare the results obtained in both analyses and to visualize how this pathology affects this structure see in Fig. 21.5.

Once the necessary modification has been made to the model to simulate the osteomyelitis, the same process is carried out as with the first model, in order to obtain the corresponding analysis, see Fig. 21.6.

Fig. 21.4 Boundary conditions and load applied to the model

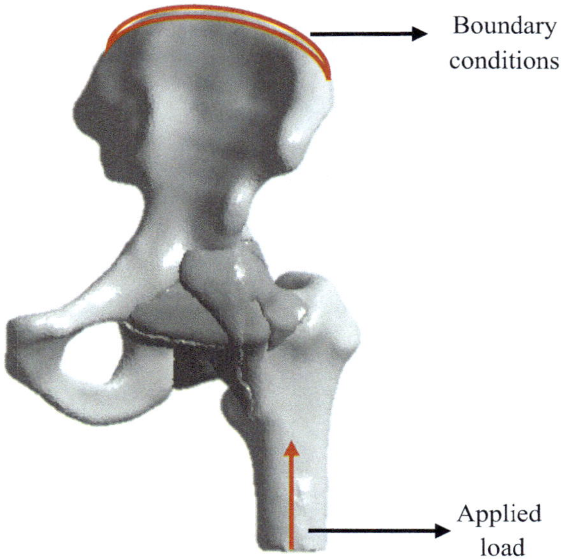

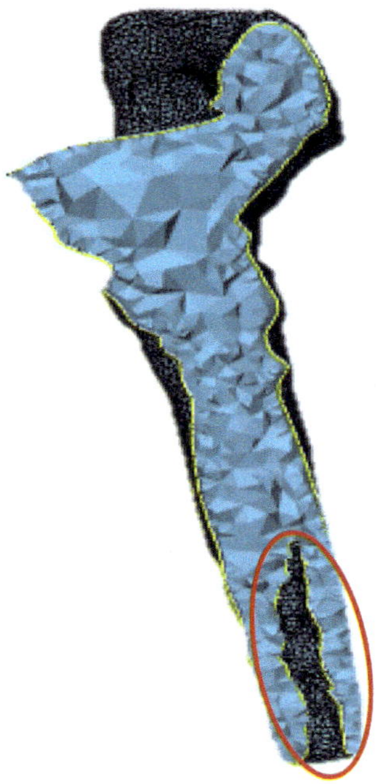

Fig. 21.5 Modification made to the trabecular bone of the femur

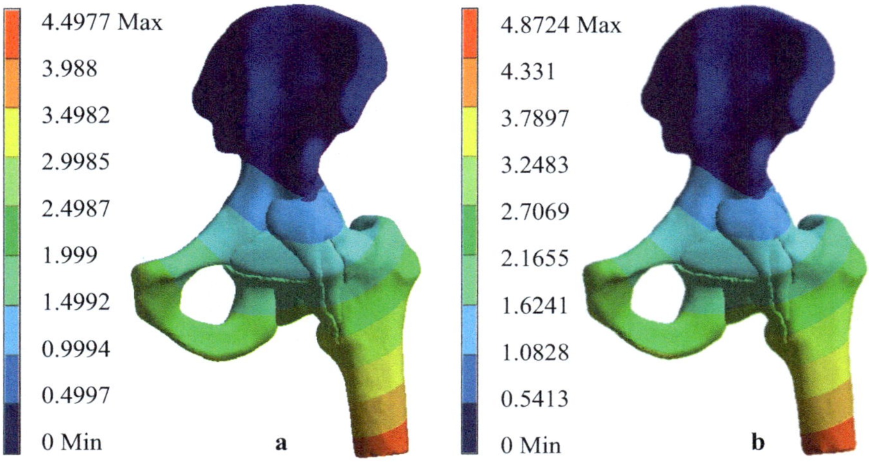

Fig. 21.6 Results obtained to the total deformation in the analyses in mm. **a** Model in healthy clinical conditions. **b** Model with osteomyelitis

21.4 Conclusion

Although the images obtained in both analyses are very similar and the changes that have occurred cannot be seen with the naked eye, when observing the results of the total deformations in both, it can be seen that they increase slightly in the model in which the osteomyelitis was implemented, which indicates that the structure has undergone a change due to this pathology. Other analyses such as strain or stress can be performed to obtain a better interpretation of the results and to know how and in which part of the structure is most affected. Any biological structure can be simulated with the help of *FEM*, which has been used in the area of orthopedics for several years to evaluate any bone structure that has suffered some kind of damage and, thus, to offer timely treatment to different pathologies, as well as for the design of different orthopedic devices.

Acknowledgements The authors thanks to the Instituto Politécnico Nacional and COFAA for the given support for the presentation of this investigation.

References

Ashman RB, Cowin SC, Van Buskirk WC, Rice JC (1984) A continuous wave technique for the measurement of the elastic properties of cortical bone. Biomechanics 17(5):349–361. https://doi.org/10.1016/0021-9290(84)90029-0

D'Amico DS, Cisilino AP, Sammartino MR, Capiel C (2005) Modelado computacional de estructuras óseas utilizando el método de los elementos finitos y tomografías computarizadas: análisis de la estabilidad de un implante gleno-humeral. *Mecánica Computacional* 24:1895–1911

Guede D, González P, Caeiro JR (2013) Biomecánica y hueso (I): conceptos básicos y ensayos mecánicos clásicos. *Rev Osteoporos Metab Miner* 5(1):43–50. https://doi.org/10.4321/S1889-836X2013000100008

Lafita-Tejedor J (2003) Fisiología y patología ósea. An Sist Sanit Navar 26(3):7–15

López-Liévano A, López-Liévano DR, Caicedo Ortiz HE, González-Rebattú A (2017) Biomodelado de los componentes del oído medio humano haciendo uso de imágenes de resonancia magnética. *Científica* 21(1):3–8

Rao N, Ziran BH, Lipsky, B.A (2011) Treating osteomyelitis: antibiotics and surgery. *Cirugía Plástica y Reconstructiva* 177–187. https://doi.org/10.1097/PRS.0b013e3182001f0f

Serna-Serna W, Trujillo-Lemus JP, Rivera-Piedrahita JH (2010) Descripción del estándar DICOM para un acceso confiable a la información de las imágenes médicas. *Scientia Et Technica* 16(45):289–294

Chapter 22
Entropy and Complexity in the Analysis of the Neuromechanical Pattern of the Vocal Folds Affected by Smoking

Xóchitl Siordia-Vásquez, Luz Yazmín Villagrán-Villegas, Luis Héctor Hernández-Gómez, Eduardo Hernández- Márquez, and José Omar Millán-Tejeda

Abstract It has been demonstrated that the act of smoking can irritate the mucous membrane, thereby stimulating the secretion of mucus. This protective mechanism serves to safeguard the vocal folds from potential damage caused by biomechanical friction. Furthermore, it has been shown that these alterations influence the sensory feedback data that the brain employs to encode a motor control pattern, which in turn regulates the neuromechanical processes of the vocal folds during the vocal production process. The purpose of these modifications is to establish a coupling between the trigeminal innervation and the neuromechanical control pattern, thereby enabling the vibratory rhythm of the vocal folds to be self-regulating in accordance with the physiological characteristics of each individual's mucosa. This will facilitate the generation of cycle-to-cycle oscillations, thereby maintaining the movements that produce the voice over time. The aim of this study is to analyse the complexity of neuromechanical patterns using an entropy estimator to quantify the decoupling deficit between vocal fold movement rhythms in search of differences between larynxes not exposed

---

X. Siordia-Vásquez (✉)
School of Electronics and Communications Engineering, Universidad Veracruzana, Veracruz, México
e-mail: xsiordia@uv.mx

L. Y. Villagrán-Villegas
School of Mechanical and Electrical Engineering, Universidad Veracruzana, Veracruz, México
e-mail: yvillagran@uv.mx

L. H. Hernández-Gómez
School of Mechanical and Electrical Engineering, Instituto Politécnico Nacional, Zacatenco, Ciudad de México, México

E. H.-. Márquez · J. O. Millán-Tejeda
Mechatronics Engineering Department, Tecnológico Nacional de México, Campus Poza Rica, Poza Rica, Veracruz, México
e-mail: eduardo.hernandez@itspozarica.edu.mx

J. O. Millán-Tejeda
e-mail: jose.millan@itspozarica.edu.mx

A. Öchsner and H. Altenbach (eds.), *Engineering Design Applications VII*, Advanced Structured Materials 230, https://doi.org/10.1007/978-3-031-84346-4_22

to nicotinic toxicants. To do this, we use an entropy-based prediction model that provides a graphical approximation of the neuromechanical pattern and an attractor that characterizes the deficit in coordination in the rhythm of vocal fold movements and the inefficiency of motor control to sustain them over time. To analyse the variability and complexity of biological signals, a one-factor ANOVA and Tukey's test were performed on three groups of individuals with an irritant factor affecting mucus secretion: smokers with laryngeal reflux disease (ERGE), non-smokers with stress, and non-smokers with viral conditions. The results showed clear differences between the neuromechanical coupling deficits of smokers and non-smokers. Smokers had the lowest values for neuromechanical pattern entropy and high values for coupling deficits suggesting that there is less complexity associated with a lower adaptive and functional capacity of the vocal folds compared to non-smokers.

Keywords Neuromechanical patters · Complexity · Entropy in biological systems · Asymmetry in vocal cords · Voice biomarkers · Smoking disorders · Motor control processing

22.1 Introduction

The main alkaloid or psychoactive substance responsible for the addictive nature of smoking is nicotine, an organic compound contained in the tobacco leaf from the Solanaceae family of the genus Nicotiana Tabacum mainly in the Virginia and Purpurea species, (Martín et al. 2004). The studies (Castro et al. 2023) reveal that smoking is a combustion process that converts nicotine into toxic gases and, in this composition, it crosses the smoker's blood–brain barrier within 9–19 s and is not metabolized in the brain, becoming a neurotoxic agent that is absorbed by the lungs in 70–90% of cases (Pérez 1995). In other cases 50% of this poison is absorbed by the mucous membranes that protect the pharynx, the sublingual plexus and the vocal cords as the smoker inhales it through the mouth to reach the respiratory system. When this neurotoxin is trapped in the vocal fold membranes, chronic inflammation and increased mucus secretion occur, resulting in thickening of the collagen and elastin fibres.

This leads to asymmetric contact between the two vocal folds, as illustrated in Fig. 22.1, section A (e). This asymmetry results in a neuromechanical deficit in the larynx of smokers, as the motor control that regulates the tension and vibration of the vocal folds is not maintained throughout the phonatory cycle. This phenomenon allows for the distinction between healthy larynxes (see Fig. 22.1, section (B) and those exposed to tobacco toxins, with similarities between male smokers (see Fig. 22.1, section (C) and female smokers (see Fig. 22.1, section D).

Evidence suggests that changes in mucosal secretion alter sensory feedback and lead to a deficit in neuromechanical coupling. (Kelleher et al. 2013) Research shows that after an initial linear interval, the stiffness of the vocal fold tension increases as the deformation of elastin and collagen fibres in the mucous membrane increases.

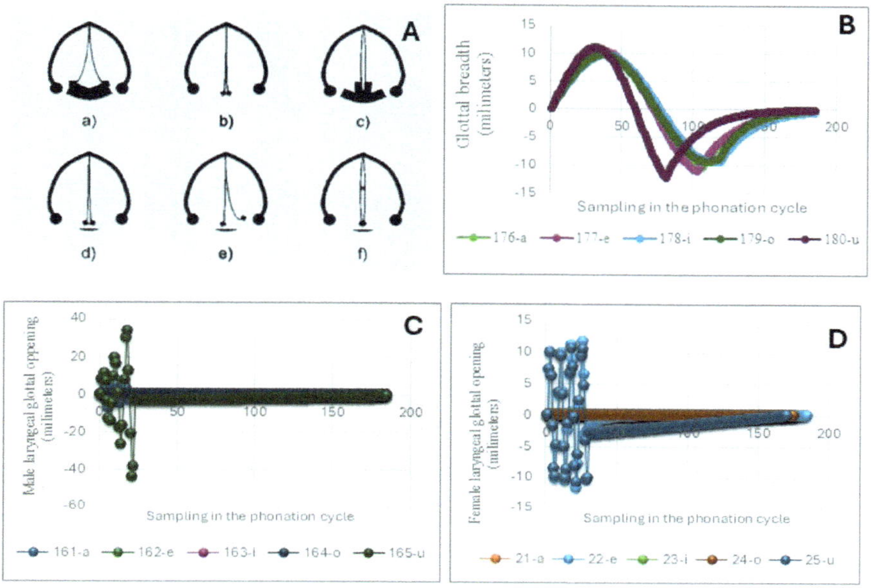

Fig. 22.1 Alteration of the vocal cord membrane by smoking

The greater the change in mucous flow secretion in the membrane, the greater the stiffness of the vocal fold tension, leading leading to a deficit in the neuromotor coordination coupling of the vocal folds that is perceived in the vocal timbre signal (Taiz 2020; Zhanga 2016; Cardoso 2021). As shown in Fig. 22.2, this phenomenon is reflected in the laryngeal dynamics and acoustic waveform of the speech signal y, has been studied by Chan and Rodríguez (2008), Kazemirad et al. (2014), and Zhang (2016), they found that the coupling deficit correlates with vocal fold stiffness and membrane disruption, as the system tends to balance in an attempt to achieve efficient voice production.

On the other hand, Siordia et al. (2021) demonstrated that in the presence of vocal fold rigidity, a very long pre-phonatory preamble is established, during which the system that regulates the motor control of the larynx tries to find a neuromechanical pattern that gradually self-adjusts the speed of the rhythm and the coordination of the movements according to the level of rigidity of the vocal folds. These two factors also relate to the neuromechanical pattern and the coupling deficit, both of which are physiological phenomena influenced by a purposeful and intentional nervous system.

This deficit in neuromechanical coupling has been commonly referred to as noise or interference and has been associated with the fundamental frequency, F_O, (Byeon and Cha 2020), a parameter related to the rate at which the vocal folds vibrate during voice production. Thus, Glover and Duhamel (2023) suggest analysing the neuromechanical coupling deficit using parameters that quantify the perturbation of the fundamental frequency, *Fo*, in a parameter known as jitter (Zealouk and Satori 2018), or by

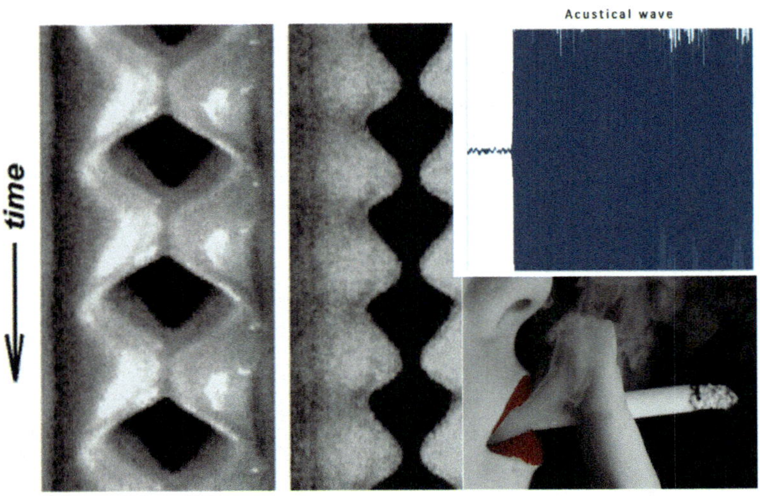

Fig. 22.2 The acoustic waveform of the speech signal reflects changes in the neuromechanical pattern

the harmonic-to-noise ratio, (NHR) (Lustosa Pinto et al. 2014) however, these parameters are the result of models based on linear approximations that do not represent the natural phenomenon that generates them (Fagherazzi et al. 2021; Nieto 2001). Although they manage to quantify the disturbance in the acoustic wave (Lustosa Pinto et al. 2014), it is not clear how they relate to the deficit in neuromechanical coupling or how to quantify it.

Since the physiological conditions of the mucous membranes of smokers have been shown to differ, it is also possible that there are differences in the attractors of neuromechanical coupling deficits in the vocal folds affected by smoking (Glover and Duhamel 2023). Therefore, there is a growing interest in exploring new models to predict neuromechanical deficits and using entropy-based methods to analyse the nature, variability and complexity of this signal (Harakannanavar 2019) and to identify possible biomarkers that differentiate the neuromechanical pattern of smokers.

This paper uses an entropy-based neuromechanical deficit prediction model to derive an attractor that characterizes smoking-induced vocal fold inefficiency. The analysis of the variability of these biological signals was performed by evaluating an entropy factor quantified from the attractor. One-factor ANOVA analysis was used for complexity analysis, as it can help to identify significant differences between the different neuromechanical coupling deficits caused by irritants present in the mucous membrane: laryngeal pharyngeal reflux (ERGE) in smokers and stress and viral conditions in non-smokers.

22.2 Materials and Methods

22.2.1 Experimental Approach

This study uses voice signal recordings to extract the neuromechanical pattern from the vowel acoustic wave, using a bank of Mexican speakers provided by the Nacional Rehabilitation Institute, Guillermo Ibarra Ibarra, in Mexico City. The recordings correspond to the vowels a, e, i, o, and u, held for three minutes, and come from men and women, smokers and non-smokers, aged between 25 and 45 years.

The time series in Fig. 22.3 shows that the heterogeneity or unevenness in the variability of the signal is not simply noise but represents an entropy that seems to define the self-regulating capacity of the biological system when the physiological conditions of the internal environment of the mucous membrane change, and when the membrane is exposed to irritating factors, it tends to equilibrate and corresponds to a less adaptive neuromechanical system. (Naranjo and de la Cruz 2010).

The database consists of 981 vocal recordings, grouped into three classes and previously diagnosed for the presence of a mucous membrane irritant. In order to balance the number of classes, each group is composed of the same number of samples, as shown in Table 22.1.

Each of the samples of the experimental corpus was recorded in an acoustic room with a professional Shure microphone, model SM58, with noise cancellation, placed 10 cm from the speaker's mouth. The recordings were stored in WAV format with a sampling frequency of 22,050 bits/sec. in a monophonic channel with a resolution of 16 bits, to reduce the computational cost of processing.

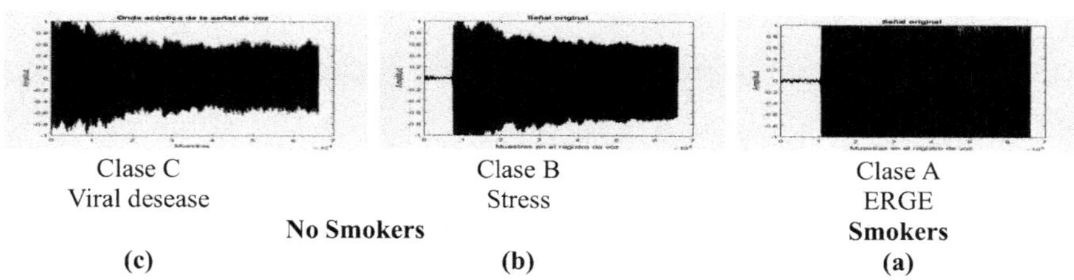

Fig. 22.3 Acoustic waves in their time series: normal coupling (**a**), non-nicotine entropy (**b**), and nicotine entropy (**c**)

Table 22.1 Description of the databank and its classes

Class	Description of records	Number of samples
A	Smokers with ERGE	327
B	No smokers stress	327
C	No smokers with viral disease	327
	Total	981

22.2.2 Neuromechanical Deficit Prediction Model Based on Entropy

The neuromechanical deficit prediction model (Siordia Vásquez and Jorge 2016) has a public domain AMDF predictor that uses a magnitude difference function to capture the complex dynamics of the vocal folds during the production process of the vocal act. This requires two opening and closing movements and regulates the coordination of the rhythms between these movements, giving rise to the phonatory cycle, *CF*. Since the phonatory cycle consists of two successive closures, $CF = 2To$, the first step is to estimate the duration of the fundamental period, *To*.

In order to quantify the neuromechanical deficit, a dynamic function with sliding windows is included, with a size equal to the number of samples that make up the *CF* phonation cycle; the windows are designed with an entropy function, see Fig. 22.4, that passes through the acoustic wave cycle by cycle to obtain a normalized envelope signal that also has a periodicity that corresponds to the physiological characteristics that identify the mucous membrane of each individual in a complete vocal cycle. The envelope is a graphical approximation of the neuromechanical pattern, see Eq. (22.1).

$$T_o = AMDF(n) = \frac{1}{N}\sum_{j=m}^{m+N-1} |s(j) - s(j+n)| \tag{22.1}$$

Then, using an entropy predictor, see Eq. (22.2), the envelope of the neuromechanical pattern is run cycle by cycle to quantify the deficit of each phonation cycle, CF, to form a vector of data with which a graphical approximation of an attractor is reconstructed, characterizing the variability of the dynamic peculiarities that model

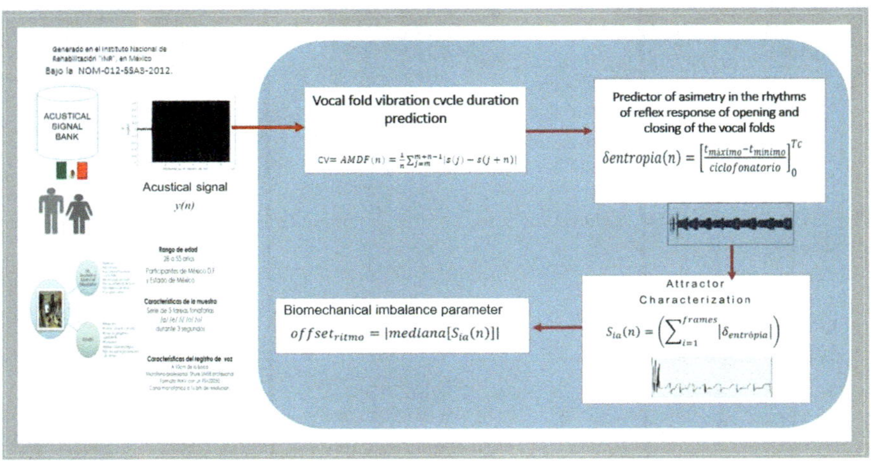

Fig. 22.4 Entropy-based model of neuromechanical deficit prediction

the complexity of the deficit of the neuromechanical coupling that particularizes the vocal fold efficiency of each individual.

$$\delta entropia(n) = \left[\frac{t_{maximo} - t_{minimo}}{ciclofonatorio}\right]_{0}^{Tc} \tag{22.2}$$

Once the attractor of the neuromechanical coupling deficits was predicted, the entropy factor of the system was calculated using a dynamic function described in Eq. (22.3).

$$S_{ia}(n) = \left(\sum_{i=1}^{frames} \left|\delta_{entropia}\right|\right) \tag{22.3}$$

This parameter represents the inefficiency of the vocal folds to regulate the rhythm in its movement and to maintain it in time, see Eq. (22.4)

$$offsetritmo = mediana[S_{ia}(n)] \tag{22.4}$$

The set of algorithms that make up the model was implemented at version 2022 in MATLAB software on a Lenovo ThinkPad computer with an Intel Core i7 processor. This computer was also used to analyse the 981 acoustic waves of the speech signals corresponding to each of the three classes that make up the database: acoustic signals from smokers with gastroesophageal reflux disease (ERGE) (A), non-smokers with stress (B), and non-smokers with a viral illness or flue (C).

22.3 Results

The results show that the model is robust to obtain the graphical approximation of the neuromechanical patterns that show significant differences between each of the three factors analysed. Figure 22.5 shows that even when evaluating the same factor, there are significant differences in the neuromechanical patterns of the vocal folds affected by smoking, which determines that the system tends to equilibrate in order to maintain vocal production without collapsing, despite the changes to which its mucous membrane and vocal folds are subjected and the thousands of interactions that occur within them.

Figure 22.5 also shows that entropy quantifies the regularity of the system, independent of father and gender. Studies show that when the system tends towards equilibrium, this type of series is less complex and more predictable, corresponding to a less adaptive neuromechanical pattern, so lower entropy values are expected compared to the other two groups.

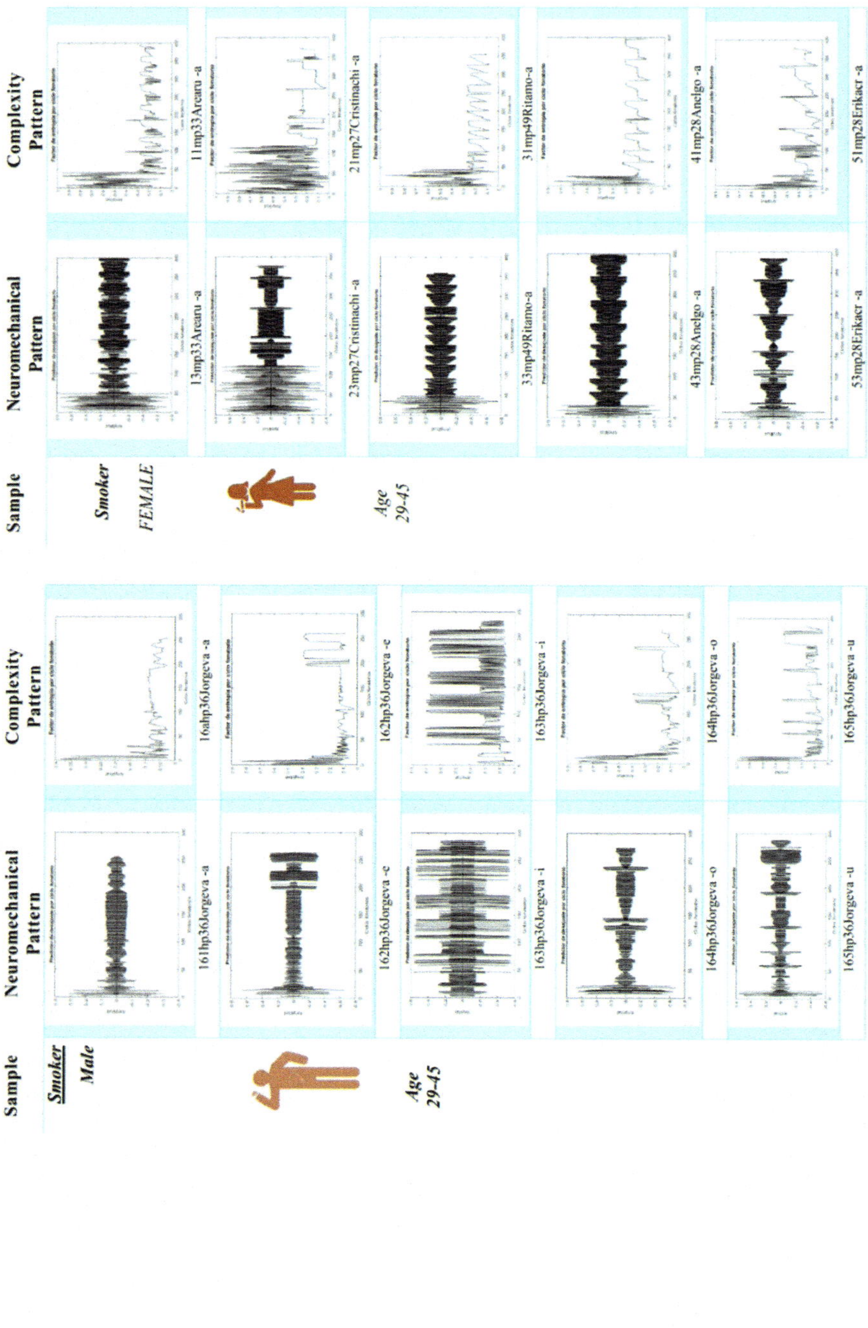

Fig. 22.5 Neuromechanical patterns and the coupling deficit in smokers' vocal cords

This is associated with a loss of flexibility and adaptability of the vocal folds, creating rigid patterns that tend to balance the system in an attempt to maintain function despite changes in the internal environment, as shown in Fig. 22.5. Figure 22.6 shows that in the non-smoking groups the neuromechanical patterns and attractors have more complex series and greater variability, which is synonymous with health since maximum complexity is associated with greater variability in the regulation of vocal fold movements. It is noted that due to the complex nature and variability of the physiological conditions of the mucosa, it may be possible to find significant differences between neuromechanical patterns even within the same individual (Fig. 22.7).

Figure 22.8 demonstrates the cause-effect relationship between vocal fold stiffness caused by changes in mucus flow secretion resulting in lower entropy values, but higher coupling deficit values, which could imply that the damage caused by smoking affects the ability of the vocal folds to function in a coordinated and efficient manner. The analysis of the variability of the neuromechanical patterns is shown in Fig. 22.8, where it is observed that the entropy values of the vocal folds affected by smoking (ERGE) are lower, which physically demonstrates the inefficiency of the motor control to regulate the rhythm of the vocal fold movements and to maintain them over time. Figure 22.8 shows that the vocal folds of non-smokers show greater variability in the neuromechanical pattern as their entropy values are lower, which is associated with the quality of the mucous membrane.

To assess the complexity and determine if there are significant differences in the neuromechanical pattern due to the three different factors irritating the mucous membrane, a single factor analysis of variance ANOVA is performed as it can help us to compare the changes occurring in the attractor patterns using the entropy factor obtained from the three study groups.

Table 22.2 shows that smokers with GERD have different neuromechanical patterns than those with stress or viral disease. These patterns may indicate a risk factor for vocal cord disease, which could help prevent early vocal cord cancer (Tables 22.3 and 22.4).

22.4 Conclusions

In this study, the complexity and variability of vocal fold neuromechanical patterns were analysed using one-factor ANOVA analysis. The results showed an *F-value* = 19.72 and a *p-value* of less than 0.05, indicating that a level of statistical significance was reached. These results suggest that there are significant differences in the entropy and complexity of the neuromechanical patterns in the vocal folds of smokers compared to the other groups analysed. The statistical significance obtained supports the alternative hypothesis and highlights the importance of quantifying the neuromechanical deficit in the context studied. These results provide a solid basis for demonstrating that the neurotoxic agent nicotine affects the neuromechanical performance of the vocal folds and provide a basis for future research that could

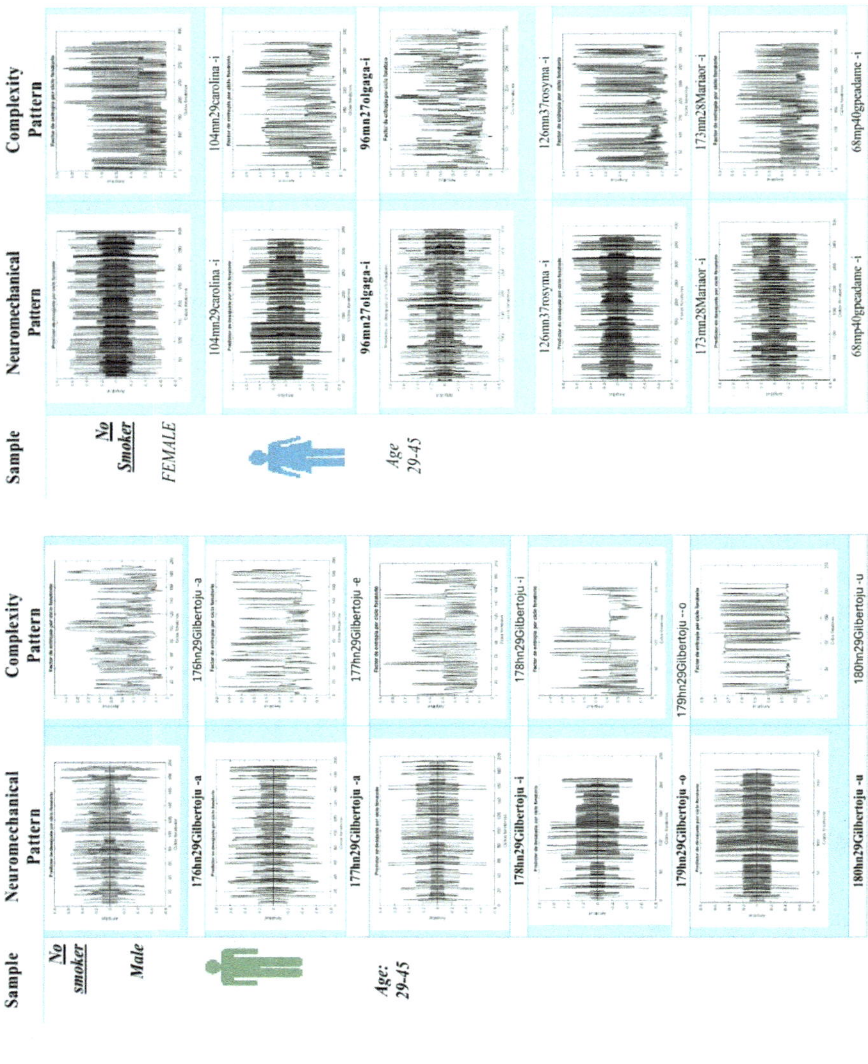

Fig. 22.6 Neuromechanical patterns and the coupling deficit in vocal folds of non-smokers

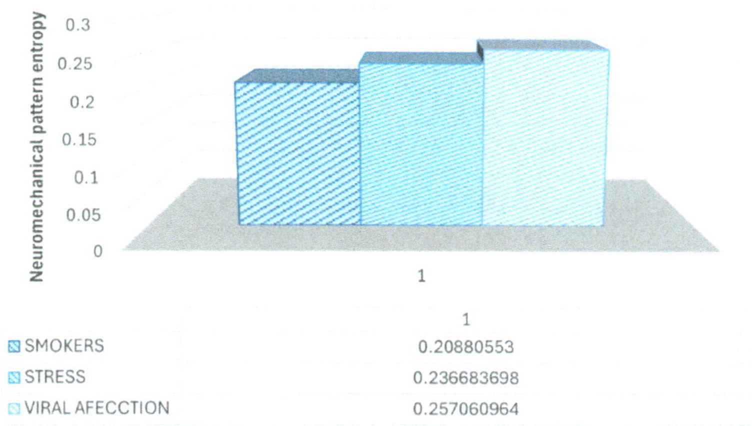

Fig. 22.7 Smoker and non-smoker average entropy values

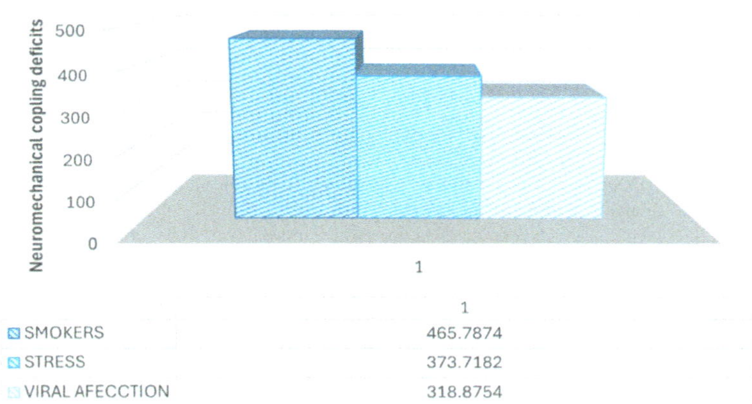

Fig. 22.8 Average values of the coupling deficit of neuromechanical patterns

Table 22.2 Data variability

	Groups	Counting	Sum	Mean	Variance
Smokers	366.66	396	184,550.94	466.037727	129,588.708
Stress	930.61	396	147,435.52	372.311919	80,942.0757
Viral affection	741.72	396	125,851.82	317.807626	132,359.574

have important implications for the development of digital biomarkers useful in the detection of vocal fold cancer risk patterns.

Table 22.3 Analysis in one-factor ANOVA

Origin of the variations	Sum of square	Grados de libertad	Promedio de cuadrados	F	Probability	F critical value
Between groups	4,452,017.842	2	2,226,008.921	19.476	4.7627E-09	3.0033184
Within the groups	135,441,691.5	1185	114,296.7861			
Total	139,893,709.3	1187				

Table 22.4 Analysing the difference between coupling deficits using Tukey's test

Population difference	Sampling difference	Difference
Smokers—stress	92.069	Significant
Smokers—viral affection	146.91	Significant
Stress—viral affection	54.82	No significant

References

Bhattacharya P, Siegmund T (2015) The role of glottal surface adhesion on vocal folds biomechanics. Biomech Model Mechanobiol 14:283–295

Byeon H, Cha S (2020) Evaluating the effects of smoking on the voice and subjective voice problems using a meta-analysis approach. Sci Rep 10(4720)

Cardoso L (2021) Análisis biomecánico de la voz en el estudio del ciclo de vibración de los pliegues vocales y su patología. Madrid, España

Castro EM, Shahrdad L, Frances ML (2023) Nicotine on the developing brain. Pharmacol Res 1–28

Chan R, Rodríguez M (2008) A single shear rheometer for linear viscoelastic characterisation of vocal fold tissues at phonation frequencies. J Acoust Soc Am 124(2):1207–12019

Fagherazzi C, Fischer A, Muhannad I, Despotovic V (2021) El uso de biomarcadores vocales desde la investigación hasta la práctica clínica. Digit Biomark 78–88

Glover M, Duhamel M-F (2023) Assessment of two audio-recording methods for remote collection of vocal biomarkers indicative of tobacco smoking harm. Acoust Aust 51:39–52

Kanabur V, Harakannanavar S, Torse D (2019) An extensive review of feature extraction techniques, speech recognition. Int J Image Graph Signal Process 11(5):1–12

Kazemirad S, Bakhshaee H, Mongeau L, Kost K (2014) Non-invasive in vivo measurement of the shear modulus of human vocal fold tissue. J Biomech 47(5):1173–1179

Kelleher J, Siegmund T, Du M, Naseri E, Chan R (2013) Medidas empíricas de la anisotropía biomecánica de la lámina propia de las cuerdas vocales humanas. Biomech Model Mechanobiol 555–567

Lustosa Pinto GA, Nubiato Crespo A, Figueiredo Mourão L (2014) Influence of smoking isolated and associated to multifactorial aspects in vocal acoustic parameters. Braz J Otorhinolaryngol 80(1):60–67

Martín RA, Rodríguez GI, Rubio C, Revert C, Hardisson A (2004) Efectos tóxicos del tabaco. *Revista de toxicologia* 21(2):64–71

Naranjo O, de la Cruz T (2010) La entropia y la irreversibilidad multiescala en el anáisis de sistemas complejosen fisiologia humana. *Revista Andaluza de medicina del deporte* 3(1):29–32

Nieto A (2001) Chaos and nonlinear dynamics in the vibration of the vocal cords. [En línea]. https://www.medwave.cl/puestadia/cursos/3160.html

de la Ossa BA, Uribe Giraldo N, Ibarra Jiménez GM (2019) La disfunción de las cuerdas vocales como diagnóstico diferencial en pacientes con tos y sibilancias respiratorias. Revista de ASMA 4(3):92–96
Pérez T (1995) Naturaleza del humo del tabaco. *Arch Broncomumol* (31):101–108
Petrov AA (2019) Vocal cord dysfunction: the spectrum across the ages. Immunol Allergy Clin N Am 39(4):547–560
Pizarro M, Centeno D, Loreto NC (2022) Laryngeal neuropathy and paradoxical vocal fold motion in the pediatric. *Rev. Otorrinolaringol. Cir. Cabeza Cuello* 82:338–342
Rand M (1989) Neuropharmacological effects of nicotine in relation to cholinergic mechanisms. de Prog Brain Res 3–11. Elseiver
Siordia VX () Método no Invasivo para caracterizar la fuente laringea. México Patente Mx/a/2015/018053
Siordia Vásquez X, Jorge CC (2016) Parametros biomecanicos de la membrana epitelial:Un biomarcador de riesgo del cáncer de laringe. *Revista de analisis cuantitativo y estadistico* 3(9):17–30
Taiz A (2020) Procesamiento motor vocal en tareas de lenguaje,» Universidad de Buenos Aires, Buenos Aires
Zealouk O, Satori H (2018) Vocal parameters analysis of smoker using Amazigh language. Int J Speech Technol 21:85–91
Zhang Z (2016) Cause-effect relationship between vocal fold physiology and voice production in a three-dimensional phonation model. J Acoust Soc Am 139(4):1493–1507
Zhanga Z (2016) Mechanics of human voice production and control. J Acoust Soc Am 140(4):2614–2635

Chapter 23
Two Modeling Approaches for an Epidemic Disease Spreading Process

Douglas Ferraz Corrêa, Marcelo Cargnelutti Rossato, David A. Pelta, Cláudio Fabiano M. Toledo, João Frederico C. A. Meyer, Antônio José da Silva Neto, and Luiz Bevilacqua

Abstract This work presents two different approaches for an epidemic disease spreading process based on the already established SIR model, the first approach uses a cellular automata approach to simulate an epidemy, the second approach uses a biflux spatial diffusion equation, based on the model developed by Bevilacqua and Galeão to simulate the same phenomena. The results are shown in figures and briefly commented. The goal is to present those two approaches that might allow for a better representation of the phenomena.

D. Ferraz Corrêa (✉) · A. J. da Silva Neto
Instituto Politécnico, Universidade do Estado do Rio de Janeiro, Nova Friburgo, RJ, Brazil
e-mail: douglas.correa@iprj.uerj.br

A. J. da Silva Neto
e-mail: ajsneto@iprj.uerj.br

M. Cargnelutti Rossato · J. F. C. A. Meyer
Instituto de Matemática, Estatística e Computação Científica, Universidade Estadual de Campinas, Campinas, SP, Brazil
e-mail: m264880@dac.unicamp.br

J. F. C. A. Meyer
e-mail: joni@ime.unicamp.br

D. A. Pelta
Department of Computer Science & A.I, Universidad de Granada, Granada, Spain
e-mail: dpelta@ugr.es

C. F. M. Toledo
Instituto de Ciências Matemáticas e de Computação, Universidade de São Paulo, São Paulo, SP, Brazil
e-mail: claudio@icmc.usp.br

L. Bevilacqua
Universidade Federal do Rio de Janeiro, Rio de Janeiro, RJ, Brazil
e-mail: bevilacqua@coc.ufrj.br

A. Öchsner and H. Altenbach (eds.), *Engineering Design Applications VII*, Advanced Structured Materials 230, https://doi.org/10.1007/978-3-031-84346-4_23

23.1 Introduction

Different models have been proposed to better understand the transmission mechanism of infections and in the most common models the population is divided into categories, usually: Susceptible, infected, and recovered/removed, this model is known as SIR model. There are several approaches that based on those categories tries to simulate an epidemic scenario. The regular SIR model formulation does not take in consideration the movement of the individuals in the population of study, some studies (Cai et al. 2019; Wang et al. 2020; Zhu et al. 2018) have successfully applied the diffusion-reaction equation to model this phenomena in a SIR model.

In this work, we present two different ways of modeling an epidemic disease spreading process based on the SIR model: a discrete process with the aid of a cellular automata and a continuum approach with the biflux anomalous diffusion equation.

It is stated in the literature that geographic features such as rivers and mountains have been implicated as physical barriers to a raccoon population movement, which means a potential impediments to epidemic rabies (Bevilacqua et al. 2011, 2013; Smith et al. 2002) have developed an analytical formulation for anomalous diffusion, the BG model, in which temporary retention that may be caused by a physical barrier, is taken into consideration, resulting in a fourth-order partial differential equation, besides the retention phenomena a secondary flux of recovering can be represented by the subsidiary flux of this model (Bevilacqua 2021).

Up to our knowledge, there are no previous implementation of the BG model in modeling a disease spreading phenomena, due to the relevance of a retention phenomena in this context and the need for understanding the infection evolution we aim to propose a model based on biflux anomalous diffusion that we hope can be used to better represent this very broad and important subject.

23.2 Cellular Automata

A cellular automaton consists of a model in which space, time, and states are discrete (Mistro 1998). In one dimension, it can be understood as a vector with a finite set of possible values that can vary over time iterations, and in two dimensions it can be understood as a matrix with the same characteristics.

Therefore, we will define a SIR model through an $N \times M$ automaton where each cell represents an individual, who can be susceptible, infected, or recovered. For the one-dimensional model, we consider $M = 1$. The possible state transitions are described on Fig. 23.1.

For each iteration, all individuals interact with v other individuals who are up to d units away from them horizontally for the one-dimensional model and horizontally and vertically for the two-dimensional model. In each interaction between an infected

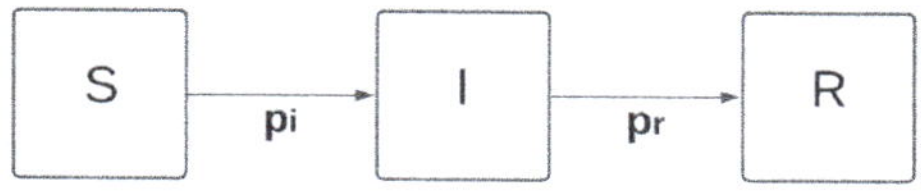

Fig. 23.1 Diagram with the possible states and transitions

and a susceptible individual, the latter can become infected with probability p_i. At the end of each iteration, infected individuals can recover with probability p_r.

23.3 Biflux Anomalous Diffusion

For the sake of simplicity, in this section we shall assume the discrete and symmetric case of redistribution.

Let's consider a region in space that contains a high concentration of particles, represented by cell i in Fig. 23.2. In a particle spreading process with retention, a portion (α) of the particles is retained, and the non-retained portion (β) is redistributed into neighboring cells. This process is represented in Fig. 23.2, where $\beta = 1 - \alpha$. When it comes to the diffusion of an infectious agent in a population β can also be related to the probability of a state transition.

This phenomenon in a continuous medium and with the possibility that the redistribution has a spatial dependence is governed by the following partial differential equation (Bevilacqua et al. 2011):

$$\frac{\partial p(x,t)}{\partial t} = K_2 \frac{\partial}{\partial x}\left[\beta(x)\frac{\partial p(x,t)}{\partial x}\right] - K_4 \frac{\partial}{\partial x}\left[\beta(x)(1-\beta(x))\frac{\partial^3 p(x,t)}{\partial x^3}\right] \quad (23.1)$$

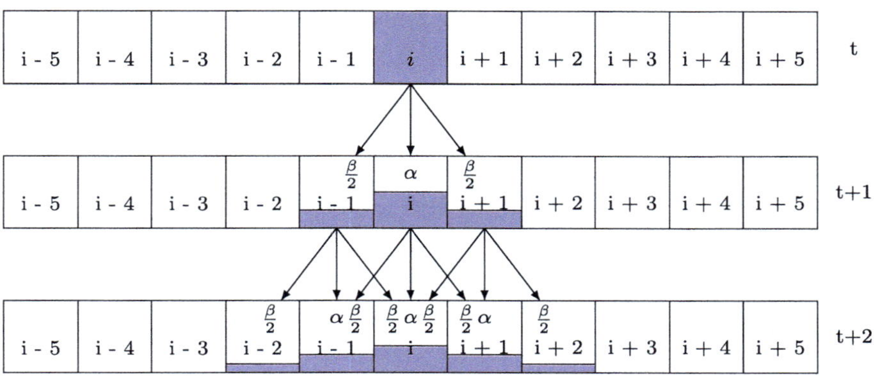

Fig. 23.2 The discrete distribution as a function of time in the BG model for biflux anomalous diffusion with constant β

where $p(x, t)$ represents the concentration, K_2 is the diffusion coefficient, and K_4 is the reactivity coefficient.

Equation (23.1) is called the biflux anomalous diffusion equation, or BG model for anomalous diffusion and it is the base of development of the more general equation. For a review of how Eq. (23.1) was obtained we recommend reading the analytical development of this model in Bevilacqua et al. (2011, 2013).

23.3.1 Biflux Theory in the Disease Spreading Context

Let's focus now on the phenomenological characteristics of the BG model applied in an epidemic disease spread context, which is one of the approaches proposed in the present study.

According to Bevilacqua (2021) the subsidiary flux can represent a flux of state transition, that is for example, from the infected state to the recovered state. Based on that we start by stating the following hypothesis

1. There's an evolution in space of the agent state;
2. There is a constant rate of recovering;
3. Rate of infected is also affected by a secondary flux of recovering;
4. A recovered agent never leaves this state, i.e., once recovered it becomes immune to the infection.

hypothesis (i) means we are representing the phenomena with a spatial diffusion formulation, (ii) means the main recovering mechanism is not the secondary flux, (iii) sets the implementation of the secondary flux as a state transition flux, and (iv) simplifies the studied phenomena not accounting for deaths in the population with the exception for deaths due to the boundary conditions.

With the above hypothesis the proposed model is given by the following system of partial differential equations

$$\frac{\partial S}{dt} = \frac{\partial}{\partial x}\left(D_S \frac{\partial S}{\partial x}\right) - rSI \tag{23.2}$$

$$\frac{\partial I}{dt} = d_i \frac{\partial}{\partial x}\left(\beta(x)\frac{\partial I}{\partial x}\right) + rSI - \alpha I - K_4 \frac{\partial}{\partial x}\left[\beta(x)(1-\beta(x))\frac{\partial^3 I(x,t)}{\partial x^3}\right] \tag{23.3}$$

$$\frac{\partial R}{dt} = \frac{\partial}{\partial x}\left(D_R \frac{\partial S}{\partial x}\right) + aI + K_4 \frac{\partial}{\partial x}\left[\beta(x)(1-\beta(x))\frac{\partial^3 I(x,t)}{\partial x^3}\right] \tag{23.4}$$

here $S(x, t)$, $I(x, t)$, and $R(x, t)$ denote the densities of susceptible, infected, and recovered individuals at location x and time t. D_S, d_I, and D_R are positive diffusion coefficients, α is the rate of recovering, and r is the rate of disease transmission.

Two boundary conditions are used to generate results with the proposed model, the first is the Dirichlet boundary condition given by

$$S(x,t) = S_0 \tag{23.5}$$

$$I(x,t) = I_0 \tag{23.6}$$

$$R(x,t) = R_0 \tag{23.7}$$

at $x = 0$ or $x = L$. With $S_0 = I_0 = R_0 = 0$ an environment where the boundary of the region is hostile for the survival of the population is represented. Due to the characteristics of the phenomena $S(x,t)$, $I(x,t)$, and $R(x,t)$ are all ≥ 0 for $\{x \in R | 0 < x < L\}$ and with the divergence theorem it is trivial to show that the population decays to 0 when $t \to \infty$.

The second is the Neumann boundary condition, given by

$$\left.\frac{\partial S(x,t)}{\partial x}\right|_{x=0,L} = 0 \tag{23.8}$$

$$\left.\frac{\partial I(x,t)}{\partial x}\right|_{x=0,L} = 0 \tag{23.9}$$

$$\left.\frac{\partial R(x,t)}{\partial x}\right|_{x=0,L} = 0 \tag{23.10}$$

in this case the population number remains constant since there's no flux on the boundaries. The second boundary condition for Infected is defined as no subsidiary flux at the boundary.

23.4 Results

23.4.1 Cellular Automata

A summary of all parameters used in the simulations is shown in Table 23.1.

To simulate the one-dimensional model, we consider $N = 1000$ and $M = 1$. As for the two-dimensional model, we take $N = M = 100$. In both cases, we let the

Table 23.1 Parameters of the SIR model via automaton

Parameter	Meaning
N	Number of automaton lines
M	Number of automaton columns
v	Number of interactions per individual
d	Maximum interaction distance
p_i	Probability of infection
p_r	Probability of recovery

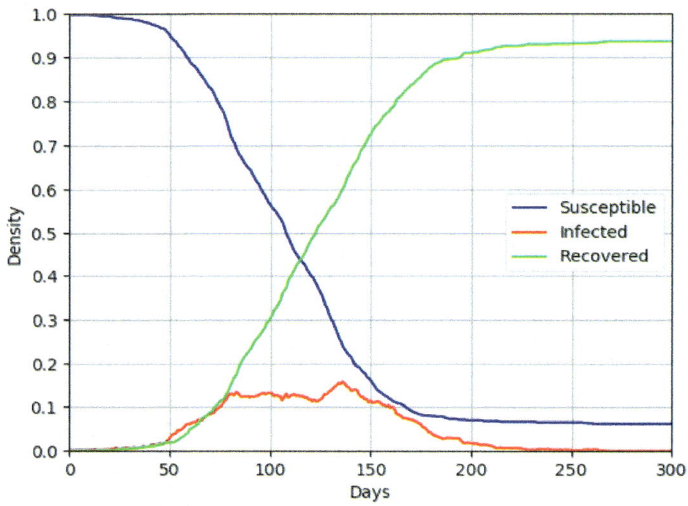

Fig. 23.3 One-dimensional simulation results with $v = 3$ and $d = 50$

infection and recovery probabilities $p_i = 0.03$ and $p_r = 0.06$ fixed, changing only the number of interactions with neighbors v and the maximum interaction distance d.

In Fig. 23.3, the result of 300 iterations of the one-dimensional model with $v = 3$ and $d = 50$ is presented.

It should be noted that the value of d must be much smaller in the two-dimensional model than in the one-dimensional model. For individuals far enough from the frontier, while in the one-dimensional model an individual has $2d$ neighbors, in the two-dimensional model this number grows to $(2d + 1)^2 - 1$.

Therefore, while in the one-dimensional model for $d = 50$ each individual had 100 neighbors, for $d = 3$ and $d = 5$ in the two-dimensional model, each individual has 48 and 120 neighbors, respectively. The result of the simulations of 400 iterations of the two-dimensional model with $d = 3$ and $d = 5$ for $v = 3$ are presented, respectively, in Figs. 23.4 and 23.5.

23.4.2 BG Model

The results were generated with the following parameters presented in Table 23.2 and the initial condition is given by Wang et al. (2020)

$$I(x, 0) = 0.01e^{-(x-10)^2} \tag{23.11}$$

$$S(x, 0) = 1 - I(x, 0) \tag{23.12}$$

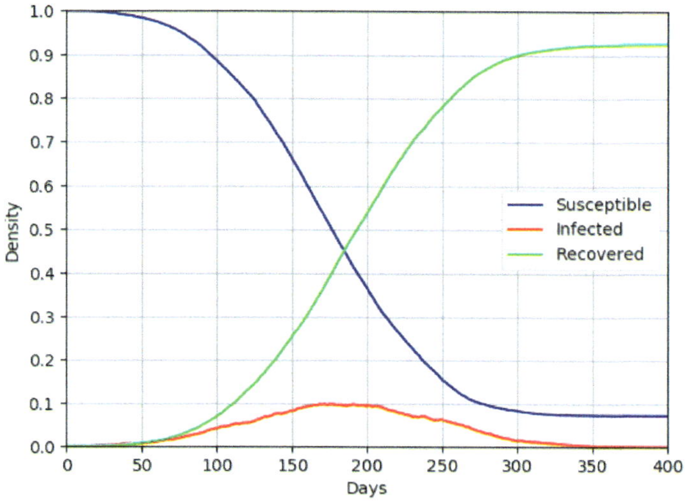

Fig. 23.4 Two-dimensional simulation results with $v = 3$ and $d = 3$

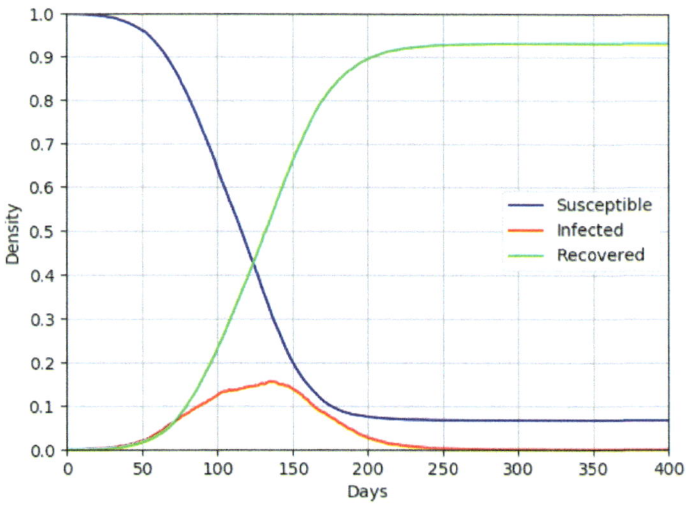

Fig. 23.5 Two-dimensional simulation results with $v = 3$ and $d = 5$

$$R(x, 0) = 0 \tag{23.13}$$

Figure 23.6 shows the numerical solution of Eqs. (23.3), (23.4), and (23.2), with the parameters presented in Table 23.2 and the no flux at the boundaries condition.

Figure 23.7 shows the result with the deadly boundary condition, this implies the population number decays with time. Interestingly, in this case the infected peak

Table 23.2 Parameters of the BG model

Parameter	Meaning	Value
$D_r = D_s = d_i$	Diffusivity coefficient	0.02
r	Rate of infection transmission	0.03
α	Rate of recovery	0.06
K_4	–	10^{-4}

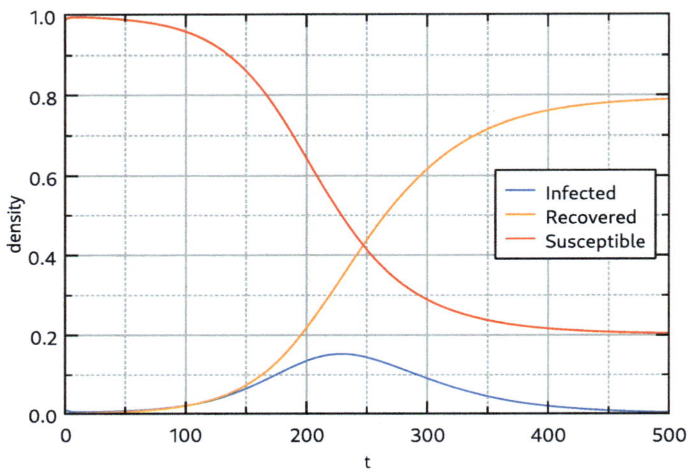

Fig. 23.6 Solution of the proposed BG model for a disease spreading process with $\beta = 0.5$ and constant population

happens faster. The actual reasoning for this still needs more research. The computational cost of the BG model implementation on a i7-12700H with NVIDIA RTX 3060 was around 16 ms, which is a reasonable time considering the relevance of the topic.

Comparing Figs. 23.3 and 23.6 we can see two main differences, the first one is the density of people in the recovered and susceptibles categories are not the same, this happens because no correlation between the models parameters were used and both were chosen based on previous published work in the literature for each model. The second difference is the fact that the cellular automata method was able to predict some degree of variations, almost like random category transfer happening during the simulation.

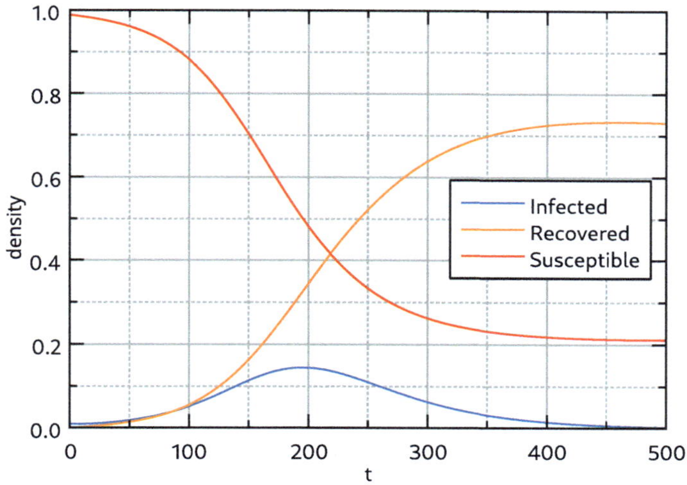

Fig. 23.7 Solution of the proposed BG model for a disease spreading process with $\beta = 0.5$ and decaying population

23.5 Conclusions

The present work shows two different approaches to modeling a disease spreading process, the first approach uses a SIR model with cellular automata while the second approach includes a subsidiary flux to represent a transition between agent state in a novel approach that we hope will allow for a better refinement of the phenomena, both approaches were able to represent the disease spreading process. When faster results are needed the BG model proved itself less computational costly and with the capacity of representing a secondary flux of state transition, while the cellular automata showed that it is a method capable of capturing small and random variations in the state transition that makes a difference when the population number is small, the difference becomes unnoticeable in the 2D scenario due to the bigger number of agents involved.

For future works we aim to introduce the retention effect on the cellular automata, while also evolving the theory behind the application of the BG model in an epidemic process and, hopefully, find a correlation of parameters between both methods.

Acknowledgements This study was financed in part by the Coordenação de Aperfeiçoamento de Pessoal de Nível Superior—Brasil (CAPES)—Finance Code 001, Conselho Nacional de Desenvolvimento Científico e Tecnológico (CNPq) and Fundação Carlos Chagas Filho de Amparo à Pesquisa do Estado do Rio de Janeiro (FAPERJ). D. Corrêa acknowledges CAPES also for the scholarship for his stay in the University of Granada (Grant CAPES/PrInt No. 88887.717186/2022-00). D. Pelta acknowledges support from projects PID2020-112754GB-I00, MCIN /AEI/ 10.13039 /501100011033 and FEDER/Junta de Andalucía-Consejería de Transformación Económica, Industria, Conocimiento y Universidades / Proyecto (BTIC-640-UGR20).

References

Bevilacqua L (2021) Processos de difusão de partículas da mesma natureza em dois estados de energia diferentes. Revista Brasileira de Ensino de Física 43
Bevilacqua L, Galeão ACNR, Costa FP (2011) A new analytical formulation of retention effects on article diffusion process. An Acad Bras Cienc 83:1443–1464
Bevilacqua L, Galeão ACNR, Simas JG, Doce APR (2013) A new theory for anomalous diffusion with a bimodal flux distribution. J Braz Soc Mech Sci Eng 35(4):1–10
Cai Y, Lian X, Peng Z, Wang W (2019) Spatiotemporal transmission dynamics for influenza disease in a heterogenous environment. Nonlinear Anal Real World Appl 46:178–194
Mistro DC (1998) Modelos para dispersão de abelhas africanizadas: um zoom matemático. 182 f. Tese (Doutorado em Matemática Aplicada) – Unicamp, Campinas
Smith David L, Lucey Brendan, Waller Lance A, Childs James E, Real Leslie A (2002) Predicting the spatial dynamics of rabies epidemics on heterogeneous landscapes. Proc Natl Acad Sci 99:3668–3672
Wang J, Xie F, Kuniya T (2020) Analysis of a reaction-diffusion cholera epidemic model in a spatially heterogeneous environment. Commun Nonlinear Sci Numer Simul 80:104951
Zhu D, Ren J, Zhu H (2018) Spatial-temporal basic reproduction number and dynamics for a dengue disease diffusion model. Math Methods Appl Sci 41(14):5388–5403

Chapter 24
Innovations in Biomedical Engineering and Surgical Techniques for Improved Outcomes in Young Patients with Floating Hips: A Multidisciplinary Approach

Juan Alfonso Beltrán-Fernández and Elías Humberto Hermida-Ochoa

Abstract Reconstructing acetabular fractures in young patients poses significant orthopedic challenges, particularly regarding implant longevity and functional outcomes. This chapter explores state-of-the-art innovations in biomedical engineering and surgical techniques aimed at enhancing patient outcomes. Leveraging advanced computer-aided design tools such as ANSYS and SolidWorks, in conjunction with 3D printing and finite element analysis, this paper investigates the impact of personalized implants, biomechanical simulations, and robotic-assisted surgery. Comparative studies demonstrate notable enhancements in implant durability, joint stability, and postoperative recovery, particularly among young, physically active patients. Furthermore, the interdisciplinary approach outlined in this study offers an adequate proposal for optimizing the clinical management of complex orthopedic cases in young patients.

Keywords Floating hip · Biomedical engineering · CAD · Finite element analysis · Robotic-assisted surgery · 3D printing · Implant design · Orthopedic surgery · Surgical navigation · Biomechanics

---

J. A. Beltrán-Fernández (✉)
Escuela Superior de Ingeniería Mecánica y Eléctrica, Instituto Politécnico Nacional–Sección de Estudios de Posgrado E Investigación Edificio 5, 2Do Piso, Unidad Profesional Adolfo López Mateos "Zacatenco" Col. Lindavista, Ciudad de México, México
e-mail: jbeltran@ipn.mx

E. H. Hermida-Ochoa
Hospital German Diaz Lombardo–Centro de Investigación y Laboratorio Biomecánico, Carmen #18, CDMX, México
e-mail: drelias@hermidagroup.com

A. Öchsner and H. Altenbach (eds.), *Engineering Design Applications VII*, Advanced Structured Materials 230, https://doi.org/10.1007/978-3-031-84346-4_24

24.1 Introduction

Reconstructing floating hips in young patients is one of the most challenging orthopedic problems, requiring multidisciplinary intervention. Young patients, such as a 34-year-old woman, present specific demands for implant durability, mobility, and quality of life, emphasizing the need for customized implant designs, advanced materials, and precise surgical techniques. With advancements in biomedical engineering, particularly in computer-aided design (CAD) software such as ANSYS, SolidWorks, and PTC Creo Parametric, the approach to implant design, preoperative planning, and postoperative recovery has evolved significantly. This article explores these innovations, emphasizing the role of 3D printing, biomechanical simulations, and robot-assisted surgery in improving outcomes for young patients with floating hips.

Floating hips, characterized by their instability and increased risk of dislocation, represent a significant challenge in orthopedic surgery, particularly in young patients who require mobility for their active lifestyles. As the incidence of hip disorders continues to rise, driven by factors such as increased physical activity, obesity, and trauma, the need for effective surgical interventions becomes paramount. These demographic developments demand not only surgical success but also a focus on enhancing postoperative quality of life and functional outcomes.

Recent advancements in biomedical engineering offer promising solutions to these challenges. Innovative materials and technologies have transformed the landscape of hip reconstruction, allowing for the design of implants that are not only robust but also tailored to the unique anatomical and physiological needs of young patients. For instance, the use of biocompatible materials such as titanium alloys and ceramics provides the necessary strength while minimizing the risk of adverse reactions within the body (Rivera et al. 2024; Ali et al. 2024).

Moreover, the advent of computer-aided design tools and 3D printing technology enables the customization of implants to achieve optimal fit and functionality. This personalization is crucial in reducing surgical complications and improving patient outcomes, particularly in young individuals whose mobility and activity levels are essential for their overall well-being (Göktaş et al. 2022; Teng et al. 2023).

In conjunction with advanced surgical techniques, including robotic-assisted procedures and navigational systems, orthopedic surgeons can achieve unprecedented precision in implant placement and orientation. These innovations not only enhance surgical accuracy but also contribute to shorter recovery times and improved postoperative rehabilitation outcomes (Anthony and Tarun 1438).

Figure 24.1 illustrates a CAD model of an implant designed specifically for floating hip reconstruction, highlighting the importance of customization in implant design.

The main objective of this research is to analyze the advancements in biomedical engineering and surgical techniques utilized in the reconstruction of floating hips in young patients. The emphasis is placed on how the integration of advanced technologies such as computer-aided design and biomechanical simulations using the

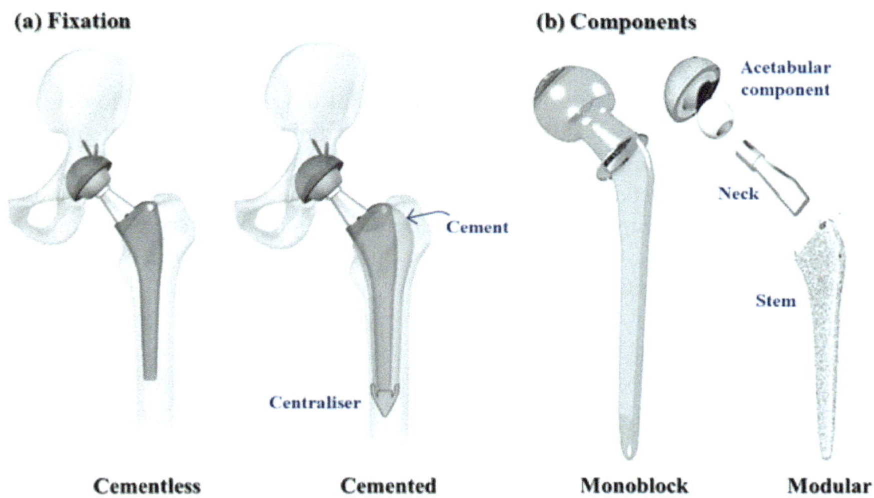

Fig. 24.1 CAD model of a titanium alloy implant designed for floating hip reconstruction, showcasing optimized geometry for better load distribution (Dias et al. 2024) Reprinted with permission from Springer Nature publishers

finite element method (FEM). These outcomes encompass enhanced implant durability, joint stability, and a reduction in postoperative complications. Furthermore, the chapter addresses the disparities in access to these technologies between developed and developing countries, with a specific focus on the current situation in Mexico.

24.2 Methodology

This chapter for designing a custom floating hip prosthesis for a 32-year-old female patient involves several critical steps:

1. **CT scan acquisition and segmentation**:
2. **3D model generation and refinement**:
3. **Biomechanical simulation**:
4. **Custom implant design**: Based on biomechanical analysis, a titanium alloy implant is designed, optimized for the patient's anatomy, and manufactured using the 3D printing technology. Specific attention is given to the implant's geometry to promote osseointegration.

It is important to note that this case will not cover preoperative planning and robotic assistance but will include a brief overview.

24.2.1 Analysis of a Study Case: 32-Year-Old Female Patient

The study focuses on a 32-year-old female patient diagnosed with a floating hip. A computed tomography (CT) scan was used for detailed anatomical imaging. The raw scan data were segmented using Scan IP and RadiiAnt, two software tools that allowed for precise visualization of the hip structure. After segmentation, the Meshmixer software was employed to clean and refine the model, ensuring accuracy for implant design. Figures 24.2 and 24.3 present the segmentation and Meshmixer cleaning process, respectively.

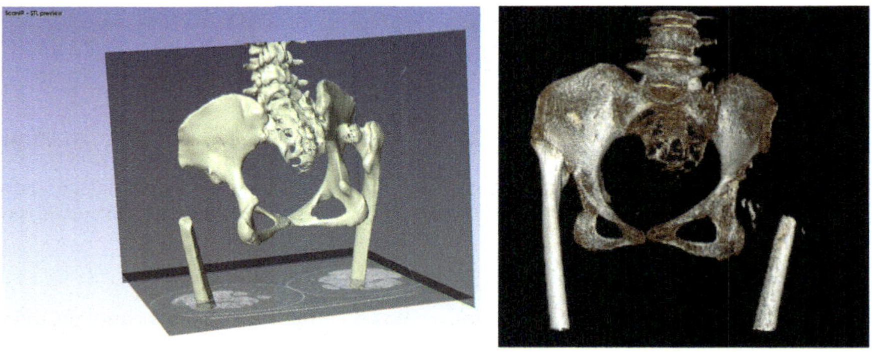

Fig. 24.2 Segmentation of CT data using the Scan IP © and Radiiant software ©

Fig. 24.3 Meshmixer process for refining segmented data

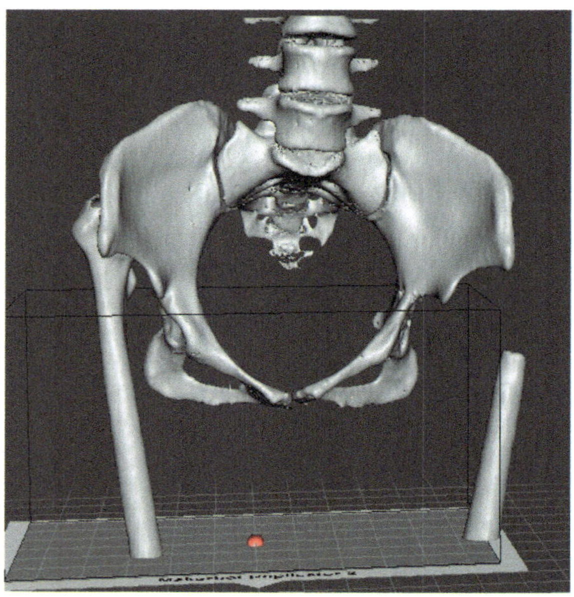

CT Scan Acquisition and Segmentation

The first step uses CT images of the hip region, segmented with the Scan IP software to obtain a precise model of the patient's bone structures. This segmentation isolates the pelvic and femoral regions, critical for accurate modeling.

The segmentation of medical images using Scan IP is crucial for developing precise 3D models for complex surgical interventions like floating hip reconstruction. In the case of our 32-year-old female patient, Scan IP accurately segmented the CT scan, isolating bone structures necessary for designing the custom implant. Scan IP's advanced capabilities ensure that even the smallest anatomical details are captured, leading to more individualized treatment plans. By precisely segmenting the pelvic bones and surrounding tissues, we can create 3D models that accurately represent the anatomical conditions, which is particularly important for younger patients. This level of customization is essential to optimize implant fit, prevent complications, and improve the biomechanical outcomes of the surgery. Detailed segmentation also facilitates the simulation of mechanical stresses on the reconstructed joint, aiding in the selection of suitable materials and design for the implant to enhance durability and functional performance.

3D Model Generation and Refinement

The segmented data are exported as an STL file. The file is refined using **the** Meshmixer software to smooth and optimize the model for further analysis. This step ensures a clean and manufacturable model, improving the fidelity of the design.

After the segmentation process, the Meshmixer software was utilized as a crucial tool for refining the segmented 3D models. The STL file obtained from Scan IP segmentation is often found to contain irregularities or surface noise, which needs to be corrected before the model is deemed ready for biomechanical simulation or 3D printing. In our approach, the use of Meshmixer allows bone surfaces to be smoothed and artifacts to be removed from the STL file, resulting in a cleaner, more accurate model that better represents the patient's anatomy. This refinement step is considered crucial for finite element analysis (FEA), as the model's accuracy directly influences the reliability of the simulation outcomes. Furthermore, the model is prepared for 3D printing through Meshmixer, ensuring that the implant design is not only anatomically precise but also manufacturable, thereby reducing the risk of errors during the printing process. By utilizing these digital tools, highly tailored implants can be created that perfectly conform to the patient's hip structure, significantly improving the surgical results and long-term functionality of the reconstructed joint.

Biomechanical Design and Simulation

Using a finite element method software (Solidworks and ANSYS), stress distributions and load-bearing capacities of potential implant designs are analyzed. This phase is essential for validating the structural integrity and fit of the prosthesis.

The process of designing an orthopedic implant starts with creating a solid model. The initial step is to develop a device that can be adjusted within the femur, based on the requirements and observations of the orthopedic surgeon. This involves designing a stem and reconstructing the acetabular cup. Once these considerations are made, the design of the device is initiated, taking into account the necessary cuts and angles for proper positioning of the stem. The angle between the stem and the bone's vertical diaphysis should range from 125° to 130°, referencing a healthy femoral head or considering the greater and lesser trochanters. Proposals for the stem and femoral head are designed using templates, which help maintain the appropriate geometry, as this is a customized prosthesis tailored specifically to the patient's size (Fig. 24.4).

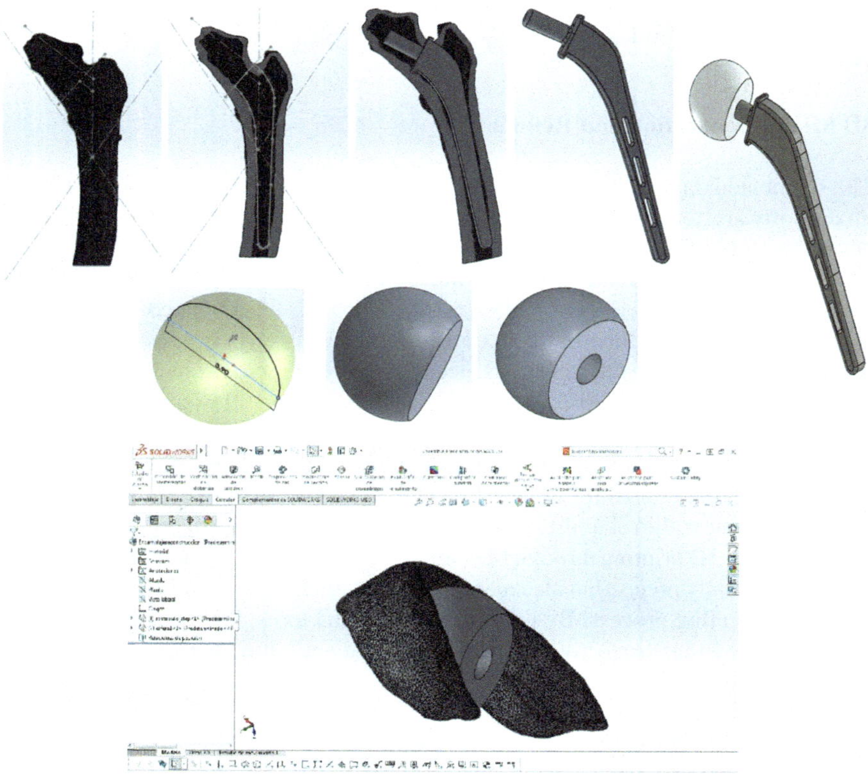

Fig. 24.4 Orthopedic implant design process

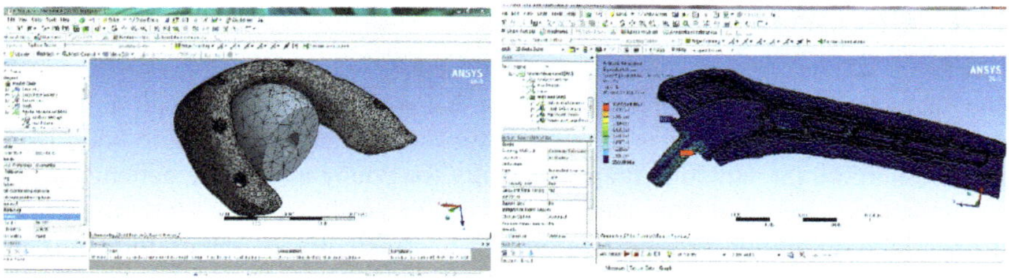

Fig. 24.5 FEM simulation analyzing stress distribution within the hip joint

The final step involved exporting the segmented and cleaned model into a CAD software for further analysis and implant design. Using ANSYS, a biomechanical simulation based on the finite element method was performed to assess the implant's behavior under mechanical load. This simulation helped predict stress points within the hip joint and allowed for adjustments to the implant design. Figure 24.5 demonstrates the FEM simulation process.

Surgical navigation systems provided real-time guidance during the procedure, ensuring that the implant was placed precisely according to the preoperative plan.

24.3 Results

In the finite element analysis performed, the primary components of the floating hip prosthesis, including the stem and the acetabular bone reconstruction, were evaluated to determine the von Mises stress distribution. The data obtained are presented in Table 24.1.

These results highlight the mechanical strength of the design under simulated load conditions, demonstrating each component's capacity to withstand anticipated forces and loads during the patient's physical activity. Figure 6 shows the stress distribution in the stem and the distribution in the reconstructed acetabular bone.

Table 24.1 Results of the mechanical strength of the design under simulated load conditions

Component	Average von Mises stress (MPa)
Prosthesis stem	339
Acetabular bone reconstruction	28.1

24.4 Discussion

The analysis shows that the prosthesis stem has a high resistance, with an average stress of 339 MPa. This meets the design standards required to endure body loads and stress in the implantation area, especially for young and active patients. These results indicate that both the material and design of the stem offer the necessary stiffness and durability for long-term use of the prosthesis, thereby minimizing the risks of fracture or deformation.

In contrast, the stress measured in the reconstructed acetabular bone was 28.1 MPa, indicating a good distribution of load and a low risk of structural failure. This measurement suggests that the design helps maintain the integrity of the surrounding bone, which promotes better integration of the implant and reduces the likelihood of future complications, such as prosthesis loosening.

The comparison between the two components highlights the significance of customized prosthesis designs in young patients. These designs optimize load distribution and improve joint biomechanics, both of which are essential for the long-term functionality of the joint.

While robotic-assisted surgical techniques and patient-specific hip implants have demonstrated efficacy in improving postoperative outcomes, significant disparities persist in access to these innovations across countries. In particular, developing countries, including Mexico, face financial, logistical, and infrastructural challenges in implementing these advanced technologies on a broader scale. As reported in recent studies, healthcare systems in low- and middle-income countries often lack the funding and trained personnel needed for consistent integration of robotic systems and personalized 3D-printed implants, which are generally available in specialized centers only in urban areas (e.g., Mexico City). This gap limits the potential benefits for many patients in rural regions who would otherwise benefit from advanced surgical options. Consequently, this technological divide has implications for patient outcomes, particularly among young, active patients who may require frequent revisions if they do not receive optimal interventions initially. Addressing this disparity is critical to reducing long-term healthcare costs and improving the quality of life for patients in these regions (Mohanty et al. 2022; Gellad et al. 2011).

24.5 Conclusions

The analysis of the floating hip prosthesis design, tailored to meet the specific anatomical and functional needs of a young female patient, demonstrates significant biomechanical advantages. The results of the finite element analysis, particularly the average von Mises stress values recorded for both the prosthesis stem (339 MPa) and the acetabular bone reconstruction (28.1 MPa), indicate that the customized design effectively manages stress distribution. This management is critical for ensuring the stability and longevity of the implant. The balance between high resistance in the prosthetic stem and minimized stress on the surrounding bone structure suggests a robust strategy for reducing the risks of complications such as fracture, deformation, or loosening, which are key concerns for high-demand patients.

Floating hip cases are complex due to their unique biomechanical requirements and often necessitate a meticulous, interdisciplinary approach. The customized design process in this study integrates patient-specific anatomical data with advanced software, highlighting the clinical significance of personalizing implant designs. Furthermore, the statistical prevalence of floating hip cases in Mexico emphasizes the importance of such innovations for local healthcare settings, where a standardized solution may not sufficiently address individual patient needs. The use of sophisticated software like Scan IP for segmentation and Meshmixer for STL refinement enhances this precision-based approach, which could improve clinical outcomes for similar cases both nationally and globally.

This work underscores the need for continued investment in biomechanical research and personalized prosthetic solutions to optimize surgical outcomes for patients with complex hip conditions. Such advancements not only enhance the quality of life for young, active patients but also represent significant progress in the application of biomedical engineering to orthopedic surgery. Future research should focus on broadening the dataset of clinical cases to refine implant design parameters and evaluate long-term outcomes, contributing to a robust, evidence-based framework for floating hip reconstructions.

Although this chapter does not address the application of robotics in surgery, it is essential to recognize that advancements in robotic-assisted surgical techniques are vital for the future of surgical practices. These techniques have shown significant improvements in postoperative outcomes compared to traditional methods, particularly in reducing surgical complications and accelerating recovery in hip reconstruction procedures (Sweet et al. 2021).

Additionally, rehabilitation strategies tailored for younger, more active patients are essential to ensure long-term success and functionality after hip reconstruction. Specific protocols will need to be designed to meet the physical demands and expectations of this demographic (Michelle et al. 2024).

Acknowledgements The authors wish to express their gratitude to the National Council of Humanities, Science, and Technology (CONAHCYT), to the National Polytechnic Institute for its ongoing support of this work, and additionally, the authors acknowledge the support provided by IPN SIP ID: 20242695, "Assessment of Tumor Extension in Breast Cancer Using Computed Tomography Scans and 3D Printed Models" and 20241141, "Implementation of Biomechanical Studies and Characterization of the Effects on the Hepatic Liver by Neural Networks and Genomics".

References

Ali F, Kalva SN, Koc M (2024) Advancements in 3D printing techniques for biomedical applications: a comprehensive review of materials consideration, post processing, applications, and challenges. Discov Mater 4(53). https://doi.org/10.1007/s43939-024-00115-4

Anthony L, Sabatini, Goswami T (2008) Hip implants VII: finite element analysis and optimization of cross-sections. Mater Des 29:1438–1446. https://doi.org/10.1016/j.matdes.2007.09.002

Dias JM, da Silv FSCP, Gasik M et al (2024) Unveiling additively manufactured cellular structures in hip implants: a comprehensive review. Int J Adv Manuf Technol 130:4073–4122. https://doi.org/10.1007/s00170-023-12769-0

Gellad WF, Gibbons MM, Miake-Lye IM, et al. (2011) A Comparison of Joint Replacement Disparities in VA and Non-VA Settings: A Systematic Review. https://www.ncbi.nlm.nih.gov/books/NBK82596/

Göktaş H, et al. (2022) Optimization of hip implant designs based on its mechanical behaviour. In: Hadamus A, Piszczatowski S, Syczewska M, Błażkiewicz M (eds) Biomechanics in Medicine, Sport and Biology. Lecture Notes in Networks and Systems, vol 328. Springer, Switzerland. https://doi.org/10.1007/978-3-030-86297-8_4

Michelle LF, Paul GJ, MacMillan L, Carsen S (2024) Rehabilitation and return to play following hip arthroscopy in young athletes. J Pediatr Soc North Am 7. https://doi.org/10.1016/j.jposna.2024.100051

Mohanty A, Lombardo AM, Judge C, Gundeti MS (2022) Are there disparities in access to robot-assisted laparoscopic surgery among pediatric urology patients? US institutional experience. Int J Urol 29(7):661–666. https://doi.org/10.1111/iju.14866

Rivera RJ, Karasavvidis T, Pagan C, et al. (2024) Functional assessment in patients undergoing total hip arthroplasty. Bone Jt J 106-B(8):764–774. https://doi.org/10.1302/0301-620X.106B8.BJJ-2024-0142.R1

Sweet MC, Borrelli GJ, Manawar SS, Miladore N (2021) Comparison of outcomes after robotic-assisted or conventional total hip arthroplasty at a minimum 2-year follow-up: a systematic review. J Bone Jt Surg 9(6). https://doi.org/10.2106/JBJS.RVW.20.00144. PMID: 34130305

Li Teng, Badre A, Alambeigi F, et al. (2023) Robotic systems and navigation techniques in orthopedics: a historical review. Appl Sci 13(17). https://doi.org/10.3390/app13179768

Chapter 25
Innovative Solutions in Surgical Approaches for Joint Prosthesis Installation and Reconstruction

Elías Humberto Hermida-Ochoa and Juan Alfonso Beltrán-Fernández

Abstract This chapter explores the latest advancements in surgical techniques for joint prosthesis installation and reconstruction, highlighting the significant role of biomedical engineering. It emphasizes the evolution of surgical methods, material science, and advanced imaging technologies in the field of orthopedics. By enhancing prosthesis design, prolonging their lifespan, and refining surgical accuracy, these innovations are transforming patient outcomes. The integration of artificial intelligence in medical imaging and tissue engineering presents new possibilities for personalized therapies and regenerative medicine. The analysis focuses on the global adoption of these developments, with a specific emphasis on the current contributions and challenges within Mexico. It also highlights a groundbreaking, globally patented therapy for cartilage regeneration, showcasing its potential in treating joint damages.

Keywords Surgical techniques · Biomedical engineering · Orthopedic innovations · Artificial intelligence in medicine · Regenerative therapies

25.1 Introduction

Advancements in biomedical engineering have led to significant improvements in joint prosthesis installation. By combining engineering principles with medical science, new horizons have opened in orthopedics, resulting in enhanced surgical techniques and material use. The demand for durable prostheses that offer long-term

E. H. Hermida-Ochoa
Hospital German Diaz Lombardo, Centro de Investigación y Laboratorio Biomecánico, Chimalistac San Ángel Carmen #18, México, CDMX, México
e-mail: drelias@hermidagroup.com

J. A. Beltrán-Fernández (✉)
Instituto Politécnico Nacional-Escuela Superior de Ingeniería Mecánica y Eléctrica-Sección de Estudios de Posgrado E Investigación Edificio 5, Unidad Profesional Adolfo López Mateos "Zacatenco" Col. Lindavista, 2Do Piso, Ciudad de México, México
e-mail: jbeltranf@hotmail.com

A. Öchsner and H. Altenbach (eds.), *Engineering Design Applications VII*, Advanced Structured Materials 230, https://doi.org/10.1007/978-3-031-84346-4_25

functionality while minimizing the risks of implant failure has prompted researchers to explore new approaches, from innovative materials to AI-assisted surgery. Prostheses are widely used to restore function in patients suffering from conditions such as osteoarthritis, traumatic injuries, or congenital deformities. The integration of advanced materials like polyetheretherketone (PEEK), known for its strength and bio-inertness, as well as biodegradable polymers like polylactic acid (PLA), shows great promise in improving biocompatibility and mechanical performance in various applications. The focus is on evaluating these materials alongside surgical innovations that improve outcomes in joint replacement surgery (Beltrán-Fernández et al. 2007; Urriolagoitia et al. 2020).

In addition, custom-made prosthetics produced using 3D printing methods are transforming orthopedic surgeries by creating implants that are designed to fit each patient's unique anatomical structures (Urriolagoitia et al. 2020). This not only enhances the precision of joint reconstruction, but also improves postoperative results, as personalized implants distribute mechanical loads more effectively.

The general objective of this research is to review and analyze the current state of research and development in advanced materials for joint prostheses and to identify potential areas for further investigation and improvement.

25.2 Methodology

Joint prostheses are now increasingly being made with materials tailored to meet the specific needs of the patient. These needs depend on factors such as anatomical structure, mechanical demands, and biological integration. Among the most prominent materials under investigation are PEEK and PLA, each offering distinct advantages. PEEK has shown high mechanical resistance, low wear rates, and excellent biocompatibility. On the other hand, PLA is biodegradable and useful in temporary applications where scaffolds support tissue regeneration. Both materials are being explored in the context of their application in prostheses designed for long-term stability and integration with surrounding tissues (Smith and Johnson 2018).

The preoperative planning stage is crucial in ensuring optimal results in joint replacement surgeries. Medical imaging techniques, such as 3D reconstructions from CT scans, allow for detailed anatomical modeling. The use of AI in analyzing these models has enhanced the surgeon's ability to make precise decisions, particularly in the fitting of prostheses to the patient's unique anatomical structure (Fritz and Fritz 2022). Figure 25.1 demonstrates the use of AI-enhanced segmentation of joint imaging, allowing for a more detailed preoperative plan and minimizing the risk of misalignment post-surgery.

Tissue engineering is essential for improving the durability and success of joint prostheses. Methods using biomaterials such as hydroxyapatite, along with stem cells, are used to create cartilage and bone tissues. These techniques have the potential for regeneration, allowing for both the mechanical reconstruction of joints and biological repair. Figure 25.2 illustrates the process of implanting tissue-engineered

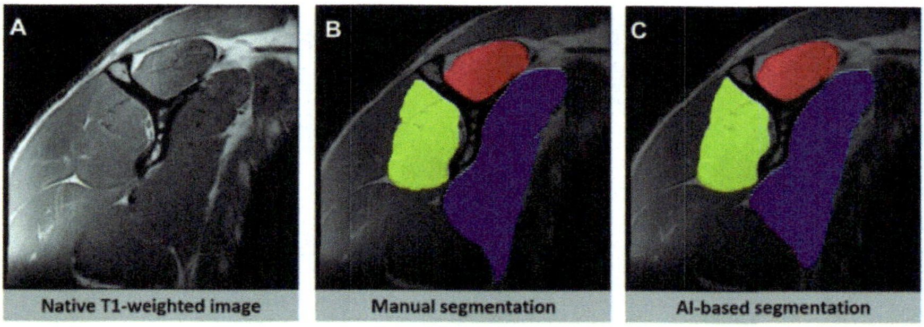

Fig. 25.1 The use of AI-enhanced segmentation of joint imaging (Fritz and Fritz 2022)

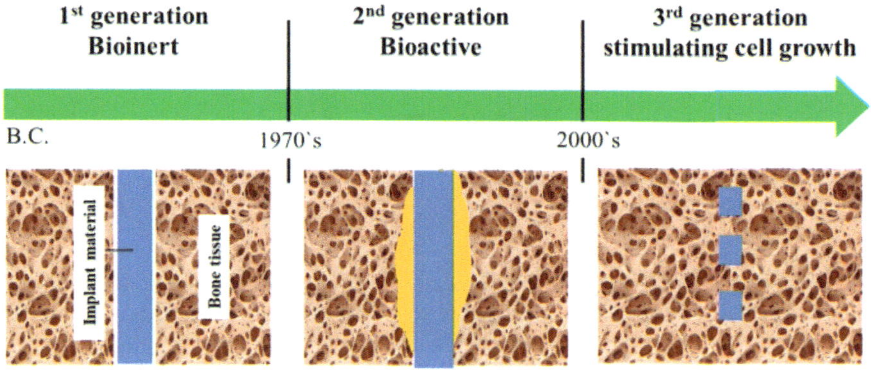

Fig. 25.2 The process of tissue-engineered cartilage implantation in a preclinical model (Farag 2023)

cartilage in a preclinical model, showing the early stages of integration within the host tissue (Farag 2023).

25.3 Results

The implementation of PEEK and PLA in custom-designed prostheses has demonstrated significant advantages in terms of mechanical performance and biocompatibility. Comparative mechanical testing between PEEK and titanium implants shows that PEEK has a slightly lower compressive strength but significantly reduced wear under physiological loading, making it a suitable material for long-term joint prostheses (Guner et al. 2024). Figure 25.3 compares mechanical stress distribution between titanium and PEEK-based prostheses under physiological loads, indicating a 15% lower wear rate for PEEK implants.

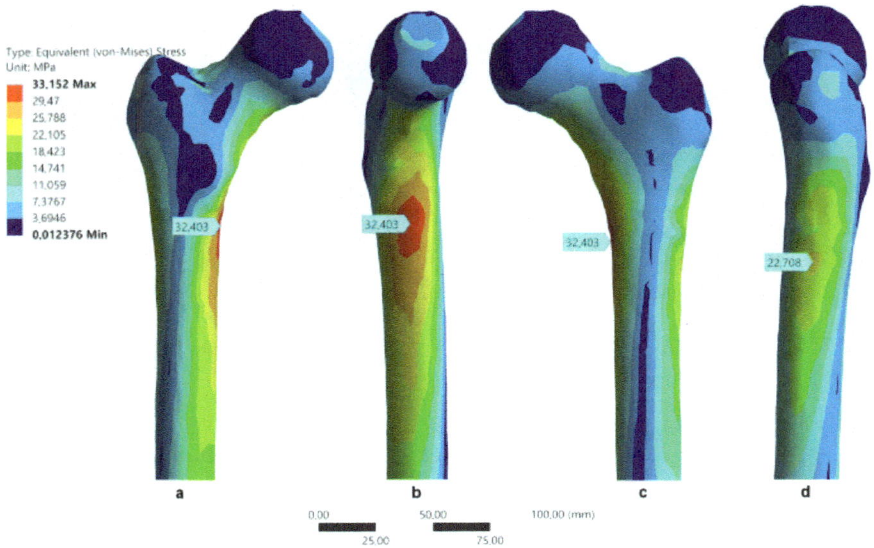

Fig. 25.3 Mechanical stress distribution between titanium and PEEK-based prostheses under physiological loads (Guner et al. 2024)

AI-driven imaging tools, when integrated into preoperative planning, have reduced the incidence of postoperative misalignment by 20% compared to traditional methods (Wu et al. 2019). As shown in Fig. 25.4, the accuracy of implant placement was significantly improved in AI-assisted procedures, with deviations from the planned angle reduced to less than 0.5°.

Cartilage regeneration therapies using biomaterials and stem cells have shown encouraging results in preclinical trials. Specifically, the patented cartilage regeneration therapy has demonstrated a 70% improvement in cartilage thickness in patients with osteoarthritis compared to untreated controls (Yang et al. 2024).

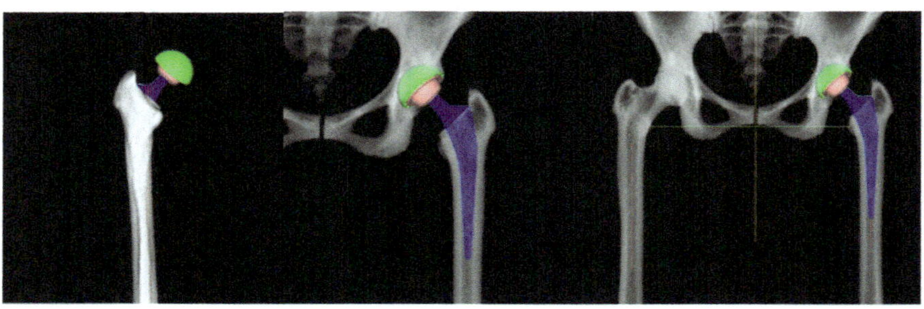

Fig. 25.4 AI-assisted procedures into preoperative planning (Wu et al. 2019)

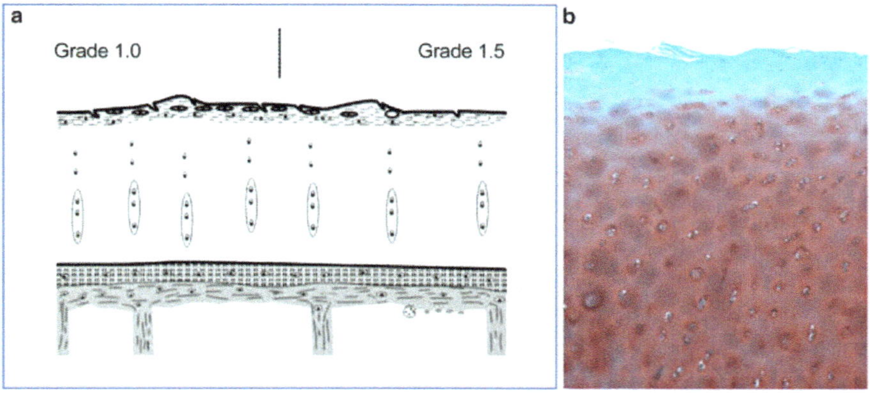

Fig. 25.5 Histological evidence of cartilage regeneration using biomaterials and stem cells (Yang et al. 2024)

Figure 25.5 provides histological evidence of cartilage regeneration three months after implantation, showcasing restored cartilage morphology.

Table 25.1 summarizes global findings in joint prosthesis installation, comparing material properties, failure rates, and biomechanical stability in clinical trials conducted in Mexico, the U.S., Germany, Japan, and France.

25.4 Discussion

The combination of PEEK and PLA materials in custom-designed prostheses has not only enhanced the mechanical properties but also opened new pathways for bio-integration and long-term functionality. With the integration of AI tools into preoperative planning, the accuracy of implant placement has dramatically improved. This reduction in surgical errors correlates with better postoperative recovery and lower rates of revision surgery (Pritzker and Gahunia 2020).

The global landscape for these innovations is rapidly evolving, with countries like Mexico making significant strides in adopting these technologies, as shown in Table 25.1. The focus on cost-effective, durable materials has made these technologies accessible to a wider population, ensuring that patients benefit from cutting-edge solutions. Figure 25.6 illustrates the integration of tissue-engineered implants in a clinical setting, showing promising early-stage results in osseointegration and functional restoration (Yamamoto et al. 2017).

The patented therapy for regenerating cartilage has the potential to revolutionize treatments for joint prostheses. As shown in Fig. 25.7, the regenerated cartilage

Table 25.1 Comparative biomechanical performance of joint prostheses in Mexico vs. global standards

Study	Materials	Outcome focus	Mechanical strength (MPa)	Wear rate (mm^3)	Fatigue resistance (Cycles)	Current global trends	Cite
Study A (Mexico)	PEEK	Osseointegration	120 MPa	0.8 mm^3	1.5 million cycles	Personalized prostheses	Farag (2023)
Study B (USA)	PLA	Reduced subsidence rates	85 MPa	1.2 mm^3	1.1 million cycles	Long-term implant stability	Guner et al. (2024)
Study C (Germany)	Ti-alloys	Compressive strength	150 MPa	0.5 mm^3	2 million cycles	Minimally invasive surgery	Wu et al. (2019)
Study D (Japan)	PEEK	Flexibility and bioresorbability	130 MPa	0.7 mm^3	1.7 million cycles	Biomechanical integrity	Yang et al. (2024)
Study E (France)	PLA + hydroxyapatite	Regenerative capacity	200 MPa	0.3 mm^3	2.5 million cycles	Tissue engineering	Pritzker and Gahunia (2020)

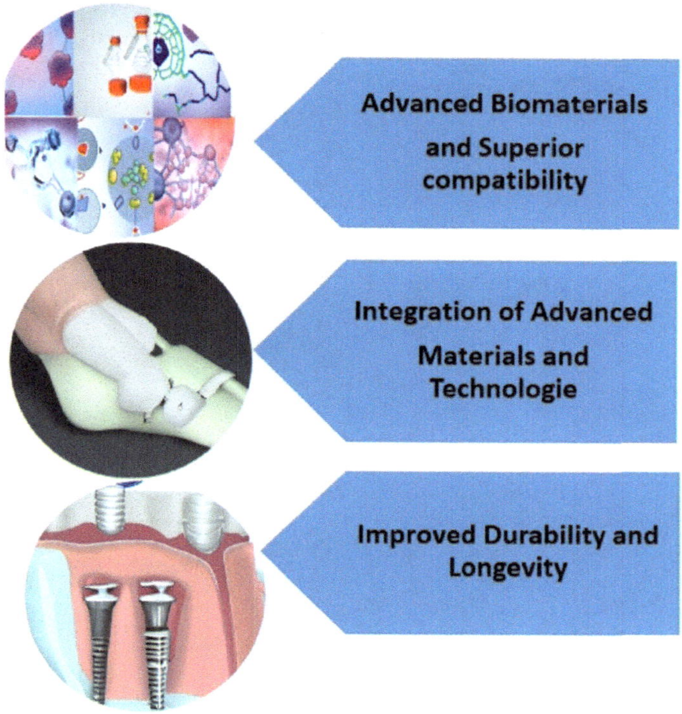

Fig. 25.6 Integration of tissue-engineered implants in a clinical setting (Yamamoto et al. 2017)

resulting from stem cell treatment has demonstrated greater durability and functionality compared to traditional prosthetic cartilage treatments. This discovery highlights the necessity for continued exploration of the long-term clinical applications of these techniques.

25.5 Conclusions

The evolution of surgical approaches for joint prosthesis installation and reconstruction is redefining the field of orthopedics. Innovations in material science, surgical techniques, and tissue engineering are collectively contributing to more durable, functional, and biocompatible implants. The implementation of AI in preoperative planning has minimized risks and improved surgical outcomes, leading to reduced failure rates and longer-lasting prostheses.

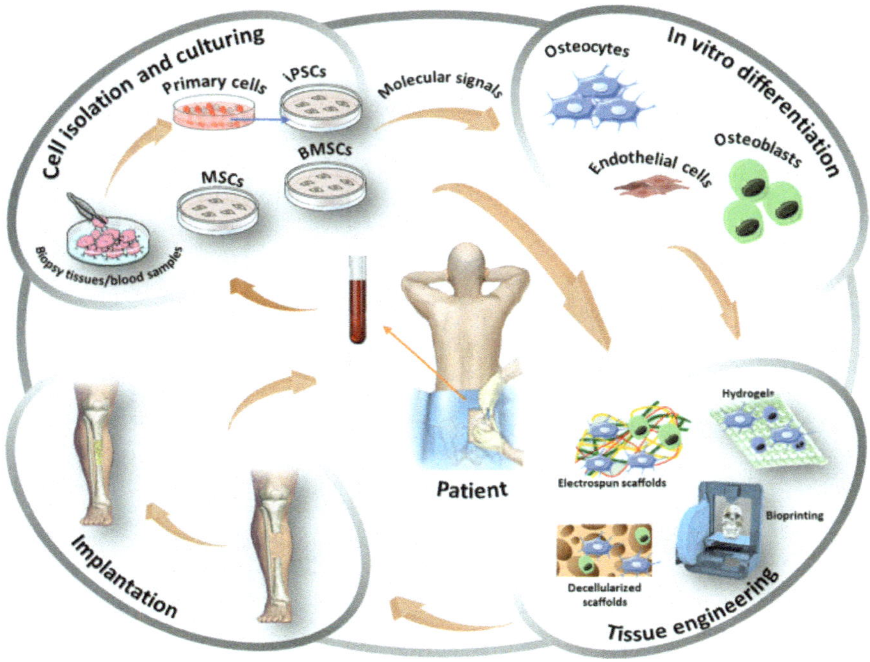

Fig. 25.7 The integration of tissue-engineered implants in a clinical setting (Kulkarni et al. 2024)

Mexico is emerging as a key player in these developments, showing a strong adoption of innovative solutions tailored to local needs. The patented cartilage regeneration therapy holds great promise for patients with cartilage injuries, potentially reducing the need for invasive surgeries and improving overall joint function.

Future research should focus on large-scale clinical trials to assess the long-term outcomes of these innovations and further refine the integration of AI and regenerative therapies into everyday orthopedic practice.

Acknowledgements Instituto Politécnico Nacional for all the contributions and support in the development of the study.

Hospital German Diaz Lombardo and CILAB for all the facilities.

References

Augustine R, Gezek M, Nikolopoulos VK et al (2024) Stem cells in bone tissue engineering: progress, promises and challenges. Stem Cell Rev and Rep 20:1692–1731. https://doi.org/10.1007/s12015-024-10738-y

Beltrán-Fernández JA, Hernández-Gómez LH, Rodríguez-Cañizo RG, Urriolagoitia-Calderón G, Urriolagoitia-Sosa G, González-Rebatú A, Dufoo-Olvera M (2007) Mechanical behavior of a

calcium phosphate ceramic bone graft used in the rehabilitation of a C4 human vertebra. Appl Mech Mater 7–8:101–106. https://doi.org/10.4028/www.scientific.net/amm.7-8.101
Farag MM (2023) Recent trends on biomaterials for tissue regeneration applications: review. J Mater Sci 58:527–558. https://doi.org/10.1007/s10853-022-08102-x
Fritz B, Fritz J (2022) Artificial intelligence for MRI diagnosis of joints: a scoping review of the current state-of-the-art of deep learning-based approaches. Skeletal Radiol 51:315–329. https://doi.org/10.1007/s00256-021-03830-8
Guner AT, Kocak S, Meran C (2024) Mechanical analysis of a PEEK titanium alloy macro-composite hip stem by finite element method. J Braz Soc Mech Sci Eng 46:338. https://doi.org/10.1007/s40430-024-04939-2
Kulkarni PG, Paudel N, Magar S et al (2024) Overcoming challenges and innovations in orthopedic prosthesis design: an interdisciplinary perspective. Biomed Mater & Devices 2:58–69. https://doi.org/10.1007/s44174-023-00087-8
Pritzker KPH, Gahunia HK (2020) Histopathology evaluation of cartilage disease and repair. In Gahunia H, Gross A, Pritzker K, Babyn P, Murnaghan L (eds) Articular cartilage of the knee. Springer, New York, NY. https://doi.org/10.1007/978-1-4939-7587-7_15
Smith P, Johnson M (2018) The use of 3D printing in custom joint prostheses: benefits and challenges. J Orthop Surg 45(3):345–367. https://doi.org/10.1177/0022034518804247
Urriolagoitia G, Hernández-Gómez LH, Rodríguez-Cañizo RG, Urriolagoitia-Calderón G, Dufoo-Olvera M (2020) Biomechanical evaluation of PLA and PEEK in custom orthopedic implants. J Biomater Appl 34(6):1–15. https://doi.org/10.1177/0885328220908202
Wu W, Lu X, He B (2019) Advances in biomaterials for orthopedic implants: trends and future directions. J Mater Sci-Mater Med 30(4):1–15. https://doi.org/10.1007/s10856-019-6253-6
Yamamoto T, Murakami T, Shibata Y (2017) The role of AI in improving surgical outcomes in joint replacement: early evidence from clinical trials. J Med Syst 41(9):132. https://doi.org/10.1007/s10916-017-0786-z
Yang W, Gao T, Liu X et al (2024) Clinical application of artificial intelligence-assisted three-dimensional planning in direct anterior approach hip arthroplasty. Int Orthop (SICOT) 48:773–783. https://doi.org/10.1007/s00264-023-06029-9

Chapter 26
Development of Biocompatible Implants Using 3D Printing Technology for Anterior Cervical Fusion

Eduardo Tenorio-Monzalvo, Juan Alfonso Beltrán-Fernández, Erik Francisco Rodríguez-Piñón, Héctor Sánchez de la Vega-Covarrubias, Elías Humberto Hermida-Ochoa, Pamela López-Miranda, Andrea Hernández-Benitez, and Julieta Aimee Zilli-González

Abstract This research analyzes the application of 3D-printed implants made from biocompatible materials for anterior cervical fusion procedures. Anterior cervical discectomy and fusion is a common surgical intervention for various cervical spine conditions, which requires the use of implants to achieve spinal stability and promote fusion. However, traditional implants may not be customizable to individual anatomical requirements and may not be biocompatible, leading to limitations in patient outcomes. The development and application of 3D printing technology to create patient-specific implants tailored to individual anatomical requirements is explored in this research. Using 3D printing, implants can be customized precisely to match

E. Tenorio-Monzalvo · E. F. Rodríguez-Piñón · H. S. de la Vega-Covarrubias · E. H. Hermida-Ochoa
Facultad Mexicana de Medicina, Universidad La Salle México, Hospital German Diaz Lombardo, Centro de Investigación y Laboratorio Biomecánico, Carmen, CDMX, México
e-mail: drtenoriotrauma@gmail.com

H. S. de la Vega-Covarrubias
e-mail: hector.sanchez0@uabc.edu.mx

E. H. Hermida-Ochoa
e-mail: drelias@hermidagroup.com

J. A. Beltrán-Fernández (✉) · P. López-Miranda · A. Hernández-Benitez · J. A. Zilli-González
Sección de Estudios de Posgrado E Investigación, Escuela Superior de Ingeniería Mecánica y Eléctrica, Instituto Politécnico Nacional, Unidad Zacatenco, "Unidad Profesional Adolfo López Mateos", Colonia Lindavista, Ciudad de México, México
e-mail: jbeltran@ipn.mx

P. López-Miranda
e-mail: pamela.01lop@gmail.com

A. Hernández-Benitez
e-mail: andreaahb21@gmail.com

J. A. Zilli-González
e-mail: julietaaimee.zilliglez@gmail.com

A. Öchsner and H. Altenbach (eds.), *Engineering Design Applications VII*, Advanced Structured Materials 230, https://doi.org/10.1007/978-3-031-84346-4_26

patient anatomy, optimize fit, and enhance osseointegration. Furthermore, biocompatible materials ensure that the implants are compatible with surrounding tissues and minimize the risk of adverse reactions while promoting bone healing. This study aims to demonstrate the feasibility of 3D-printed biocompatible implants for anterior cervical fusion through computational modeling, and experiment. The use of 3D-printed biocompatible implants for anterior cervical fusion offers potential benefits in terms of surgical outcomes, patient satisfaction, and long-term spinal health.

Keywords 3D-printed · Implants · Cervical fusion · Peek · PLA · Biomechanics

26.1 Introduction

Additive manufacturing, particularly 3D printing, has revolutionized the customization of medical implants, significantly improving outcomes in procedures like cervical discectomy and fusion (ACDF) (Sheoran et al. 2020). In ACDF, the implant serves a dual purpose: maintaining disc height and promoting vertebral fusion following discectomy. Traditional implants, however, often fail to adapt to the patient's unique anatomy, resulting in mechanical instability and insufficient osseointegration (Mobarak et al. 2023). By leveraging patient-specific designs, 3D-printed implants offer the potential for a more precise anatomical fit and improved biological integration (Wang and Wu 2023; Sharma et al. 2023) (Fig. 26.1).

The limitations of conventional implants, including mechanical mismatch and suboptimal biological performance, highlight the need for advancements. Titanium implants, despite their strength, tend to induce stress shielding, hindering bone formation around the implant (Patel 2018; Bai et al. 2021). PEEK, while offering better biological performance than titanium, still presents challenges in fully promoting

Fig. 26.1 3D-printed cervical implant tailored from patient-specific computed tomography (CT) data, illustrating its precise anatomical conformity

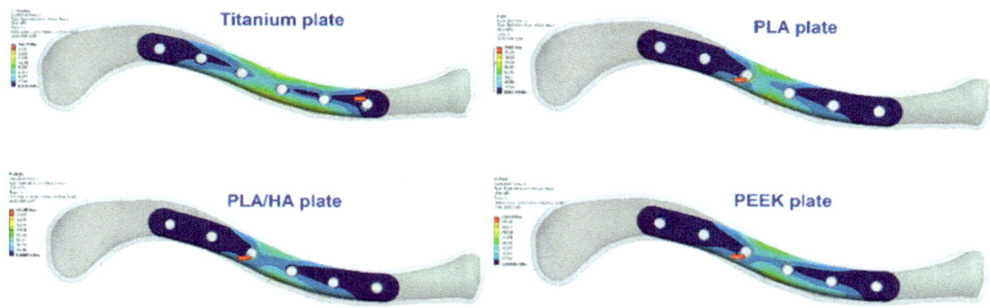

Fig. 26.2 A finite element analysis (FEA) simulation comparing the stress distribution in PEEK and polylactic acid (PLA) implants under physiological loading conditions

vertebral fusion (Tabrizi and Fadaei 2023). 3D printing enables the customization of implant geometry, aligning it more closely with the patient's anatomical structure and enhancing biomechanical performance. The use of biocompatible materials, such as hydroxyapatite, further improves osseointegration by mimicking the natural bone environment (Yurteri et al. 2024) (Fig. 26.2).

The removal of a cervical disc and the placement of an implant to maintain the disc space and support vertebral fusion are involved in ACDF. The success of the procedure is contingent on the biomechanical properties of the implant. Conventional implants may create localized stress concentrations, which may impede the fusion process. In contrast, 3D-printed implants, designed to match the patient's anatomy, distribute stress more evenly, thus promoting better long-term stability and bone growth (Cheung and Chang 2023; Sadiq and Muthusamy 2023) (Fig. 26.3).

26.2 Materials and Methods

The methodology of this work is shown in Fig. 26.4, as follows:

The methodology presented includes everything necessary to address the clinical case discussed in this work. It includes the use of computerized axial tomography, a computer design program, the selection of mechanical properties, and the application of the finite element method.

26.2.1 Implant Design and Finite Element Analysis (FEA)

The implants and cervical spine were designed using Onshape CAD, Radiiant, and Meshmixer software, allowing for detailed customization based on patient-specific computed tomography (CT) data (see Fig. 26.5). The designs were 3D printed in PLA and PEEK to be evaluated in image correlation technique. This experimental

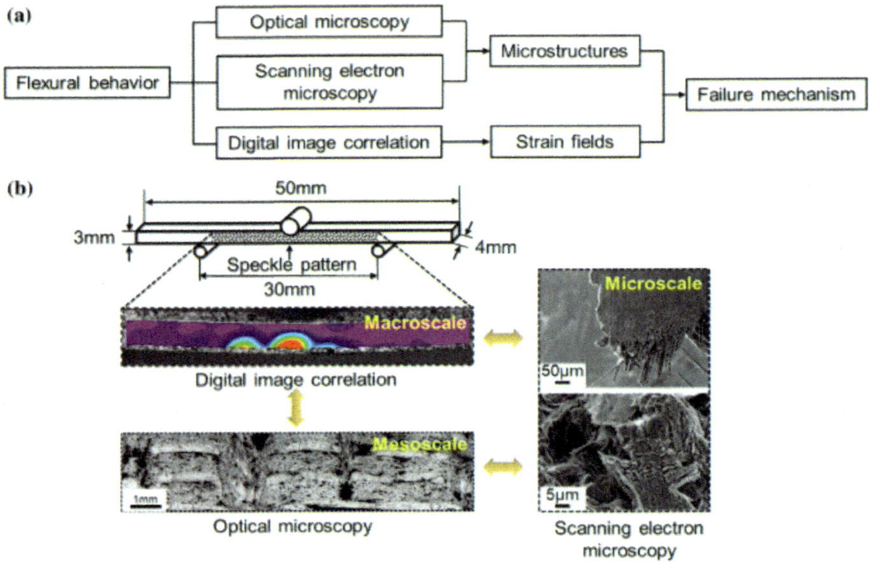

Fig. 26.3 Image correlation analysis used during mechanical testing of 3D-printed implants, indicating improved strain distribution compared to traditional implants (Xu et al. 2022). Reprinted with permission from SpringerNature publishers

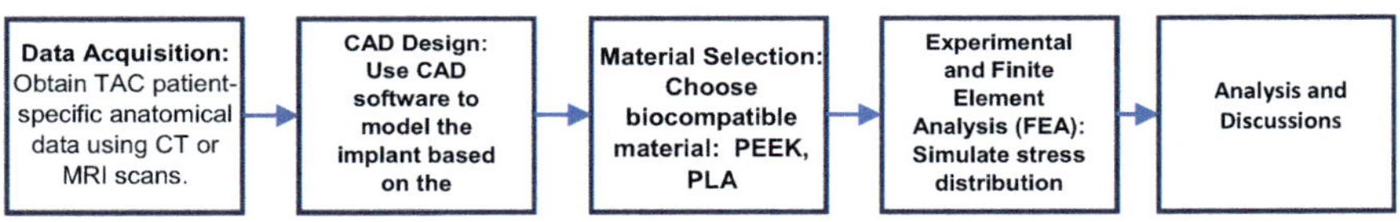

Fig. 26.4 Methodology

testing allowed to understand the mechanical behavior of the implants, particularly strain and stress distribution under specific loading conditions.

A parametric model of the inter somatic cage was developed to be assembled with the cervical numerical model, as illustrated in Fig. 26.6.

In the same line, a segmented model of the cervical spine with PEEK implants was developed to be assembled with the inter somatic cages and to compare mechanical performance under the same loading conditions, as illustrated in Fig. 26.7.

26.2.2 *Experimental Testings*

Mechanical testings were conducted using both standardized and custom 3D-printed implants to evaluate their mechanical stability. The Kinovea software was used to

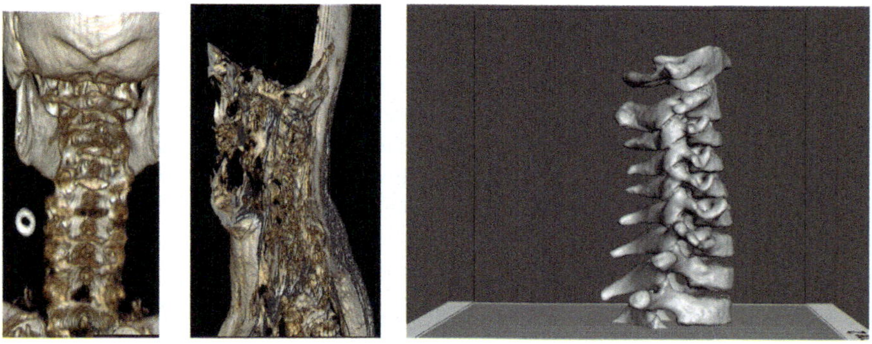

Fig. 26.5 Segmentation of TAC using Radiiant© and Meshmixer© software

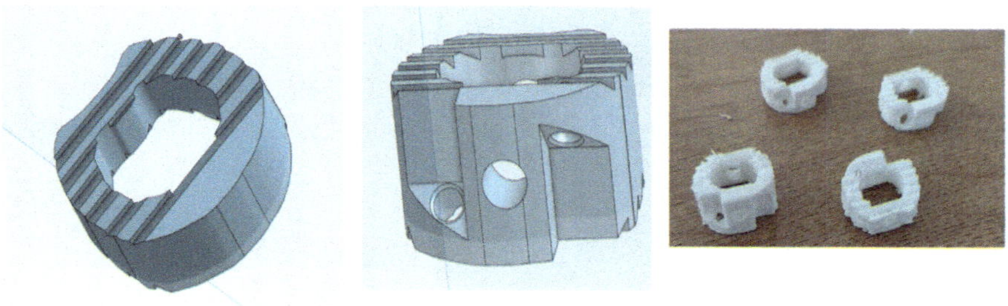

Fig. 26.6 CAD and manufactured implants

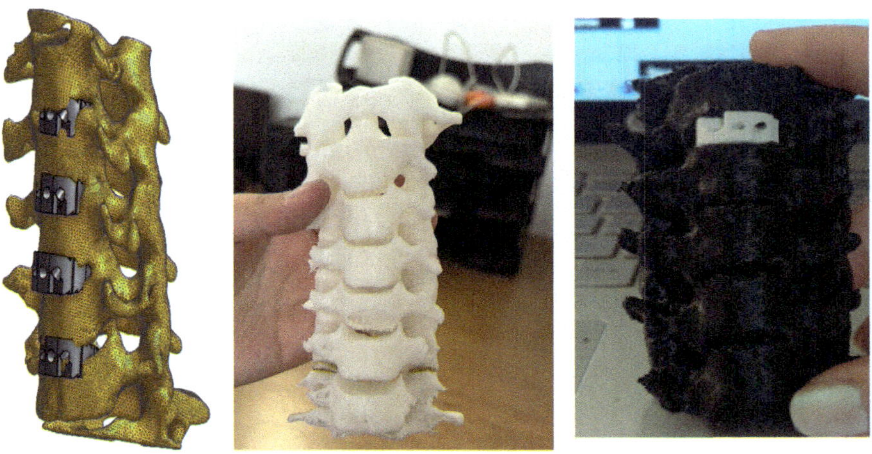

Fig. 26.7 Assembly of the implants with cervical model

determine the Cobbśangle and the GOM software was employed for digital image correlation (DIC) (Fig. 26.8).

About the image correlation technique, precise strain distribution during compression tests was measured by this testing. Compression loading conditions were applied by these mechanical tests. The implants were tested using 3D printed cervical vertebrae (human cervical spine). This allowed a direct comparison between the compressive performance of 3D-printed implants made from PEEK and polylactic acid (PLA) (Fig. 26.9).

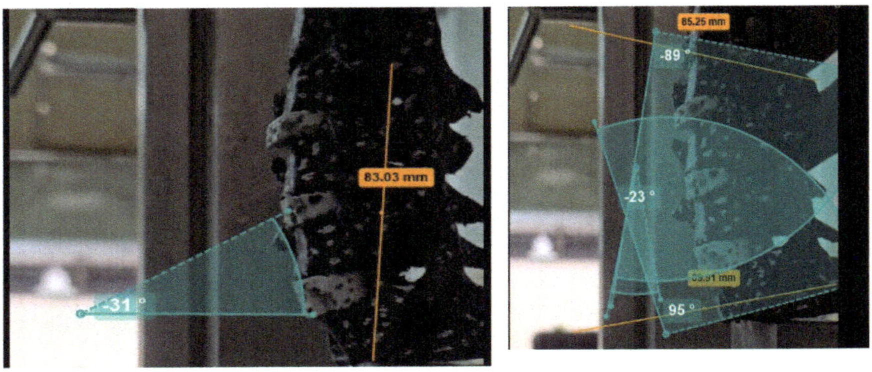

Fig. 26.8 Cobbś angle using Kinovea software

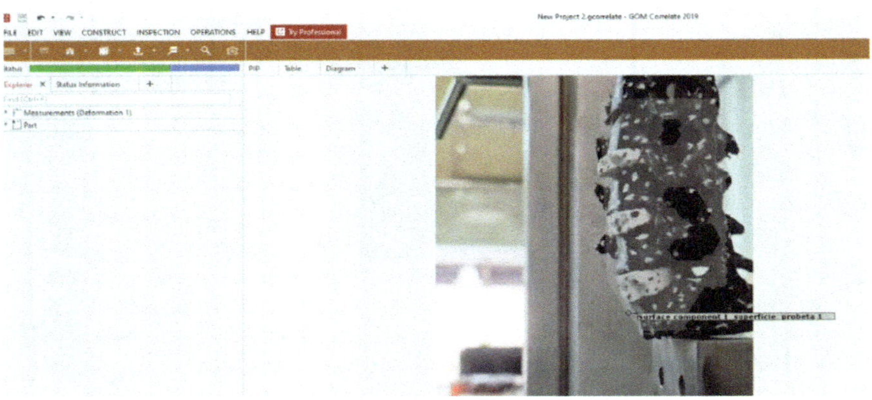

Fig. 26.9 Experimental technique using GOM correlate software

26.3 Results

In the analysis of the results, the Cobb angle was found to be within the range of 10° to 25°, indicative of mild scoliosis. This classification is consistent with the criteria established by Cobb (1948). A mild scoliosis typically does not require aggressive intervention but is an important indicator for monitoring spinal stability based on these parameters (Cobb et al. 1948) (Fig. 26.10).

Furthermore, the strain analysis performed using GOM Correlate demonstrated low strain in both PLA and PEEK implants, as measured during mechanical testing. The reduced strain values suggest vertebral stability, which aligns with the Müller stability scale for vertebral assessment. This outcome highlights the mechanical robustness of both implant materials, supporting their clinical viability for spinal fusion procedures. The results were corroborated by the findings of Beltrán-Fernández et al. (2007) (Fig. 26.11).

Mechanical performance under physiological loading conditions was demonstrated by the low strain levels observed in the image correlation analysis, highlighting the materials' effectiveness in maintaining vertebral stability. These results validate the capability of the 3D-printed implants, both in PLA and PEEK, to preserve vertebral alignment and ensure efficient load distribution. Such mechanical characteristics are indicative of the implants' potential to contribute to stable and favorable post-operative outcomes.

26.4 Conclusions

In this work, a specific analysis of the mechanical behavior of 3D-printed spinal implants made from PEEK and PLA was presented, with a specific focus on their application in cervical spine fusion. Utilizing advanced design tools such as Onshape CAD, the implants were tested to meet patient-specific anatomical requirements. The integration of these personalized designs with modern 3D printing technologies has been shown to offer significant advantages over traditional implant materials, particularly in terms of improving load distribution and minimizing subsidence risks during physiological conditions. The biomechanical testing conducted using GOM Correlate provided important insights into the strain behavior of both PEEK and PLA implants. These results show that both materials can maintain vertebral alignment and withstand biomechanical loads, even during critical phases of post-operative recovery. The low strain levels observed in both materials, measured against the Muller scale, confirm their suitability for spinal fusion applications. The maintenance of vertebral alignment within the Cobb angle range of 10° to 25° further confirms the effectiveness of these implants in correcting mild scoliosis cases, ensuring both short-term recovery and long-term spine stability.

The enhanced performance of 3D-printed implants in stress distribution compared to traditional materials is particularly significant. Image correlation analysis revealed

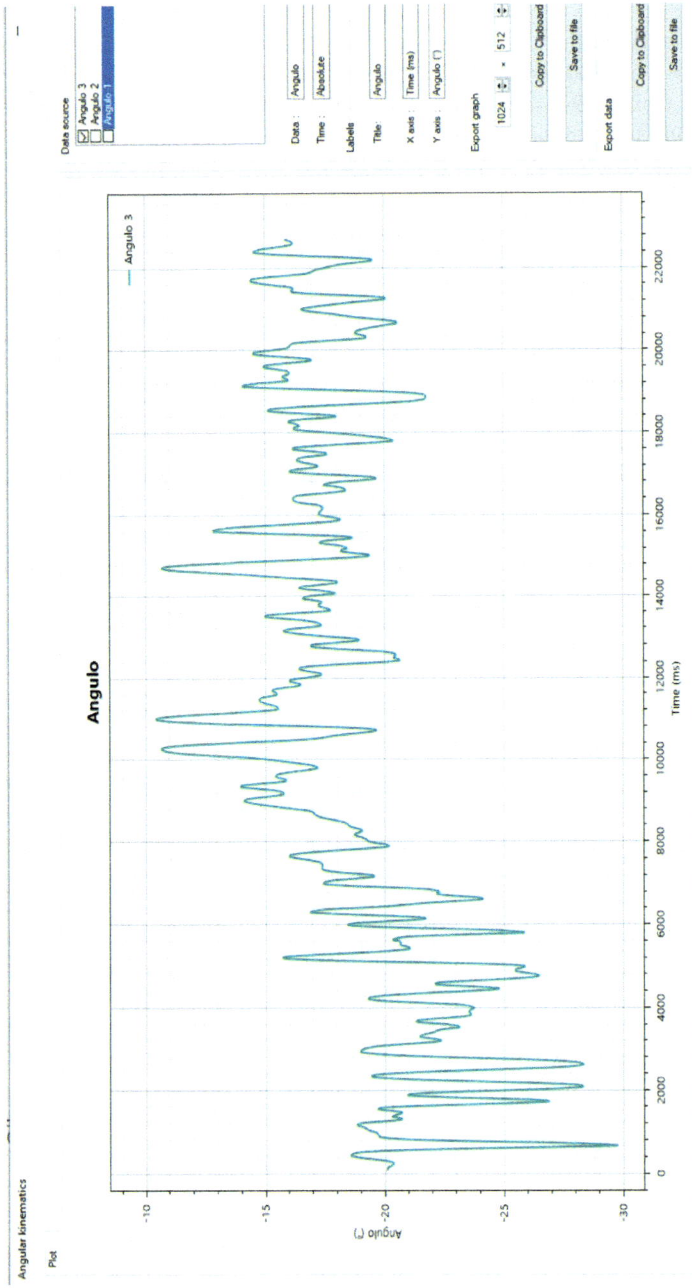

Fig. 26.10 Dynamic measurement of Cobb's angle using Kinovea software

Fig. 26.11 Strain measuring using GOM correlate software

that custom-fitted implants result in a more uniform strain distribution under loading conditions. This is crucial for reducing implant subsidence and promoting successful osseointegration. This observation is consistent with previous studies that highlight the significance of patient-specific designs in improving clinical outcomes in spinal surgeries.

The results highlight the potential of 3D-printed, custom spinal implants. These implants not only closely match anatomical structures but also demonstrate strong mechanical properties. When rigorously designed and tested, PEEK and PLA implants show the ability to improve post-operative spinal stability, decrease failure risks, and contribute to the long-term functionality of the implant. Further research is needed to explore the long-term clinical implications of these findings, especially in diverse patient populations and more complex scoliosis cases, to fully use the potential of 3D printing in spinal surgery.

Acknowledgements The authors would like to express their sincere gratitude to the National Council of Humanities, Science, and Technology (CONAHCYT), the National Polytechnic Institute, and CILAB–Hospital Germán Díaz Lombardo for their continuous support in this research.

References

Bai J, Liu H, Zhang Y, Wang B, Zhao J (2021) Titanium interlayer-mediated hydroxyapatite-coated polyetheretherketone cage in transforaminal lumbar interbody fusion surgery. BMC Musculoskelet Disord 22(1). https://doi.org/10.1186/s12891-021-04691-2

Beltrán-Fernández JA, Urriolagoitia Calderón G, Urriolagoitia Sosa G, González Rebatú A et al (2007) Mechanical behavior of a calcium phosphate ceramic bone graft used in the rehabilitation of a C4 human vertebra. AMM. 7–8:101–106 (2007). https://doi.org/10.4028/www.scientific.net/amm.7-8.101

Cheung M, Chang C (2023) Material properties and biomechanical performance of PEEK versus titanium for spinal implants: a comparative study. JSS 9(1):45–53 (2023). https://doi.org/10.21037/jss-23-12

Cobb JR, Cobb J, Cobb JL (1948) Outline for the study of scoliosis

Mobarak MH, Islam MA, Hossain N, Al Mahmud MZ, Rayhan MT, Nishi NJ, Chowdhury MA (2023) Recent advances of additive manufacturing in implant fabrication–a review. Appl Surf Sci Adv 18:100462. https://doi.org/10.1016/j.apsadv.2023.100462

Patel M et al (2018) Stress shielding in titanium spinal implants. J Biomed Eng

Sadiq A, Muthusamy P (2023) Enhanced osseointegration of PEEK spinal implants: a review of the current strategies and their efficacy. Spine J. https://doi.org/10.1016/j.spinee.2023.05.002

Sharma S, Pahuja S, Gupta V et al (2023) 3D printing for spine pathologies: a state-of-the-art review. Biomed Eng Lett 13:579–589. https://doi.org/10.1007/s13534-023-00302-x

Sheoran AJ, Kumar H, Arora PK, Moona G (2020) Bio-medical applications of additive manufacturing: a review. Procedia Manuf 51:663–670. ISSN 2351–9789. https://doi.org/10.1016/j.promfg.2020.10.093

Tabrizi R, Fadaei F (2023) Comparison of PEEK and titanium in spinal fusion surgery: a systematic review and meta-analysis. Becker's Spine Review. https://www.beckersspine.com/spine/item/52504-peek-vs-titanium-coated-peek-implants-for-spinal-fusion-5-key-findings

Wang R, Wu Z (2023) Recent advancement in finite element analysis of spinal interbody cages: a review. Front Bioeng Biotechnol 11:1041973. https://doi.org/10.3389/fbioe.2023.1041973

Xu H, Li L, Li G et al (2022) In situ characterization of the flexural behavior and failure mechanism of 2D needle punched carbon/carbon composites by digital image correlation. J Mater Sci 57:11077–11091. https://doi.org/10.1007/s10853-022-07272-y

Yurteri A, Mercan N, Uğur L (2024) Comparison of the use of biocompatible materials and titanium in the treatment of midshaft clavicle fractures with a patient-specific plate: a finite element analysis study. Arch Orthop Trauma Surg 144:3255–3266. https://doi.org/10.1007/s00402-024-05449-5

Chapter 27
Aerodynamic Analysis and Application of High-Pressure Air Gun

Hong-Bin Chang, Uzu-Kuei Hsu, Yao-Yu Tsai, Chen-Yu Wu, and Cheng-Hsien Tai

Abstract To maintain ecological conservation and species balance, it is necessary to utilize shooting sterilization vaccines to manage the issue of overpopulation among species. Therefore, exploring the application of shooting devices is crucial. The key to effective deployment lies in designing these devices to achieve an effective range while minimizing the distress caused to animals during vaccination. This study explores the use of high-pressure air for shooting device applications, utilizing the differential pressure between high-pressure air and atmospheric pressure to provide an initial flight velocity to a syringe loaded with vaccine. The force with which the syringe is fired is affected by variations in pressure and barrel length, which are related to the internal energy of high-pressure air and the gas expansion ratio during the expansion stroke. The research involves conducting multiple simulations and experiments with various barrel lengths and storage pressures to determine the optimal length and pressure relationship under fixed volume conditions. Additionally, unmanned rotorcrafts are utilized as tools for air-to-ground operations. To reduce the payload weight on unmanned aerial vehicles, resonant waves are also integrated into the study. Resonance within the barrel can minimize the use of barrel length and gas pressure, achieving task completion with minimal equipment resources. The

H.-B. Chang · Y.-Y. Tsai
General Research Service Center, National Pingtung University of Science and Technology, Pingtung County, ROC, Taiwan
e-mail: changhb@mail.npust.edu.tw

Y.-Y. Tsai
e-mail: og73230153@mail.npust.edu.tw

U.-K. Hsu
Bachelor Program in Artificial Intelligence and Mechatronics, National Pingtung University of Science and Technology, Pingtung County, ROC, Taiwan
e-mail: ukhsu@mail.npust.edu.tw

C.-Y. Wu
Department of Vehicle Engineering, National Pingtung University of Science and Technology, Pingtung County, ROC, Taiwan

C.-H. Tai (✉)
College of Biomedical Engineering and Technology, Tzu Chi University, Hualien, ROC, Taiwan
e-mail: chtai0109@gms.tcu.edu.tw

A. Öchsner and H. Altenbach (eds.), *Engineering Design Applications VII*, Advanced Structured Materials 230, https://doi.org/10.1007/978-3-031-84346-4_27

application of shock wave shooting devices can potentially be extended to various fields in the future, such as ecological diagnostics of trees, real-time environmental testing, and cold launches in military engineering.

Keywords Rotorcraft applications · Air pressure shooting · Resonant waves · Computational fluid dynamics

27.1 Introduction

The sika deer population in Kenting, Taiwan, has gradually developed into a stable group, impacting the local ecosystem and agricultural production. The activities of the population have especially altered the vegetation structure and ecological diversity of their habitat, affecting the living conditions of other wildlife. Birth and death rates are the primary natural factors influencing the size of the sika deer population, forming a complex dynamic balance. When the mortality rate increases, the birth rate also increases to compensate for the reduced population density; the reverse is also true.

Using direct hunting or euthanasia as a method of population control has very low acceptance at both public and policy levels. Furthermore, due to reproductive compensation, the population density may quickly recover after such measures are taken. Traditional surgical contraception methods are no longer suitable at this stage. Contraceptive vaccine technology offers an economical and rapid reproductive control strategy that can be applied on a large scale. Developing appropriate injection techniques is crucial to ensure that the vaccine coverage and continuous contraceptive effect meet management goals. Currently, traditional methods such as capture or remote injection using drug injection guns are employed to administer vaccines. To achieve effective population control, it is necessary to administer contraceptive vaccines on a large scale, with at least 80% of female individuals effectively contracepted.

Due to Taiwan's hot and humid climate and human activity factors, sika deer primarily move at night and have excellent horizontal vision and perceptual abilities, making it extremely difficult to administer contraceptive vaccines manually at close range. As of 2023, the number of wild sika deer in Kenting National Park, Taiwan, has exceeded 2,000 due to the lack of natural predators. Traditional capture or remote injection methods are inadequate to vaccinate a large number of female individuals within three months (the period when males are distinguishable by their antlers is from June to September).

In recent years, innovations and advancements in drone technology have had extensive applications in the ecological industry, including ecological surveys, ecological restoration, and wildlife healthcare. Field experiments using drones to fly over sika deer have shown that sika deer have relatively weak vision and perception abilities in the vertical direction and are not sensitive to aerial threats. They typically do not exhibit escape responses or only move away slowly. Moreover, sika deer in Kenting, Taiwan, mainly gather in open grassland environments at the forest edge during dawn and dusk, making drone technology a potential solution (Hampton et al. 2016).

27.2 Research Methods

27.2.1 Research Process

This study aims to design a shooting device that can be mounted on a drone, capable of firing syringes loaded with animal vaccines. The design of the device will take into account practical challenges, such as the payload limitations of the drone, shooting distance, and shooting stability. Therefore, the design must focus on lightweight construction while ensuring sufficient shooting power.

This study begins by establishing an initial model based on preliminary concepts to conduct actual shooting experiments and measure the initial bullet velocity. Using computational fluid dynamics (CFD) methods, flow field analysis of the high-pressure air launching device is conducted (Aldert Johan Cilliers 2021; Anderson 2009; Ferziger and Perić 2002). By varying the pressure and volume of the air storage tank, the process of gas expansion and bullet ejection is examined to determine the optimal volume and pressure ratio (Gkritzapis et al. 2007; Hainz and Costello 2005).

In addition to using compressed air alone, this study also explores the use of shock waves and resonant waves as energy sources for shooting. The principle of energy superposition of shock waves is used to investigate achieving the same or better results with a shorter barrel and less air storage volume (Tai et al. 1999, 2006, 2005; Lin 2012; Mandella et al. 1985; Neumann 1943).

The research will compare the shooting effects of three energy sources: air pressure, shock waves, and resonant waves. For each shooting method, the effects of barrel length, resonant mass block weight, and spacing will be examined. Numerical analysis and post-processing results will be compared, and a final procedure will be proposed to establish the most effective shooting device.

27.2.2 Experimental Methods

The device design utilizes high-pressure gas as the energy source for shooting, primarily due to its advantages of being lightweight and quieter during firing compared to gunpowder. However, using high-pressure gas for shooting requires an investigation into the impact of gas expansion speed on the shooting performance. Therefore, the study will involve testing with components capable of rapid gas exhaust.

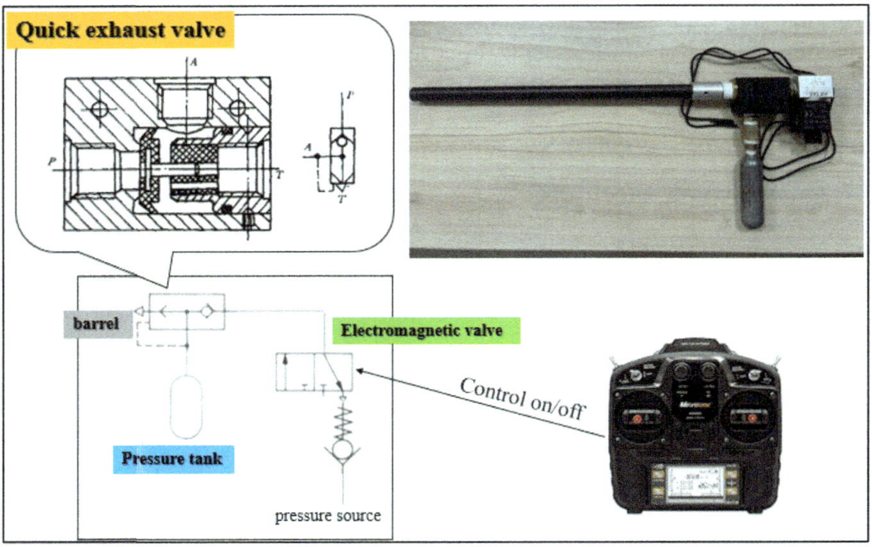

Fig. 27.1 Prototype of air pressure shooting device

The prototype of the shooting device used in this study is shown in Fig. 27.1. A CO_2 gas cylinder is used as an air storage tank to store high-pressure air. The high-pressure air is quickly released through the rapid exhaust valve, generating thrust to fire the bullet. Since the shooting device is mounted on a drone, a solenoid valve is used as a control medium to operate the rapid exhaust valve.

To determine the minimum allowable distance for the shooting device when deployed, we conducted field flight tests prior to the experiment to observe the drone's flight over sika deers and assess its impact on them. The tests showed that when the drone flew at an altitude of 15 meters above the sika deer, the noise and downward airflow from the drone caused the deer to start searching for environmental threats. When the altitude decreased to 10 meters, the deer began to move, as shown in Fig. 27.2. Therefore, our operational height was set within the range of 10 to 15 meters, with experiments conducted at the maximum height of 15 meters to test the bullet's initial velocity and accuracy. Through the experiments, we determined the minimum allowable bullet velocity to be 24 m/s, as shown in Fig. 27.3.

27.2.3 Calculation Model and Boundary Definition

In this study, the velocity of the bullet exiting the muzzle is taken as the research objective. The calculation model is established with the storage space for high-pressure gas, the bullet, the barrel, and the external flow field, as shown in Fig. 27.4. To reduce computational costs, simulations are performed using axisymmetric modeling. Fully

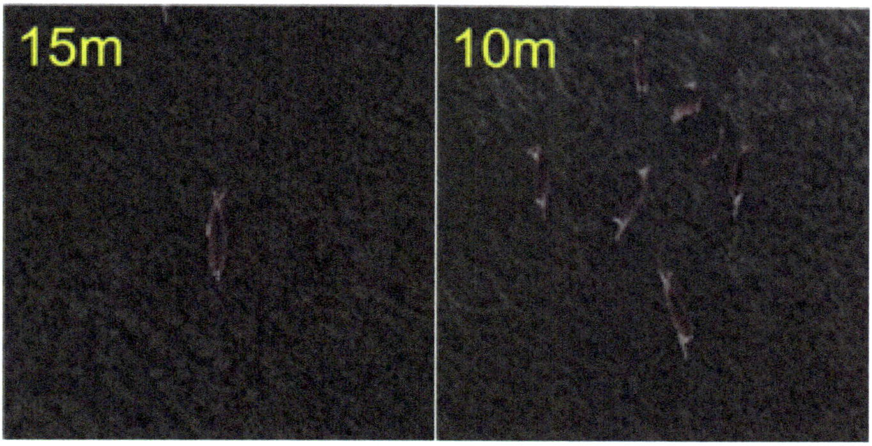

Fig. 27.2 Impact of drone flight altitude on sika deer

Test number	Velocity (m/s)
1	24.3
2	24.1
3	23.9
4	24.3
5	24.5

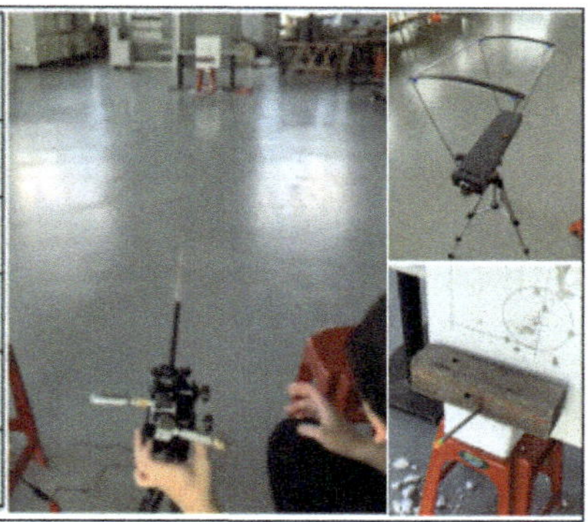

Fig. 27.3 Target initial shooting velocity experiment

orthogonal grids are used for the mesh setup. Dynamic simulations are conducted using dynamic mesh and six degrees of freedom (six DOF) methods to investigate the effect of variables on the bullet's muzzle velocity, as shown in Fig. 27.5 (Altufayl 2019; Gkritzapis and Kaimakamis 2008; Khalil et al. 2015).

The simulations will employ three methods: pressure waves, shock waves, and resonant waves. The design modifications for the pressure wave simulation will involve changing the barrel length and altering the pressure of the high-pressure air. Additionally, to accommodate the space constraints of the storage tank mounted on the drone, the shape and angle of the storage tank will be modified without changing its volume, as shown in Fig. 27.6.

The shock wave simulation involves using high-pressure air to propel a mass block. The space between the mass block and the bullet contains air at atmospheric

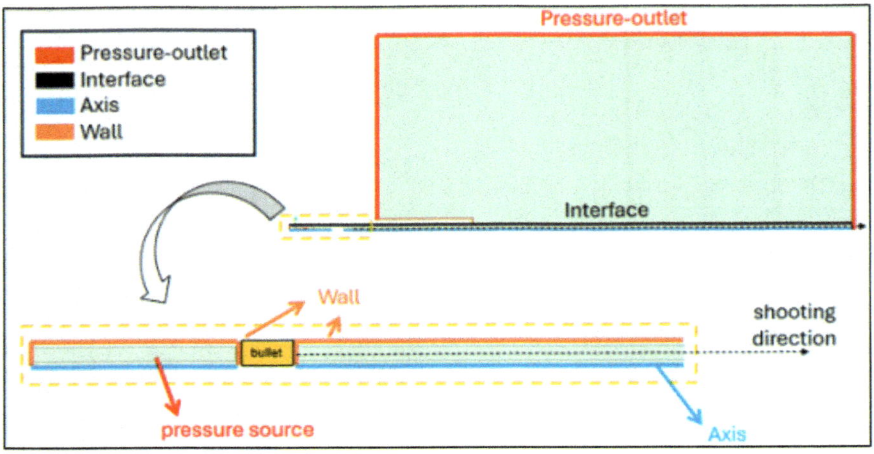

Fig. 27.4 Calculation model boundary settings

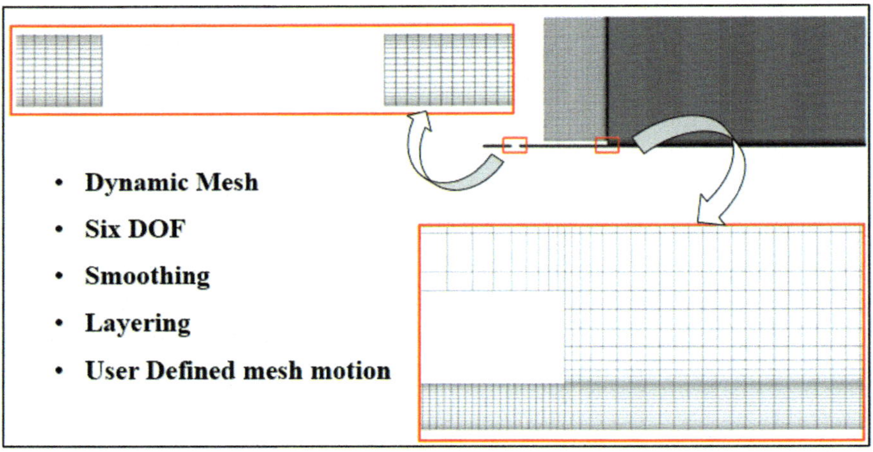

Fig. 27.5 Dynamic mesh setup

pressure. The shock wave generated by the mass block compressing the air is used to fire the bullet. Figure 27.7 illustrates the design modification factors for the shock wave simulation. The distance between the mass block and the bullet, as well as the weight of the mass block, will be varied to investigate their effects on the bullet's firing speed.

To achieve better shock wave superposition effects, the study incorporates a user-defined function (UDF) to define the displacement of the mass block, simulating it as a shock wave generator. As shown in Fig. 27.8, the design modification factors for the resonant wave will use three frequencies: 1000 Hz, 2000 Hz, and 3000 Hz. The

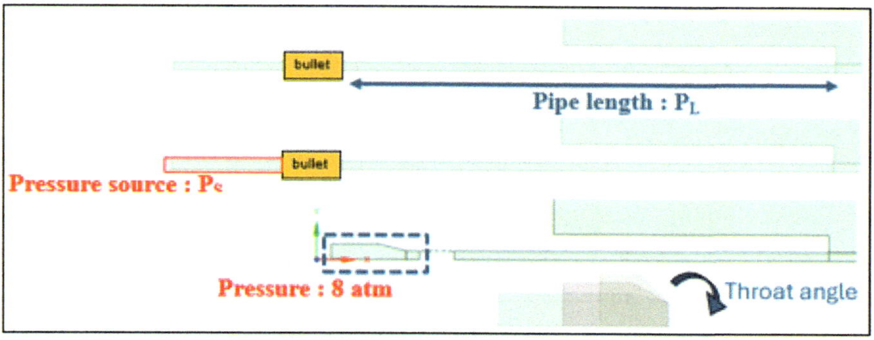

Fig. 27.6 Design variations for pressure wave simulation

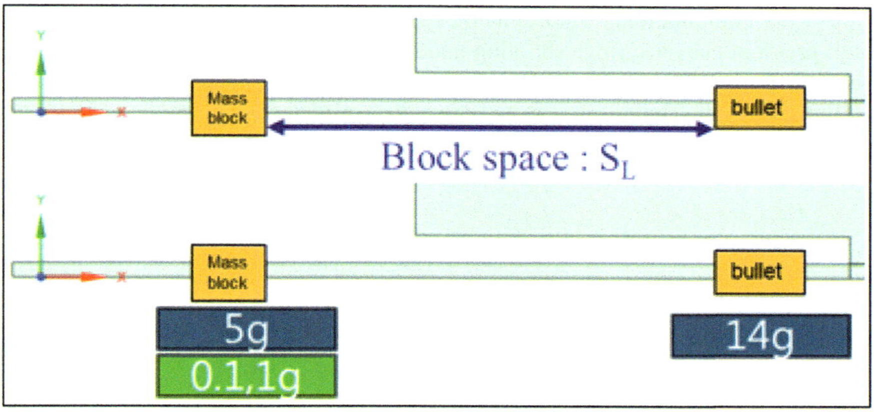

Fig. 27.7 Design variations for shock wave simulation

simulation of resonant waves also includes an investigation of the distance between the mass block and the bullet.

27.2.4 Numerical Methods

To study the effect of pressure waves on the propulsion force of the bullet in the shooting device simulation, the fluid is assumed to be compressible, and dynamic meshing is used for the bullet shooting simulation. To simplify the model, axisymmetric methods are employed to simulate three-dimensional flow phenomena in a two-dimensional form, and the friction between the bullet and the barrel is neglected. For the resonant wave section, user-defined mesh motion is used to define the displacement of the mass block.

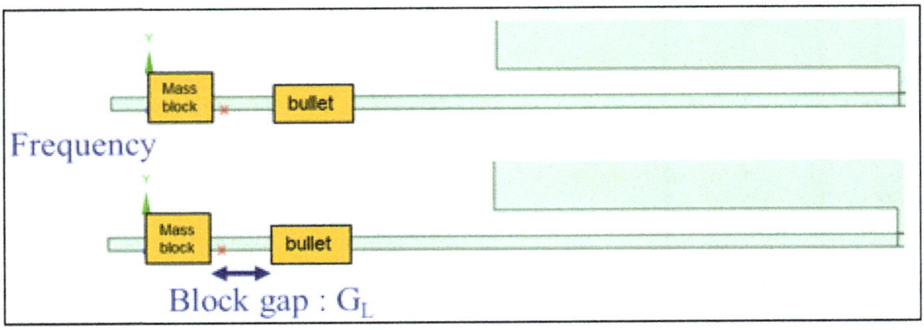

Fig. 27.8 Design variations for resonance wave simulation

This study utilizes numerical simulation methods to analyze flow field phenomena, employing computational fluid dynamics (CFD) to explore the dynamic effects of high-pressure gas on syringe shooting under different design parameters. The study needs to examine the pressure-driven situation of the bullet and the flow dynamics within the barrel. Therefore, it is necessary to consider changes in the velocity and pressure fields. The flow phenomena should satisfy both the conservation of mass and the conservation of momentum:

$$\frac{\partial \rho}{\partial t} + \nabla \cdot (\rho \overline{u}) = 0 \tag{27.1}$$

$$\frac{\partial}{\partial t}(\rho \overline{u}) + \nabla \cdot (\rho \overline{u} \overline{u}) = -\nabla p + \nabla \cdot \left(\overline{\overline{\tau}}\right) + S_i \tag{27.2}$$

In the equations, $\overline{\overline{\tau}}$ represents the stress tensor, and μ represents the dynamic viscosity. The relationship is as follows:

$$\overline{\overline{\tau}} = \mu\left[\left(\nabla \overline{u} + \nabla \overline{u}^T\right)\right] - \frac{2}{3}\nabla \cdot \overline{u} I \tag{27.3}$$

The k-ε turbulence model used in this study is the standard k-ε turbulence model. The turbulent viscosity in the standard k-ε mathematical model is defined as:

$$\mu_t = \rho C_\mu \frac{k^2}{\varepsilon} \tag{27.4}$$

where $C_\mu = 0.09$ is an empirical constant, the transport equations for the turbulent kinetic energy k and the turbulent dissipation rate ε are expressed as follows:

Turbulent kinetic energy equation (k):

$$\frac{\partial}{\partial t}(\rho k) + \frac{\partial}{\partial x_j}\left(\rho k \overline{u_j}\right) = \frac{\partial}{\partial x_j}\left[\left(\mu + \frac{\mu_t}{\sigma_k}\right)\frac{\partial k}{\partial x_j}\right] + G_k - \rho\varepsilon \tag{27.5}$$

Turbulent dissipation rate equation (ε):

$$\frac{\partial}{\partial t}(\rho\varepsilon) + \frac{\partial}{\partial x_j}(\rho\varepsilon\overline{u_i}) = \frac{\partial}{\partial x_j}\left[\left(\mu + \frac{\mu_t}{\sigma_\varepsilon}\right)\frac{\partial\varepsilon}{\partial x_j}\right] + C_{1\varepsilon}\frac{\varepsilon}{k}(G_k) - C_{2\varepsilon}\rho\frac{\varepsilon^2}{k} \quad (27.6)$$

The constants $C_{1\varepsilon} = 1.44$, $C_{2\varepsilon} = 1.92$, $\sigma_k = 1.0$, and $\sigma_\varepsilon = 1.3$, in Eqs. (27.5) and (27.6) are empirical constants. σ_k and σ_ε are the turbulent Prandtl numbers for the turbulent kinetic energy and turbulent dissipation rate, respectively. G_k is the production term of turbulent kinetic energy, defined as follows:

$$G_k = \mu_t S_s^{\ 2} \quad (27.7)$$

27.3 Results and Discussion

To achieve an effective range and identify an optimal configuration for shooting stability and lightweight design, this research aims to improve the bullet's velocity as it exits the muzzle. By increasing the initial shooting speed as the bullet leaves the barrel, better shooting stability and flight distance can be achieved.

Before conducting numerical simulations, high-pressure air shooting experiments are performed to determine the minimum allowable initial velocity, which serves as the baseline for simulations. The parameter settings in the simulation are based on experimental values, and a grid convergence analysis is performed for boundary layer definitions to ensure the accuracy and reliability of the simulation results. As shown in Fig. 27.9, the boundary layer on the barrel wall was reduced from 1 mm to 0.05 mm. The evaluation criterion was the air pressure speed at the muzzle, with a converged value of 30.3 m/s. Thus, a 0.1 mm boundary layer was deemed the ideal setting for subsequent simulations.

The study then proceeds to simulate and refine the design by comparing pressure waves, shock waves, and resonant waves. The effects of these three types of pressure waves are analyzed to determine the best direction for improvement.

27.3.1 Exploration Using a Pressure Wave Shooting Device

This study explores design modifications in a shooting device operating solely under pressure waves by varying the barrel length, the pressure of the air source, and the shape and angle of the air source.

In the experiment, the barrel length used was 300 mm. In the simulation, we increased and decreased the barrel length by 100 mm respectively. It was found that, in the absence of friction, the initial velocity of the bullet was minimally affected

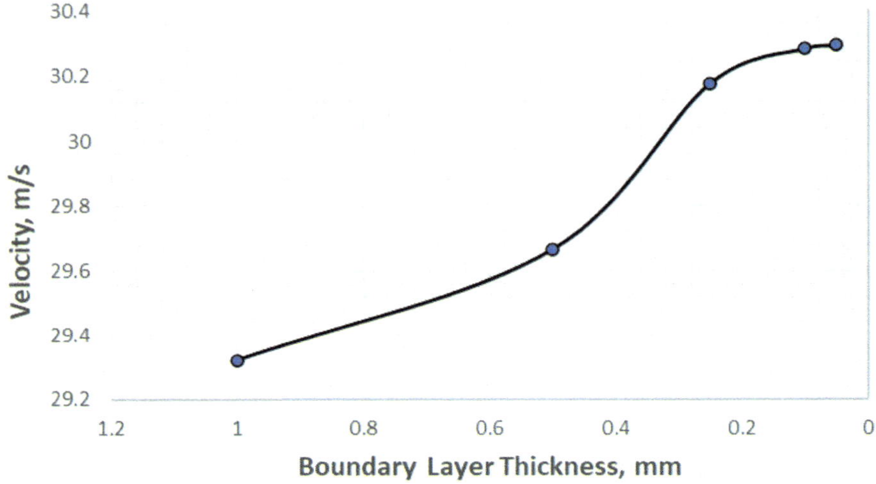

Fig. 27.9 Boundary layer mesh convergence curve

by the barrel length. The force exerted on the bullet over time showed no variation in the thrust from high-pressure air for the three different lengths, as observed in Fig. 27.10.

By changing the pressure of the air source, it is evident that higher pressure results in better initial velocity performance, as shown in Fig. 27.11.

The study also investigates the impact of shape variations on the initial velocity of the shooting device. As shown in Fig. 27.12, in an enclosed volume, as long as the pressure and volume remain constant, the shape does not affect the thrust.

As summarized in Fig. 27.13, it can be observed that changing the air pressure while keeping the volume constant is the best method to improve the initial velocity.

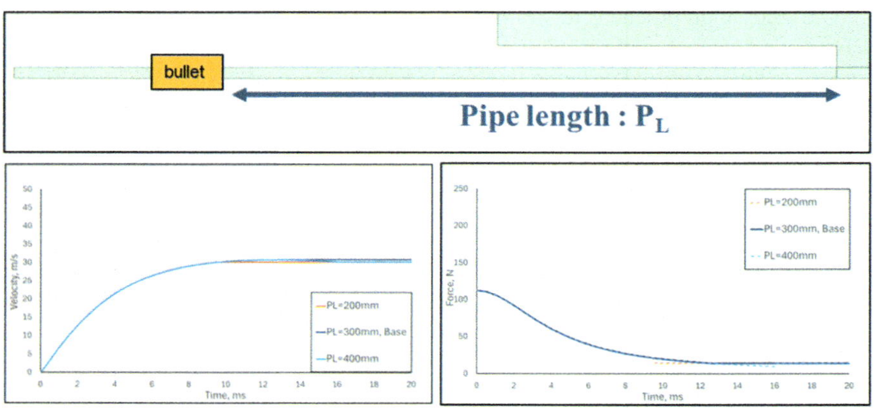

Fig. 27.10 Effect of barrel length on bullet initial velocity using air pressure wave

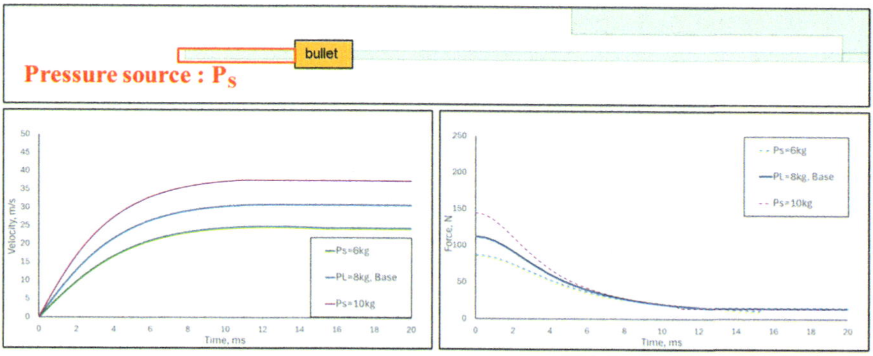

Fig. 27.11 Effect of air source pressure on bullet initial velocity using air pressure wave

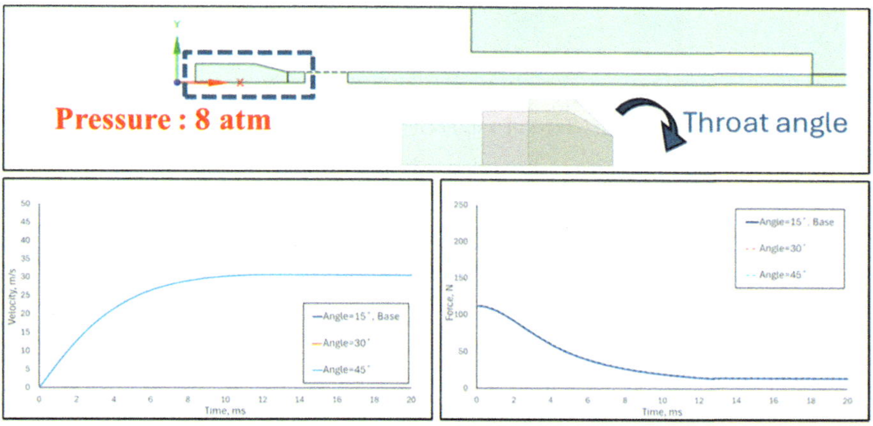

Fig. 27.12 Effect of air tank shape on bullet initial velocity using air pressure wave

27.3.2 Exploration Using a Shock Wave Shooting Device

This study primarily aims to design a shooting device suitable for field use. Analysis of pressure waves indicates that higher air pressure is needed for better shooting performance. However, in field environments, there is limited access to large energy sources, thus limiting the method of increasing air pressure. To achieve lower energy costs, this study will attempt to use shock waves for testing.

As shown in Fig. 27.14, adjusting the distance between the mass block and the bullet can effectively change the initial velocity of the bullet. Observing the thrust variation over time, it is found that the thrust curve maintains around 14 N when the bullet exits the muzzle. However, in simulations with distances SL = 125 mm, SL = 225 mm, and SL = 325 mm, a discontinuity occurs when the bullet exits the barrel. At a distance of SL = 44 mm, the bullet experiences a thrust lower than 14 N

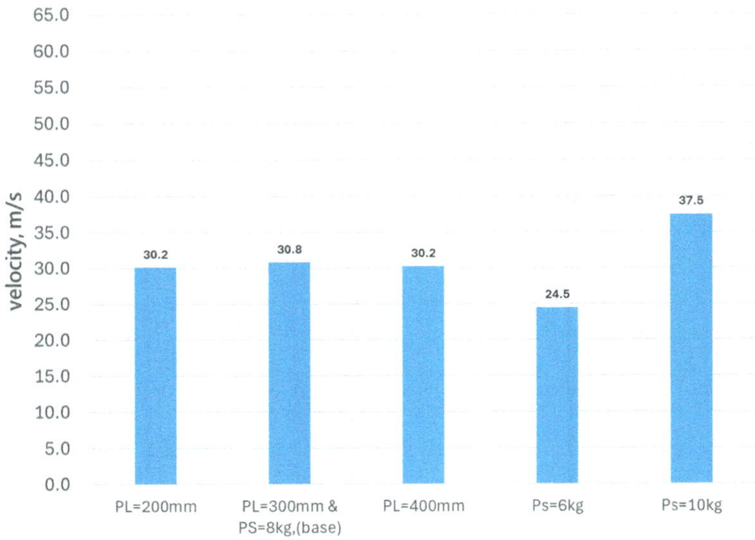

Fig. 27.13 Comparison of the effects of various modifications on bullet initial velocity using air pressure wave

upon exiting the muzzle. If 14 N is considered the baseline for maintaining balanced bullet thrust, it can be reasonably inferred that in the cases of SL = 125 mm, SL = 225 mm, and SL = 325 mm, the air pressure wave inside the barrel has not fully developed to the muzzle position. At SL = 44 mm, it is inferred that the direction of the air pressure wave's force opposes the bullet's flight direction when it exits the muzzle.

Therefore, finding an optimal distance where the position of the air pressure wave's force application coincides with the bullet's exit from the muzzle can achieve the best thrust effect, as illustrated in Fig. 27.15.

Under the shock wave configuration, the mass block becomes a crucial factor. Changing the mass of the block will affect the development of the air pressure wave between it and the bullet. As shown in Fig. 27.16, it is observed that the heavier the mass block, the faster the initial velocity of the bullet. However, the time from when the bullet starts moving to when it exits the muzzle is longer. The thrust curve also shows that different mass blocks have a relationship with the barrel length. Therefore, to purely investigate the effect of the mass block's weight, the calculation model extends the barrel to an infinite length to analyze the bullet's velocity changes without other influencing factors, as illustrated in Fig. 27.17.

From the statistical chart in Fig. 27.18, it can be observed that the effectiveness of using shock waves is inferior to that of pressure waves. However, the simulation results of shock waves reveal that the method of using shock waves is directly influenced by the barrel length and the weight of the mass block.

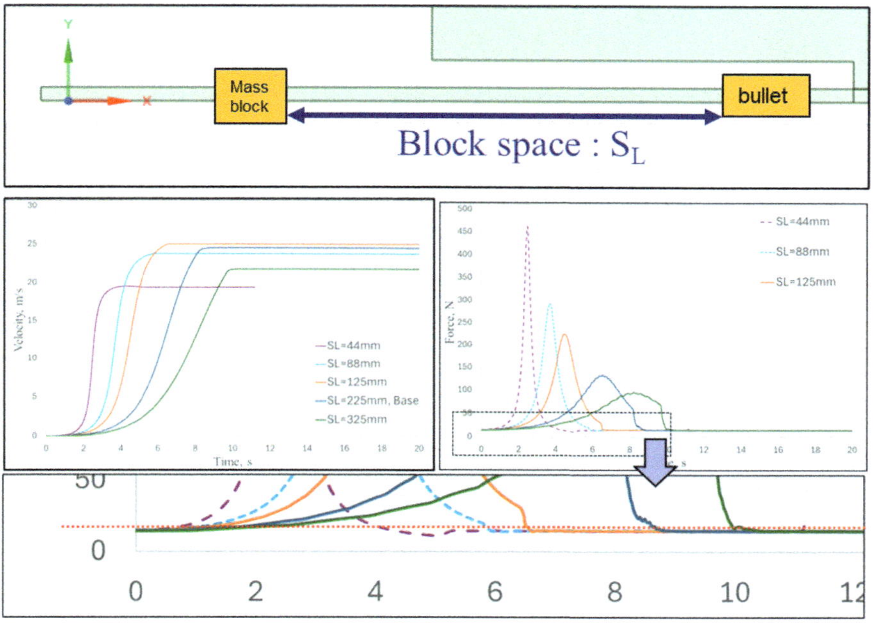

Fig. 27.14 Effect of distance on bullet initial velocity using shock waves

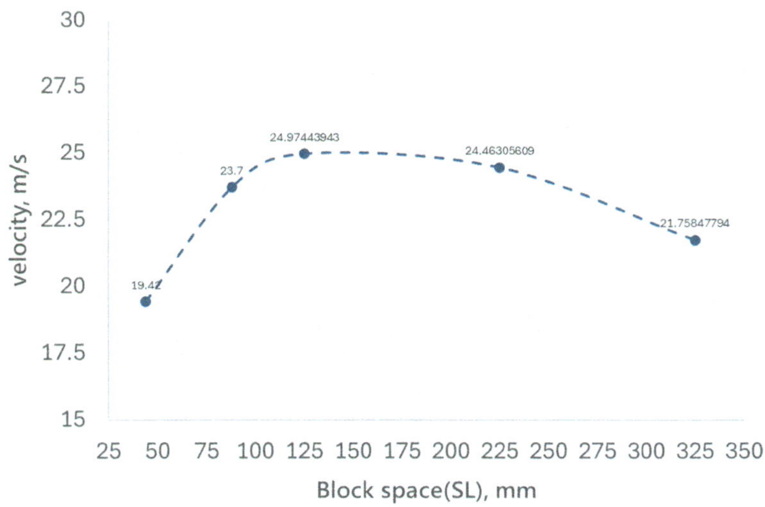

Fig. 27.15 Bullet initial velocity performance at different distances using shock waves

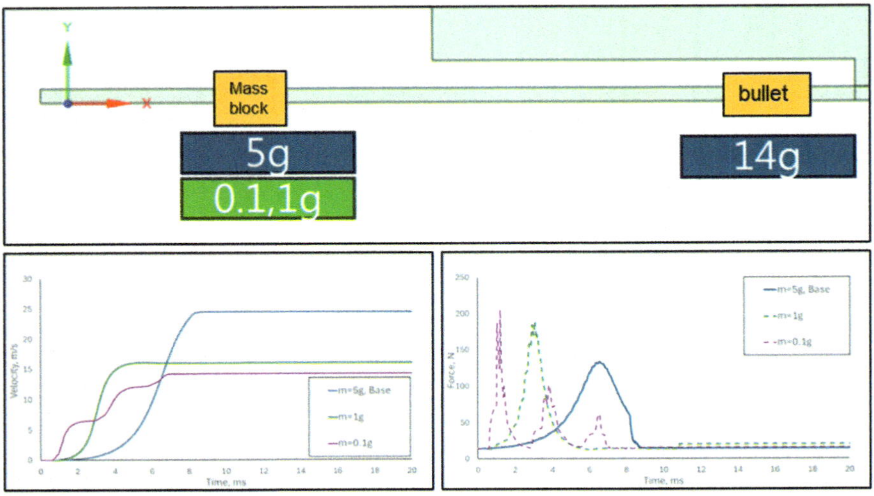

Fig. 27.16 Effect of mass on bullet initial velocity using shock waves

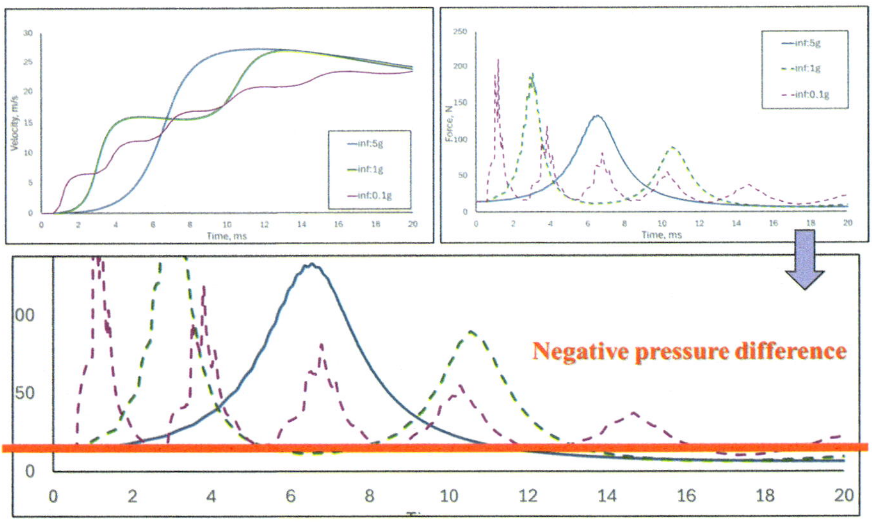

Fig. 27.17 Effect of mass on bullet initial velocity using extended barrel

27.3.3 Exploration Using a Resonant Wave Shooting Device

From the previous simulations, it was found that the performance of shock waves was inferior to that of pressure waves due to the additional shock wave working space, while the pressure source at the front remained unchanged. This space's air

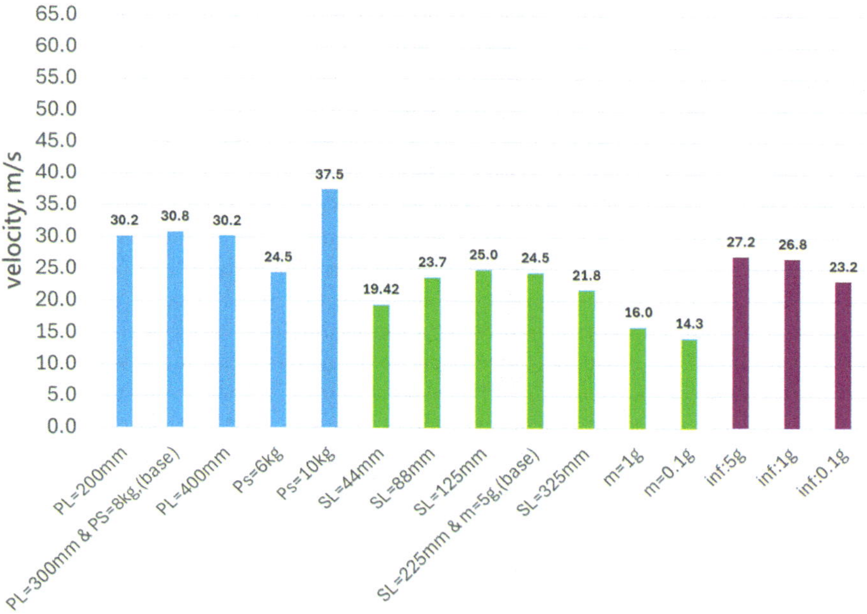

Fig. 27.18 Comparison of the effects of various modifications on bullet initial velocity using air pressure waves and shock waves

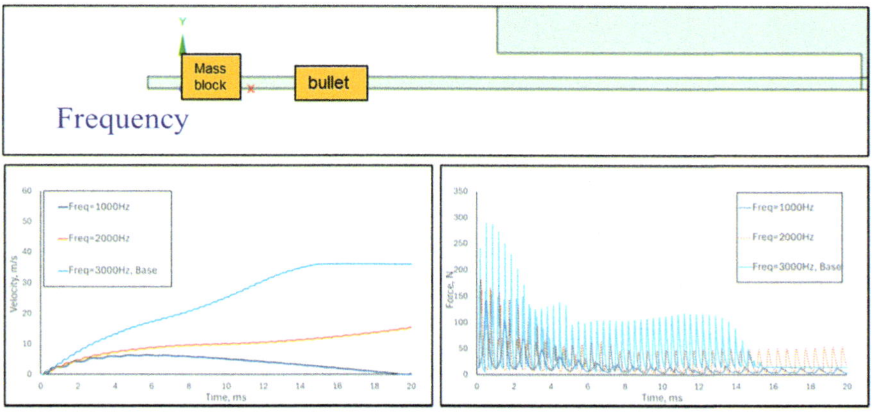

Fig. 27.19 Effect of vibration frequency on bullet initial velocity using resonant waves

absorbed the pressure energy from the front during compression. Additionally, as the bullet was gradually pushed out of the barrel by the generated shock waves, it affected the volume of the intermediate space, indirectly altering the development of the pressure waves.

To effectively develop the application of shock waves, the mass block at the front end will be replaced with an oscillator that can change over time. By continuously generating shock waves that resonate within the barrel, we aim to find the optimal ratio to achieve the best energy superposition effect.

First, we can start by changing the oscillation frequency of the mass block. Beginning with a simulation at Frq = 1000 Hz, we set the initial bullet velocity as the target to find the optimal frequency. Once this optimal frequency is identified, further research on other variable factors can be conducted. At Frq = 3000 Hz, we find a bullet initial velocity slightly greater than the experimental value.

The results of the shock wave study highlight the importance of the distance between the mass block and the bullet in the research of shock wave launching devices. Figure 27.20 illustrates the impact of varying distances on the waves generated by the mass block oscillating at 3000 Hz. Figure 27.21 shows a comparison of the initial bullet velocities produced at different distances. The overall trend indicates that neither the longest nor the shortest distance is optimal; instead, the best result is achieved at GL = 270 mm.

Comprehensively evaluating the performance of using the three types of waves as energy sources, Fig. 27.22 shows that the application of resonant waves achieves the best target values. In the trend analysis of the performance and design modifications of the three waves, we can observe how changes in model settings can yield better trends. Therefore, integrating the above studies, we can start by fixing the experimental constraints. Then, through simulations in an infinitely long barrel, we identify the optimal weight for the mass block. Finally, we adjust the barrel length and distance for analysis to find the optimal design.

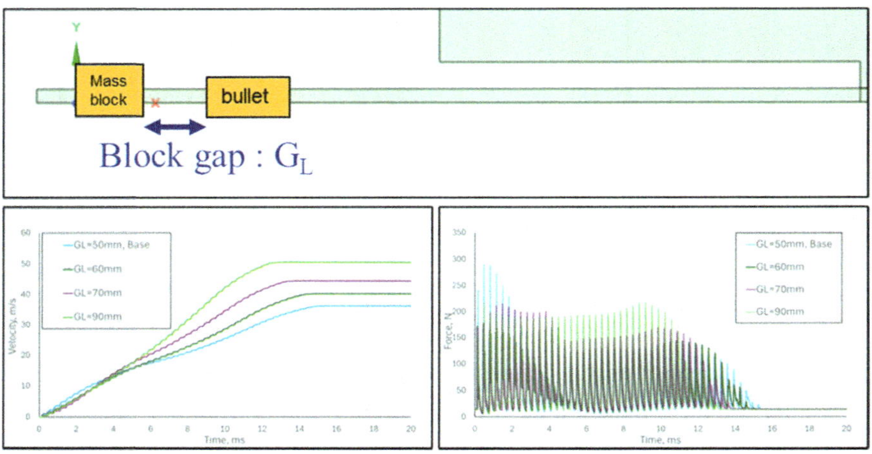

Fig. 27.20 Effect of distance on bullet initial velocity using resonant waves

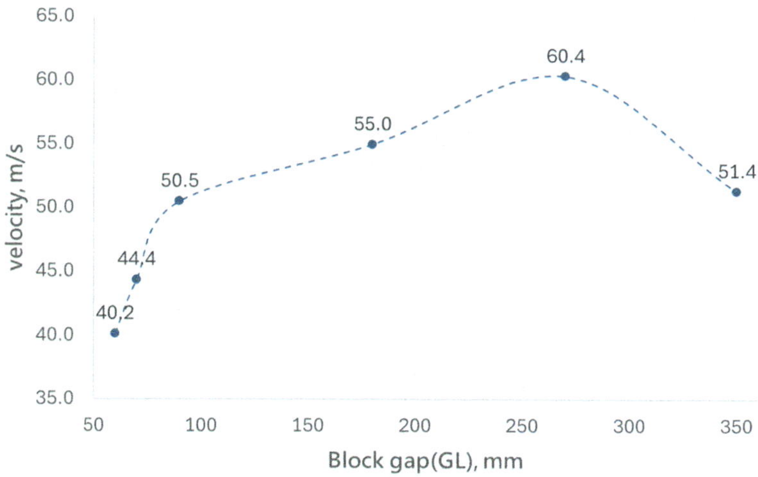

Fig. 27.21 Bullet initial velocity performance at different distances using resonant waves

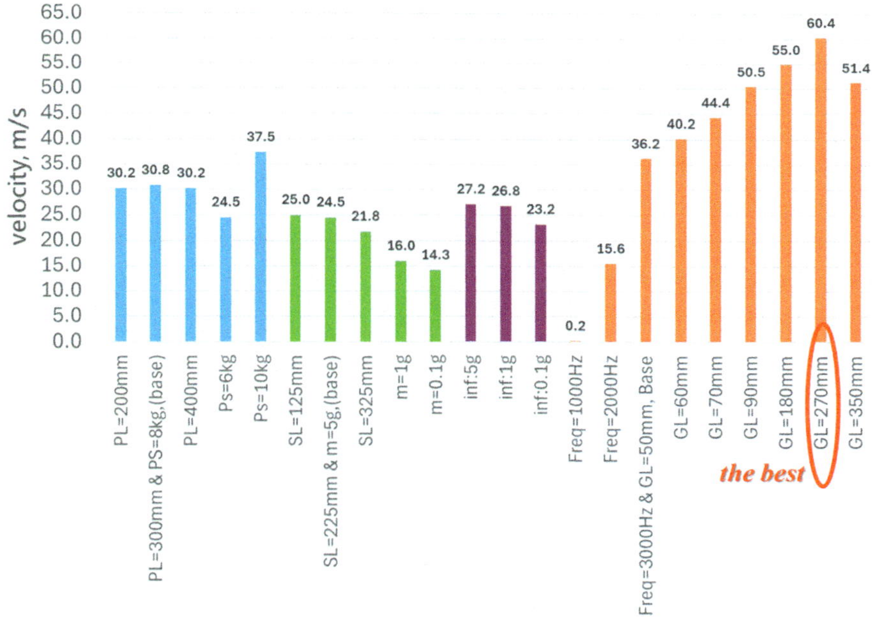

Fig. 27.22 Comparison of the effects of various modifications on bullet initial velocity using three types of wave sources

27.4 Conclusion

This study, set against the backdrop of ecological conservation, aims to design a device for launching contraceptive vaccines. Simulations and analyses of three types of pressure waves are conducted to explore how design modifications can achieve the optimal bullet velocity for stable and effective shooting operations.

In the study, the application of waves is explored. The simulation results reveal that:

(1) The muzzle velocity of the projectile can be accurately predicted using numerical modeling. Because frictional resistance cannot be considered, the simulated values are on the high side, but the trends are very consistent.
(2) Use high-pressure gas alone to push the projectile. As the air pressure increases, the internal energy increases and the muzzle velocity of the projectile also increases.
(3) The shape of the high-pressure air container has little effect on the muzzle velocity of the projectile.
(4) In order to reduce frictional resistance, shock waves can be used to push the projectile.
(5) When using a shock wave to push a projectile, the mass of the rear shock wave generator and the distance from the front projectile will have a great impact on the energy accumulated by the shock wave. The distance from the front projectile to the running mouth will also affect the use of shock wave energy. Although there are many parameters, applying numerical simulation can effectively obtain the optimal parameters.
(6) Using the electromagnetic oscillation to generate shock waves will greatly enhance the energy emitted.
(7) To effectively reduce the size and mass of the launch system, it is a feasible method to use electromagnetic oscillation to generate a series of shock waves to push the projectile.
(8) The frequency of electromagnetic oscillation is closely related to the frequency of shock wave generation. To obtain the best resonance effect, careful adjustment is required. Since the shock wave oscillation frequency of the projectile changes at any time as it accelerates forward, the most effective way is to install a pressure sensor pad at the front end of the shock wave generation block. Having a pressure sensor that senses the shock wave strength and makes a corresponding electromagnetic response will achieve the best results.
(9) This principle can apply shock wave resonance to many aspects, such as using a drone to collect samples from atmosphere and water, rapid non-destructive testing trunk health, and even cold launch of missiles or rockets.

Among the results of the three types of waves, the resonant wave showed the best performance. The relationship between the vibration frequency and the wave frequency is crucial for shock wave superposition. During the shooting process, the syringe barrel moves, causing the wave frequency inside the barrel to change over time. The next focus of the research will be on how to combine the generated shock waves with the reflected waves from the bullet impact to create a stronger energy wave.

References

Altufayl A (2019) Development of a 6-DOF trajectory simulation model for asymmetric projectiles. Ph.D. thesis, North-West University

Anderson J (2009) Fundementals of aerodynamics. 5th edn. McGraw-Hill Higher Education

Cilliers AJ (2021) Modelling the external ballistics of tranquilliser darts. Engineering at Stellenbosch University

Ferziger JH, Perić M (2002) Computational methods for fluid dynamics

Gkritzapis DN, Kaimakamis G (2008) A review of flight dynamic simulation model of missiles Hellenic army academy. pp 257–261

Gkritzapis DN, Panagiotopoulos EE, Margaris DP, Papanikas DG (2007) Computational atmospheric trajectory simulation analysis of spin-stabilized projectiles and small bullets. In: Collection of technical papers-2007 AIAA atmospheric flight mechanics conference. ISBN 1563479052. ISSN 1752–5055

Hainz LC, Costello M (2005) Modified projectile linear theory for rapid trajectory prediction. J Guid, Control, Dyn 28(5):1006–1014. ISSN 07315090

Hampton JO, Adams PJ, Forsyth DM, Cowled BD, Stuart IG, Hyndman TH, Collins T (2016) Improving animal welfare in wildlife shooting. Wildl Soc Bull 40(4):1–27

Khalil M, Rui X, Hendy H (2015) Discrete time transfer matrix method for projectile trajectory prediction. J Aerosp Eng 28(2). ISSN 0893–1321

Lin YJ (2012) Numerical studies of shock waves interaction and energy focusing. Master thesis, National Pingtung University of Science & Technology

Mandella M, Moon YJ, Bershader D (1985) Quantitative study of shock generated compressible vortex flows. In: Proceedings of 15th international shock waves and shock tubes. pp 471–477

Tai C-H, Lo S-W, Miao J-M, Yao-Kuang S (1999) Numerical study of muzzle transient flow of howitzer with muzzle brake. Chin J Mech Ser B 22(2):111–125

Tai CH, Tneg JT, Lo SW, Liu CW (2005) A three-dimensional numerical investigation into the interaction of blast waves with bomb shelters. JSME, Series B 48(4):820–829

Tai CH, Lo SW, Wu CN (2006) Numerical investigation in the interaction between shock waves. The Third Asian-Pacific Shock and Vortex Workshop, Taiwan

Von Neumann J (1943) Oblique reflection of shock, explosion research reports 12. Washington D. C.

Chapter 28
Implementation of a Monitoring System for a Hydroponic Microclimate Through the Internet of Things

Luis Héctor Hernández-Gómez, Mauricio Aarón Pérez-Romero, Juan José Muñoz-Cesar, Armando Josué Piña-Díaz, Alejandra Armenta-Molina, Diego Esquivel-Reyes, and Angélica Ruiz-Font

Abstract The production of vegetables at home, for marketing or own consumption, is possible. However, temperature and humidity, as well as the pH of the nutrient solution must be controlled. They directly affect productivity. A monitoring system was implemented, so the user can take corrective actions if necessary. As a result, information about the variables of interest is available. The implementation of a monitoring system in a hydroponic microclimate is described in this paper. The nutrient film technique was followed, and the Internet of Things was used, through an embedded system with a low-cost microcontroller. This system collects and sends the information to a database. The purpose is to provide a report on the temperature, pH, and humidity and improve productivity.

L. H. Hernández-Gómez (✉) · A. Armenta-Molina · D. Esquivel-Reyes
Escuela Superior de Ingeniería Mecánica y Eléctrica, Sección de Estudios de Posgrado e Investigación, Instituto Politécnico Nacional, Ciudad de México, México
e-mail: lhhernandez@ipn.mx

A. Armenta-Molina
e-mail: aarmentamo@ipn.mx

D. Esquivel-Reyes
e-mail: desquivelr1700@alumno.ipn.mx

M. A. Pérez-Romero · J. J. Muñoz-Cesar · A. J. Piña-Díaz
Escuela Superior de Ingeniería Mecánica y Eléctrica, Instituto Politécnico Nacional, Ingeniería en Control y Automatización, Ciudad de México, México
e-mail: maperezro@ipn.mx

J. J. Muñoz-Cesar
e-mail: jjmunoz@ipn.mx

A. J. Piña-Díaz
e-mail: apinad@ipn.mx

A. Ruiz-Font
Instituto Politécnico Nacional, CIBA-IPN, Tlaxcala, México
e-mail: afont@ipn.mx

A. Öchsner and H. Altenbach (eds.), *Engineering Design Applications VII*, Advanced Structured Materials 230, https://doi.org/10.1007/978-3-031-84346-4_28

Keywords Crops · pH control · Vegetables · Nutrient solution · Nutrient film technique

28.1 Introduction

Hydroponics is a concept that has become quite renowned today (Halim et al. 2024). Over the years, it has represented the conjunction between agriculture and technology (Tan et al. 2023). Facilities to harvest almost any plant have been developed. Consequently, the main objective is to grow plants without soil (Souza et al. 2023). However, nutrient solutions are required in addition to water (Farooq et al. 2023). Such solutions contain all the nutrients that a crop requires to grow properly. A second objective is the elimination of pests and diseases. They are developed by the earth or some unavoidable external factor. Modern hydroponics has gained popularity. It makes it easier to avoid the loss of time and money. The technology optimizes the traditional activities in this field.

In (Popkova 2022) have mentioned that hydroponics has been a great step toward the second green revolution. Its advantages have impacted horticulture, vegetable production, and gardening, among others. From this point of view, these systems are appropriate for all those who wish to start a new form of cultivation. However, the use of technology is required.

Technology can be related to the Internet of Things (IoT). It is an interconnection of everyday objects through the internet (Sneineh and Shabaneh 2023). It is an intelligent interrelation of objects, people, or processes. In this way, the obtained systems are supported by other technologies such as data analytics and artificial intelligence, among others. It is possible to make decisions without the need for people.

One of the main reasons to choose a hydroponic system, over traditional growing methods, is that food is obtained in places where the space is reduced. Sensors are used in a hydroponic system to evaluate variables and exchange information. Crops are kept in the best conditions. Besides, plants or vegetables, which are not feasible in ordinary conditions in places of interest, are produced.

Hydroponics develops the cultivation in water. However, there is not a unique procedure. The management of nutrients considers the environment, oxygen, temperature, and other parameters that help the crop survive. Diverse methods have been followed to optimize production (Monisha et al. 2023). Human consumption and economic needs are satisfied. The hydroponic systems can be divided into six general systems:

- Ebb and flow: The growth trays are temporarily flooded with a nutrient solution. After a certain period, it is pumped back into a tank. This flow is controlled with a timer, which is activated several times.

- Aeroponics: It is a system in which the roots are suspended in the air in a dark medium. They are nebulized with a nutrient solution. However, this technique has a disadvantage. The roots can be dried in each nebulization cycle.
- Floating root: The plants are in a sheet or bag that floats on the nutrient solution, so the roots are submerged in the solution. Besides, oxygen is conducted to the roots. It is one of the best techniques and is widely used in teaching activities. However, it is only suitable for lettuces and other green leaves.
- Drip system: It has a timer that controls a pump. It causes the nutrient solution to drip onto the plant. The advantages of this system are the following: the nutrient solution can be recovered, and it is easier to control the pH and concentration of nutrients.
- Wick system: This technique is simple since it does not require pumps to flow the nutrient solution from the storage container to the growing trays. The solution is dispensed through wicks. However, it is only suitable for plants that require little water.
- Nutrient film technique (NFT): A nutrient solution film is recirculated. There is a constant flow of the solution. It does not require timers and a substrate. The nutrient solution is pumped from a reservoir to the growing trays or the roots through PVC tubes. The cycle is closed when the solution returns to the reservoir. It has been widely used in domestic applications.

28.2 Statement of the Problem

Hydroponics has a great advantage over other cultivation methods, such as avoiding the use of soil, skipping the germination stage, or avoiding pests, that are generated in the soil, such as fungi, bacteria, viruses, earthworms, or nematodes. However, characteristic problems have arisen from this system. They are not only related to an excess or lack of circulating water through the roots or due to lack of time (Kailashkumar et al. 2023). Factors such as humidity, pH, and temperature, directly influence the vegetables.

A NFT hydroponic system has been designed to produce cherry tomatoes, ornamental chili, and strawberries. The crops are risen in a room 2 m wide, 2 m long, and 2.3 m high. It is covered with a 720-caliber plastic. Such plastic filters the solar rays. Besides, it prevents the burning of crops due to prolonged exposure to UV rays. There is an internal corridor, which is 0.5 m wide.

The IoT manages this system. The water circulation is kept with the nutrient solution at least 3 times a day. The roots get the necessary nutrients for their development. It is controlled with growth lamps, a recirculation pump, and a heater.

28.3 Materials and Methods

The application of automation in a greenhouse is achieved with sensors that allow monitoring of the environment necessary for vegetables. The environment is controlled, such as air heating and ventilation. Besides, there is constant irrigation (Gazieva et al. 2023).

Technical conditions: Implementing a hydroponic system that reduces the production time. In this way, an appropriate system is developed to grow vegetables. Using sensors for the monitoring of temperature, electrical conductivity, pH, and humidity, it is possible to obtain a report of the state of the crop. Accordingly, data is obtained in real time. Implementing an air temperature control strategy ensures that the vegetables grow in a suitable medium. Besides, LED lights replace the sunlight. They are programmed to operate between 8 and 16 h. A water pump recirculates the nutrient solution. In such a way the plants can absorb the required nutrients.

Monitoring of variables: Due to the constant changes in the environmental variables of hydroponic crops, repetitive tasks cannot be followed. Monitoring the variables of interest provides information on the current state of the crop. Actions are taken and correct development of the crop takes place.

Implementation of the IoT: The following proposal has been considered the most feasible. An on/off control strategy with hysteresis is being considered. The temperature is kept higher than 20 °C with a heater. It is suitable for an area of 4 m^2. Besides, solid-state relays control the centrifugal pump and the growth lamp. A timer is required to develop adequate management of the crop development.

An XAMPP server is required for the management of the database and the implementation/design of the web server. It can activate the port 80. Under these conditions, the internet is browsed with a web page, and it can be connected with other ports. The associated software can determine if there is an activation error. Figure 28.1 shows the proposed system solution.

Figure 28.2 shows the diagram of the implementation of the proposal described above.

A closed system has been designed. It is spacious and comfortable for the user. Its size is adequate and can be used at home or elsewhere for vegetables. A prototype was developed with these ideas in mind. Figure 28.3 shows a schematic arrangement.

28.4 Description of the Prototype

The temperature in a hydroponic system, which is isolated employing a plastic coating, must be kept above 21 °C (set point). Below this level for at least a couple of hours or even days, the crop growth will be slowed, or in the worst situation, the harvest is lost. The temperature sensor (TE) sends an electrical signal to the comparator. Under the results, a digital signal is sent to a solid-state relay (actuator). In this way, the operation of the heater is controlled.

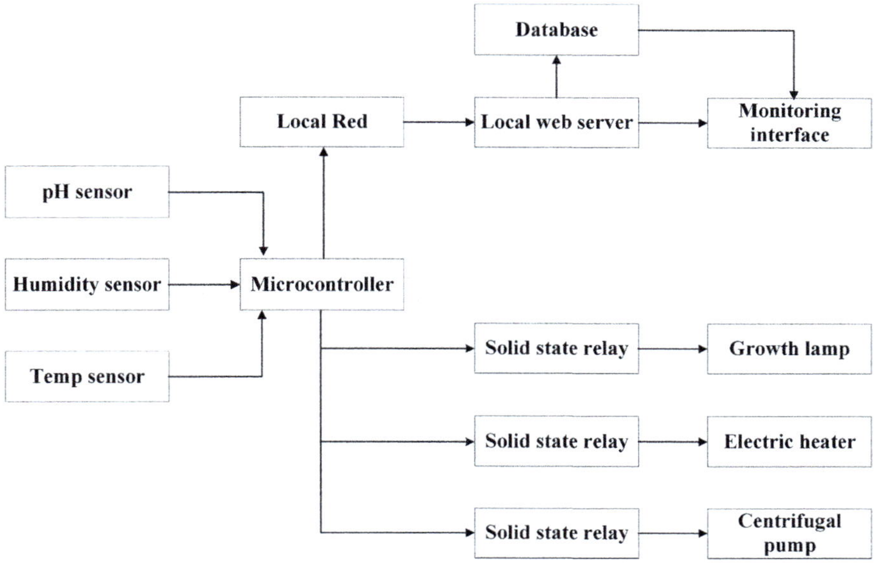

Fig. 28.1 Proposed solution

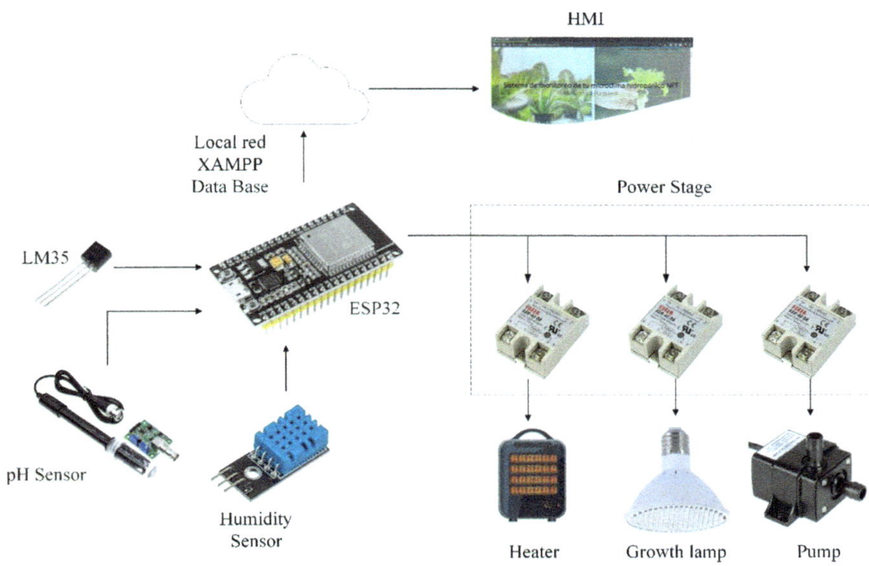

Fig. 28.2 Implementation of the proposal

The plant nutrient solution is stored in two interconnected containers. A pump is activated, for 30 minutes, every 6 hours. So, the nutrients circulate through the crop

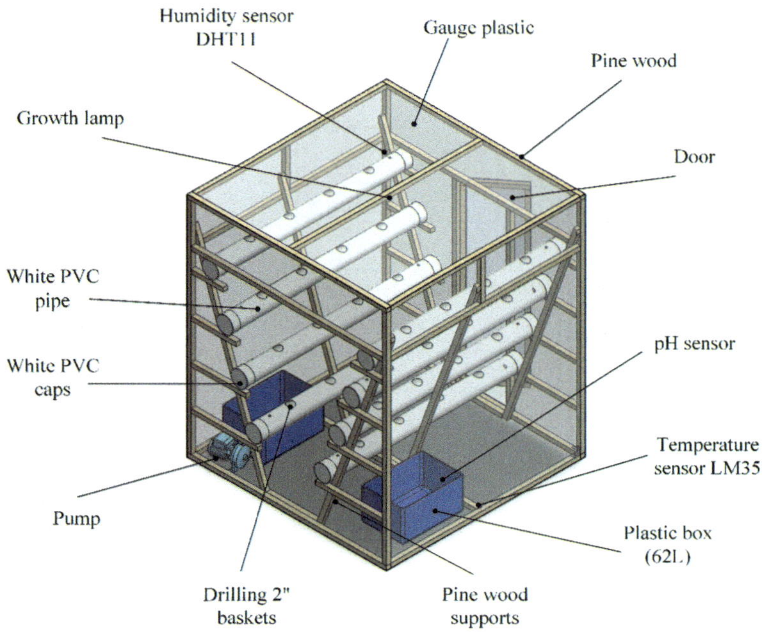

Fig. 28.3 Schematic arrangement of the prototype

lines. Besides, an ultraviolet light lamp is turned on for 6 hours. Then after, it is turned off for 6 h. This cycle is repeated constantly. The pH of the solution is measured every 10 minutes, and the humidity is evaluated every 10 min. The data obtained from these sensors is sent to the microcontroller. A flowchart of the operation of the monitoring system is shown in Fig. 28.4.

The control loops are shown in the following diagram of piping and instrumentation (Fig. 28.5). The voltage supply to the elements of the system and their connection to the microcontroller is illustrated in Fig. 28.6.

28.5 Main Characteristics of the Control Program

An on/off control system is proposed, with the controlled variable being the temperature through an electric heater. The temperature sensor sends an electrical signal to the comparator. It is evaluated if the temperature is higher or lower than the set point. The microcontroller ESP32 considers the programming of the sensors and the actuators. In this way, the microclimate is kept in operation. Besides, the user interface can be seen. The microcontroller was programmed in an Arduino environment. The activation of sensor and actuator functions are configured as parallel tasks using

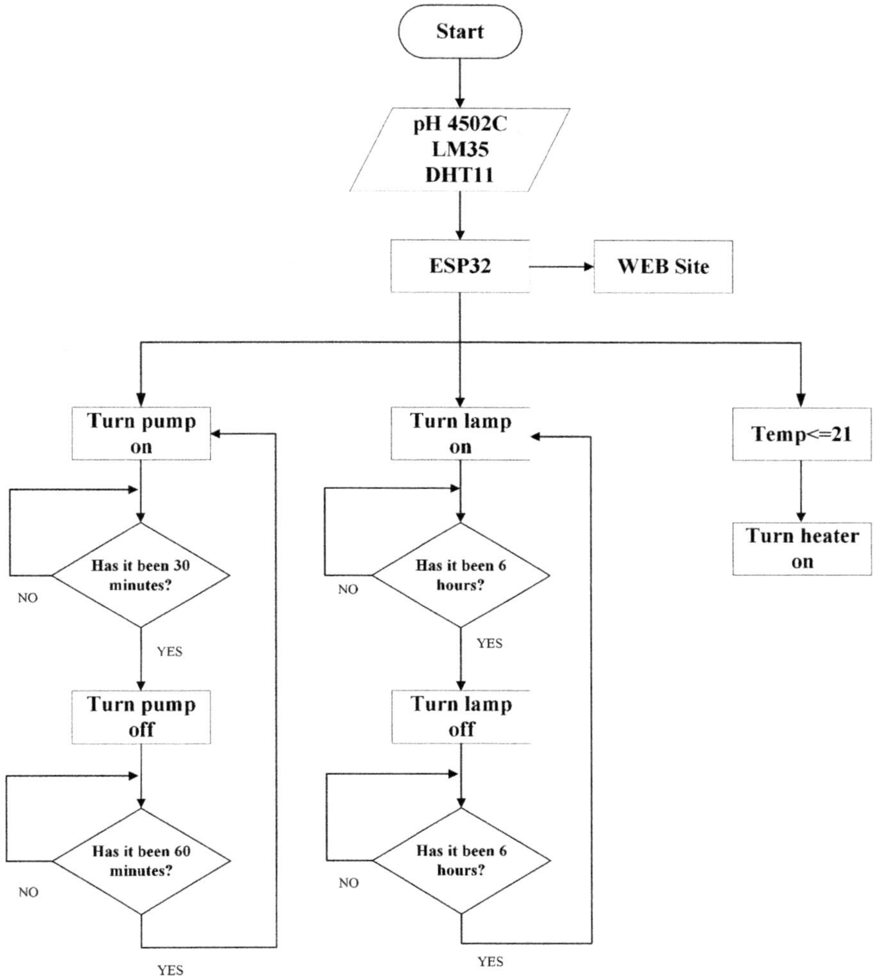

Fig. 28.4 Flowchart of the operation system

a library. Unlimited tasks can be done, as well as the configuration of memory size, priority, and even the space where they can be executed within the microcontroller.

The user interface monitors the microclimate locally and remotely. The user knows the variables and takes action. The user interface is developed with XAMPP, since this software allows the free design of web pages. The user interface is designed with PHP and HTML programming. In the case of PHP, it is possible to send information collected by the database to the web server. Besides, HTML allows the reading and graphic design of the information in the database.

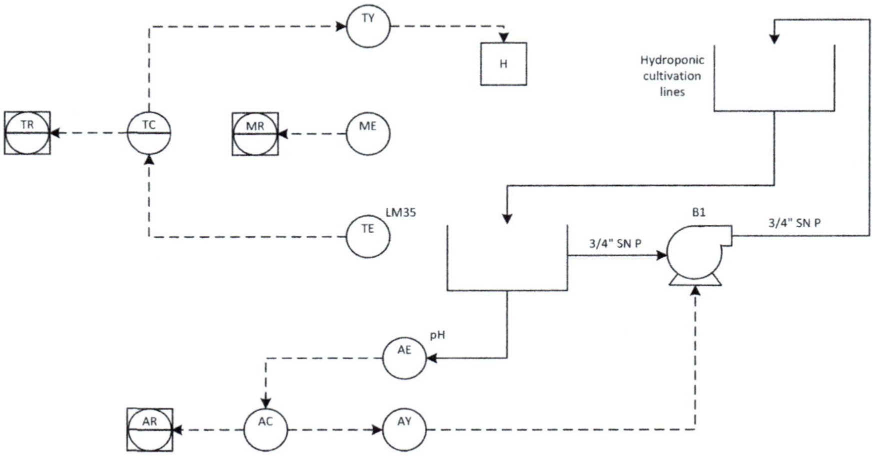

Fig. 28.5 Piping and instrumentation diagram

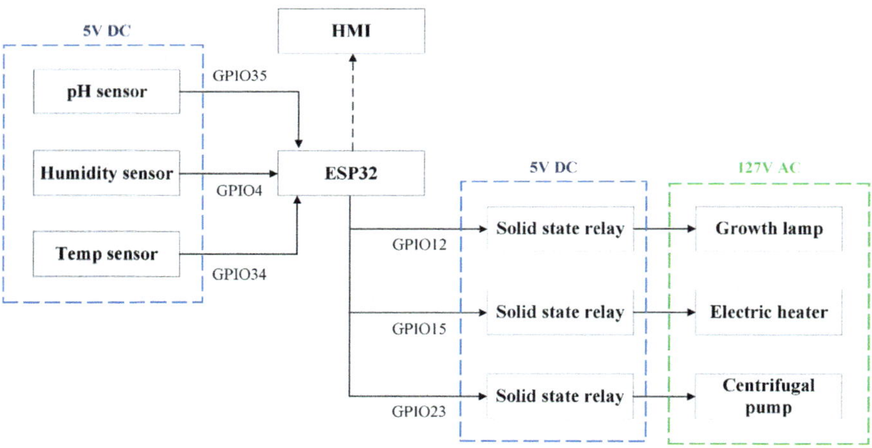

Fig. 28.6 Power supply and connection of the sensors with the microcontroller

28.6 Discussion of the Results

The temperature was compiled in 48 h. It was observed how the temperature of the microclimate was controlled. The temperature graph shows that the system was at the ambient temperature (Fig. 28.7). The temperature increase is slow and constant. It showed that the system is not affected by outdoor conditions. It gives a guideline to adjust the sampling time from 5 to 10 min. The increase in the sampling time reduces the data storage in the MySQL database. Besides, it is possible to maintain an ideal

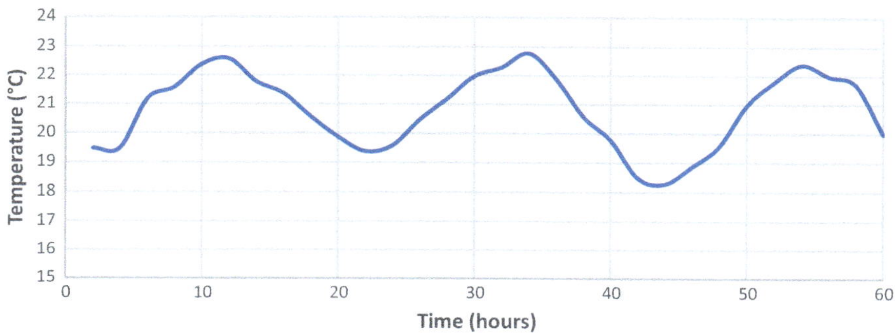

Fig. 28.7 Temperature variation

temperature in the greenhouse, and graphs of different periods can be obtained. In this way, the production of crops can be optimized.

The humidity was evaluated in the same period mentioned above. In this way, it was possible to observe the effect of the outdoor temperature on the adjustments in the microclimate conditions. The analysis of this data established the criteria of operation. In the first instance, the humidity variation is determined throughout the day (Fig. 28.8). It increases between 5 and 8 o'clock in the morning. During this period, the dew point is high (more than 90%RH). The walls, the growing channels, and the storage boxes are wet. A similar situation takes place when the outdoor temperature is high. In this scenario, a monitoring period of 10 min was established, because the variation is slow.

To obtain a satisfactory result, it is essential to consider that the humidity must be greater than 40% in the greenhouse. However, it must be lower than 60%. The vegetables can coexist under these conditions. This humidity range is an average value.

The pH determines the characteristics of the nutrient solution. Its range must be between 5.5 and 6.5. Values above 7 are not recommended. Figure 28.9 shows the

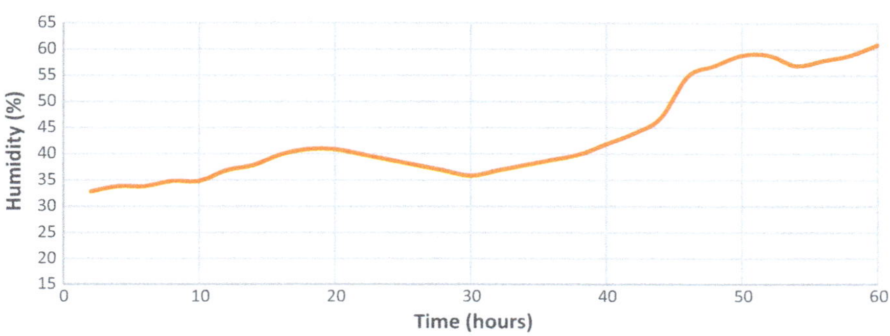

Fig. 28.8 Humidity variation

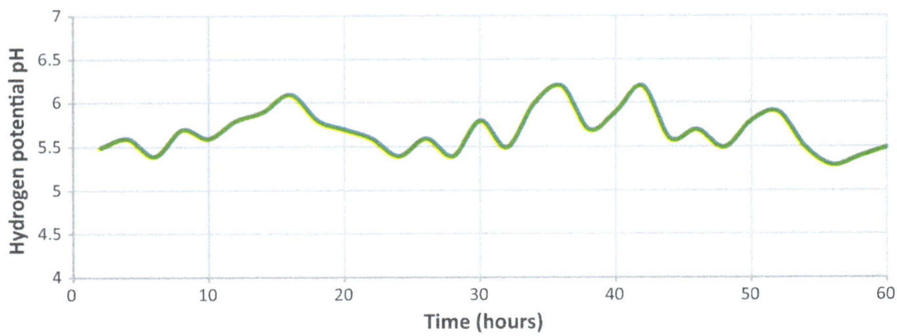

Fig. 28.9 pH variation

pH variation in 48 h. It was kept in an appropriate range and the vegetables absorbed the nutrients. Besides, the sampling time can be increased. The change in the pH is low through time. The informatic resources can be used appropriately.

28.7 Growth and Development of Vegetables

The growth process of the cherry tomato is shown in Fig. 28.10. Its development was analyzed for 2 months, from the transplanting.

The vegetable lost its leaves in the first week due to climate change. In the second week, it recovered its condition and the initial flower buds were developed. The tomatoes appeared in the third week. The sprout size was approximately 2 cm at the end of the seventh week. The vegetable lost most of its leaves and the tomato sprouts became red during the last week. The results were compared with the data reported in the open literature. The results are summarized in Table 28.1. The crop development was accelerated in the proposed system. In the case of garden at home, it was reduced by 33%. Regarding how to plant, it was 20%.

The growth and development of the strawberry differs from any other vegetable. The results will be different from those obtained by the cherry tomato and the ornamental chili. However, this does not prevent a brief analysis. The period of observation was two months. It is summarized in Fig. 28.11. Its growth began in the first week. Its flower buds appeared in the second and third weeks.

It is known that the strawberries are obtained after 5–6 months. Their maximum and minimum sizes are 5 cm and 3 cm, respectively. It can be harvested after 6–10 months. Alternatively, a 1.5 cm strawberry is obtained after 2 months in the proposed system. It is possible to harvest the strawberries after 6 months.

Regarding the ornamental chili, a 2-month growth scheme is illustrated in Fig. 28.12. A small and circular flower is developed in the first 2 weeks. There are sprouts by the third week. They are green and their size is 2 cm. By the fourth week, they are bigger. There is a change from green to yellow-brown in the sixth

Fig. 28.10 Cherry tomato growth process

Table 28.1 Cherry tomatoes under different conditions

Characteristic	Microclimate	Garden at home (garden at home: how and when to harvest cherry tomatoes 2023)	How to plant (how to plant: how to plant cherry tomatoes–care and harvesting of fruits 2023)
First fruits	2 weeks	4 weeks	4 weeks
First harvest	8 weeks	12 weeks	10 weeks

week. In the eighth week, a yellow chili is obtained. Table 28.2 compares the results obtained with those reported in the open literature. In general terms, the growth of chili is 30% faster than the traditional procedures.

A recovery plan was implemented. It was focused on those crops located outside the greenhouse with bad growth. This is the case when a plant loses its flowers and

Fig. 28.11 Strawberry growth process

leaves. Their stem began to wither. In this case, the plant was cleaned. It was left for 48 h in a tube. Afterward, it was transplanted. It was observed that the crop recovered.

The main advantages are as follows: (1) The proposed system allows a precise control of plant nutrition. (2) The irrigation systems are simplified because soil sterilization is not required and optimizes the supply of nutrients. (3) The direct contact of the roots with a nutrient solution is increased. (4) The growth of the crops is accelerated up to 30%. So, the annual production is increased. (4) High-quality vegetables are obtained when the system is managed properly. On the other hand, the disadvantages are as follows: (1) A high initial investment is required. (2) A plant care is essential. (3) It is unsuitable for all species. (4). Plants are susceptible to changes. (5) There are a limited number of experts on the subject.

Fig. 28.12 Ornato chili growth process

Table 28.2 Chile ornamental growth comparison

Characteristic	Microclimate	Garden at home (garden at home: how to germinate chili seeds 2023)	How to plant (how to plant: planting chili peppers–planting guide for this crop, care and harvest 2023)
First fruits	3 weeks	4 weeks	4 weeks
First harvest	8 weeks	12 weeks	12 weeks

28.8 Conclusions

The implementation of the proposed hydroponic system at home has proven to be efficient. For this purpose, an area of 4 m^2 is adequate. The activities can be developed without any problem. Additionally, it is useful for reviving withered plants. By adding a greenhouse-style dome, the crops were isolated from external factors, such as changes in temperature, excessive rain, and hail. There is greater protection against nocive insects. Regarding, the cases reported in this paper, the proposed system improves production. The cherry tomato crop in the microclimate is reduced by more than 20%. Besides, the ornamental chili plants have better development in the microclimate environment. The production time is reduced by 30%. A similar improvement is observed in the production of strawberries.

The graphical interface is friendly to the user. The ISA 5.5 HMI was followed. The striking colors indicate if there is a parameter outside the range of operation. Temperature, pH, and humidity were monitored in real time. A database was generated with the information obtained. The required actions were based on such information. Humidity monitoring is an example. The control of this parameter generates improvements.

An adequate pH promotes good absorption of nutrients. Fungus can be created if this parameter is not controlled properly. In an extreme situation, the crop may be lost. Besides, the temperature plays an important role. The data collected allowed the heater to maintain an adequate temperature. In this way, the impact of the external climatic conditions is avoided.

Acknowledgements The authors kindly acknowledge the grant awarded by the Instituto Politécnico Nacional and COFAA for the development of this research.

References

Farooq A, Verma AK, Hittinahalli CM, Harika N, Pai M (2023) Iron supplementation in aquaculture wastewater and its effect on the growth of spinach and pangasius in nutrient film technique-based aquaponics. Agric Water Manag 277:108126. https://doi.org/10.1016/j.agwat.2022.108126

Garden at home: how and when to harvest cherry tomatoes? (2023). https://huerto-en-casa.com/cosecha-tomate-cherry/. Accessed 10 Dec 2023

Garden at home: *how to germinate chili seeds?* (2023). https://huerto-en-casa.com/como-germinar-semillas-de-aji/. Accessed 13 Dec 2023

Gazieva L, Belyaeva E, Kosulin V (2023) Effectiveness and profitability of automation technologies in greenhouse productivity and food security. In: 2nd international conference on environmental sustainability management and green technologies. https://doi.org/10.1051/e3sconf/202345102012

Halim AHA, Taufik MA, Mahamarowi NH, Ahmad TDAS (2024) Introduction of IoT integration into smart indoor farming hydroponic systems for urban communities: Plensis system. In: 14th symposium on computer applications & industrial electronics (IEEE ISCAIE). Penang Island, Malaysia

How to plant: how to plant cherry tomatoes–care and harvesting of fruits (2023). https://como-plantar.com/jardin-comestible/frutas/tomate-cherry/#cosecha. Accessed 10 Dec 2023

How to plant: planting chili peppers–Planting guide for this crop, care and harvest (2023). https://como-plantar.com/jardin-comestible/verduras/aji/#cosecha-del-chile. Accessed 13 Dec 2023

Kailashkumar B, Priyadharshini K, Logapriya M (2023) Hydroponic cultivation: factors affecting its success and efficacy. Int J Environ Clim Change 13(10):2403–2410. https://doi.org/10.9734/IJECC/2023/v13i102905

Monisha K, Kalai Selvi H, Sivanandhini P, Sona Nachammai A, Anuradha CT, Rama Devi S, Hikku GS (2023) Hydroponics agriculture as a modern agriculture technique. J Ach Mat Man Eng 116(1):25–35. https://doi.org/10.5604/01.3001.0016.3395

Popkova EG (2022) Vertical farms based on hydroponics, deep learning, and AI as smart innovation in agriculture. In: Popkova EG, Sergi BS (eds) Smart innovation in agriculture. Smart innovation, systems and technologies, vol 264. Springer, Singapore. https://doi.org/10.1007/978-981-16-7633-8_28

Sneineh AA, Shabaneh AA (2023) Design of a smart hydroponics monitoring system using an ESP32 microcontroller and the internet of things. MethodsX 11:102401. https://doi.org/10.1016/j.mex.2023.102401

Souza V, Gimenes RMT, de Almeida MG, Farinha MUS, Bernardo LVM, Ruviaro CF (2023) Economic feasibility of adopting a hydroponics system on substrate in small rural properties. Clean Tech Env Pol 25(8):2761–2775. https://doi.org/10.1007/s10098-023-02529-9

Tan DMY, Ng WJ, Chua HS, Thing Thing G, Law FL (2023) Exploring the global hydroponic cultivation system: current and future perspectives. In: International conference on mechanical, manufacturing and process plant engineering, Putrajaya, Malaysia. Springer Nature Singapore. Adv Mater Sci Eng 26–41

Chapter 29
Universal Joint Modeling in Catia V5, Synopsis of the Kinematics: The First Step for Failure Analysis

Nader G. Zamani

Abstract Universal joints are widely used in a large number of mechanisms today. They are efficient tools for transferring motion between skewed shafts. Such components are vulnerable to material failure on a regular basis. Among other factors, it can be attributed to ignoring the kinematics behaviors altogether. The U-joint kinematic modeling capability is implemented in most CAE software but not clearly documented in the package. This expository paper concentrates on the Catia v5 code attempting to shed some light on the different tools available to the users for this joint. The basic, yet sound exposition of the joint which is attempted in the paper.

Keywords Universal joint · Kinematics · Digital mockup workbench · Cross-pin orientation · PLM

29.1 Introduction

The ingenious idea behind a universal joint is responsible for many machine dynamics advances since the industrial revolution (Budynass and Nibsett 2014; Norton 2004). This joint is primarily used for transferring motion, and therefore power, between the two shafts which are skewed at an angle relative to each other. This eliminates the need for a more expensive solution such as a gearing system. The depiction in Fig. 1a, b shows a "shiny new" U-joint and one that has been in service respectively. Notice that the axis of the two shafts is skewed. The part shown in Fig. 1c, where two U-joints are connected is known as the constant velocity joint (CV joint) which is not the focus of the present paper.

In spite of the widespread applications of such an important joint, the author finds it curious that the undergraduate engineering curriculum does not cover the fundamentals at more depth. In the author's experience, having seen it at several engineering schools, when the students are asked about the angular velocity of the two

N. G. Zamani (✉)
Mechanical Engineering, University of Windsor, Windsor, ON, Canada
e-mail: zamani@uwindsor.ca

A. Öchsner and H. Altenbach (eds.), *Engineering Design Applications VII*, Advanced Structured Materials 230, https://doi.org/10.1007/978-3-031-84346-4_29

Fig. 29.1 U-joint and the closely related CV joint

shafts, many believe that they are the same. Since some of the engineering graduates will in fact use CAE software to model such joints in real world problems, it is important to have a simple yet more complete explanation of the "Modus Operandi" of the joint. This will in turn answers the questions raised by many users when modeling these types of problems on the computer.

The U-joint assembly is shown below (Zamani and Weaver 2012; Zamani 2016). It consists of two skewed shafts connected together through a "cross-pin". One of these shafts is referred to as the "input" while the other one as the "output". It could be that the input shaft turns at a constant speed but that is not a requirement (Fig. 29.2).

Some of the key facts about this joint are described below. These statements are not all trivial mainly because of the complicated three-dimensional motion of the cross-pin. These will however be discussed in the paper at a later point.

(a) Even if the angular velocity of the input shaft is a constant, the angular velocity of the output shaft is not constant, however it will be periodic. The speed variation of the output depends only on the skew angle of the two shafts. This is shown in Fig. 29.3, for two different angles and a constant input velocity (Universal Joint).

(b) The speed variation in terms of the skew angle β is reasonable. In the case of the smallest such angle, namely $\beta = 0$, the two shafts are coaxial and turn at the same speed with no variation at all. Therefore, the smaller β, the smaller the amplitude. The entity T in the figure represents the time that takes for the input shaft to make a full revolution. Clearly, the period of turning of the output shaft

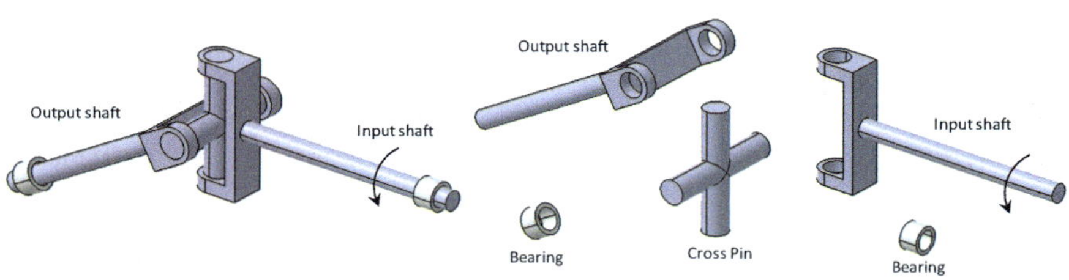

Fig. 29.2 The U-joint assembly and the exploded view

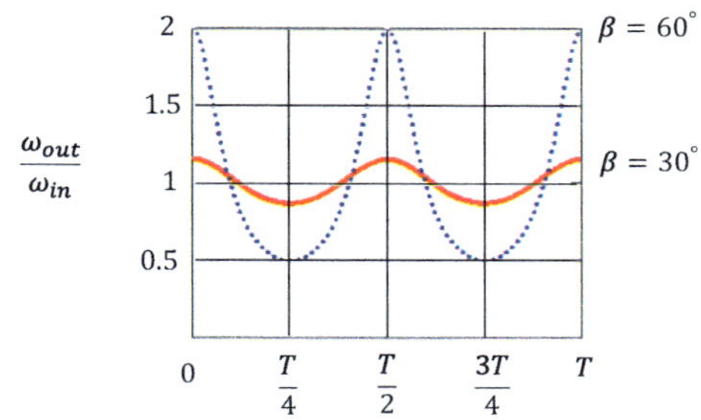

Fig. 29.3 The non-dimensional output angular velocity for two different skew angles

is half of the input shaft. This justifies the presence of two cycles for every turn of the input shaft.

(c) The plane of the cross-pin is displayed in Fig. 29.4. This plane has a complex motion in space. The normal to this plane is shown with the solid red line in the figure. At two extremes, the indicated plane is normal to the direction of the input shaft and the direction of the output shaft. The side view of these two extreme conditions are also shown in the same figure.

(d) The normal to the plane of the cross-pin part (the red line), lies on the surface of a cone emanating at the centroid of the pin and makes a complete revolution for every half a turn of the input shaft. This is consistent with the claims made in part (b), namely the frequency of the output shaft is double the frequency of the input shaft.

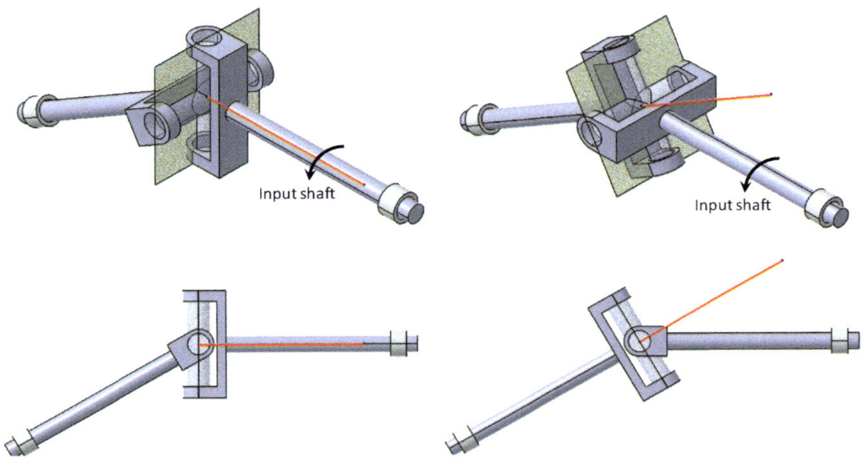

Fig. 29.4 The plane of the cross-pin and its normal shown in red

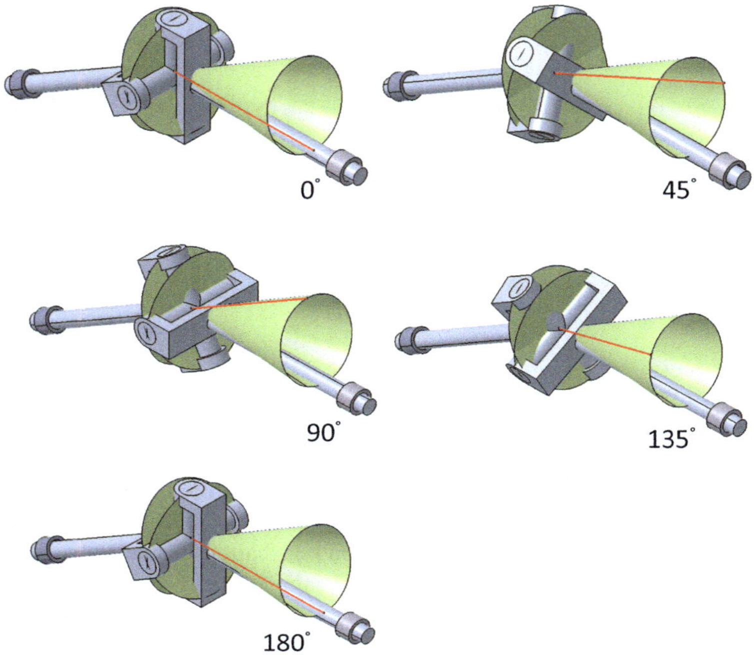

Fig. 29.5 The cone in which the normal to the cross-pin resides as it moves in space

Graphically this is shown in Fig. 29.5. The orientation of the normal to the cross-pin (the red line) is shown at five different angles. Each of these angles represents the rotation of the input shaft by the specified value. Note that after 180-degree rotation of the input shaft, the normal line has made a complete revolution and returns to the initial configuration. Two more planar surfaces (in the form of a circle) are also displayed in Fig. 29.5. The significance of these surfaces will be discussed next.

(e) In reference to the cross-pin, Fig. 29.6, it has a vertical arm which always lies in the plane perpendicular to the axis of the input shaft, the circle shown in the figure and labeled as "Cross sectional plane of the input shaft". The cross-pin also has a horizontal arm which always remains in the second circle. This second surface is referred to the "Cross sectional plane of the output shaft".

29.2 Kinematics of the Universal Joint

In this section, the kinematics governing the universal joint is derived. In order to aid the setup visually, the shafts and the cross-pin are not displayed in Fig. 29.7, however, the two fixed planes perpendicular to the axis of the input and output shafts are included. Without the loss of generality, the cross-pin shown as the double headed

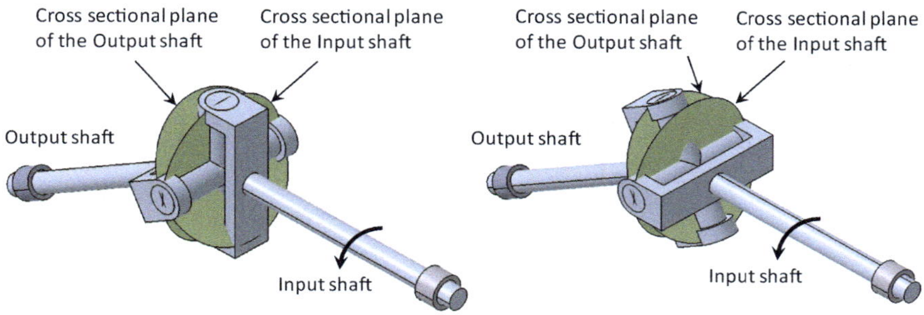

Fig. 29.6 The arms of the cross-pin always remain in the two fixed planes

dashed arrows, in that particular configuration is assumed to have a height and width of unity. Note that the vertical arm, always remains in the circle labeled "Cross sectional plane of the input shaft". Furthermore, the horizontal arm always remains in the circle labeled "Cross sectional plane of the output shaft". This was already alluded to in part (e) of the previous section but repeated due to its significance.

The two directions $\boldsymbol{X}_1$ and $\boldsymbol{X}_2$ are fixed in the space and do not change orientation. The angle between the two fixed planes, is the skew angle β which is also fixed. Although it may be obvious, β is the angle between $\boldsymbol{Proj}_{X_1}$ and the $\boldsymbol{X}_1$ direction. The vector $\boldsymbol{Proj}_{X_1}$ is the projection of the $\boldsymbol{X}_1$ onto the "Cross sectional plane of the output shaft".

As indicated earlier, the actual input and output shafts are not shown however, they are symbolized by the dashed lines in the drawing. It is worth mentioning that the

Fig. 29.7 The setup and conventions used to derive the kinematics of the U-joint

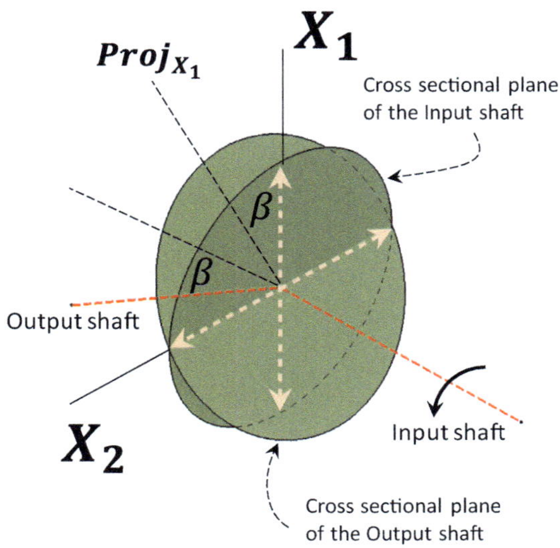

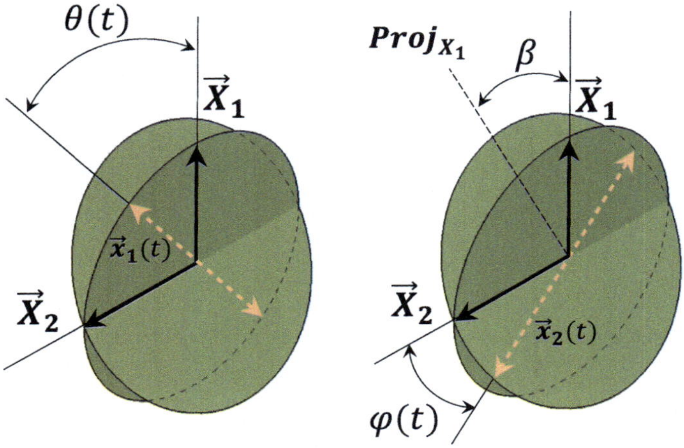

Fig. 29.8 The variables used in the kinematics formulas

input and output axis will also be referred to as the "Spin1" and "Spin2" directions as adopted by the Catia v5 software developers (Catia v5 R21 2018).

Let us assume that the vertical arm of the cross-pin rotates by the amount $\theta(t)$ in the fixed plane as shown in Fig. 29.8. This is the angle between the rotating arm $\vec{x}_1(t)$ and the fixed vector $\vec{X}_1$. This motion leads to the rotation of the horizontal arm of the cross- pin in the slanted fixed plane by $\varphi(t)$. This is the angle between the rotating arm $\vec{x}_2(t)$ and the fixed vector $\vec{X}_2$.

Based on elementary trigonometry, one can write the following expressions.

$$\vec{x}_1(t) = (\cos(\theta(t)), \sin(\theta(t)), 0) \tag{29.2.1}$$

$$\vec{x}_2(t) = (-\cos(\beta)\sin(\varphi(t)), \cos(\varphi(t)), \sin(\beta)\sin(\varphi(t))) \tag{29.2.2}$$

$$\vec{X}_1(t) = (1, 0, 0) \tag{29.2.3}$$

$$\vec{X}_2(t) = (0, 1, 0) \tag{29.2.4}$$

Since the arms of the cross-pin are perpendicular to each other, the dot product of the corresponding vectors is zero.

$$\vec{x}_1(t) * \vec{x}_2(t) = 0 \tag{29.2.5}$$

Expanding this expression, it terms of its components yields

$$-\cos(\beta)\cos(\theta(t))\sin(\varphi(t)) + \sin(\theta(t))\cos(\varphi(t)) = 0 \tag{29.2.6}$$

Rearranging the expression (29.2.6) results in the following formulas.

$$\tan(\varphi(t)) = \frac{\tan(\theta(t))}{\cos(\beta)} \tag{29.2.7}$$

$$\varphi(t) = tan^{-1}\left[\frac{\tan(\theta(t))}{\cos(\beta)}\right] = tan^{-1}\left[\frac{\sin(\theta(t))}{\cos(\theta(t))\cos(\beta)}\right] \tag{29.2.8}$$

Based on physical grounds, $\varphi(t)$ is a continuous function of time. However, the $\tan^{-1}$ function is not one-to-one. Restricting the range of $\theta(t)$ to the interval $-\pi < \theta(t) < \pi$, allows the calculation of $\varphi(t)$ correctly from the expression

$$\varphi(t) = \text{atan2}(\sin(\theta(t))/(\cos(\beta)\cos(\theta(t))) \tag{29.2.9}$$

The time differentiation of (2.9) using the chain rule, and algebra leads to the expression

$$\frac{d\varphi}{dt} = \frac{d\theta}{dt}\frac{\cos(\beta)}{1 - \sin^2(\beta)\cos^2(\theta(t))} \tag{29.2.10}$$

Or

***

$$\omega_{out}(t) = \omega_{in}(t)\frac{\cos(\beta)}{1 - \sin^2(\beta)\cos^2(\theta(t))} \tag{29.2.11}$$

$$\frac{\omega_{out}(t)}{\omega_{in}(t)} = \frac{\cos(\beta)}{1 - \sin^2(\beta)\cos^2(\theta(t))} \tag{29.2.12}$$

A further differentiation of (2.12) leads to the relationship between the angular accelerations $\alpha_{in}(t)$ and $\alpha_{out}(t)$.

$$\alpha_{out}(t) = \alpha_{in}(t)\frac{\cos(\beta)}{1 - \sin^2(\beta)\cos^2(\theta(t))} - \omega_{in}(t)\frac{\cos(\beta)\sin^2(\beta)\sin(2\theta(t))}{(1 - \sin^2(\beta)\cos^2(\theta(t)))^2} \tag{29.2.13}$$

Referring to the expression (29.2.12), the presence of $\cos^2(\theta(t))$, and the half-angle formula in trigonometry, $\cos(2\theta(t)) = 1 - 2\cos^2(\theta(t))$, clearly supports the claim (d) in Sect. 29.1.

Test Case

This is the appropriate point in the article to test the results derived above for a specific case. Suppose that the input shaft is turning at $\omega_{in} = 240$ rpm which is equal

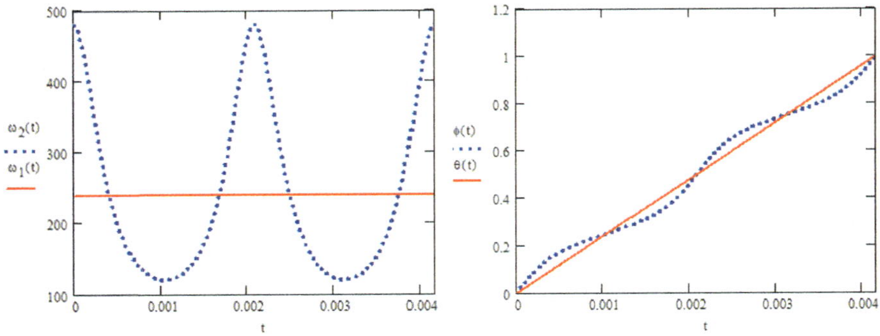

Fig. 29.9 The plots of the output shaft velocity and displacement

to $f = 4$ Hz. Therefore, the period of the input shaft is $T = \frac{1}{f} = 0.25$s. Assuming a skew angle of $\beta = 60°$, the non-dimensional output rpm is given by

$$\frac{\omega_{out}(t)}{\omega_{in}(t)} = \frac{\cos(60°)}{1 - \sin^2(60°)\cos^2(\theta(t))} \tag{29.2.14}$$

where, $\theta(t) = 2\pi ft = 8\pi t\,rad$. The angular displacement of the output shaft can be calculated from the following expression.

$$\varphi(t) = \int_0^t \omega_{out}(\tau)d\tau = \int_0^t \omega_{in}\frac{\cos(60°)}{1 - \sin^2(60°)\cos^2(\theta(\tau))}d\tau \tag{29.2.15}$$

The plots of the output shaft velocity and angular displacement are shown in Fig. 29.9. Here, the time variable is in seconds and the shaft speed (both input and output) are in rpm. Finally, the angular displacements are in radians. The red solid curve is associated with the input shaft, whereas the blue dotted curve is the output shaft results.

29.3 The Digital Mockup Module and the Catia DMU Kinematics Workbench

In this section, the universal joint creation in Catia is discussed. All joints are created within the DMU Kinematics Workbench which is part of the Digital Mockup Module of the software as shown in Fig. 29.10.

There are two approaches for modeling a universal joint. Keep in mind that the cross-pin plays no significant role in the kinematics of the problem which only depends on the skew angle β. It is this angle that determines the speed variation of the output shaft. If there is an absolute need to model the cross-pin (perhaps for

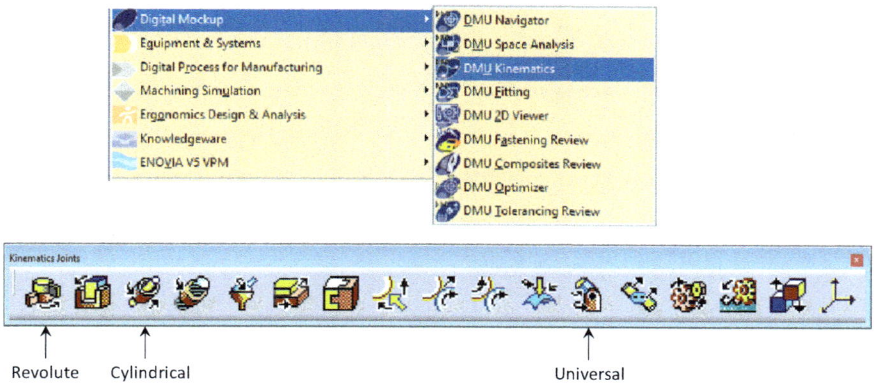

Fig. 29.10 The kinematic joints available in Catia

aesthetic reasons), it can be done with the aid of a revolute joint which drives the input shaft and three cylindrical joints which ties the assembly together. Obviously this approach does not even require the use of "U-joint"  available in the menu.

The second approach is based on using the U-joint directly available in the menu. This is also shown in Fig. 29.11. It can be done using, a universal joint, and four cylindrical one. As in the previous case, the angle driven revolute joint is responsible for turning the input shaft.

Upon selection the U-joint icon, the dialogue box shown in Fig. 29.12 appears. There are five entities three of which must be selected by the user.

The first two selections, namely "Spin 1" and "Spin 2" are rather straightforward. For now, the suggestion is to select the direction of the "input" shaft as "Spin 1" and the "output" shaft as the "Spin 2". We will discuss the implication of reversing the order of this selection later in the article. The reader should keep in mind that the "Angle driven" cylindrical joint is responsible for turning the input shaft. It turns out that one can instead make the cylindrical joint angle driven. This however will be ignored presently.

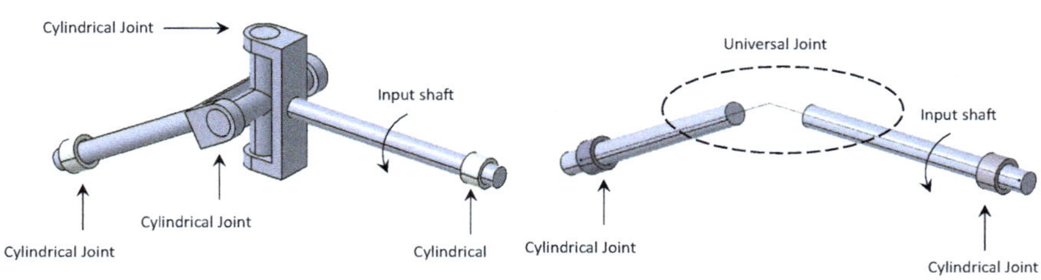

Fig. 29.11 The two scenarios for modeling a universal joint

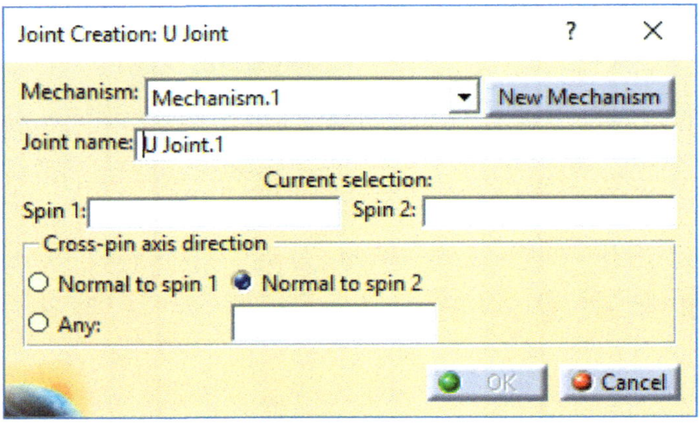

Fig. 29.12 The dialogue box associated with a U-joint creation

There are three remaining "radio buttons" that are left in the menu which needs an explanation. These radio buttons deal with the "Cross-pin axis direction" which is neither defined in the software documentation, nor the standard engineering literature (Fig. 29.13).

If the "Normal to spin 2" is selected (which is the default), it means that at the initial time $t = 0$, the cross-pin which is *not modeled* is at the indicated position shown in Fig. 29.14. The significance of this selection is that at that instant, the rotational velocity of the output shaft is minimum. In the case of constant input speed, in the selected orientation,

$\omega_{out} = \omega_{in}\cos(\beta)$ at the initial instant.

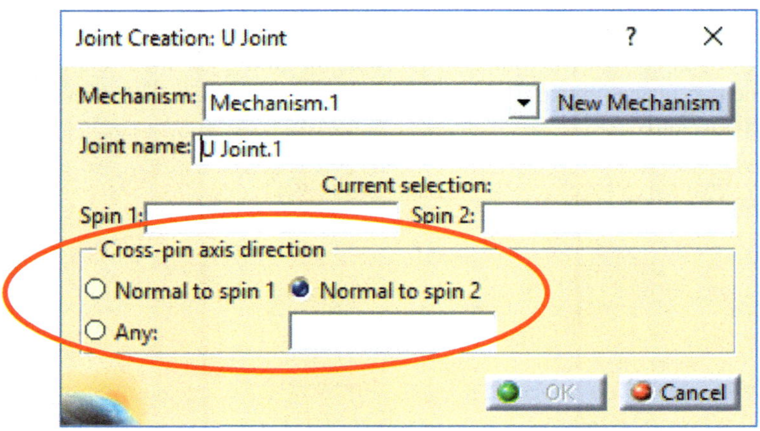

Fig. 29.13 What is the role of these "radio buttons"?

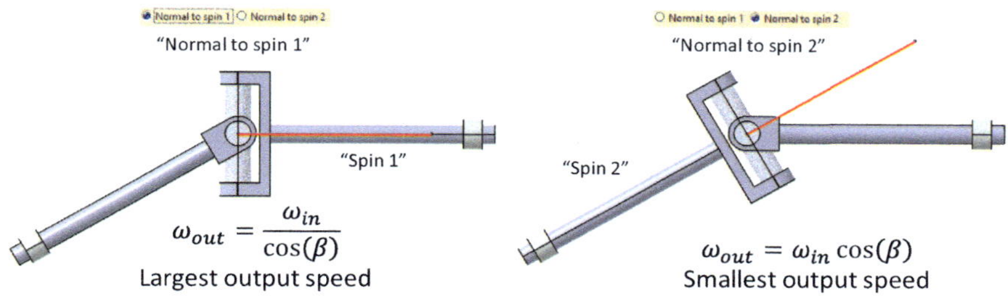

Fig. 29.14 Orientation of the cross-pin (not modeled) depending on the radio button selection

However, if "Normal to spin 1" is selected, the configuration of the virtual cross-pin (i.e. not modeled) is as shown in Fig. 29.14. At that instant, $\omega_{out} = \omega_{in}/\cos(\beta)$ and therefore it has a maximum value for a constant input shaft speed.

The only remaining possible selection is the "Any" radio button. When this is selected, the "direction" must be chosen. The situation is clarified in Fig. 29.15. It is important to keep in mind that the selected direction must be perpendicular to the "Spin 1" direction, and it represents the angle made with the vertical axis $\boldsymbol{X}_1$. The implication of this choice is that at that instant (i.e. initial time $t = 0$), the output shaft velocity is calculated from the expression below (also see Eq. (29.2.11)).

$$\omega_{out}(0) = \omega_{in}\frac{\cos(\beta)}{1 - \sin^2(\beta)\cos^2(\Omega)} \tag{29.3.1}$$

Needless to say that $\omega_{out}(0)$ is in the range $\omega_{in}\cos(\beta) < \omega_{out}(0) < \omega_{in}/\cos(\beta)$.

It was mentioned earlier that the order of the selection of "Spin 1" and "Spin 2" can be reversed. This means that the output shaft can be chosen as "Spin 1" and

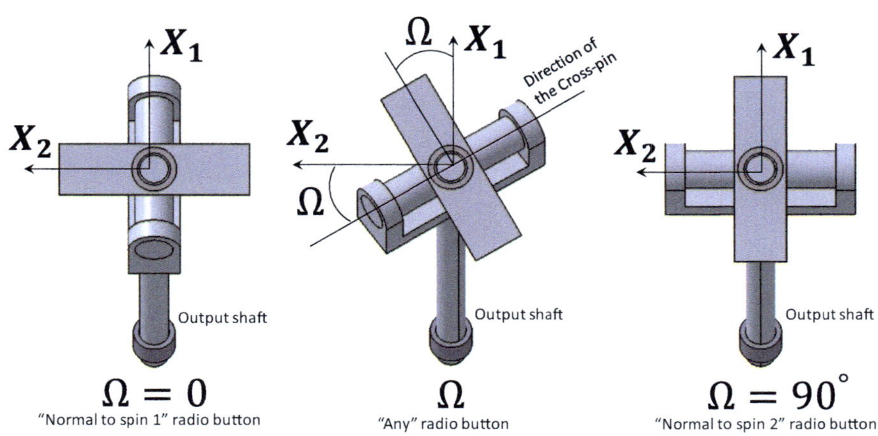

Fig. 29.15 Orientation of the cross-pin (not modeled), the input shaft is pointing out of plane

the input shaft as "Spin 2". Assuming that the revolute joint is still controlling the motion of the input shaft, the statements about the maximum and minimum output speeds must be reversed. Therefore, when the virtual cross-pin is perpendicular to the output shaft, the speed of this shaft is maximum. On the other hand, when the virtual cross-pin is perpendicular to the input shaft, the speed of the output shaft is minimum.

The author is cautioned that if the selected direction "Any" is not perpendicular to "Spin 1", the system locks and produces erroneous results.

29.4 Verification with Catia

At this point, the claims and derivations made in the previous two sections will be tested using the mechanism generated in Catia. The assembly in Fig. 29.16 has a skew angle $\beta = 60°$ and the cross-pin is not modeled. Therefore, the mechanism involves a U-joint, a revolute driving the input shaft and a cylindrical joint. The input shaft is assumed to have a constant speed $\omega_{in} = 60$ rpm.

The four directions tested are also shown in Fig. 29.17. Please note that $\Omega = 0°$ is equivalent to the selection "Normal to spin 1" and $\Omega = 90°$ is the same as "Normal to spin 2".

Before displaying the results, we refer to Fig. 29.17, the only clue in the online documentation which is clearly vague and inadequate.

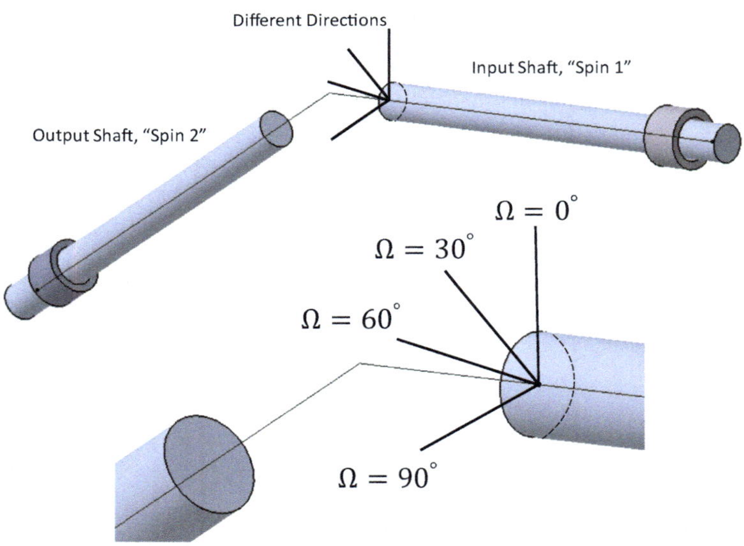

Fig. 29.16 The assembly to be tested, different cross-pin axis directions are used

7. Select the cross-pin axis direction. Select the Any option. In this case, you need to select the direction of the cross-pin axis (which has to be perpendicular with one of the two axis previously selected). In our example, select an edge of the green cylinder.

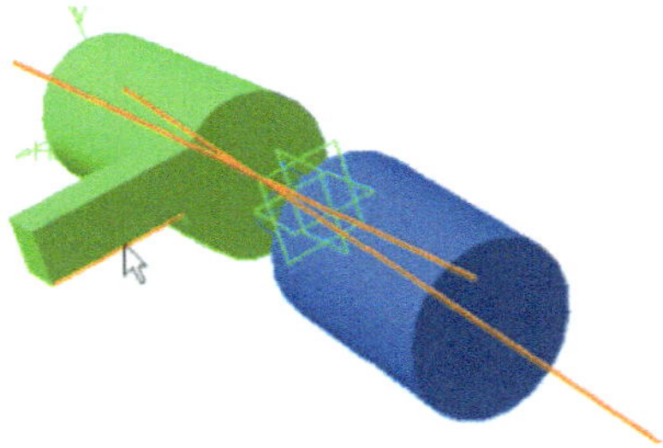

Fig. 29.17 Excerpt from the online Catia documentation (Catia v5 R21, Dassault Systemes Corporation. 2018) regarding the issue of cross-pin direction

Case 1, Ω $= 0°$

For the "Any" direction radio button, the vertical line as shown in Fig. 29.18 is selected. This also corresponds to choice "Normal to Spin 1". For this case, the angular velocity of the output shaft is maximum at the start of the simulation,

$$\omega_{out}(0) = \omega_{in}\frac{\cos(\beta)}{1-\sin^2(\beta)\cos^2(\theta(0))} = \omega_{in}\frac{\cos(\beta)}{1-\sin^2(\beta)\cos^2(\Omega)} = 120\text{rpm} \tag{29.3.2}$$

Case 2, Ω $= 30°$

For the "Any" direction radio button, the second line (at 30° with respect to the axis $\boldsymbol{X}_1$) as shown in Fig. 29.19 is selected. For this case, the angular velocity of the output shaft at the start of the simulation is calculated below.

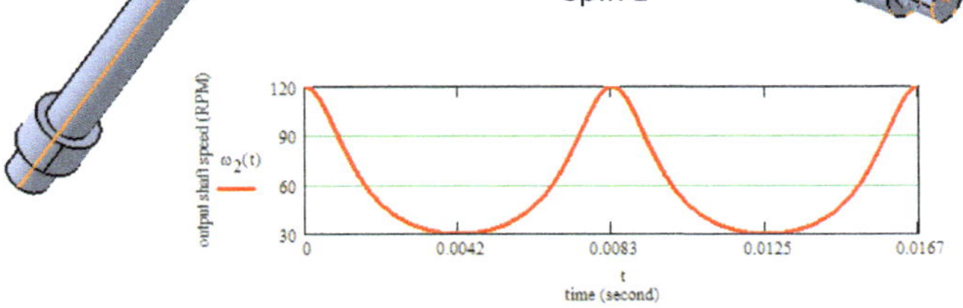

Fig. 29.18 The direction selected corresponds to $\Omega = 0°$

$$\omega_{out}(0) = \omega_{in}\frac{\cos(\beta)}{1-\sin^2(\beta)\cos^2(\theta(0))} = \omega_{in}\frac{\cos(\beta)}{1-\sin^2(\beta)\cos^2(\Omega)} = 68.6\text{rpm} \tag{29.3.3}$$

Case 3, $\boldsymbol{\Omega} = 60°$

For the "Any" direction radio button, the third line (at 60° with respect to the axis $\boldsymbol{X}_1$) as shown in Fig. 29.20 is selected. For this case, the angular velocity of the output shaft at the start of the simulation is calculated below.

$$\omega_{out}(0) = \omega_{in}\frac{\cos(\beta)}{1-\sin^2(\beta)\cos^2(\theta(0))} = \omega_{in}\frac{\cos(\beta)}{1-\sin^2(\beta)\cos^2(\Omega)} = 36.9\text{rpm} \tag{29.3.4}$$

Case 4, $\boldsymbol{\Omega} = 90°$

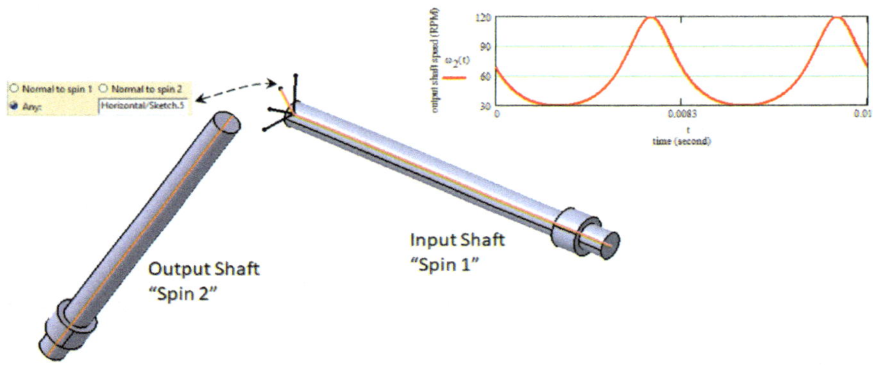

Fig. 29.19 The direction selected corresponds to $\Omega = 30°$

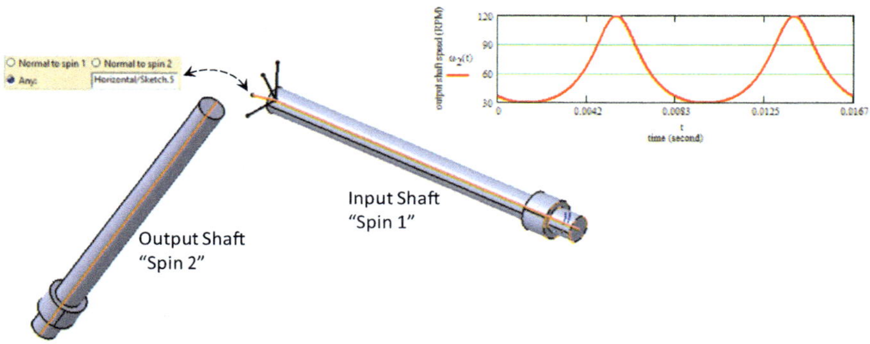

Fig. 29.20 The direction selected corresponds to $\Omega = 60°$

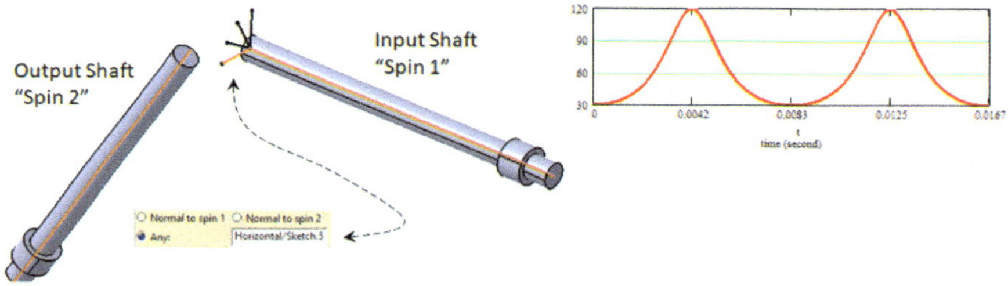

Fig. 29.21 The direction selected corresponds to $\Omega = 90°$

For the "Any" direction radio button, the vertical line as shown in Fig. 29.21 is selected. This also corresponds to choice "Normal to Spin 2". For this case, the angular velocity of the output shaft is minimum at the start of the simulation,

$$\omega_{out}(0) = \omega_{in}\frac{\cos(\beta)}{1 - \sin^2(\beta)\cos^2(\theta(0))} = \omega_{in}\frac{\cos(\beta)}{1 - \sin^2(\beta)\cos^2(\Omega)} = 30\text{rpm} \quad (29.3.5)$$

Case 5, an Ω Whose Direction $\overrightarrow{H}$ is not Perpendicular to "Spin 1"

This case deals with an arbitrary direction in space which is not perpendicular to the "Spin 1" axis. Obviously, **Ω** is the angle between the selected direction $\overrightarrow{\boldsymbol{H}}$ and $\overrightarrow{\boldsymbol{X}}_1$ which can be computed from the definition of dot product.

$$\cos(\Omega) = \frac{\overrightarrow{\boldsymbol{H}} * \overrightarrow{\boldsymbol{X}}_1}{\|\overrightarrow{\boldsymbol{H}}\| \|\overrightarrow{\boldsymbol{X}}_1\|} \quad (29.3.6)$$

Here, the system locks and Catia produces erroneous results as shown in Fig. 29.22.

Case 6, Swapping the "Spin 1" and "Spin 2" Directions

It was mentioned earlier, that if everything stays the same and only swapping the directions, very little changes. The result is that the max and min output angular speeds are swapped. To be more precise, suppose that as shown in Fig. 29.23, the output shaft is selected as "Spin 1", the input shaft selected as "Spin 2", and the default radio button "Normal to spin 2" is picked. Once the plot of the output shaft angular speed is generated, it is the opposite of case 4. Therefore, in that configuration, the output shaft starts with a maximum speed.

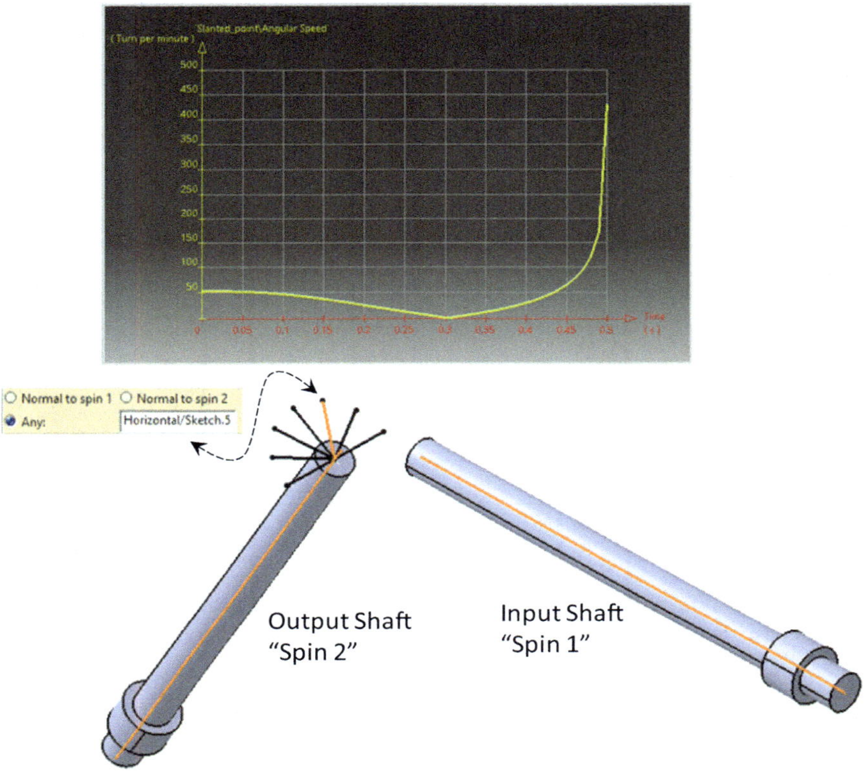

Fig. 29.22 The direction selected is not perpendicular to "Spin 1"

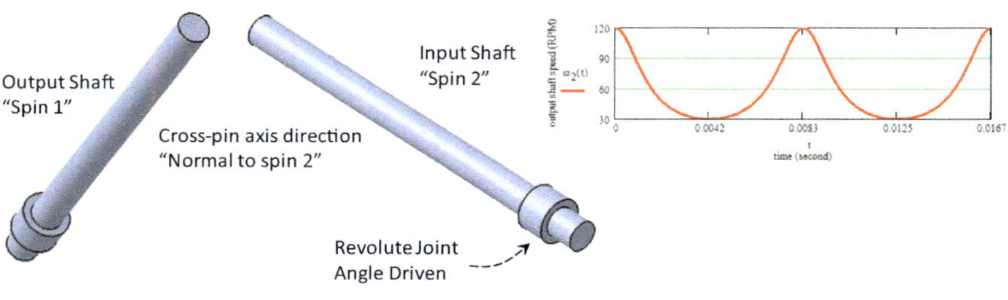

Fig. 29.23 The effect of swapping "Spin 1" and "Spin 2"

29.5 Comments on a Universal Joint

In the development presented earlier on the universal joint, it was assumed that the "spin" directions were fixed in space. The motion was factored in because of the rotation of the shafts about their own axis. That assumption however is not a

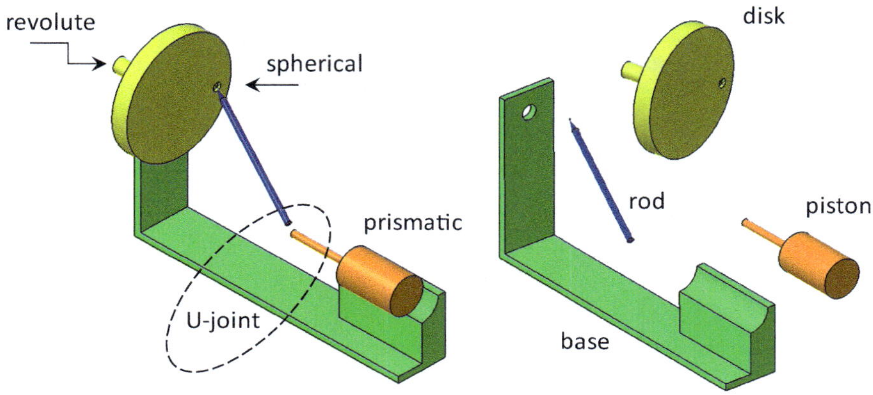

Fig. 29.24 The mechanism involving a U-joint not fixed inn space

prerequisite for utilizing U-joints. An application is depicted in Fig. 29.24. Here, the assembly consists of four parts as shown. The disk is revolving in the base through an angle driven revolute joint. The piston shown (which has a shaft) needs a prismatic joint to travel along its path. It is tempting to model this as a cylindrical joint but it will suffer from and extra degree of freedom which needs to be accounted for. The connecting rod interacts with the disk through a spherical joint as shown. Keep in mind that a spherical joint is also commonly known as a ball-socket joint. The issue that is relevant to the present paper is the kind of joint needed between the rod and the piston.

The interaction between these two parts depends on whether one decides to model the cross-pin or to use the universal joint as prescribed in the Catia program. These are both acceptable however the latter requires an additional part (cross-pin) and joints. Although for esthetic reasons this approach may be pleasing, it does not contribute additional information if the goal is to mimic the motion and the kinematics. The setup is displayed in Fig. 29.25. Note that an axis of the U-joint is not fixed and revolves (in addition to spinning) in space.

It was mentioned in the introduction that it is often mistakenly assumed that the angular speed of the shaft is the same as the input shaft. As seen earlier that is not the case. In many applications, the input shaft turns at a constant speed and the objective is to transfer the motion to an output shaft having the same speed. This can be achieved by using two universal joints as shown in Fig. 29.26, where the axis of the two shafts is parallel. The resulting joint in Catia is known as the constant velocity joint (CV) and simply requires the three spin directions.

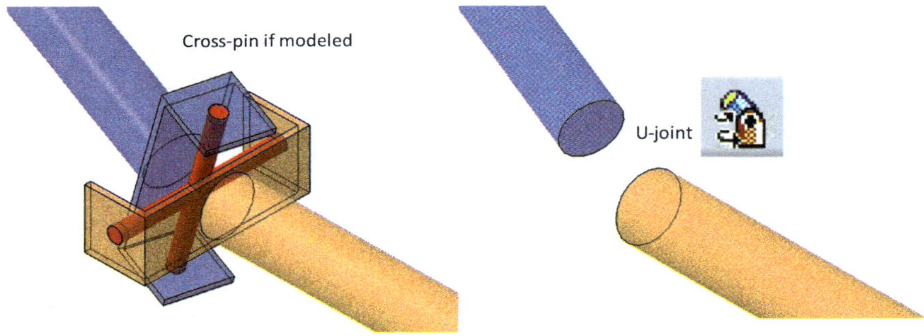

Fig. 29.25 The two approached for modeling the interaction between shafts

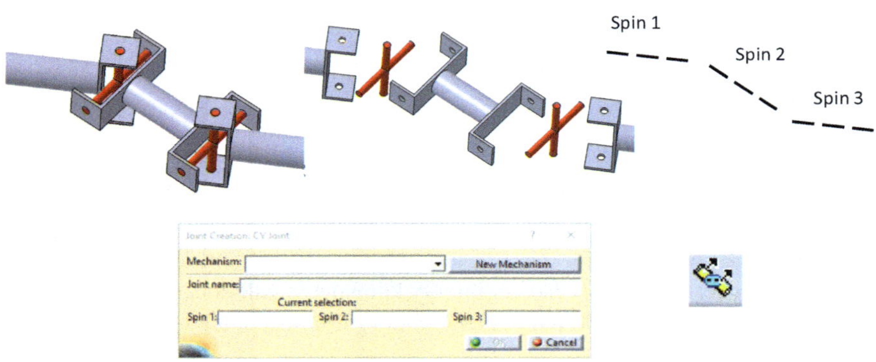

Fig. 29.26 The Constant Velocity (CV) joint

29.6 Concluding Remarks

The motivation behind this paper was to shed some light on the "Modus Operandi" of the universal joint which plays such a critical role in many mechanisms. Such a joint can be modeled in most CAE software including Catia but unfortunately the documentation of these software do not clearly spell out the different options available to the users. The basic kinematics relationships that prevail a universal joint was reviewed and were compared with U-joint modeling in Catia. Relying on the theoretical developments, some unanswered question in the Catia software were addressed. Among such issues was the role of the "Cross-pin axis direction" and its relevance which were discussed in detail. Although dynamics analysis (calculation forces) was not included in the present paper, the reader is referred to a publication discussing these issues in detail (Dynamics of Universal Joints 2012).

Acknowledgements The author gratefully acknowledges the support of the Faculty of Engineering at the University of Windsor under the internal grant 809153.

Appendix

In this appendix, the angular velocity vector components of the cross-pin are calculated. The derivation is based on elementary relative velocity formulas and selecting two point "A" and "B" where the velocity vectors are known in the magnitude and directions. These are shown in Fig. 29.27. Although the height (and width) of the cross-pin will drop out of the calculation, it is assumed to be "L". The magnitudes of the velocities at "A" and "B" are $v_A = \omega_1 L$ and$v_B = \omega_2 L$. Furthermore, in the displayed configuration, the output shaft angular velocity, is given by $\omega_2 = \omega_1/\cos(\beta)$. This was confirmed in the text of the paper. Using the relative velocity expression,

$$\vec{v}_A = \vec{v}_B + \vec{v}_{A/B} \tag{A1}$$

$$\vec{v}_A = \vec{v}_B + \vec{\omega}_{pin} \times \vec{r}_{A/B} \tag{A2}$$

Here, $\vec{\omega}_{pin}$ is the unknown angular velocity of the cross-pin.

$$\vec{\omega}_{pin} = \left[\omega_{pin,x}\ \omega_{pin,y}\ \omega_{pin,z}\right]^T \tag{A3}$$

Expanding the above expressions in the component form, and substituting the known entities results in the following system of equation.

$$\begin{pmatrix} \omega_1 L \\ 0 \\ 0 \end{pmatrix} = \begin{pmatrix} 0 \\ \omega_2 L\sin\beta \\ -\omega_2 L\cos\beta \end{pmatrix} + \begin{pmatrix} \omega_{pin,x} \\ \omega_{pin,y} \\ \omega_{pin,z} \end{pmatrix} \times \begin{pmatrix} -L \\ 0 \\ L \end{pmatrix} \tag{A4}$$

If the cross product is carried out, and both sides are divided by "L", (A4) reduces to,

$$\begin{pmatrix} \omega_1 \\ 0 \\ 0 \end{pmatrix} = \begin{pmatrix} 0 \\ \omega_2 sin\beta \\ -\omega_2 cos\beta \end{pmatrix} + \begin{pmatrix} \omega_{pin,y} \\ -\omega_{pin,x} - \omega_{pin,z} \\ \omega_{pin,y} \end{pmatrix} \tag{A5}$$

Using a matrix notation, (A5) can be rewritten as

$$\begin{pmatrix} \omega_1 \\ 0 \\ 0 \end{pmatrix} = \begin{pmatrix} 0 \\ \omega_2 \sin\beta \\ -\omega_2 \cos\beta \end{pmatrix} + \begin{bmatrix} 0 & 1 & 0 \\ -1 & 0 & -1 \\ 0 & 1 & 0 \end{bmatrix} \begin{Bmatrix} \omega_{pin,x} \\ \omega_{pin,y} \\ \omega_{pin,z} \end{Bmatrix} \tag{A6}$$

The matrix is singular, however, at the instant shown in Fig. 29.27, the cross-pin has no ability to rotate about the Z axis, therefore $\omega_{pin,z} = 0$. This enables us to solve the system for the other two knowns.

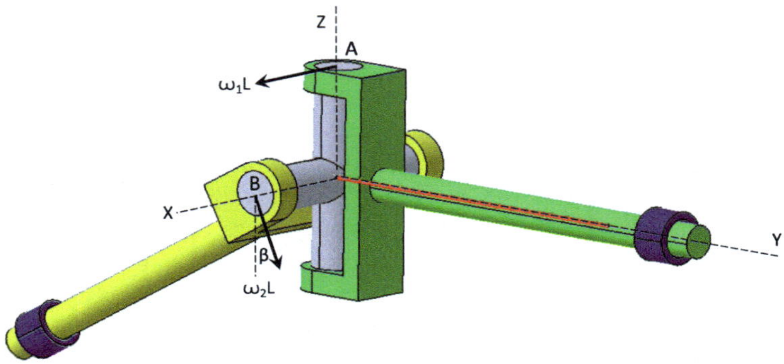

Fig. 29.27 The setup for calculating the angular velocity of the cross-pin

$$\begin{Bmatrix} \omega_{pin,x} \\ \omega_{pin,y} \\ \omega_{pin,z} \end{Bmatrix} = \begin{Bmatrix} \omega_1 \tan(\beta) \\ \omega_1 \\ 0 \end{Bmatrix} \tag{A7}$$

Note that we have used $\omega_2 = \omega_1/\cos(\beta)$ at the instant under consideration.

References

Budynass RG, Nibsett K (2014) Shigley's mechanical engineering design. McGraw Hill, New York

Catia v5 R21 (2018) Dassault systemes corporation. Velizy, France

Norton RL (2004) Design of machinery. McGraw Hill, New York

Universal Joint, Wikipedia. https://en.wikipedia.org/wiki/Universal_joint

Vesali F, Rezvani M, Kashfi M (2012) Dynamics of universal joints, its failures and some propositions for practically improving its performance and life expectancy. J Mech Sci Technol 26–8:2439–2449

Zamani NG (2016) Mechanism design essentials in 3DEXPERIENCE R2016x using Catia applications. SDC Publications, Mission Kansas

Zamani NG, Weaver JM (2012) Catia v5 tutorials in mechanism design and animation. SDC Publications, Mission Kansas

Chapter 30
Constructing a Cost-Effective Stroboscope for Calibration Laboratories Amidst Market Inflation and Export Challenges

Mirna Osama and Ahmed Y. Shash

Abstract This research explores the development of an economical stroboscope within a calibration laboratory setting, addressing the challenges of high market costs and export restrictions. The study focuses on designing and constructing a stroboscope using locally available materials and advanced yet cost-effective technology such as LED lighting and microcontroller programming. The methodology includes the integration of these components to achieve a high flash rate and sufficient illumination for effective motion analysis. Preliminary results indicate that the prototype meets necessary operational standards while significantly reducing costs. This project not only offers a practical solution to resource constraints in scientific equipment but also contributes to the broader field of applied physics by demonstrating the feasibility of low-cost, high-performance laboratory instruments.

30.1 Introduction

The stroboscope is an essential tool in various industrial and scientific applications, including motion analysis, speed measurement, and calibration of optical tachometers. However, the high cost of commercial stroboscopes and export restrictions pose significant challenges for many laboratories, particularly in developing regions. This study aims to address these challenges by developing a cost-effective stroboscope using locally available materials and advanced, yet affordable, technology (Laboratories and Inc.: Application of Strobe Application Devices).

M. Osama · A. Y. Shash (✉)
Faculty of Engineering and Materials Science, German University in Cairo, Cairo, Egypt
e-mail: ahmed.shash@cu.edu.eg; ahmed.shash@guc.edu.eg

A. Y. Shash
Mechanical Design and Production Department, Faculty of Engineering, Cairo University, Cairo, Egypt

A. Öchsner and H. Altenbach (eds.), *Engineering Design Applications VII*, Advanced Structured Materials 230, https://doi.org/10.1007/978-3-031-84346-4_30

Stroboscopes have been widely used in research and industry for decades. The importance of stroboscopes in measurement processes, preventive maintenance, and education is well-documented (IET Labs).

However, the high cost of commercial stroboscopes, which can cost upwards of $2,000, limits their accessibility (Brancalião 2020).

30.2 Methodology

30.2.1 Design and Component Selection

The Stroboscope was designed to be both cost-effective and high performing therefore the following key components include:

- LED light source: High-intensity LEDs were chosen for their efficiency and longevity as shown in Fig. 30.1.
- Microcontroller: An affordable microcontroller was used to control the flash rate and synchronize the system as shown in Fig. 30.2.
- Power supply: A standard power supply unit was selected to ensure consistent performance.
- LCD screen: An I2C LCD screen was used to show the required frequency as shown in Fig. 30.3.

Fig. 30.1 LED light source used in the developed stroboscope

Fig. 30.2 Micro-controller and circuit components

Fig. 30.3 User interface components that includes the LCD screen and buttons

30.2.2 Construction

The construction process involved assembling the selected components onto a custom-designed printed circuit board (PCB) as shown in Fig. 30.4. The LEDs were arranged to provide uniform illumination as shown in Fig. 30.5, and the microcontroller was programmed to control the flash rate accurately. Figure 30.6 displays the stroboscope's user interface. Figure 30.7 shows the assembled inside of the stroboscope and Fig. 30.8 shows the working LED lights.

30.2.3 Calibration and Testing

The stroboscope was calibrated using a reference tachometer. Ten different frequencies were tested, ranging from 1 to 1500 Hz. The corresponding tachometer readings are recorded, and the absolute and relative errors were calculated to assess accuracy.

The steps for calibrating and testing the developed stroboscope are as follows and shown in Fig. 30.9:

- Place the stroboscope on a flat surface.
- Within 12–30 cm distance align the optical tachometer facing the stroboscope's LED lights.
- Turn on the stroboscope.
- Set the light's frequency using the stroboscope's buttons.
- Record the readings of the calibrating tachometer against the stroboscope's readings.
- Repeat the last two steps with various frequencies.

30.3 Results

Table 30.1 shows the test results that were obtained using the developed stroboscope and calibrated tachometer.

To obtain the test results the following calculations were conducted:

- Mean absolute error (difference between the measured value and the true value) =

$$\Sigma\, Absolute\ Errors/No.\ of\ readings$$

  $= 3.38\ /\ 10 \approx 0.3$ Hz.

- Accuracy $= \pm$ mean relative error (ratio of the absolute error to the true value) =

$$\Sigma\, Relative\ Errors/No.\ of\ readings$$

  $= 2.08\ /\ 10 \approx \pm\, 0.2\%$

- Maximum absolute error: 1.3 Hz (at 1500 Hz)

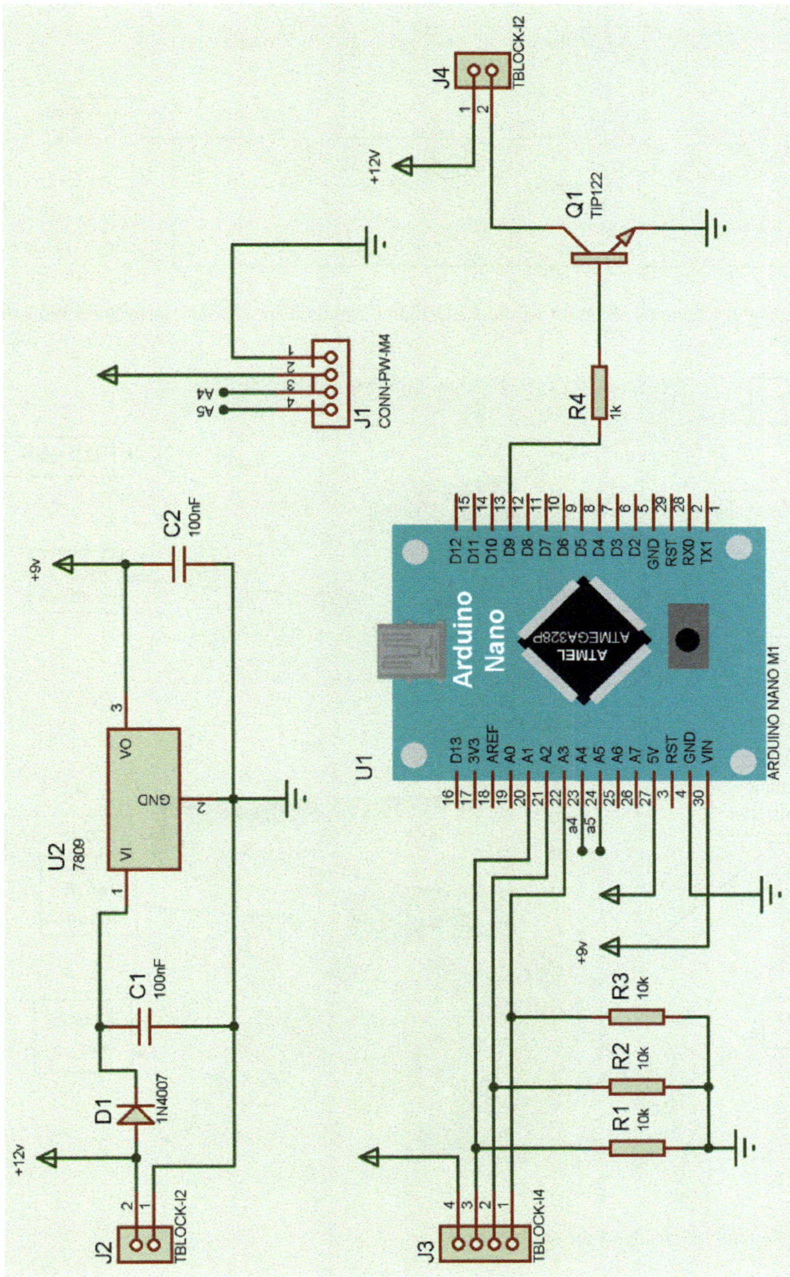

Fig. 30.4 Light schematic diagram of the developed stroboscope's circuit board

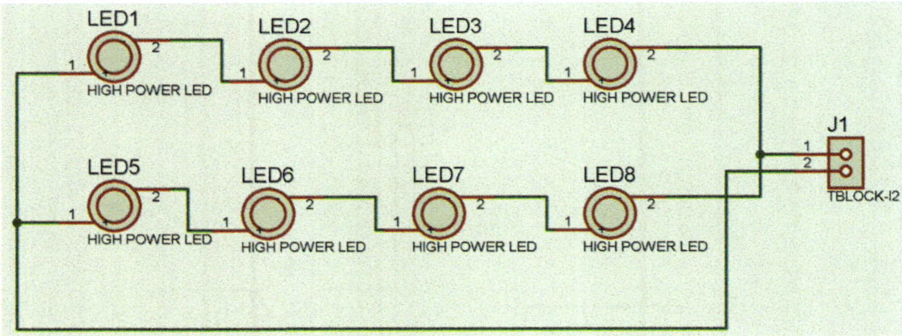

Fig. 30.5 Light schematic diagram to the LED lights arranged in a circuit to provide uniform illumination

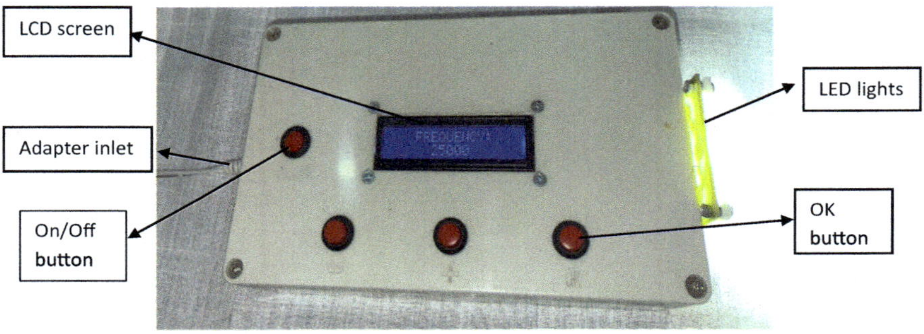

Fig. 30.6 Stroboscope's final user interface

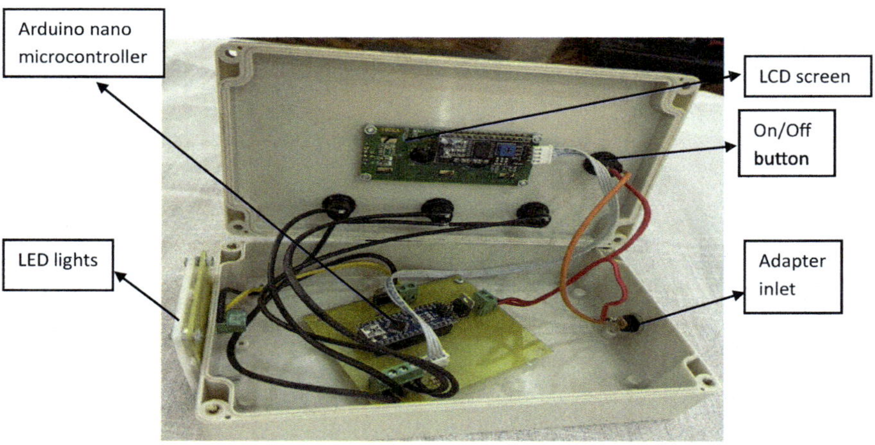

Fig. 30.7 Assembled inside of the stroboscope

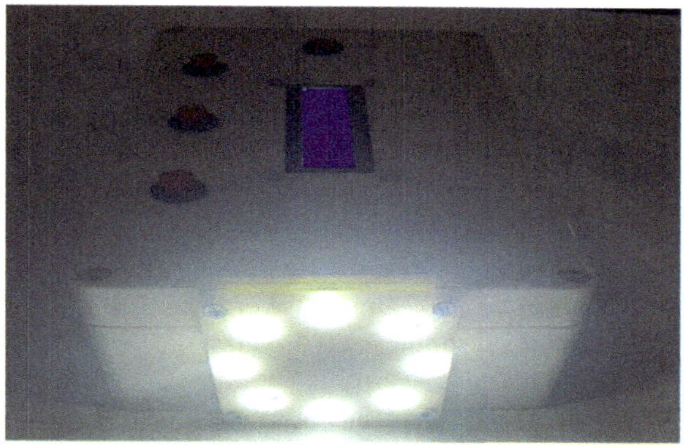

Fig. 30.8 Working LED lights

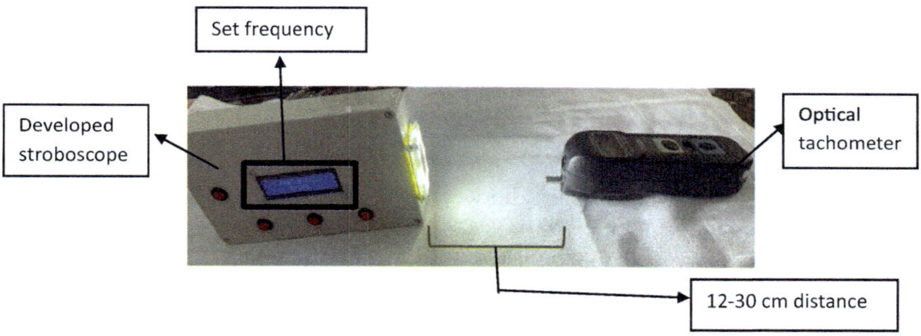

Fig. 30.9 Testing and calibrating procedure of the stroboscope

Table 30.1 Developed stroboscope test results

Stroboscope (Hz)	Tachometer (Hz)	Absolute error (Hz)	Relative error (%)
A	B	C = A–B	$D = \frac{C}{B} \times 100$
1	0.99	0.01	1.01
5	4.99	0.01	0.20
10	9.99	0.01	0.10
50	49.95	0.05	0.10
100	99.9	0.1	0.10
200	199.8	0.2	0.10
400	399.6	0.4	0.10
500	499.5	0.5	0.10
1000	999.2	0.8	0.08
1500	1498.7	1.3	0.09

- Maximum relative error: 1.01% (at 1 Hz)

30.3.1 Developed Stroboscope Specifications

- Frequency operating range: 1 Hz to 50,000 Hz
- Mean absolute error: 0.3 Hz
- Accuracy = ± mean relative error: ± 0.2%
- Maximum absolute error: 1.3 Hz
- Maximum relative error: 1.01%
- Linearity: The error increases slightly with frequency, but remains below 0.1% for most of the range

30.3.2 Comparison Between the Developed Stroboscope and the Commercial Stroboscopes

Table 30.2 shows the difference between the developed stroboscope and the commercial ones in terms of price, frequency range, accuracy and light source.

The developed stroboscope offers exceptional value for its price, with a wider frequency range than most commercial models and competitive accuracy. Its significantly competitive cost could make it an attractive option for many users, especially in markets where budget constraints are a major factor.

The low-cost, high-performance stroboscope has the potential to significantly impact calibration laboratories by reducing costs, increasing accessibility, and potentially disrupting the market for calibration equipment. This could lead to more widespread use of stroboscopes in calibration procedures and potentially inspire further innovations in cost-effective calibration solutions.

Table 30.2 Comparison between developed stroboscope and commercial stroboscopes

Stroboscope model	Price (USD)	Frequency (Hz)	Accuracy (%)	Light source
Developed Stroboscope	53	1 to 50,000	± 0.2%	LED
Shimpo DT-3015N (Shimpo Instruments: Shimpo DT-3015N Portable Stroboscope)	1150	1 to 550	± 0.05%	LED
Monarch Instrument NOVA STROBE DAX 115 (115 Xenon Stroboscope with 115 VAC Power)	1000	1 to 330	± 0.004%	LED
Fluke 820–2 LED stroboscope (Fluke Corporation: Fluke 820–2 LED Stroboscope)	2000	1 to 5000	± 0.02%	LED

30.4 Conclusion

This study successfully developed a cost-effective stroboscope using locally available materials and advanced technology. The stroboscope achieved high accuracy and performance while significantly reducing costs. This project offers a practical solution to resource constraints in scientific equipment and contributes to the broader field of applied physics by demonstrating the feasibility of low-cost, high-performance laboratory instruments.

References

Brancalião L (2020) Prototyping and calibration of a low-cost stroboscope. Polytechnic Institute of Bragança. https://bibliotecadigital.ipb.pt/bitstream/10198/21588/1/Brancali%C3%A3o_Laiany.pdf

Fluke corporation: fluke 820–2 LED stroboscope. https://www.fluke.com/en-us/product/mechanical-maintenance/vibration-analysis/fluke-820-2

IET labs: handbook of stroboscopy. https://www.ietlabs.com/pdf/application_notes/Handbook_Stroboscopy.pdf

Monarch instruments: monarch nova-strobe DAX 115 xenon stroboscope with 115VAC power. Mitchell instrument company. https://www.mitchellinstrument.com/dax115-strobe-stroboscope-ac-line-powered.html

Shimpo instruments: shimpo DT-3015N portable stroboscope. https://www.shimpo-direct.com/product/shimpo-dt-3015n-portable-stroboscope

Sugawara laboratories Inc.: application of strobe application devices. https://www.sugawara-labs.co.jp/en/application_strobe

Chapter 31
Optimizations of an Ergonomic Postural Auxiliary System for Industrial Applications

Kenia Madeleyn Ortiz-Morales, Juan Alfonso Beltrán-Fernández, Ricardo González-Romano, Juan Atonal-Sánchez, and Verónica Guzmán-Mercado

Abstract The optimization process for the design of an industrial postural support system aimed at mitigating muscular and postural fatigue during prolonged shifts is described in this study. CAD modeling and numerical testing were employed to ensure the system was dimensioned with an appropriate safety factor, reducing the risk of failure. Additionally, the system helps to minimize muscular effort, benefiting user health and quality of life. The research includes an investigation of various exoskeleton systems, focusing on their geometric characteristics, materials, application areas, and associated commercial costs. The results enable a comparative analysis of the prototype with existing commercial systems, evaluating its advantages and limitations based on performance.

Keywords Postural system · Optimization · Muscle fatigue · Working posture · Muscle work reduction

---

K. M. Ortiz-Morales · J. A. Beltrán-Fernández (✉) · J. Atonal-Sánchez · V. Guzmán-Mercado
Instituto Politécnico Nacional, Sección de Estudios de Posgrado E Investigación, Escuela Superior de Ingeniería Mecánica y Eléctrica, Unidad Zacatenco, "Unidad Profesional Adolfo López Mateos", Ciudad de México, Mexico
e-mail: jbeltran@ipn.mx

K. M. Ortiz-Morales
e-mail: kmadeleeyn@gmail.com

J. Atonal-Sánchez
e-mail: jatonals@ipn.mx

V. Guzmán-Mercado
e-mail: vero_guzmanm94@hotmail.com

R. González-Romano
Transformadora De Metales, Mexico City, Mexico
e-mail: transfmetal@prodigy.net.mx

A. Öchsner and H. Altenbach (eds.), *Engineering Design Applications VII*, Advanced Structured Materials 230, https://doi.org/10.1007/978-3-031-84346-4_31

31.1 Introduction

The postural system is responsible for maintaining an individual in an upright position, ensuring balance despite external forces exerted on the body, both in static conditions and during movement. This balance is achieved through the correct positioning of the center of gravity and the support polygon. The postural system ensures that ergonomic posture is maintained and that the body can adapt to varying environments and circumstances (Zhang and Zhang 2021).

The postural system relies on external sensory information, musculoskeletal feedback, and data from the central nervous system. Proper postural alignment promotes physiological and biomechanical efficiency, reducing stress and overload caused by gravity (ELIAS et al. 2014; NTP 1162). During prolonged standing, strategies such as ankle oscillation, hip movement, and alternating weight distribution between the legs are employed to maintain balance. However, various personal, mechanical, or environmental disturbances may lead to muscle fatigue, reducing stability and potentially causing chronic pathologies if untreated (NTP 1162; Haddadin and Albu-Schäffer 2018; Kumar and Saini 2019; Farina and Holobar 2017).

In Mexico, institutions such as the Mexican Institute of Social Security (IMSS) have reported musculoskeletal disorders as a significant occupational health issue, particularly among industrial workers. Common conditions include pain in the neck, back, shoulders, and waist, often linked to repetitive motions, manual labor, or prolonged standing and sitting. These risk factors contribute to the development of occupational diseases. Globally, back and cervical pain are recognized as leading causes of disability, affecting physical abilities, productivity, and family life. This further impacts on the economy by increasing healthcare costs and reducing workforce productivity (Gobierno; Investigación and y Desarrollo en Ciencias de la Salud 09 Diciembre 2018).

31.2 Materials and Methods

A study of exoskeleton systems used as postural support was conducted, as shown in Fig. 31.1, focusing on their primary characteristics, materials, applications, and market costs. Based on the collected data, the advantages and disadvantages of each system were identified and compared to the developed prototype (Ries and Ross 2016).

Exoskeletons are defined as wearable frame systems designed to assist individuals in physically demanding tasks. Their primary function is to enhance physical capabilities, providing additional strength and reducing the muscular effort required for certain activities (Goujon and Gosselin 2020).

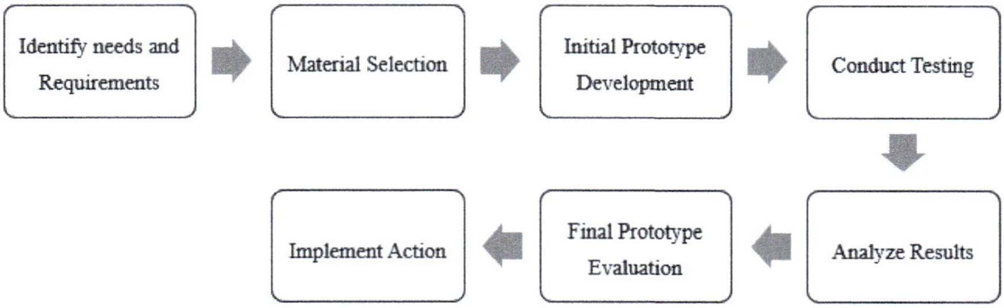

Fig. 31.1 Methodology

31.3 Initial Prototype

Mechanical exoskeletons are designed to operate without external power sources for moving parts or components. Instead, the energy produced by the user's own movements is harnessed, stored, and redistributed to adjust the physical load exerted on various parts of the body. These exoskeletons are primarily employed to reduce localized loads, particularly in the lumbar region, alleviating strain on the user's back. This load reduction is achieved by redistributing forces more effectively, which decreases muscle fatigue and prevents overuse injuries (Bogue 2020; Rossi and Vitiello 2015).

The initial prototype of this project was developed by Ricardo González at Transformadora de Metales Company, as shown in Fig. 31.2. However, limitations were identified in the prototype that required optimization for effective implementation. A key challenge was the reduction of the mechanism's weight, as the original prototype, weighing 4 kg, proved uncomfortable for prolonged use throughout a typical workday.

To assess the potential improvements of the initial prototype, a comprehensive modeling of its components and assembly was conducted. This was followed by a stress analysis using the finite element method (FEM). As illustrated in Fig. 31.3, the von Mises stress distribution reveals a maximum stress of 232.3 MPa, which is

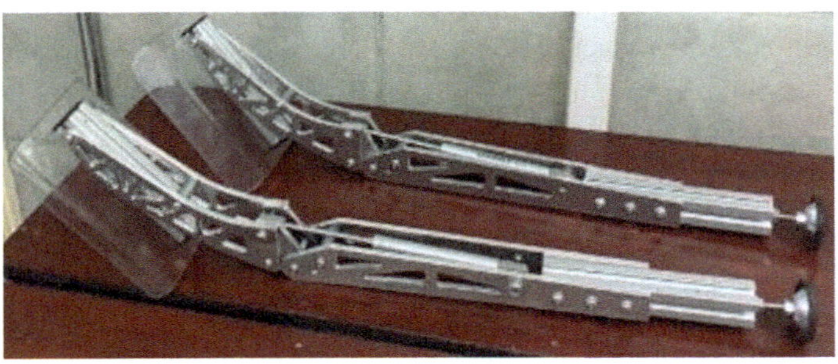

Fig. 31.2 Initial prototype

Inital Prototype

Aluminum 6061 T6

Force: 588.6 N

Coefficient of thermal expansion:	2.4e-05 /Kelvin
Name:	6061-T6 (SS)
Yield Strength:	275 MPa
Elastic Modulus:	69 GPa
Poisson's Ratio:	0.33
Density:	2700 kg/m^3
Shear Modulus:	26 GPa

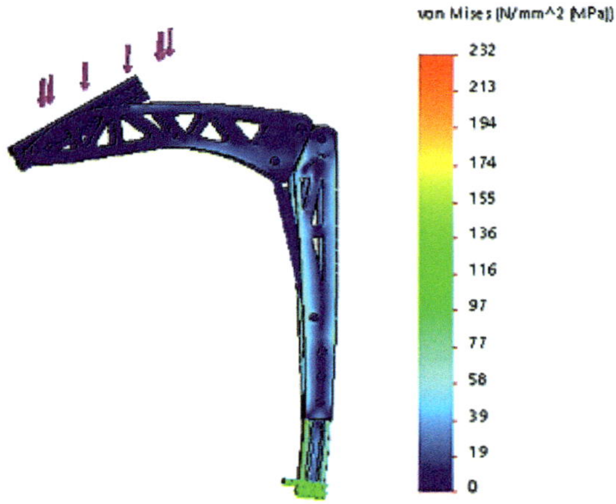

Fig. 31.3 Von Mises stress analysis of the initial prototype

below the yield strength of 275 MPa for 6061 T6 aluminum. This results in a safety factor of 1.18.

As illustrated in Fig. 31.4, the primary critical point of the assembly is identified at the junction of the two links, where a stress of 232 MPa is generated. Although this stress is below the yield strength of the material, the safety factor of 1.18 suggests that the prototype may be prone to failure for users with higher body weights. Therefore, it is recommended that this mechanism be utilized only by individuals who meet the weight specifications established in this study or are below this threshold. To address this limitation, it is proposed that the geometry of the prototype be modified to reduce stress concentrations and improve the safety factor.

To initiate the optimization process, an analysis was conducted on the average weight and height of workers in the metropolitan area, which ranges from 165 to 170 cm. Consequently, the prototype was designed to accommodate this height range. Despite the World Health Organization (WHO) guidelines recommending a maximum weight of 72 kg for a male of 170 cm based on Body Mass Index (BMI) tables, data from the National Institute of Statistics and Geography (INEGI) indicate that men in Mexico with a height of 165 cm typically weigh around 75 kg.

Moreover, similar products currently available are designed to support weights between 100 and 120 kg. Therefore, the prototype was specifically engineered to accommodate a maximum weight of 120 kg.

The adjustment of link dimensions to match the average height was carried out using the anthropometric measurement method proposed by Drillis and Contini in 1966. This process determined that the upper link should measure approximately 32.8 cm, while the lower link should measure 40.3 cm. Additionally, it was recommended that both the upper gluteal support and the floor support be adjustable by 5 cm to accommodate the average height range of the Mexican population, allowing for customization based on individual user needs.

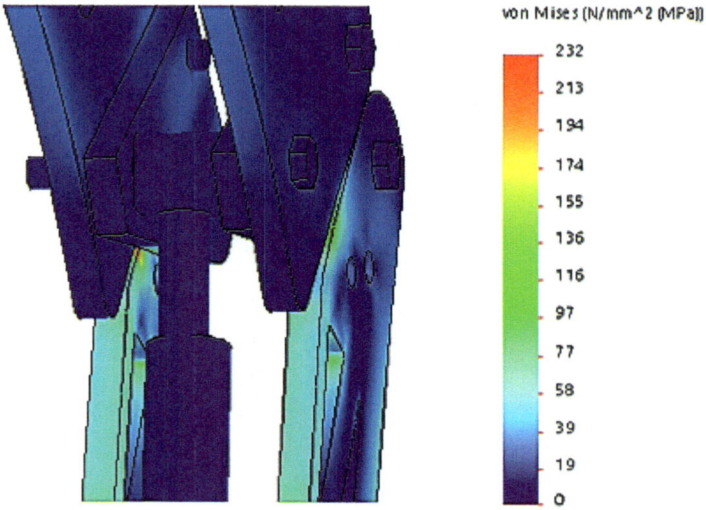

Fig. 31.4 Critical point of the initial prototype

After examining the material properties, a new design was evaluated using the material from the first prototype: Aluminum 6061 T6. The "T6" designation refers to the temper or hardness level, which is achieved through precipitation hardening. This grade provides an excellent strength-to-weight ratio, making it ideal for projects that require a lightweight yet strong material. Additionally, it boasts exceptional workability, weldability, and corrosion resistance (Edwards et al. 1930) (Fig. 31.5).

Figure 31.6 shows the stress analysis results for the redesign proposal. The maximum stress was reduced to 88 MPa from the initial 232 MPa observed in the original prototype. This reduction resulted in a safety factor of 3.12, ensuring that the mechanism can support more than the previously specified maximum capacity of 120 kg.

Substantial improvements in both structural performance and user comfort were demonstrated through the comparative analysis of the initial and redesigned prototypes, as shown in Table 31.1. The initial prototype, constructed from aluminum 6061 T6 and weighing 4 kg, was found to exhibit a maximum stress of 232 MPa, resulting in a safety factor of 1.18. Although these parameters met the preliminary safety criteria, significant discomfort during prolonged use was reported due to the prototype's weight.

In response, a redesign was implemented, incorporating a thinner profile of the same material. This adjustment led to a 30.75% reduction in weight to 1.23 kg. Consequently, a substantial decrease in maximum stress to 88 MPa was achieved, and the safety factor was increased to 3.12, representing a 164% improvement over the initial design. The enhanced safety factor suggests that the new design is capable of supporting greater loads and offers improved durability. Additionally, an adjustable

Fig. 31.5 CAD assembly of the redesigned proposal

Redesign	
Aluminum 6061 T6	
Force: 588.6 N	
Coefficient of thermal expansion:	2.4e-05 /Kelvin
Name:	6061-T6 (SS)
Yield Strength:	275 MPa
Elastic Modulus:	69 GPa
Poisson's Ratio:	0.33
Density:	2700 kg/m^3
Shear Modulus:	26 GPa

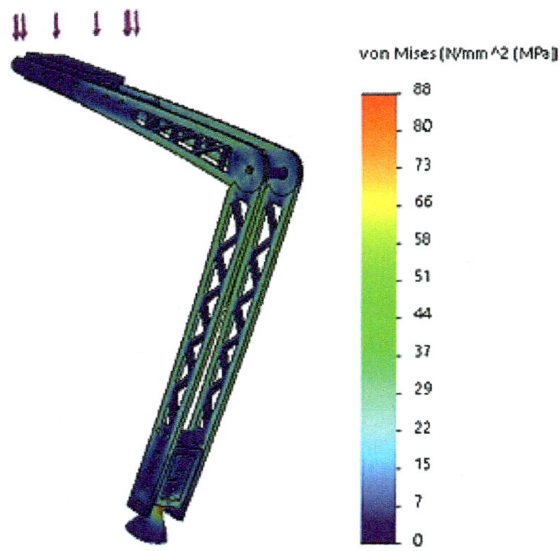

Fig. 31.6 Von Mises stress analysis of the redesigned proposal

Table 31.1 Comparative analysis of the initial and redesigned prototypes

Parameter	Initial prototype	Redesigned prototype	Improvement
Material	Aluminum 6061 T6	Aluminum 6061 T6	Same material, reduced thickness
Weight	4.00 kg	1.23 kg	30.75% reduction
Maximum stress	232 MPa	88 MPa	62.1% reduction
Safety factor	1.18	3.12	Improved by 164%
Height adjustment range	Not adjustable	5 cm	Added adjustability
Design modification	Initial design	Updated geometry and thickness	Enhanced for better ergonomics
Usability range	Limited to initial specifications	Expanded to 165–170 cm	Broader usability range

mechanism was integrated into the redesigned prototype to accommodate a height range of 165–170 cm, thus broadening its applicability to a wider user demographic.

The modifications not only improve structural integrity but also increase ergonomic comfort. By integrating adjustable supports for the gluteal and floor areas, users can achieve a customized fit that enhances their overall experience. Ongoing research and development are essential to further refine the prototype, focusing on adding more ergonomic features and materials to boost user support and comfort.

31.4 Conclusions

The stress analysis of the initial model revealed that a maximum stress of 232 MPa, close to the yield strength of Aluminum 6061 T6, was present. This resulted in a safety factor of 1.18, indicating that the prototype met the safety specifications established by Eng. Ricardo González concerning weight, material, and geometry. However, the mechanism's usability range in terms of weight and height was expanded, and efforts were made to reduce its weight. The initial prototype, weighing 4 kg, was found to be cumbersome for users despite fulfilling its intended purpose.

A new design, featuring updated geometry, was proposed. Although constructed from the same material, the design utilized a thinner profile, resulting in a weight reduction of 1.23 kg—a 30.75% decrease compared to the initial prototype. This redesign achieved a maximum stress of 88 MPa and provided a safety factor of 3.12, representing an improvement over the initial prototype's factor of 1.18. The new design incorporated an extension mechanism for both the gluteal and floor supports, allowing for a 5 cm adjustment and accommodating a height range of 165–170 cm. Ongoing research is focused on enhancing the ergonomic aspects of the prototype, including material selection and the incorporation of additional features to improve user comfort. Additionally, efforts are being made to integrate lumbar support into the exoskeleton design.

Acknowledgements Gratitude is extended to the institutions that were essential to the development of this work:

- Instituto Politécnico Nacional (IPN)
- Comisión de Operación y Fomento de Actividades Académicas del IPN (COFAA)
- Consejo Nacional de Humanidades, Ciencia y Tecnologías (CONAHCyT).

References

Bogue R (2020) Exoskeletons: A review of current developments and applications. Ind Robot: Int J 47(6):673–683. https://doi.org/10.1108/IR-08-2020-0182

Dunai L, Lengua I, Peris-Fajarnés G, Defes-García B (2019) Design of a lower-limb exoskeleton. Universidad Politécnica de Valencia (España). Centro de Investigación de Tecnologías Gráficas

Edwards JD, Frary FC, Jeffries Z (1930) The aluminum industry, vol 2. McGraw-Hill, New York

ELIAS LA, et al (2014) Spinal mechanisms may provide a combination of intermittent and continuous control of human posture: predictions from a biologically based neuromusculoskeletal model. PLOS Comput Biol 10

Farina D, Holobar A (2017) Wearable technology for muscle activity monitoring and feedback. IEEE Trans Biomed Eng 64(8):1884–1891. https://doi.org/10.1109/TBME.2016.2624700

Gobierno. Secretaria del Trabajo y Previsión Social: Actualización Tabla de Enfermedades de Trabajo. gob.mx (www.gob.mx)

Goujon A, Gosselin C (2020) Design and optimization of exoskeletons: a review. Robot Auton Syst 125:103331. https://doi.org/10.1016/j.robot.2019.103331

Haddadin S, Albu-Schäffer A (2018) Robotics for rehabilitation: A review of exoskeletons and their applications. IEEE Trans Rob 34(6):1249–1266. https://doi.org/10.1109/TRO.2018.2875866

IMC, talla y peso IMSS admayor_imc.pdf (imss.gob.mx)

Centro de Investigación y Desarrollo en Ciencias de la Salud 09 Diciembre 2018. El 30% de las personas que padecen lumbalgia en México requieren incapacidad – CIDICS – UANL

Kumar S, Saini A (2019) Effectiveness of ergonomic interventions in reducing musculoskeletal disorders: a review. Int J Ind Ergon 71:102924. https://doi.org/10.1016/j.ergon.2019.102924

NTP 1162: Exoesqueletos I: Definición y clasificación. (insst.es)

Ries JD, Ross SA (2016) Materials and design considerations for lightweight exoskeletons. J Mech des 138(12):121404. https://doi.org/10.1115/1.4033766

De Rossi SMM, Vitiello N (2015) Wearable robots: a review. IEEE/RSJ Int Conf Intell Robot Syst 320–327. https://doi.org/10.1109/IROS.2015.7353323

Zhang X, Zhang Q (2021) Biomechanics of human posture and its effect on exoskeleton design. J Biomech 123:110474. https://doi.org/10.1016/j.jbiomech.2021.110474

Chapter 32
Light Curves Variability in High Mass X-Ray Binaries

Leticia Corral Bustamante

Abstract In this work, light curves were analyzed through plots of the light intensity of fifteen celestial objects called high mass X-ray binary systems, HMXRBs, as a function of time. The light curves were obtained from Period04 from the High Energy Astrophysics Science Archive Research Center of NASA's HEASARC, National Aeronautics and Space Administration Goddard Space Flight Center, Rossi X-ray Timing Explorer (RXTE), Guest Observer Facility—NASA and INTEGRAL Data Archives—ESA's Cosmos databases. HMXRBs are stellar binary systems composed by an O-B star and a compact object (neutron star or black hole), systems with a super-giant companion, systems with a main-sequence companion, weird systems, unveiled by Roche-lobe overflow, wind-fed accretion, circumstellar disk, supergiant fast X-Ray transients and homogeneous stellar wind. The objective was to analyze the variability of light curves of X-ray sources, and thus, to provide relevant information on the temporal behavior of these astrophysical objects through their physical and/or geometrical parameters in order to understand the processes experienced by these binaries. From the results obtained, it was found that the type of variability of HMXRBs in some of the sources is due to orbital motion, neutron star rotation and periodic motions of material. Periodic and aperiodic light curves were found and their study allowed us to provide information on the physical processes that produce them and obtain results similar to those of other authors who have studied this type of binary using other techniques, such as variability that can give rise to the system's superorbital period, as well as a possible linear relationship. The command line tool called Period04 was used to perform Fourier temporal analysis obtaining the periodogram of each source. The calculation of the periods and frequencies was executed with the corresponding error of both subdividing the light curve into time intervals or without subdividing according to its appearance, if there are bursts, etc. and with this, the variability of the sources was obtained. The contribution of this work to the area of high-energy astrophysics focuses on giving information on phenomena that produce variability in X-rays through the qualitative analysis of light curves of the sources that involve its physical processes.

L. C. Bustamante (✉)
Tecnológico Nacional de México, Instituto Tecnológico de Cd. Cuauhtémoc, Chihuahua, México
e-mail: rcorral@itcdcuauhtemoc.edu.mx

A. Öchsner and H. Altenbach (eds.), *Engineering Design Applications VII*, Advanced Structured Materials 230, https://doi.org/10.1007/978-3-031-84346-4_32

Keywords Light curves · Variability · HMXRBs · BeXRBs · Sources · Frequency · Period · Amplitude · Phase diagram · Fourier transform · Fourier temporal analysis · Periodogram · High energy astrophysics

32.1 Introduction

The temporal analysis of a light curve of an X-ray source involves the analysis of the number of photons detected in a given time interval. With this, we can measure the X-ray brightness of a source. The variability presented by the source involves physical phenomena that occur in the universe, among which (a) the orbital movements, (b) the rotation of neutron stars and (c) the change in the physical conditions of the emitting material stand out. The first produces variability when the X-ray conditions produced change along the orbital movement; the second involves the pulsations in X-rays caused by the material falling on the magnetic poles of the neutron star as the magnetic axis is not aligned with the axis of rotation of the star and, the third, is referred to the changes undergone by the expanding material expelled by a supernova that, when cooled, causes the amount of energy emitted in the X-ray range to decrease.

In this work, fifteen sources were studied, whose data have been obtained from the ASM (RXTE Guest Observer Facility 2017; ASM Light Curves Overview 2023), NASA's HEASARC High Energy Astrophysics Science Archive Research Center (Aeronautics et al. 2023) and the INTEGRAL Data Archives (Integral Data Archives 2023) sources catalogue. The tool Period04 (Lenz and Breger 2005) was used and, to know the characteristics of the sources under study, Simbad (Astronomical and Database—CDS (Strasbourg). 2023, 2017) was consulted.

HMXRBs are bright X-ray sources whose high-energy emission is produced by the accretion of matter (from either the strong stellar wind in binaries with super-giant companions or the circumstellar disk in Be/X-ray binaries) from the massive companion onto a neutron star.

It is known that one of the sources we analyzed has a period of 90 min. From the results obtained and described in Table 32.1, it can be seen that the source referred to by Blay et al. (2006), corresponds to source 4u2206 + 54, for which Fourier analysis was found for the adjusted frequency, which has a period of 92.6901 min.

The temporal behavior of the sources studied coincides with that reported by some authors (Blay et al. 2006; Farrell et al. 2008; Naik et al. 2013; Laplace et al. 2017; Reig and Coe 1998; Stella et al. 1986; Okazaki and Negueruela 2001; Coe et al. 1988; Wilson et al. 2002, 2008; Corbet and Levine 2006; Klochkov et al. 2007; Baykal et al. 2008, 2005; Kühnel et al. 2012; Ge et al. 2020; he0309–2057 2024; Townsend and Charles 2020; Ackermann et al. 2013; Chernyakova et al. 2012; Stanek et al. 2017; Nesci 2017; Liu et al. 2023; Fürst et al. 2014; Reig et al. 2016; Rangelov et al. 2016; Camero et al. 2007; Analysis and for SMC X-1 2023; Paradijs and Kuiper 1984; Vela X-1 History from 2024; Grinberg et al. 2017; Kretschmar et al. 2019, 2021; Pacciani et al. 1917; Dugair et al. 2013; Archer et al. 2016; Müller et al. 2013; x0115 2023;

Table 32.1 Fourier analysis with results of highlighted frequencies of the studied sources that are represented in the periodogram of Period04 in Figs. 32.1, 32.2, 32.3, 32.4, 32.5, 32.6, 32.7, 32.8, 32.9, 32.10, 32.11, 32.12, 32.13, 32.14, 32.15

Name of the high energy source	Frequency ± Error $f \pm E_f$ (days^{-1})	Period ± Error $P \pm E_p$ (days)	Fig. number/Light curve time period (days)
4u2206 + 54	17.98e-05 ± 1.95e-07 s^{-1}	5561.25 ± 6.02 s (92.69 ± 0.10 min)	1/0–220,000 s
x2206 + 543	54.87e-05 ± 7.77e-06	1822.48 ± 25.82	2/800–6200
	45.56e-05 ± 1.27e-05	2195.16 ± 61.26	2/700–4300
	9e-03 ± 5.6e-4	110.70 ± 6.88	2/6200–6575
exo2030 + 375	21.70e-03 ± 4.81e-06	46.09 ± 0.01	3/700–3000
	21.84e-03 ± 5.22e-06	45.78 ± 0.01	3/5000–6000
groj1008-57	40.50e-03 ± 7.39e-06	247.02 ± 0.45	4/800–4000
	40.50e-03 ± 7.39e-06	247.02 ± 0.45	4/800–6000
	8e-03 ± 11.54e-06	124.81 ± 17.98e-02	4/4000–6000
he0309-2057	30.97e-04 ± 2.10e-05	322.94 ± 2.19	5/700–5800
ks1947 + 300	23.00e-03 ± 7.88e-05	43.48 ± 0.15	6/800–2500
	25.90e-05 ± 1.92e-06	3860.90 ± 28.65	6/800–6500
psr j0437 – 4715	91.89e-03 ± 2.59e-05 s^{-1}	10.88 ± 0.0031 s	7/700–6500
saxj2103.5 + 4545	28.83e-05 ± 5.18e-06	3468.49 ± 62.32	8/700–5800
smcx1	17.92e-03 ± 1.96e-06	55.82 ± 0.0061	9/700–6250
velax1	12.74e-05 ± 5.83e-06	7849.592 ± 359.4	10/700–6400
x0114 + 650	86.23e-03 ± 7.04e-06	11.60 ± 0.00095	11/700–6250
	78.3e-03 ± 4.2e-04	12.76 ± 0.07	11/ 6250–6600
x0115 + 634	54.51e-04 ± 1.85e-06	183.44 ± 0.06	12/700–6400
	95.83e-05 ± 4.35e-05	1043.54 ± 47.41	12/700–1750

(continued)

Table 32.1 (continued)

Name of the high energy source	Frequency ± Error $f \pm E_f$ (days^{-1})	Period ± Error $P \pm E_p$ (days)	Fig. number/Light curve time period (days)
xper	2.92e-05 ± 1.68e-07	34,248.42 ± 197.46	13/5000–6000
	17.47 e-05 ± 1.52e-06	5722.93 ± 49.89	13/0–5500
xtej1739-302	63.25e-03 ± 2.22e-05	15.81 ± 0.0056	14/700–6200
xper	103.30 ± 1.82e-05 s^{-1}	0.0097 ± 1.71e-09 s	15/less than 54,000–54,800 s

xper 2017; Romano et al. 1961; Grundstrom et al. 2007) who used different methods for their study.

The general appearance of the light curve was observed before the calculation of the periodogram, the possible nature of the period of the highlighted frequencies that results from subdividing or not the light curve into time intervals and/or on the entire curve and Fourier analysis were examined.

The HMXRBs have a very rich phenomenology, presenting scenarios in which high-energy emissions occur and which are also useful for optics, thus accentuating the need to use all possible ranges of the spectrum to be able to know in more detail what is happening in the system. These are the few systems in which the mass of the compact object can be determined by being two objects in orbit, a massive star that through stellar models can predict an accurate value of its mass. For compact objects such as black holes and neutron stars of this type of binaries, by studying their motions with Kepler's laws, the relationship between the masses of the massive star and the compact object can be obtained.

HMXRBs are compact objects that have a massive optical companion, that is, more than eight times the mass of the sun, while LMXRBs are of low mass of the solar type or with masses less than that of the Sun. X-ray emission occurs when there is an exchange of mass from the optical companion to the compact object as a common scenario, the exchange changes from one system to another according to the associated phenomenology and the type of processes that can be obtained from them through the sources. We study systems with optical companions of the O and O-B type, in which intense stellar winds occur and even accretion discs are produced, discs around the optical star that flow outwards and somehow the stellar wind is driven towards the equator of the star and there it accumulates forming a disc-shaped structure.

The three basic scenarios that may correspond to the sources whose data are used in this work, or combinations of them, refer to interaction systems with compact object that we can find in X-ray binary systems with massive companion: (a) by overflow of the Roche lobe. It may be that the optical companion is so large that it fills the Roche lobe that is its gravitational zone of action. The Roche lobes of the two objects meet at the Lagrange point. If the massive star is so large that it fills the Roche lobe at the Lagrange point, an overflow of material falling towards the compact object may occur. (b) By stellar wind. It may also be that the compact object is moving in the area of dense stellar wind, therefore, it is capturing material from that wind and in that way accretion occurs and, (c) per circumstellar disk. Combinations of cases of HMXRBs are: by stellar wind, for example, that the compact star has a more elliptical orbit and that it enters and leaves the dense stellar wind zone, and, by circumstellar disk. The compact star may be far away and not come very close.

In the case of X-rays with a low-mass companion LMXRBs, accretion almost always occurs due to overflow of the Roche lobe. The variability when there are large extensions of material is, for example, the precession of a disk around this type of system that includes a star that forms an accretion disk. That disk is not static, it is moving and probably processing, pulling due to the gravitational interaction of the two objects. This precession can produce observable variability, they can produce superorbital periods because they are longer than the orbital period and are clearly distinguishable (Townsend and Charles 2020; Ackermann et al. 2013; Chernyakova et al. 2012), such a process occurs in the source ks1947 + 300 analyzed in this work (Townsend and Charles 2020). Farrell et al. (2008), conducted a study showing that one of the sources studied here, x0114 + 650, has a superorbital period. The results obtained show that, for most of the X-ray sources analyzed, they present variability over a large range of time scales, in the shortest pulsation periods and in the longest orbital periods in days.

The objective of analyzing light curves from X-ray sources using Period04 (Lenz and Breger 2005), the time series analysis tool, is met.

32.2 Methodology

The sequence used in Period04 (Lenz and Breger 2005) to perform the Fourier analysis on the sources taking the criterion of division of time intervals of the light curve or on the complete curve without taking into account peak of the bursts, includes: import time string (mag) → import data → detection of time and observed value → observation of the light curve → subdivision by gaps (special → subdivide time string) or subdivide by fixed intervals.

Use of the Fourier transform to obtain the periodogram. Database for all the sources on Table 32.1, except 4u2206 + 54, psr j0437 – 4715 and XPer, we indicate the maximum frequency to facilitate the calculation. Whereas, for the database of this exception of binaries, we use the Nyquist frequency, which is the maximum frequency of data analysis, which are shown in Table 32.1.

Step rate indicates the accuracy to be used and returns a frequency with a certain amplitude.

Display graph. Observation of the periodogram graph to be analyzed.

Improve the setting by selecting the frequency (fit → improve all) to draw a phase diagram.

Phase plot indicates the frequency that the software calculated. data → observed (time string plot).

In Table 32.1, the measurement of periods is in seconds (s) and frequencies in Hertz (Hz) of the binaries 4u2206 + 54, psr j0437 – 4715 and XPer, while, in all other binaries, measurement of periods is in days and frequencies in days^{-1}.

The frequency is that provided by the periodogram of Period04 as a function of the amplitude in magnitudes, mag. For a given frequency, f, the corresponding period is calculated as

$$P = \frac{1}{f}. \tag{32.1}$$

Calculation of the period error, E_P, obtaining the significance of the detection of the peak at the frequency given by the relationship

$$E_P = \frac{E_f}{f^2}. \tag{32.2}$$

where E_f is the error in frequency. The final values to be provided are the period and its error, $P \pm E_P$, and the frequency and its error, $f + E_f$.

In summary, the methodology includes:

- Load the light curve.
- Observe the appearance of the light curve.
- Execute the Fourier transform.
- See if there is a peak.
- If that peak has a certain frequency, we take it as the valid periodicity.
- Calculate the error.
- See if a good phase diagram comes out or not, etc.

32.3 Results and Discussion

This section presents and discusses the results of the work carried out by studying the light curves of fifteen X-ray sources with Period04 (Lenz and Breger 2005). In Simbad (Astronomical and Database—CDS (Strasbourg) 2023; Astronomical and Database—CDS (Strasbourg) 2017) the possible scenario of each source was consulted.

The general appearance of the sources, the nature of the period and the subdivision of the light curve over periods of time or without subdivision, the Fourier analysis of the light curves and phase diagrams with binning 10 can be seen in Figs. 32.1, 32.2, 32.3, 32.4, 32.5, 32.6, 32.7, 32.8, 32.9, 32.10, 32.11, 32.12, 32.13, 32.14, 32.15.

Period04 (Lenz and Breger 2005) was used to perform temporal analysis by Fourier analysis for a light curve subdividing it into time intervals or not subdivided, according to its appearance, whether there are bursts, etc. taking the criterion of not

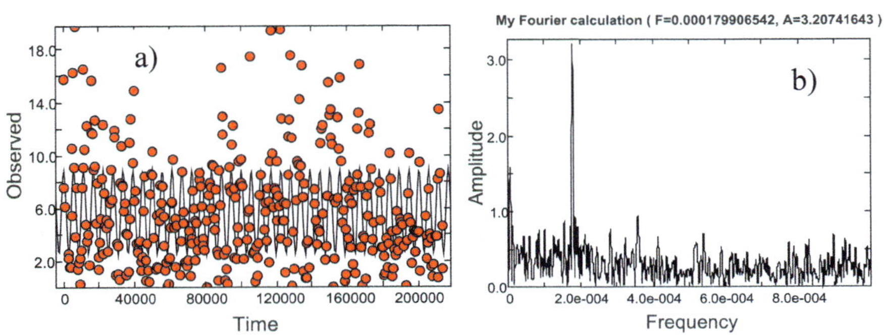

Fig. 32.1 Source 4u2206 + 54 in Period04 (Lenz and Breger 2005): **a** light curve, **b** amplitude as a function of frequency of the Fourier calculation. Note a peak at the frequency of 1.7991e–4 Hz

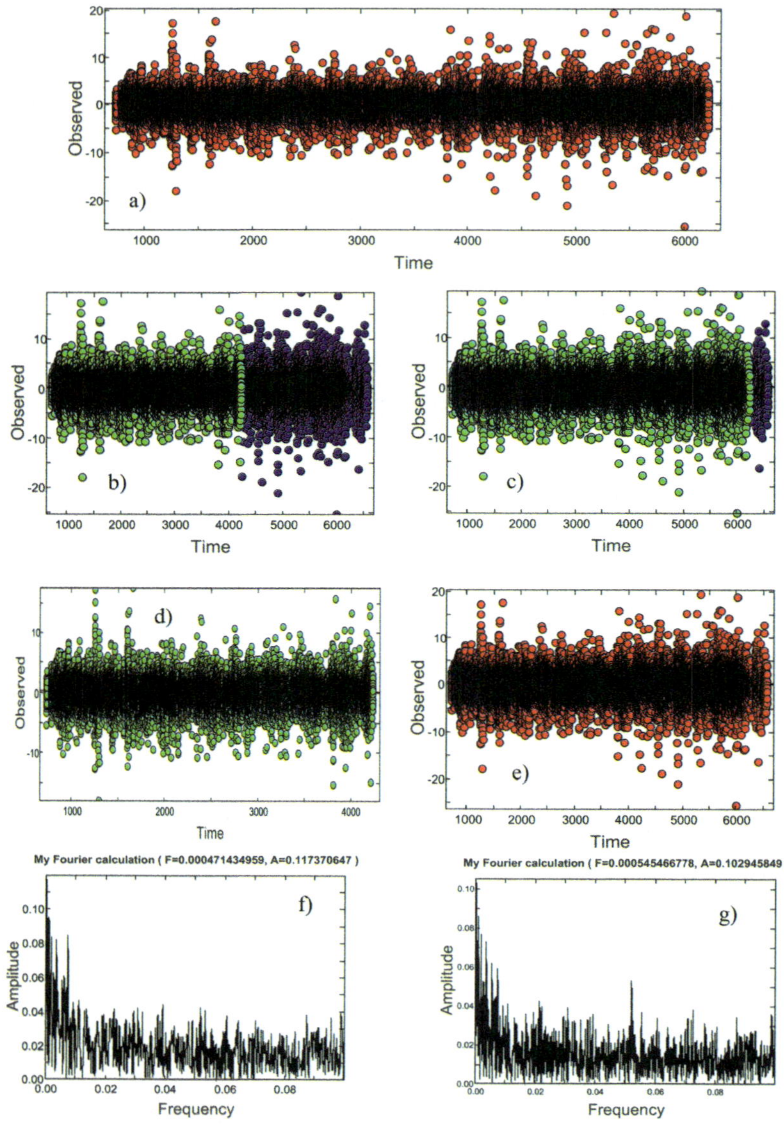

Fig. 32.2 Source x2206 + 543 in Period04 (Lenz and Breger 2005): **a** complete light curve, **b** light curve divided over two period of time, the first of 700 to 4300 days, **c** light curve divided over two period of time: 800–6200 and 6200–6575 days to be analyzed, **d** light curve of 700–4300 days, **e** light curve of 800–6200 days, **f** periodogram of the curve of (**d**), **g** periodogram of the light curve of (**e**). Continuation of source x2206 + 543 in Period04 (Lenz and Breger 2005): **h** phase diagram of (**e**) using a period of 2195 days (see Table 32.1), **i** division over a period of time of 6200–6575 of the light curve to be analyzed, **j** Fourier calculation graph for the frequency of the peak of 0.0083 days^{-1} and amplitude of 0.4891 mag—periodogram of the light curve of (**i**), and **k** phase diagram with binning 10 for the data of (**i**) using a period of 110.70 days (see Table 32.1)

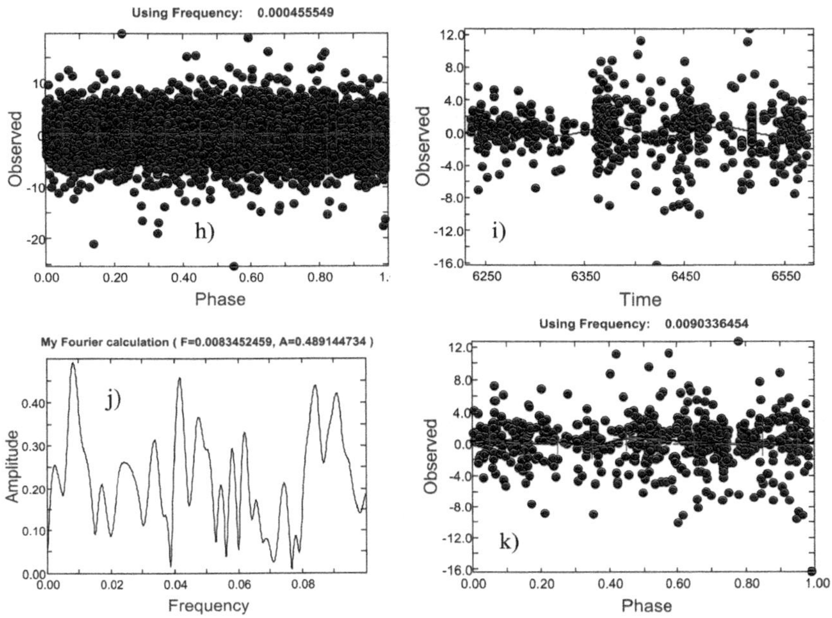

Fig. 32.2 (continued)

analyzing burst peaks that occur once or aperiodically, but those time intervals with an approximately periodic behavior.

Table 32.1 presents the calculations made in Period04 (Lenz and Breger 2005) of frequency and period with their corresponding errors for light curves divided over periods of time or completes.

Due to their appearance, the light curves of sources 4u2206 + 54, psr j0437 – 4715 and XPer were not divided into time intervals (Figs. 32.1, 32.7 and 32.15, respectively). Next, the results obtained for each source are presented.

4u2206 + 54

This is an object of spectral type HMXRB, O9.5Vep, subtype 9.5 from 0 to 9, where zero denotes the hottest star of the class and V designates the coolest star of the luminosity class (Astronomical and Database—CDS (Strasbourg) 2023; Astronomical and Database—CDS (Strasbourg) 2017).

It is noted that, in the case of the study of the counterpart of this source of Blay et al. (2006), the star BD + 53 2790, that there is no clear pattern of variability of the He lines (see Fig. 32.1a), therefore, the light curve was not subdivided.

The periodogram exhibits a major peak of greater than 3.0 mag amplitude for a frequency of 1.799e-04 s^{-1} (Fig. 32.1b). The period obtained for the light curve for this source is 5561.25 $\pm$ 6.02 s = 92.69 $\pm$ 0.10 min, which is in accordance with Blay et al. (2006).

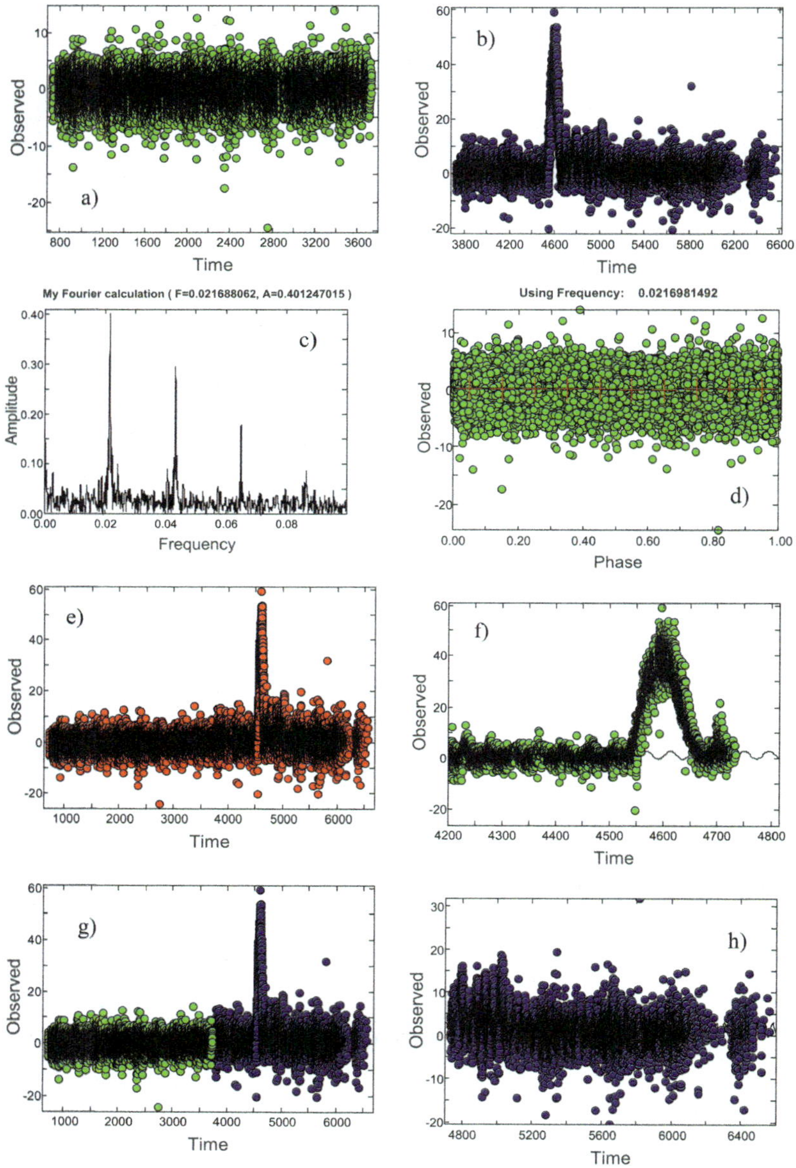

Fig. 32.3 Light curve of exo2030 + 375: **a** divided over 700–3000 days, **b** light curve in time interval of $3800 \leq t \leq 6600$ days, **c** graph of the Fourier calculation of the frequency of the peak of 0.0217 Hz, **d** phase diagram of (**a**) using a period of 46.09 days, **e** complete light curve, **f** peak of the light curve, **g** light curve divided over two period of time, **h** 5000–6500 days of the curve to be analyzed. Continuation of the source exo2030 + 375 in Period04 (Lenz and Breger 2005): **i** phase diagram of **h** using a period of 45.78 days and **j** periodogram of the Fourier calculation of the frequency 0.0218 Hz

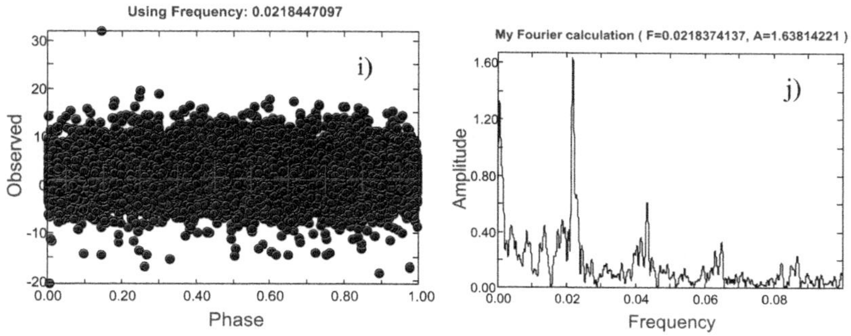

Fig. 32.3 (continued)

x2206 + 543

From the Fourier analysis shown in Fig. 32.2, dividing the light curve into two intervals of 800–6200 days and 6200–6575 days as in Fig. 32.2c, it can be seen in Table 32.1 that there is variation in the frequency and the frequency error increases in the interval at the right end of the curve that is separated from that curve from 54.87e-05 $\pm$ 7.77e-06 days^{-1} until 9e-03 $\pm$ 5.6e-4 days^{-1}, for the above-mentioned light curve time intervals, respectively.

The period of the two intervals differ by 1712 $\pm$ 18.94 days, so the light curve has periodic variations. The frequencies in Fig. 32.2f and g indicate the variability before, after and during the prominent emission peaks in Fig. 32.2d and e, respectively.

By its appearance, we could compare Fig. 32.2 with Fig. 32.11. By analyzing the two time intervals that exhibit the light curve of x0114 + 650 in Fig. 32.11, it can be seen in Table 32.1 that the frequency does not change dramatically (7.88e-03 days^{-1}), although the error of the frequency calculated by Period04 (Lenz and Breger 2005) increases from the first to the second interval, from 7.04e-06 to 4.2e-04 days^{-1}. While the period suffers a variation of 1.1673 days and the period error increases by 6.7e-02 days.

From the analysis of x2206 + 543 (Fig. 32.2) and x0114 + 650 (Fig. 32.11), it is evident that there is no pattern of periodic variability. The latter source exhibits superorbital period according to Farrell et al. (2008). It is also observed that the frequency errors of x0114 + 650 differ in inverse relation to the frequency, which is not the case for x2206 + 543. Besides, it can be seen that the frequency error is larger for the second interval of both sources, which is isolated from the rest of their respective curves.

For x0114 + 650 the period between intervals does not change much, but the period error increases in the second interval of the curve. For the source x2206 + 543 the period is much larger in the first time interval and so is the error (Fig. 32.2). From the Fourier analysis of Figs. 32.2 and 32.11, it is concluded that each source is unique and that the fact that it has a small isolated part at the right end of its light curve does not mean that it exhibits a similar behavior in a different source.

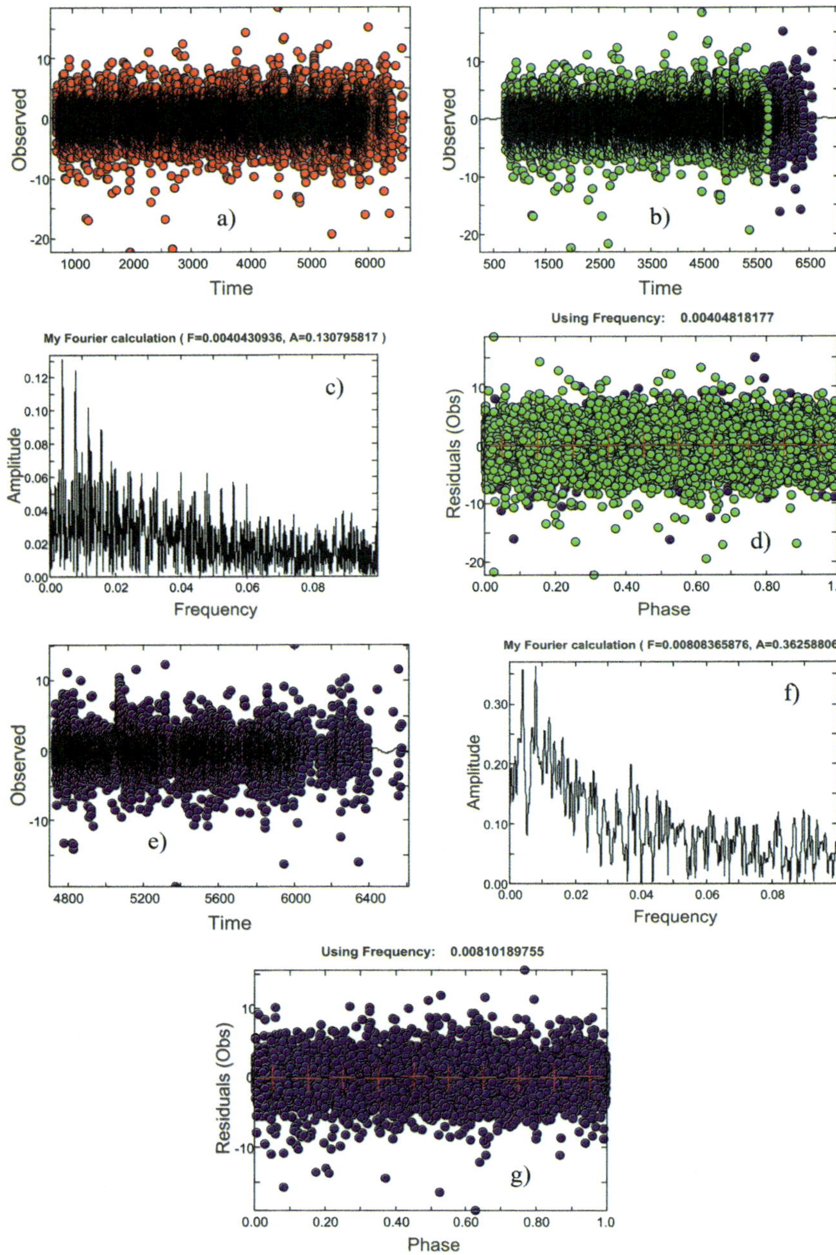

Fig. 32.4 Source groj1008-57: **a** full light curve showing a more prominent peak, **b** light curve divided over two period of time, **c** Fourier transform calculation graph for peak frequency and amplitude or periodogram, **d** 734.29–6571.41 days' phase diagram of (**b**) using a period of 247.02 days. **e** curve to be analyzed, **f** graph after Fourier transform calculation for specified peak frequency and amplitude, periodogram, and **g** phase diagram of (**e**) using a period of 124.81 days

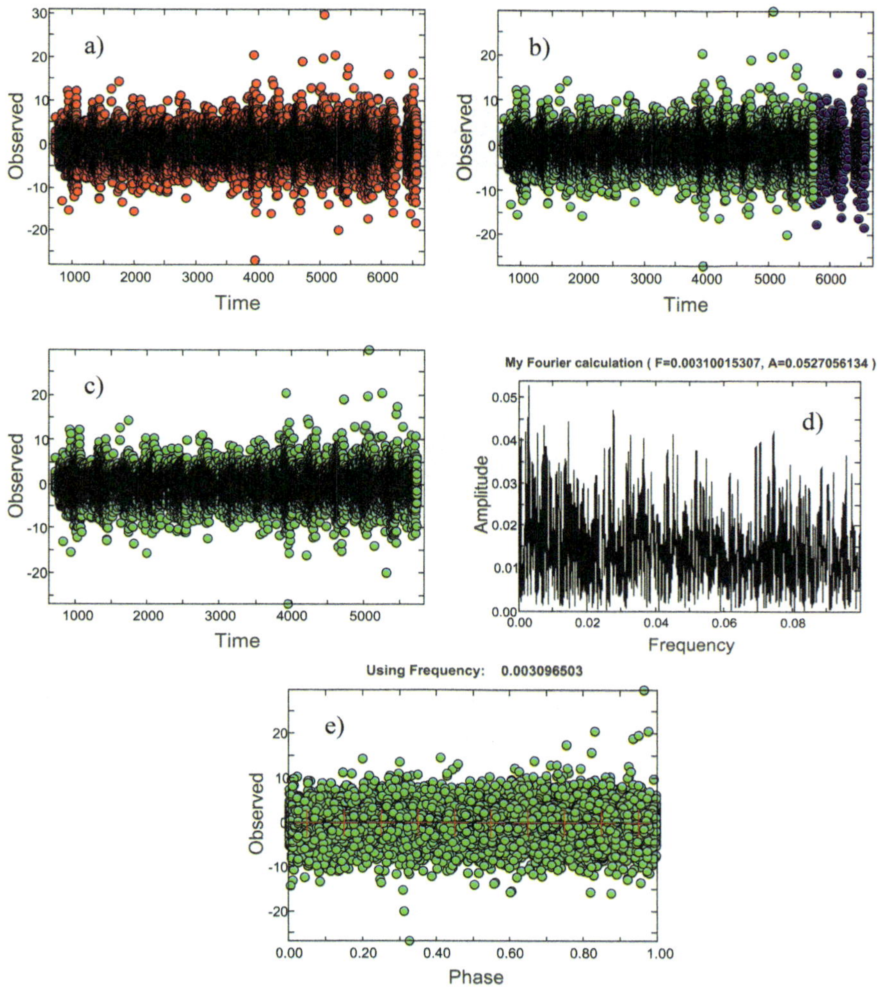

Fig. 32.5 he309-2057 complete in Period04 (Lenz and Breger 2005): **a** light curve, **b** light curve divided into two periods of time, **c** curve to be analyzed, **d** periodogram. Graph after Fourier transform calculation for the peak frequency of 0.0031 Hz and amplitude of 0.0527 mag, and **e** phase diagram with binning 10 of (**c**) using a period of 322.94 days

The general appearance of a light curve by visualization, prior to Fourier analysis, may provide a different perception of periodic variation and emission epochs.

For the source in Fig. 32.2, Table 32.1 shows frequencies adjusted in days^{-1} of 0.00055, 0.00046 and 0.00903 (which was adjusted to 7.8e-2 to obtain the periodogram), which correspond to periods in days of 1822.4792, 2195.1559 and 110.70 for the light curve intervals on days of 800–6200, 700–4300 and 6200–6575 of Fig. 32.2e, d and i, and the periodograms of Fig. 32.2g, f and j, respectively. There

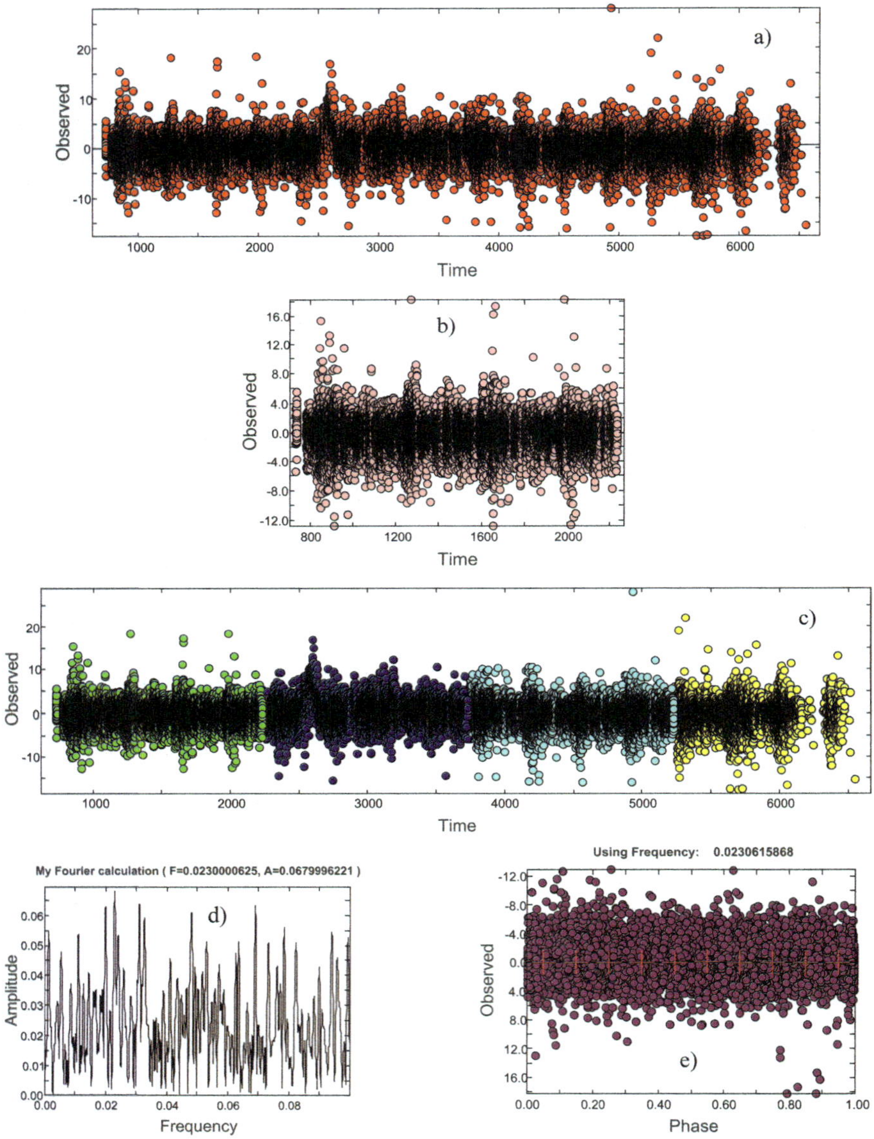

Fig. 32.6 Period04 sequence (Lenz and Breger 2005) for the time interval of ks1947 + 300 with possible superorbital period: **a** Appearance of the light curve showing a prominent peak at approximately 2600 days, **b** light curve divided over a period of time of **c** to be analyzed, **d** Periodogram. Graph after Fourier transform calculation for the peak frequency of 0.0230 Hz and amplitude of 0.0680 mag, and **e** phase diagram with binning 10 of (**b**) using 43.48 days. Continuation. Sequence Period04 (Lenz and Breger 2005) for ks1947 + 300. **f** Aspect of the light curve in an interval of time, **g** orbital phase from 0.0 to 0.2, **h** orbital phase from 0.20 to 0.70, **i** orbital phase from 0.75 to 0.89, **j** periodogram. Graph after Fourier transform calculation for the peak frequency of 0.0004335 Hz and amplitude of 0.2454 mag and **k** phase diagram of (**a**) with binning 10 using a period of 3860.90 days with a frequency of 0.000259007236 Hz

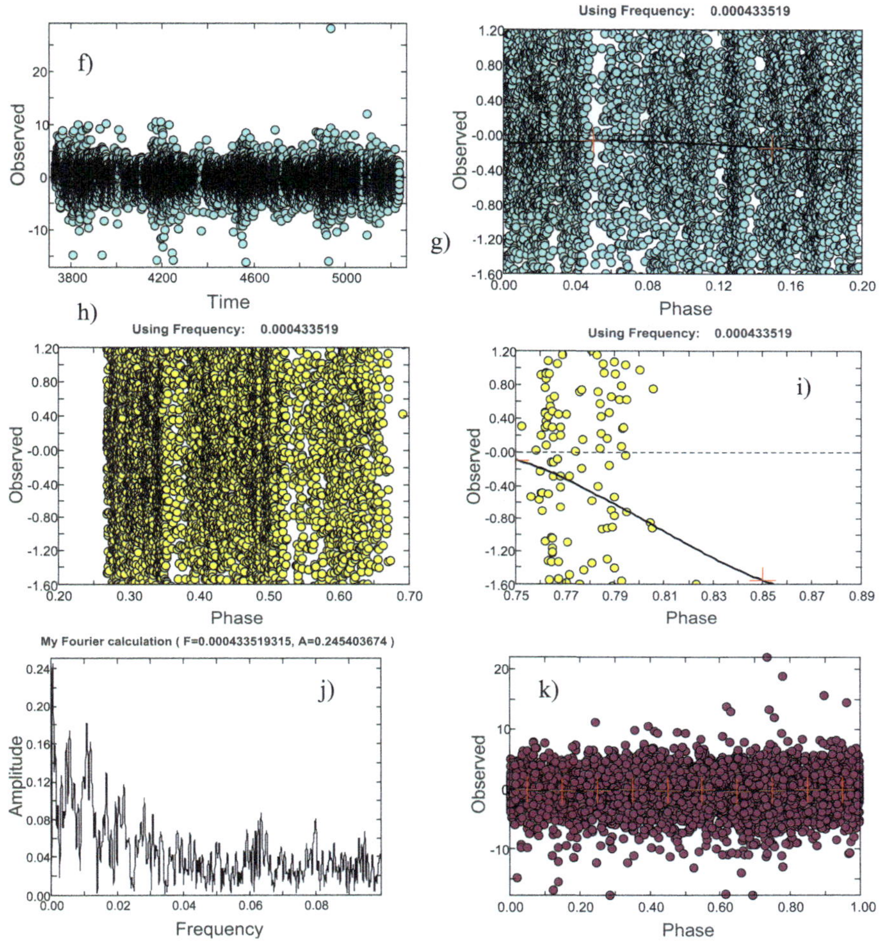

Fig. 32.6 (continued)

appears to be a repeating pattern probably due to rotational or orbital motions. The frequencies show periodic variability before the prominent emission peaks (see Fig. 32.2a). The light curves of Fig. 32.2b and c were subdivided in two different time intervals.

If we compare the 800–6200-day interval in Fig. 32.2e with respect to the 700–4300-day interval in Fig. 32.2d, we can see different periodograms for both in Fig. 32.2g and f, respectively, indicative of periodic variability in the light curve. The periods in Table 32.1 are: 1822.48 ± 25.82 and 2195.16 days, respectively, and the period of 110.70 days is from the interval of days 6200–6575 of Fig. 32.2i (see Table 32.1).

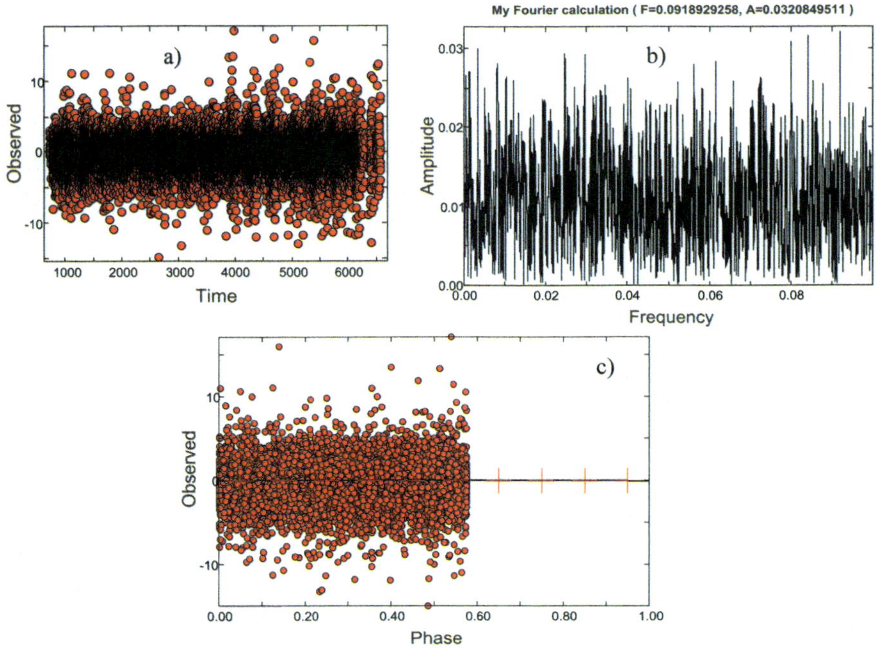

Fig. 32.7 Sequence Period04 (Lenz and Breger 2005) for the entire time interval of psrj0437 – 4715: **a** Appearance of the light curve, **b** periodogram. Graph after Fourier transform calculation for the peak frequency of 0.0919 Hz and amplitude of 0.0321 mag and **c** phase diagram of (**a**) using a period of 10.88 days

Two intervals are presented for two periodograms. For the interval in Fig. 32.2e the peak has an amplitude greater than 0.1 mag at a frequency of 5.45e-04 days^{-1}, while in the interval of Fig. 32.2d we can see a peak of almost 0.12 mag for a frequency of 4.71e-04 days^{-1} (Fig. 32.2g and f, respectively). The subdivisions of the curve into two intervals, twice, Fig. 32.2b and c that give rise to an analysis of Fig. 32.2d and e, respectively, the periods of 700–4300 days and 800–6200 days, give two different frequencies of the order of 0.00055 and 0.00046 days^{-1}.

exo2030 + 375

This binary is an object type HMXRB galactic variable star, spectral type B0Ve (Astronomical and Database—CDS (Strasbourg) 2023; Astronomical and Database—CDS (Strasbourg) 2017) Fig. 32.3a–j.

This source presents a constant intensity, although it has a peak, a sudden very large emission with peaks that are apparently repeated periodically. The source presents two types of variations: an orbital variation that occurs in the area after the peak and the variation of this peak or bursts, explosions in which suddenly the flux increases dizzily and then return to its usual level. A burst is one of these variations that occurs due to changes in the accumulation of material around the system.

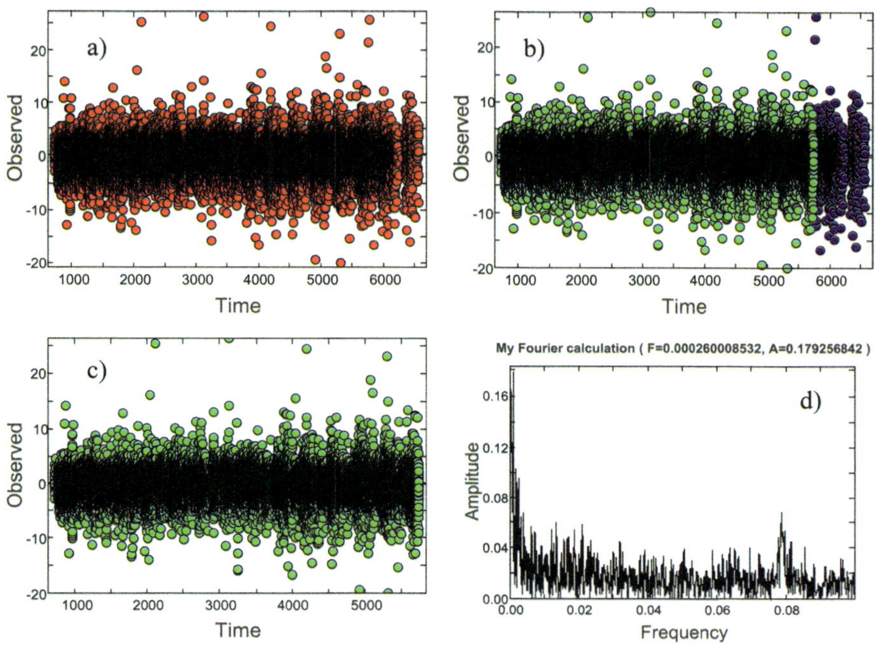

Fig. 32.8 Sequence Period04 (Lenz and Breger 2005) for the entire time interval of saxj2103.5 + 4545: **a** appearance of the light curve, **b** light curve subdivided over two period of time, **c** curve divided over a period of time, and **d** periodogram. Graph after Fourier transform calculation for 0.00026 Hz peak frequency and 0.1793 mag amplitude

In this case we have a Be star (it is a B-type star and the e indicates that it has emission lines that are produced in the circumstellar disk). It develops a disk around it which, like any disk, is usually unstable and the moment arrives when it falls apart and is ejected; at that moment, the amount of material available for accretion is enormous because the whole disk is ejected to the outside and a peak is produced. It recovers its normal course with time, as soon as this burst passes, the peaks of orbital variation are recovered.

The curve was subdivided into two time intervals (Fig. 32.3g) between 700–3000 days of Fig. 32.3a and 5000–6000 days in Fig. 32.3h to avoid prominent curve bursting. The intervals in Fig. 32.3g and h ($\approx$700–3000 days and $\approx$4800–6400 days, respectively) were analyzed.

Before and after the outburst of this light curve between 4550 and 4700 days (Fig. 32.3b, e, f and g), the curve reveals a repetitive pattern, probably due to rotational or orbital motions.

The periodogram for the 700–3000-day interval of Fig. 32.3g presents four peaks at approximately 0.00, 0.022, 0.044 and 0.065 days^{-1} of amplitudes of 0.16, 0.40, 0.30, 0.16 mag, respectively, which are shown in Fig. 32.3c. While the interval of days 5000–6000 in Fig. 32.3h with four peaks with the same frequency values as the

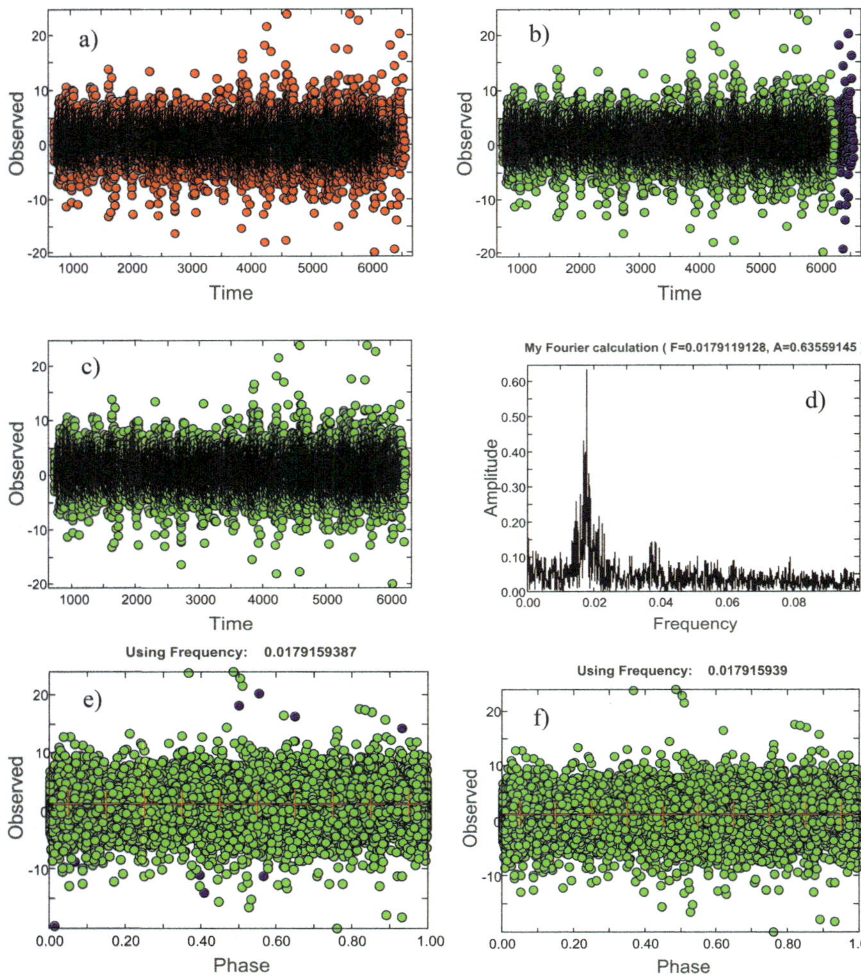

Fig. 32.9 Period04 sequence (Lenz and Breger 2005) for the entire time interval of smcx1: **a** appearance of the light curve, **b** light curve subdivided over two period of time, **c** period of time of the curve to be analyzed, **d** periodogram. Graph after Fourier transform calculation for the peak frequency of 0.0179 Hz and amplitude of 0.6356 mag showing the prominent peak, **e** phase diagram with binning 10 of (**b**) using a period of 55.82 days and **f** phase diagram with binning 10 of (**a**)

previous time interval, but with respective amplitudes of $\approx$1.36, 1.60, 0.60 y 0.30 (see Fig. 32.3j).

Thus, the peaks of both intervals are located at the same frequencies, but with different amplitudes, therefore, there are periods of more intense emission than others. The source shows a period of 46 days before and after the outburst. The respective periods and frequencies are presented in Table 32.1.

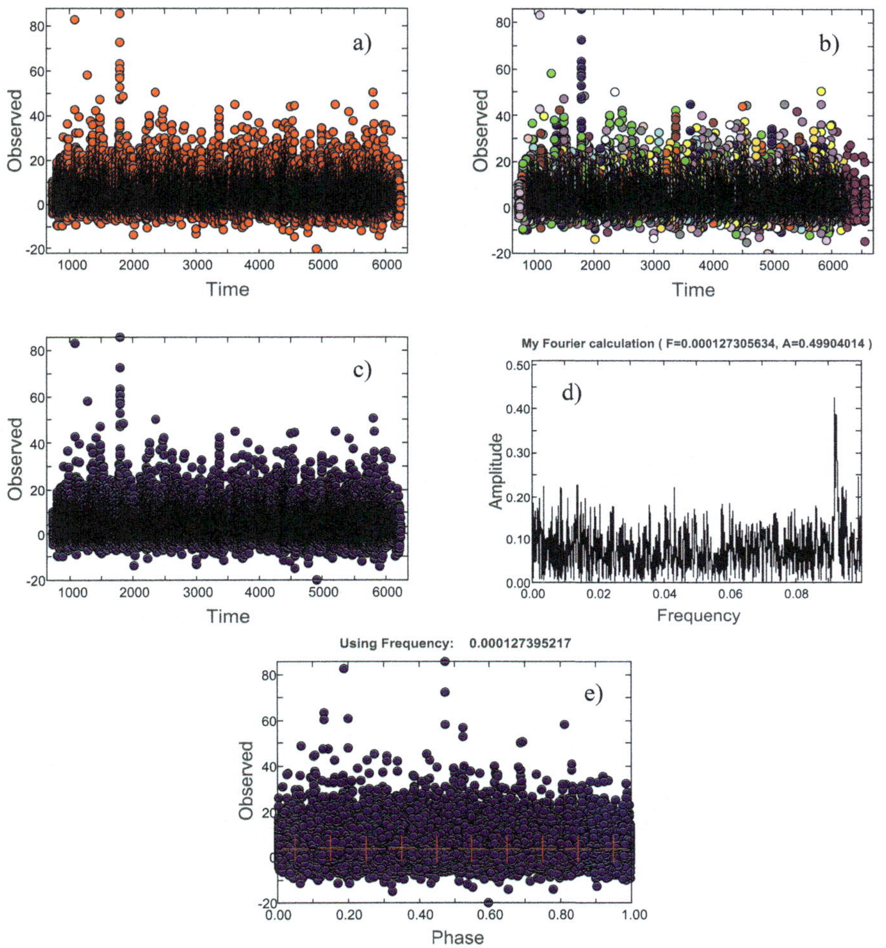

Fig. 32.10 Sequence Period04 (Lenz and Breger 2005) for the entire time interval of velax1: **a** aspect of the light curve, **b** light curve subdivided over two period of time, **c** period of time of the curve to be analyzed, **d** Periodogram. Graph after Fourier transform calculation for the peak frequency of 0.00013 Hz and amplitude of 0.4990 mag, **e** phase diagram with binning 10 of (**c**) using period of 7849.59 days

Naik et al. (2013), presented results from a study of pulsations with a period of 41.41 s and strong energy dependent pulse profiles, presence of prominent dips at several phases in the profiles which gradually disappeared and the profile appeared single-peaked. Several low energy emission lines are detected in the pulsar spectrum and the presence of dips up to higher energies.

Laplace et al. (2017) inform that this binary presents a last giant outburst in 2006 and regular X-ray outbursts (type I) were detected every periastron passage. Finally,

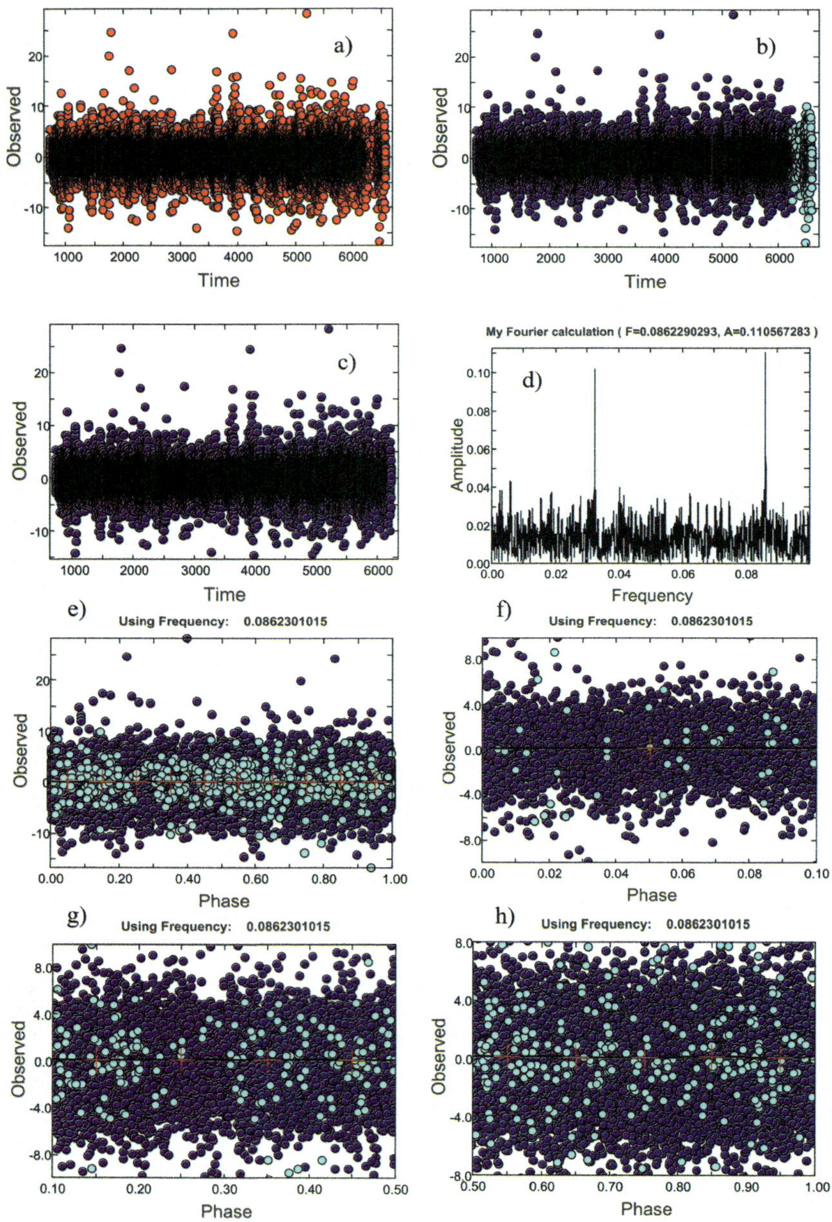

Fig. 32.11 Sequence Period04 (Lenz and Breger 2005) for the entire time interval of x0114 + 650: **a** full light curve, **b** light curve subdivided over two periods of time, **c** light curve to be analyzed, **d** periodogram. Fourier calculation plot for the peak frequency of 0.0862 Hz and amplitude of 0.1106 mag, **e**–**h** four views of the phase diagram of (**b**) with binning 10 to see the superorbital period

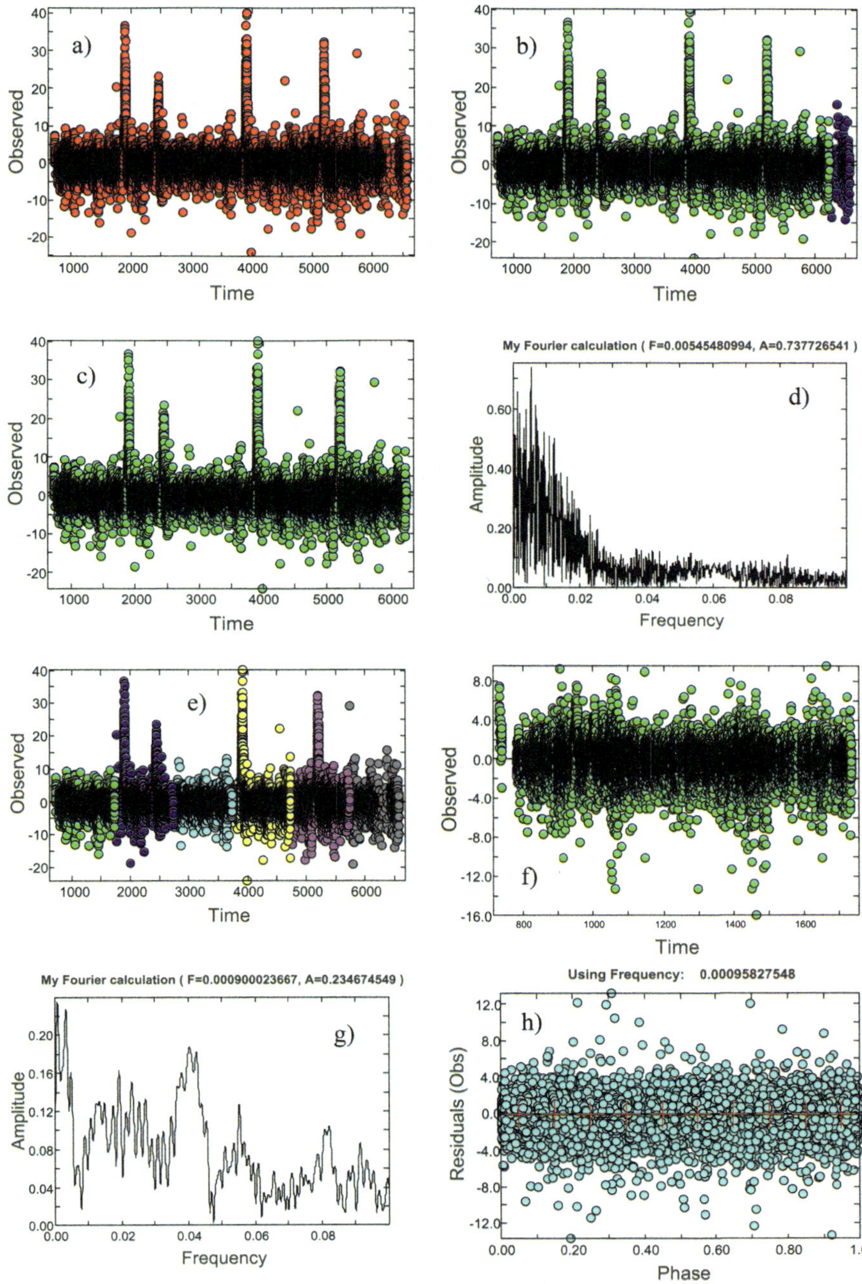

Fig. 32.12 x0115 + 634 source: **a** light curve with one peak at 3920 days, **b** light curve subdivided over two periods of time, **c** curve to be analyzed, **d** periodogram of (**c**), **e** light curve subdivided over six periods of time, **f** curve to be analyzed, **g** periodogram of (**f**). **h** phase diagram of (**f**) using a period of 1043.54 days

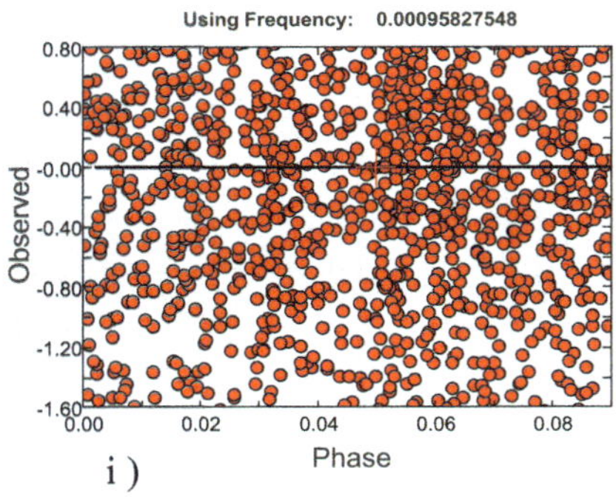

i)

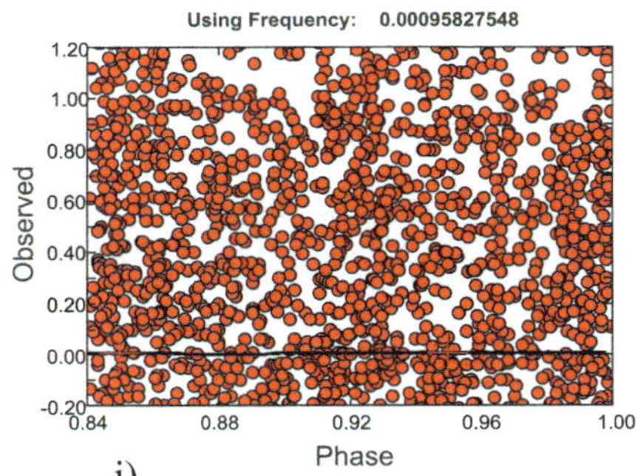

j)

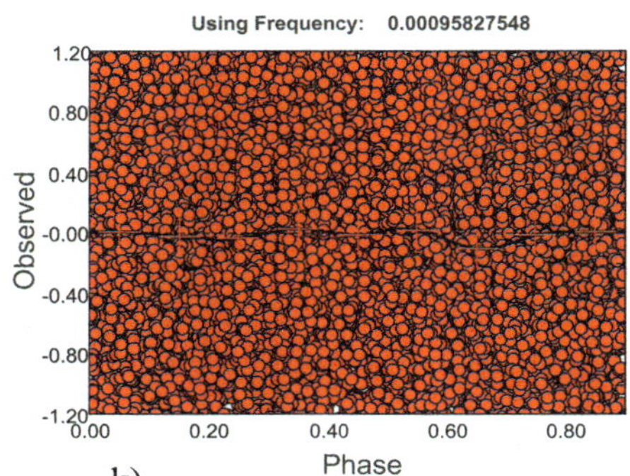

k)

Fig. 32.12 (continued)

the source started to show X-ray flux decreased significantly and type I outbursts were missed in several cases. Simultaneously, the spin frequency of the neutron star, which had been increasing steadily since the end of the 2006 giant outburst, reached a plateau.

Very recent observations indicate that the source is now starting to spin down. Just before 21 yr ago, the source displayed a sudden orbital phase shift of the outburst peak (1995) which occurred between the two giant outbursts (1985 and 2006). These phenomena suggest the system to have an underlying periodicity of 10.5 yr between orbital phase shifts and/or giant outbursts which may reflect some long-period dynamics in the circumstellar disk of the Be star.

Be X-ray binaries (BeXRBs) are binary star systems composed of a neutron star and a (O9-B2)-type star from which emission lines have been observed, known as a Be star. These stars to harbor large circumstellar disks, which provide the X-ray radiation (Reig and Coe). Most BeXRBs exhibit a characteristic transient X-ray radiation, so-called outbursts, during which the X-ray luminosity increases until it reaches a maximum and fades away again, on a timescale from a week to several months. Two main types of X-ray outbursts are commonly distinguished: type I, also known as normal outbursts which typically occur periodically at a period equal to the

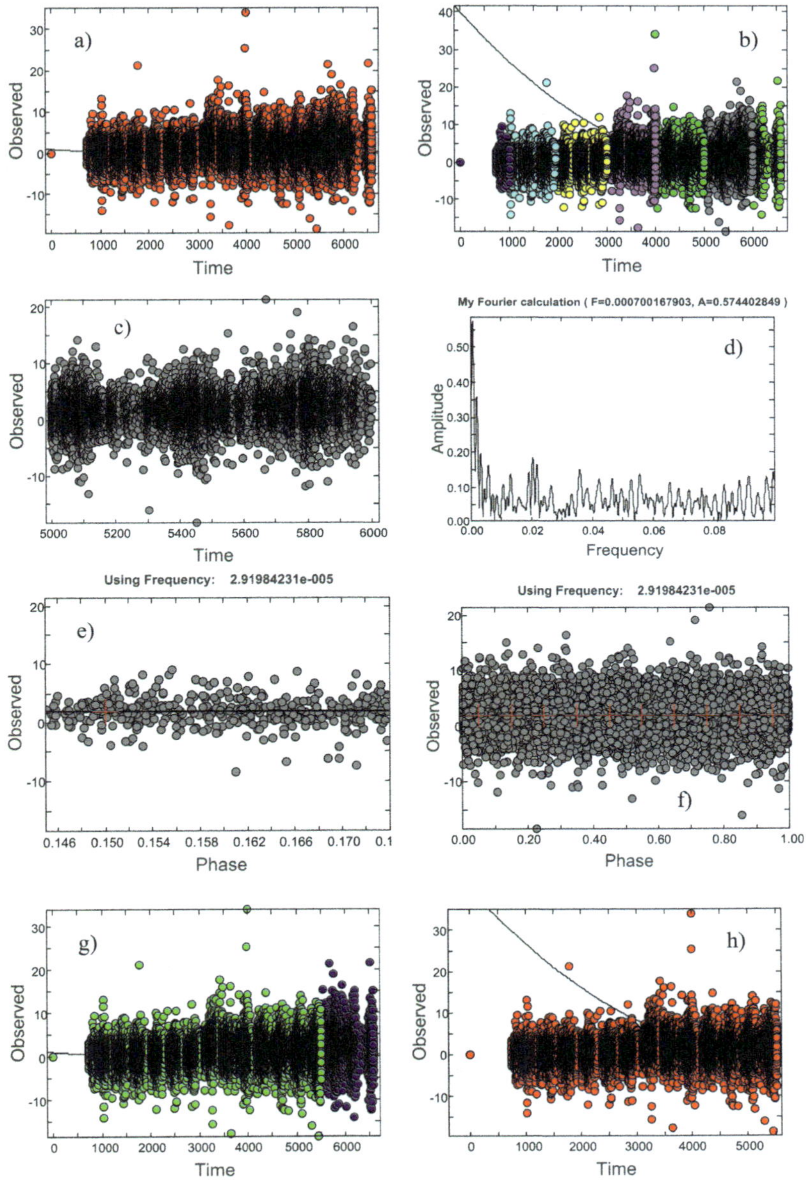

Fig. 32.13 Sequence Period04 (Lenz and Breger 2005) for xper: **a** appearance of the light curve, **b** light curve subdivided over seven periods of time, **c** period of time of the curve to be analyzed, **d** periodogram of (**c**), **e** phase diagram of (**c**) using, **f** phase diagram of (**c**) a period of 34,248.42 days, **g** light curve subdivided over two periods of time, **h** period of time of the curve to be analyzed. Continuation. xper source: **i** periodogram of (**h**). Fourier calculation graph for 0.00015 Hz peak frequency, **j** phase diagram of (**h**) a period of 5722.93 ± 49.89 days and **k** phase diagram of (**b**) with binning 10

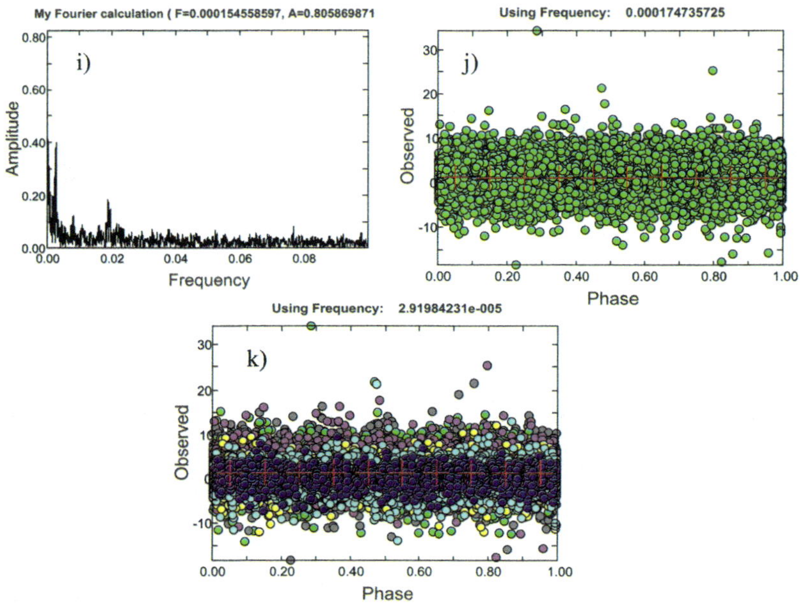

Fig. 32.13 (continued)

orbital period; and type II or giant outbursts. Giant outbursts are assumed to occur randomly (Stella et al. 1986; Okazaki and Negueruela 2001).

The origin of the X-ray radiation from type I outbursts is due by the interaction between the Be-star disk and the neutron star. If the neutron star approaches the Be star and its disk closely enough, accretion can occur, resulting in the emission of X-ray radiation.

EXO2030 + 375 is a BeXRB, composed of a magnetized neutron star and a B0Ve companion (Coe et al. (1988)). The source is characterized by an orbital period of 46.021 days (46.09 ± 0.01 y 45.78 ± 0.01 days measured in this work) and by being the BeXRB with the largest number of observed type I outbursts (over 150), occurring almost every periastron passage.

The source experienced an orbital phase jump of the type I outbursts in 1995 (Reig and Coe 1998; Wilson et al. 2002). A second giant outburst was observed in 2006 (Corbet and Levine 2006; Klochkov et al. 2007; Baykal et al. 2008; Wilson et al. 2008).

An orbital phase shift occurred in EXO2030 + 375 (Laplace et al. 2017). The outburst reached a peak at an orbital phase of 0.015.

groj1008-57

This source is an object type HMXRB, spectral type B0e, where B type 10,000–30,000 K and subtype 0 source denotes that it is the hottest star in the class (Astronomical and Database—CDS (Strasbourg) 2023; Astronomical and Database—CDS

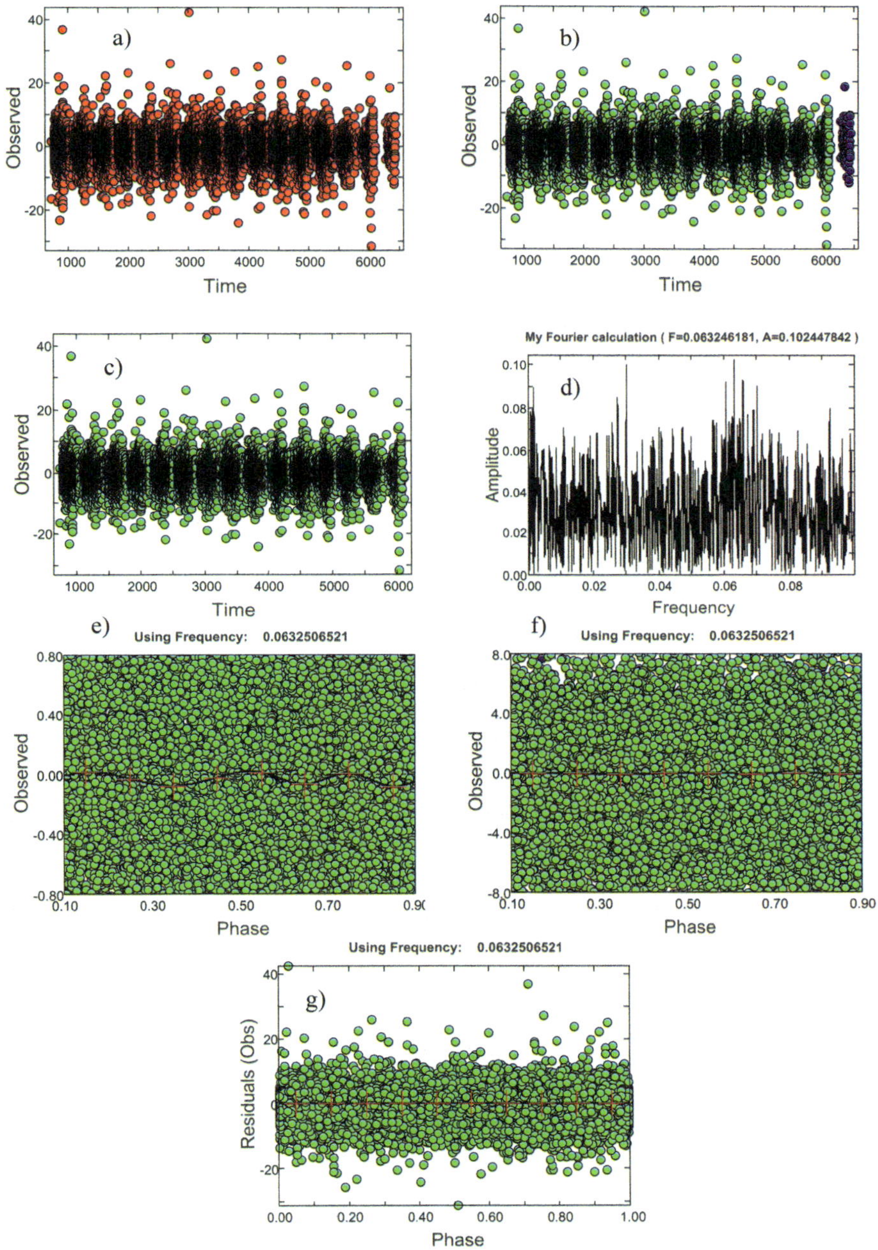

Fig. 32.14 Sequence Period04 (Lenz and Breger 2005) for xtej1739-302: **a** appearance of the complete light curve, **b** light curve subdivided over two periods of time, **c** period of time of the curve to be analyzed, **d** periodogram. Fourier calculation graph for the peak frequency of 0.0632 Hz, **e–g** three views of the phase diagram with binning 10 of (**c**) using a period of 15.81 days

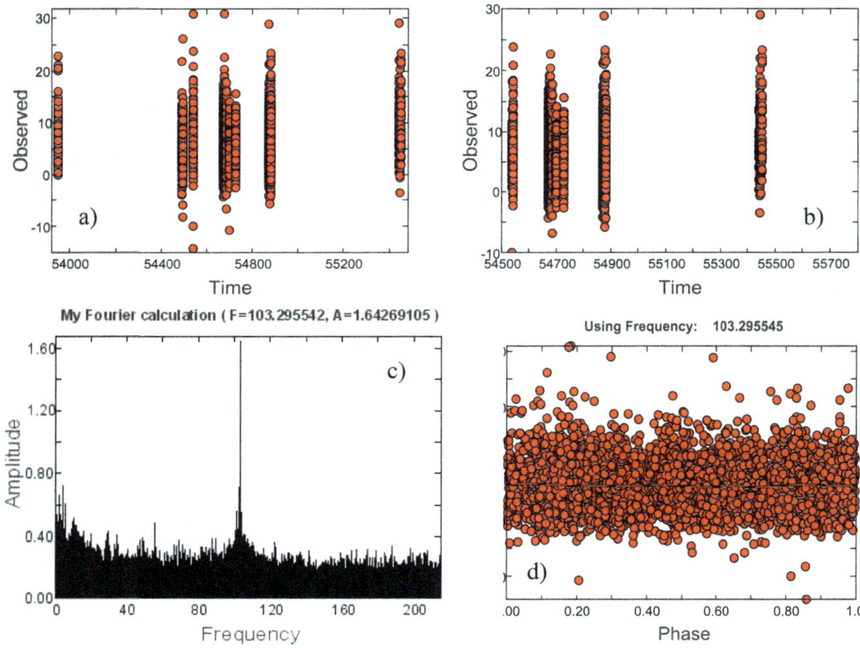

Fig. 32.15 Period04 sequence (Lenz and Breger 2005) for XPer: **a** aspect of the complete light curve, **b** aspect of the light curve at a certain time interval, **c** periodogram. Fourier calculation graph for the peak frequency of 103.2955 Hz, **d** phase diagram of (**a**) using a period of 0.00968096 s

(Strasbourg) 2017) (Fig. 32.4a–g). There is apparently a light outburst at 5100 days, otherwise, the curve exhibits a repetitive pattern (see Fig. 32.4a).

This curve was subdivided into the interval of days 4800–6400 (Fig. 32.4e) and was also analyzed over the entire time interval, i.e., 800–6000 days (Fig. 32.4a). The curve shows the slight outburst (see Fig. 32.4a and b).

The periodograms in Fig. 32.4c and f of the first time intervals of Fig. 32.4b and e, respectively, differ in frequency and amplitude, so the light curve has periodic variation and more intense emission epochs than others.

The period in Table 32.1 is 247.02 $\pm$ 0.45 days for intervals of 800–6000 (Fig. 32.4b) and 800–4000 days as expected, since the former encompasses the latter, however, the period of 124.81 days was obtained for the 4000–6000-day interval.

The periodograms of days 800–6000 exhibits peaks from highest to lowest, the highest being 0.13 mag for 4e-03 days^{-1} in Fig. 32.4c, while for 4800–6400 days we have 0.36 mag and 8e-03 days^{-1} (see Fig. 32.4e and f).

The period obtained for the light curve frequency for the first interval of Fig. 32.4b and its periodogram of Fig. 32.4c is almost twice that obtained from Fig. 32.4e and its periodogram of Fig. 32.4f, as shown in Table 32.1.

During the decay of the burst of this binary transient Be/X-ray J1008-57 (ATel # 4319), the source showed a 1-day long burst (Fig. 32.4a, b and e) show the prominent peak between 5080 and 5150 days). During the 2–3 days before and after this flare, the flux does not decay further, but remained consistent (Kühnel et al. 2012).

Ge et al. (2020) observed the accreting pulsar GRO J1008-57 performed by Insight-Hard X-ray Modulation Telescope (HXMT) at the peak of the source's 2017 outburst. Pulsations were detected with a spin period of 93.283(1) s. The pulse profile shows double peaks at soft X-rays, and only one peak above 20 keV.

he0309-2057

This source is an object type Seyfert 1 Galaxy, AGN, Seyfert1, morphological type Sy1.5 with swift name of SWIFTJ0311.5–2045 (Astronomical and Database—CDS (Strasbourg) 2023; Astronomical and Database—CDS (Strasbourg) 2017) shown in Fig. 32.5a–e.

Except for the end of the curve that has a separation between 6300 and 6400 days according to Fig. 32.5b, the light curve shows periodic variations (see Fig. 32.5a).

The periodogram of Fig. 32.5d is different in frequency and amplitude, so the light curve shows periodic variation and has more intense emission epochs than others.

The frequency of this source for 5700–6600 days is 0.0349 days^{-1} and the amplitude of 0.1993 mag. The period, frequency and amplitude obtained for the 700–5800-day interval of Fig. 32.5b shown in Fig. 32.5c and in Table 32.1 these are 322.94 $\pm$ 2.19 days, 0.0031 days^{-1} and 0.0527 mag, respectively (see Fig. 32.5d).

There is similarity of Fig. 32.5a with he0309-2057 of he0309–2057 (2024). Other results in RXTE Guest Observer Facility (2017).

ks1947 + 300

This object type HMXRB (Fig. 32.6a–k), spectral type B0Ve, has a peak at $\approx$2630 days (see Fig. 32.6a). The curve was divided into intervals of 1500 days resulting in total four intervals as in Fig. 32.6c.

The periodograms in Table 32.1 show differences in frequency and amplitude and, therefore, periodic variation and periods of emission that are more intense than others.

The 800–2500-day interval in Fig. 32.6b exhibits lines ranging from 0.00 to $\approx$0.068 mag over the entire time interval in Fig. 32.6d, with the highest peak at 0.068 mag for a frequency of 2.3e-02 days^{-1}, also shows a period of 43.48 $\pm$ 0.15 days; however, the period of 3860.90 $\pm$ 28.65 days for the whole curve was made because it was uncertain that this source had a superorbital period (orbital phase of Fig. 32.6a in k approximately between 2550 and 2650 days).

We can observe the sequence of Period04 images (Lenz and Breger 2005) for the whole time interval to analyze a possible superorbital period in the orbital phases of three intervals of Fig. 32.6g–i (Townsend and Charles 2020; Ackermann et al. 2013; Chernyakova et al. 2012). The periodogram indicates a decrease in amplitude that starts to be approximately constant from 0.025 days^{-1} onwards in Fig. 32.6j.

Long period variability that suggests a possible Psup around 3861 days is presented (Table 32.1). Using this value, we plot ks1947 + 300 in Fig. 32.1 of Townsend and Charles (2020) and show that it sits precisely amongst the other BeXRBs, thus confirming the ~3 861 days' variation as the true super-orbital period in this system. Then, the relationship in Fig. 32.1 of Townsend and Charles (2020) has predictive potential for long-term variability studies (Stanek et al. 2017; Nesci 2017). Authors like Liu et al. (2023) did the same.

Townsend and Charles (2020) have examined properties of ULXs pulsars and NS-HMXBs about orbital and super-orbital periods and showed all ULX pulsars follow a tight linear relationship between their orbital and super-orbital periods (approximately as in Fig. 32.6g–i in this work). This linear relationship can be used as a predictive potential in ULXs and HMXBs. It provides evidence that the superorbital periods generated in ULXs are an outcome of the same mechanisms at work in HMXBs, and that at the extreme end of the accretion scale many ULX pulsars may be HMXBs.

Fürst et al. (2014), present a spectral analysis of three simultaneous observations of Nuclear Spectroscopy Telescope Array and Swift/XRT of this neutron star Be ks1947 + 300 transient binary, taken during its outburst in 2013/2014 to study the evolution of its spectrum. They found that this source has a magnetic field strength that is at the lower end of the known cyclotron line sources.

Reig et al. (2016) provide long-term variability results for the line Hα in HMRXBs. They characterized the optical variability times and studied the interaction between the neutron star and the accreting material. Their results show that Be/X-ray binaries with narrow orbits are more variable than systems with long orbital periods; a Keplerian distribution of gas particles describes their disks as in classical Be/X-ray stars; a decrease in the equivalent width of Hα after major X-ray bursts that correlates with disk radius; most of them with asymmetric double-peaked profiles indicating that density inhomogeneities are a common property in their disk; the time scales of the profile variability are shorter and their Hα width are smaller than in isolated Be stars and their disk is denser than in classical Be stars. They concluded that the interaction between the compact object and the Be-type star works in two directions: the massive companion provides the source of matter for accretion, affecting the environment of the compact object and the continuous revolution of the neutron star around the optical counterpart also produces the truncation of the equatorial disk of the Be star (see Fig. 32.6g–i and k).

psr j0437 – 4715

This is an object type pulsar whose light curve in Fig. 32.7a shows a repetitive pattern without peaks for analysis as a whole.

The periodogram shows amplitude peaks ranging from 0.00 to 0.032 mag over the entire time interval. For the 0.032 mag amplitude there is a peak at a frequency of $\approx 0.092\ s^{-1}$ in Fig. 32.7b.

Rangelov et al. (2016) detected an arc shock from this pulsar-like object in the far ultraviolet of the nearest millisecond obtained with the Hubble Space Telescope. The FUV emission could be produced by interstellar medium matter in shock, or, perhaps by relativistic electrons in the pulsar's wind. They also reported X-ray emission ahead of the pulsar, possibly a pulsar wind nebula created by shocked pulsar wind (see Fig. 32.7a–c).

saxj2103.5 + 4545

For this object type HMXRB, spectral type B0Ve, in Fig. 32.8a it can be seen that the light curve exhibits a repetitive pattern. The curve was subdivided into the two time intervals in Fig. 32.8b. The periodogram in Fig. 32.8d is different in frequency and amplitude, therefore, there is periodic variation in the light curve with periods of more intense emissions than others.

For 5800–6600 days the frequency is 0.0093 days^{-1} and the amplitude 0.1521 mag, whereas for 700–5800 days, the frequency is 0.00026 days^{-1}, amplitude 0.1793 mag (Lenz and Breger 2005) and the period 3468.49 $\pm$ 62.32 days (only period reported in Table 32.1).

Camero et al. (2007) release the first long-term pulse profile study of this source, demonstrating the correlation of pulse shape with luminosity, time and energy. It was observed between 1999 and 2004 by RXTE PCA and from 2002 to 2005 by INTEGRAL. These authors report a pulse shape is complex and highly variable and that the strength of almost all the peaks increases with luminosity. They found single, double, triple and quadruple peaks pulse profile structure was obtained. The orbital solution obtained was of 12.6653 $\pm$ 0.00051 days These authors report a 12-day period, unlike the one obtained in this work of 3468.49 $\pm$ 62.32 days for 700–5800 days of the Fig. 32.8c using the database of 2017 (RXTE Guest Observer Facility 2017; ASM Light Curves Overview 2023; Aeronautics et al. 2023).

smcx1

In this source (Fig, 9a–f), the light curve appears uniform, without prominent peaks and with a lack of definition at the right end as shown in Fig. 32.9a and b, so it was analyzed as a curve with a repetitive pattern with emission epochs.

The periodogram in Fig. 32.9d indicates a different amplitude and therefore, the light curve shows periods with more intense emission, from the visualization of the frequency, it can be concluded that there is no periodic variability. For 700–6250 days in Fig. 32.9c and Table 32.1, the highest peak is at 0.0179 days^{-1} and 0.6356 mag in Fig. 32.9d. The period is 55.82 $\pm$ 0.0061 days (RXTE Guest Observer Facility 2017; Analysis and for SMC X-1 2023).

The frequency of this source for 6250–6550 days is 0.0179 days^{-1} and the amplitude is 1.1290 mag (Lenz and Breger 2005). This amplitude is greater than the highest peak of the periodogram in Fig. 32.9d for the time interval of the light curve of the Fig. 32.9c. van Paradijs & Kuiper (1984) present the results of photometric observations (Walraven system) of Sk 160, the optical counterpart of the massive X-ray binary SMCX-1 obtained between 1979 and 1982. They report an orbital period of 3.89233 + or – 0.00008 days and the double-wave optical light curve of Sk 160 is the result of ellipsoidal variations, and X-ray heating of the primary. Also, this authors inform about the minimum in the light curve near superior conjunction is symmetric and coincides with the times of X-ray eclipses (phase 0.0), that near inferior X-ray conjunction is asymmetric and comes early by about 0.02 orbital cycles, and also that SMCX-1 contains a luminous accretion disk.

velax1

For this source, Fig. 32.10a–e, the general appearance of the light curve shows a low burst as shown in Fig. 32.10a, which shows a uniform behavior over the entire time interval. The curve was divided into 700–6250 days and 6250–6600 days in Fig. 32.10b.

The period of 7849.592 $\pm$ 359.40 days was obtained for 700–6400 days (Table 32.1). The frequency of this source for 6250–6600 days is 0.0977 days^{-1} and the amplitude 2.0139 mag, whereas in Table 32.1 and Fig. 32.10d for 700–6250 days, the frequency was 0.00013 days^{-1} and the amplitude 0.4990 mag.

The periodogram in Fig. 32.10d has two peaks, one after the process has started, with an amplitude of 0.45 mag, and the other after 0.09 days^{-1}, of $\approx$0.44 mag. Except for these two peaks where the amplitude is practically the same, Fig. 32.10d appears to exhibit periodic variation and epochs of stronger emission at the other points of the plot. Rapid velax results are provided by the ASM/RXTE equipment at (RXTE Guest Observer Facility 2017; ASM Light Curves Overview 2023; Vela X-1 History from 2024).

The high-mass X-ray Binary Vela X-1 bright and eclipsing was studied for Grinberg et al. (2017). In Chandra-HETGS spectroscopy at orbital phase ~ 0.25, the authors observed changes in overall spectral shape on timescales. They demonstrated that this spectral variability is caused by changes in overall absorption and that such strongly variable absorption cannot be caused by unperturbed clumpy winds of O/B stars. The authors detected line features from high and low ionization species of silicon, magnesium and neon that imply a co-existence of cool and hot gas phases in the system that they interpreted as a highly variable, structured accretion flow close to the compact object as in simulations of wind accretion in high-mass X-ray binaries.

Kretschmar et al. (2019), have informed about Vela X-1 is as an eclipsing high mass X-ray binary (HMXB) consisting of a 283 s accreting X-ray pulsar in a close orbit of 8.964 days around the B0.5Ib supergiant HD77581 at a distance of just 2.4 kpc.

They discuss the properties and the use of the data of the system as laboratory to test developments in modelling the accretion process in HMXRBs from descriptions of the wind acceleration to modelling of the structure of the flow of matter close to the neutron star and its variations.

Kretschmar et al. (2021), suggests that improved magnetohydrodynamic models as a key factor for understanding the variability of the X-ray flux and the torques applied to the neutron star. A full characterization of the X-ray emission from the accretion column remains another so far unsolved challenge.

x0114 + 650

The light curve of this source exhibits a uniform periodic appearance (Fig. 32.11a). The behavior is one of periodic variations with uniform emissions.

From the subdivision of Fig. 32.11a into 700–6250 days in Fig. 32.11b and c and 6250–6600 days of Fig. 32.11b a periodic variation of $\approx$1.16 days in the light curve is observed from Period04 (Lenz and Breger 2005). The periods of the time intervals are 11.60 ± 0.00095 days and 12.76 ± 0.07 days (Table 32.1).

The periodogram of the first interval shows two peaks, the largest of which has an amplitude of 0.1106 mag and a frequency of 0.0862 $days^{-1}$ (Fig. 32.11d). For the 6250–6600-day interval of Fig. 32.11b, the frequency and amplitude are 0.0784 $days^{-1}$ and 0.7291 mag, respectively (calculated in Period04, see Table 32.1). As the amplitudes of the periodogram are different, the light curve shows periods of more intense emission than other ones. Farrell et al. (2008) obtained results of the spectral and temporal behavior of this source examined on the orbital and superorbital pulse of time scales, covering ~ 2 ciclos of the 30.7 days of superorbital modulation, concluding that the observed superorbital modulation is linked to the variability in the mass accretion rate. The nature of the periods obtained here for both intervals of the light curve may be attributed to the superorbital period of this source.

Pacciani et al. (1917) report enhanced X-ray emission from the pulsar 3A0114 + 650 slowly spinning in 2009 over a time period of three AGILE orbits ($\approx$5 h). In this work, we report a period of 11.60 ± 0.00095 days in 700–6250 days and ditto a period of 12.76 ± 0.07 days in 6250–6600 days with the 2017 database of (RXTE Guest Observer Facility 2017; ASM Light Curves Overview 2023; Aeronautics et al. 2023) (see Table 32.1).

Farrell et al. (2008) expose HMXRB results 2S011 + 650 onboard the RXTE on orbital and superorbital pulsing. The authors observed variability of the neutral hydrogen column density during the orbital period, which they attributed to line-of-sight motion of the neutron star through the dense circumstellar environment, as well as a reduction in the power-law photon index during the orbital maximum. No significant column density variability was observed during the superorbital period, indicating that variable dimming by a precession warp in an accretion disk is not the mechanism behind the superorbital modulation. In contrast, a significant increase in the power-law photon rate was observed during the superorbital minimum. They conclude that the observed superorbital modulation is linked to variability in the mass accretion rate (see Fig. 32.11a–e).

x0115 + 634

This object type HMXRB, spectral type B0.2 V, galactic variable star presents a light curve with more intense emission epochs (Astronomical and Database—CDS (Strasbourg) 2023; Astronomical and Database—CDS (Strasbourg) 2017), Fig. 32.12a–k. Four burst peaks are observed at approximately 1900, 2460, 3920 and 5210 days (see Fig. 32.12a–c and e). The light curve in Fig. 32.12a was divided into two time intervals in Fig. 32.12b. The 700–6400-day interval in Fig. 32.12a, b, and e has four peaks, three of which are nearly equally intense. The 700–1750-day interval in Fig. 32.12a–c, and e contains none of the peaks.

The difference in periods of 183.44 ± 0.06 days for 700–6400 days of Fig. 32.12e with respect to 1043.54 ± 47.41 days for 700–1750 days of Fig. 32.12f of Table 32.1, therefore, this is a curve with periodic variability.

Dugair et al. (2013), offer the detection of quasi-periodic oscillation (QPO) in this high-mass transient binary pulsar Be/X-ray 4U0115 + 634.

The periodogram in Fig. 32.12d shows a peak with an amplitude of 0.74 mag and a frequency of 5.45e-03 days^{-1}. While the interval in Fig. 32.12f shows peak amplitude ≈0.2347 mag for a frequency of 9e-04 days^{-1} in Fig. 32.12g.

The periodogram in Fig. 32.12d exhibits a behavior that indicates frequency decrease.

It was uncertain whether this source had a superorbital period (see orbital phases in Fig. 32.12i–k), and in this case, the nature of the period is of a more noticeable variability at the two frequencies: it varies with the orbital period of the system, 183.44 ± 0.06 days, but, it could also reveal a super-orbital period of variation at all these frequencies of about 2.86 years (1043.54 ± 47.41 days in Table 32.1) since the 700–6400-day interval encompasses the interval of days 700–1750. Archer et al. (2016) and Müller et al. (2013) have conducted studies of the impact factor of bursts from this source. Neither these authors nor Dugair et al. (2013), emphasize that this source presents a superorbital period.

Graphics of the Fig. 32.12i–k show a curved behavior, unlike a tight linear relationship between their orbital and superorbital periods showing in Fig. 32.6g–i according with authors as Townsend & Charles (2020).

x0115 + 634 binary system (x0115+634 19. 6327 + 63.7400 2023) o X0115 + 634 (4U0115 + 63) is the most active of Be/X-ray transients, showing very bright X-ray outbursts at least once every three years.

This source exhibits large X-ray bursts because transient X-ray binaries produce larger bursts in which the X-ray flux can increase at the resting level by very large factors (Archer et al. 2016). The system 4U0115 + 634 underwent explosions in 2015 although these authors did not detect gamma rays in the period of highest X-ray activity.

Müller et al. (2013), report a flash of the high X-ray mass of this binary in 2008, as observed with RXTE and INTEGRAL.

Dugair et al. (2013), presents the detection of quasi-periodic oscillation (QPO) at ~ 41 mHz in this transient binary pulsar HMXRB, using data from the RXTE explorer, during observations of the X-ray bursts of this pulsar. There are three RXTE/PCA observations where multiple QPOs were detected in the power density spectrum. Nevertheless, during the three X-ray bursts, no systematic variation was observed in any of the spectral parameters other than the anti-correlation between cyclotron absorption energy and luminosity. The presence of any systematic variation of QPO frequency and QPO rms with source flux was also investigated, yielding negative results.

Baykal et al. (2005) show the change of the optical light curve suggests a precession of the Be star disc of around a few hundred days. They also investigate the periodic signatures from the archival RXTE-ASM (Rossi X-ray Timing Explorer – All Sky Monitor) light curve founding a significant orbital modulation in the ASM light curve during the type I X-ray outburst.

xper

In this source, Fig. 32.13a-k, the light curve has an appearance that exhibits time intervals similar in shape, except in size, separated from each other, which show periodic variations with slightly more intense emission epochs in some fragments of the time interval (see Fig. 32.13a and b).

The periodogram in Fig. 32.13d show different frequencies and amplitudes, which describe a curve with periodic variation and more intense emission epochs of Fig. 32.13c. From the calculation done in Period04, the frequency of this source for 5500–6600 days is 0.0015 days^{-1} and the amplitude is 0.5633 mag; whereas for 0–5500 days, the frequency was 0.000155 days^{-1} and the amplitude, 0.8059 mag (Fig. 32.13i). In Table 32.1 we can see that the respective frequencies are of the order of 2.92e-05 $\pm$ 1.68e-07 days^{-1} for 5000–6000 days of the light curve of Fig. 32.13c and 17.47 e-05 $\pm$ 1.52e-06 days^{-1} for 0–5500 days of the light curve of Fig. 32.13h, with respective periods of 34,248.42 $\pm$ 197.46 days and 5722.93 $\pm$ 49.89 days.

The light curve was subdivided into seven time intervals of approximately 1000 days (Fig. 32.13b) and the 5000–6000-day interval (Fig. 32.13c) was analyzed based on the appearance of the light curve. Since the phase diagram exhibits only the 0.146 to 0.174 interval of the phase (Fig. 32.13e), we chose to subdivide the curve into two intervals from 0 to 5500 days and from 5500 to 6500 days (Fig. 32.13g) and analyzed the 0-to-5500-day interval (Fig. 32.13h) which yields the 0.0 to 1.00 phase diagram (Fig. 32.13j).

The periodograms of the two analyzed time intervals show similar behaviors, but less dense for the periodogram of Fig. 32.13d for 5000–6000 days of Fig. 32.13c with respect to the interval from 0 to 5500 days of Fig. 32.13h (periodogram Fig. 32.13i), showing a peak close to time zero and with amplitudes of 0.57 mag (Fig. 32.13d) and 0.40 mag (Fig. 32.13i) respectively, for frequencies of 7e-4 days^{-1} and 1.5e-04 days^{-1}. This leads to periods mentioned above, respectively.

Filters for ASM quick-look data for xper in xper (2017) comparable with the Fig. 32.13.

xtej1739 – 302

This source is an object with spectral type O8.5Iab(f), double star, morphological type HMXB (Fig. 32.14a–g). The light curve of this binary has repetitive time intervals separated into approximately equal time intervals (Fig. 32.14a). It was subdivided into two intervals of 700–6200 days and 6280–6460 days in Fig. 32.14b and the 700–6200 day intervals was analyzed (Fig. 32.14c).

The periodogram in Fig. 32.14d was observed for a frequency of 0.063 days^{-1}and 0.1024 mag. The frequency of this source for 6280–6460 days of Fig. 32.14b is 0.0559 days^{-1} and amplitude of 1.6856 mag calculated with Period04. The interval of 700–6200 days of Table 32.1 shows a period of 15.81 $\pm$ 0.0056 days.

The periodogram exhibits approximately equal lines with slight peaks at 0.03 and 0.065 days^{-1}, but the average behavior of the lines ranges between 0.00 and 0.065 mag, i.e., 0.0325 mag. The highest peak has amplitude 0.1024 mag at a frequency of 0.0632 days^{-1} (see Fig. 32.14d). The periodogram in Fig. 32.14d has slight periodic variations.

Romano et al. (1961) reports that the Swift Burst Alert Telescope (BAT) light curve activated on the fast X-ray transient supergiant xtej1739-302 provides a light curve without any significant structure with a rate that remained approximately constant during the observation of 1600 s.

xper

The appearance of the light curve of this source is restricted to non-periodic vertical lines as in Fig. 32.15a and b.

The light curve in Fig. 32.15a and b was subdivided into two intervals from less than 54,000 to 54,800 s and from 54,800–55,430 s and the first interval where most of the emissions are concentrated was analyzed. The results are shown in Table 32.1.

The approximately symmetrical periodogram of the entire light curve yields a peak with frequency and amplitude of 103.2955 s^{-1} and 1.6427 mag, respectively (Fig. 32.15c).

The frequencies found for the first time interval and for the complete time interval of the light curve show that they are approximately equal, with slightly larger amplitudes before the outburst. Apparently, this source does not offer periods of more intense emission. In Fig. 32.15a for the interval of ≈54,000–54,800 s a period of 0.0097 ± 1.71e-09 s was found.

Grundstrom et al. (2007) presented results from an H-alpha monitoring campaign of the X Per. Their observations indicate that its circumstellar disk has recently grown to near record proportions, and concurrently the system has dramatically increased in X-ray flux, possibly the result of enhanced mass accretion from the disk. They found that the H-alpha half-maximum emission radius of the disk surrounding XPer reached a size about six times larger than the stellar radius, a value that is well below the minimum separation between the Be star and neutron star.

32.4 Conclusions

From the light curves of the fifteen HMXRBs studied, we can conclude that, from the results obtained from Period04 about discrete Fourier transformation and the possibility to calculate amplitude and/or phase variations, the objective of analyzing the variability of light curves of the sources was met, providing information on their temporal behavior due to orbital and/or superorbital movement, the rotation of the compact object and the periodic movements of the material.

The periodicity of the sources through Fourier temporal analysis allowed us to contribute in the area of high-energy astrophysics in the state of the art of variability in X-rays that coincides with results presented by other authors using different techniques and time data and that significantly affects the physical processes of the binaries in space–time in which they have been studied according to NASA and ESA databases.

Acknowledgements This research has made use of the SIMBAD database, operated at CDS, Strasbourg, France 2000, A&AS,143,9, "The SIMBAD astronomical database", Wenger et al.

References

Ackermann M, Ajello M, Ballet J, Barbiellini G, Bastieri D, Bellazzini R, Bonamente E, Brandt TJ, Bregeon J, Brigida M, Bruel P, Buehler R, Buson S, Caliandro GA, Cameron RA, Caraveo PA, Casandjian JM, Cavazzuti E, Cecchi C, Chekhtman A (2013) Associating long-term gamma-ray variability with the superorbital period of LS I 61°303, 2013. Am Astron Society ApJL. https://doi.org/10.1088/2041-8205/773/2/L35

Archer A, Benbow W, Bird R, Bourbeau E, Buchovecky M, Buckley JH, Bugaev V, Byrum K, Cerruti M, Connolly MP, Cui W, Errando M, Falcone A, Feng Q, Fernandez-Alonso M, Finley JP, Fleischhack H, Flinders A, Fortson L, Furniss A, Griffin S, Grube J, Hutten M, Hanna D, Hervet O, Holder J, Humensky TB, Johnson CA, Kaaret P, Kar P, Kelley-Hoskins N, Kertzman M, Kieda D, Krause M, Kumar S, Lang MJ, Lin TTY, Maier G, Moriarty P, Mukherjee R, Nieto D, O'brien S, Ong RA, Park N, Pohl M, Popkow A, Pueschel E, Quinn J, Ragan K, Reynolds PT, Richards GT, Roache E, Rousselle J, Rovero AC, Sadeh I, Schlenstedt S, Sembroski GH, Shahinyan K, Staszak D, Telezhinsky I, Tyler J, Wakely SP, Wilcox P, Wilhelm A, Williams DA (2016) Very high energy observations of the binaries V 404 Cyg and 4U 0115+634 during giant X-ray outbursts. Astrophys J. https://doi.org/10.48550/arXiv.1608.06464

ASM Light Curves Overview. (2023). https://www.sao.ru/cats/doc/RXTE.html. Accessed 6 Dec 2023

SIMBAD astronomical database—CDS (Strasbourg). http://simbad.u-strasbg.fr/simbad/ (2017). Accessed 8 Jul 2017

SIMBAD Astronomical Database—CDS (Strasbourg). http://simbad.cds.unistra.fr/simbad/ (2023). Accessed 5 Dec 2023

Baykal A, Kızıloğlu Ü, Kızıloğlu N, Beklen E, Özbey M (2008) Recent RXTE/ASM and ROTSEIIId observations of EXO 2030+375 (V2246 Cygni). A&a 479:301–306. https://doi.org/10.1051/0004-6361:20078721

Baykal A, Kızıloğlu Ü, Kızıloğlu N, Balman S, İnaM SÇ (2005) X-ray outburst of 4U 0115+634 and ROTSE observations of its optical counterpart V635 Cas. A&A 439 1131–1134. https://doi.org/10.1051/0004-6361:20052968

Blay P, Negueruela I, Reig P, Coe MJ, Corbet RHD, Fabregat J, Tarasov AE (2006) Multiwavelength monitoring of BD+53 2790, the optical counterpart to 4U 2206+54. Astron Astrophys 446:1095–1105. https://doi.org/10.1051/0004-6361:20053951

Camero AA, Wilson CA, Finger MH, Reglero V (2007) Long-term pulse profile study of the Be/X-ray pulsar SAXJ2103.5+4545. Astron Astrophys 473:551–559. https://doi.org/10.1051/0004-6361:20077398

Chernyakova M, Neronov A, Molkov S, Malyshev D, Lutovinov A, Pooley G (2012) Superorbital modulation of X-ray emission from gamma-ray binary LSI +61 303. ApJL. https://doi.org/10.1088/2041-8205/747/2/L29

Coe MJ, Payne BJ, Longmore A, Hanson CG (1988) The optical/IR counterpart to the newly-discovered X-ray source EXO 2030+375. MNRAS 232:865. https://doi.org/10.1093/mnras/232.4.865

Corbet RHD, Levine AM (2006) Recent RXTE/ASM and ROTSEIIId observations of EXO 2030+375 (V2246 Cygni). A&a 843:301–306. https://doi.org/10.1051/0004-6361:20078721

Dugair MR, Jaisawal GK, Naik S, Jaaffrey SNA (2013) Detection of a variable QPO at $\sim$ 41mHz in the Be/X-ray transient pulsar 4U 0115+634. MNRAS.434.2458D—Mon Not R Astron Soc 434:2458–2464. https://doi.org/10.1093/mnras/stt1187

Farrell SA, Sood RK, O'neill PM, Dieters S (2008) Un estudio detallado de 2S 0114 + 650 con el Rossi X-ray timing explorer. MNRAS 389:608–628. https://doi.org/10.1111/j.1365-2966.2008.13588.x

Fürst F, Pottschmidt K, Wilms J, Kennea J, Bachetti M, Bellm E, Boggs SE, Chakrabarty D, Christensen FE, Craig WW, Hailey CJ, Harrison F, Stern D, Tomsick JA, Walton DJ, Zhang W (2014) NuSTAR discovery of a cyclotron line in ks 1947+300. ApJL. https://doi.org/10.1088/2041-8205/784/2/L40

Ge MY et al (2020) Insight-HXMT firm detection of the highest-energy fundamental cyclotron resonance scattering feature in the spectrum of GRO J1008–57. ApJL 899:L19. https://doi.org/10.3847/2041-8213/abac05

Grinberg V, Hell N, El Mellah I, Neilsen J, Sander AAC, Leutenegger M, Fürst F, Huenemoerder DP, Kretschmar P, Kühnel M, Martínez-Núñez S, Niu S, Pottschmidt K, Schulz NS, Wilms J, Nowak MA (2017) The clumpy absorber in the high-mass X-ray Binary Vela X-1. A&A. https://doi.org/10.48550/arXiv.1711.06743

Grundstrom ED, Boyajian TS, Finch C, Gies DR, Huang W, McSwain MV, O'Brien DP, Riddle RL, Trippe ML, Williams SJ, Wingert DW, Zaballa RA (2007) Joint H-alpha and X-ray observations of massive X-ray binaries. III. The be X-ray binaries HDE 245770 = A 0535+26 and X Persei. Astrophys J 660:1398–1408. https://doi.org/10.48550/arXiv.astro-ph/0702283

he0309–2057. https://www.bass-survey.com/ (2024). Accessed 14 Oct 2024

Integral Data Archives. https://www.cosmos.esa.int/web/integral/integral-data-archives (2023). Accessed 11 Dec 2023

Klochkov D, Horns D, Santangelo A, Staubert R, Segreto A, Ferrigno C, Kretschmar P, Kreykenbohm I, La Barbera A, Masetti N, McCollough M, Pottschmidt K, Schönherr G, Wilms J (2007) INTEGRAL and Swift observations of EXO 2030+375 during a giant outburst. A&a 464:L45–L48. https://doi.org/10.1051/0004-6361:20066801

Kretschmar P, Martínez-Núñez S, Fürst F, Grinberg V, Lomaeva M, El Mellah I, Manousakis A, Sander AAC, Degenaar N, van den Eijnden J (2019) Vela X-1 as a laboratory for accretion in High-Mass X-ray Binaries. Mem S.A.It.. https://doi.org/10.48550/arXiv.1905.08578

Kretschmar P, El Mellah I, Martínez-Núñez S, Fürst F, Grinberg V, Sander AAC, van den Eijnden J, Degenaar N, Maíz Apellániz J, Jiménez Esteban F, Ramos-Lerate M, Utrilla E (2021) Revisiting the archetypical wind accretor Vela X-1 in depth. Case study of a well-known X-ray binary and the limits of our knowledge. A&A. https://doi.org/10.1051/0004-6361/202040272

Kühnel M, Müller S, Kreykenbohm I, Wilms J, Pottschmidt K, Krimm H, Fürst F, Rothschild RE, Caballero I, Schoenherr G, Klochkov D, Staubert R (2012) Unexpected flaring activity of the Be/X-ray binary GRO J1008–57. The Astronomer's Telegram. https://astronomerstelegram.org/?read=4355. Accessed 14 Oct 2024

Laplace E, Mihara T, MoritanI Y, Nakajima M, Takagi T, Makishima K, Santangelo A (2017) Possible regular phenomena in EXO 2030+375. A&A 597. https://doi.org/10.1051/0004-6361/201629373

Lenz P, Breger M (2005) Period04 user guide. Comm. in Asteroseismology 146:1–82. http://www.period04.net/files/p04manual_v1.0.0.pdf

Liu W, Reig P, Yan J, Zhang P, Li X, Gao B, Xiao G, Liu Q (2023) Multiwavelength observations of KS 1947+300. A&A. https://doi.org/10.48550/arXiv.2311.13844

Müller S, Ferrigno C, Kühnel M, Schoenherr G, Becker PA, Wolff MT, Hertel D, Schwarm F-W, Grinberg V, Obst M, Caballero I, Pottschmidt K, Fuerst F, Kreykenbohm I, Rothschild RE, Hemphill P, Nunez SM, Torrejon JM, Klochkov D, Staubert R, Wilms J (2013) No anticorrelation between cyclotron line energy and X-ray flux in 4U 0115+634. A&A. https://doi.org/10.1051/0004-6361/201220359

Naik S, Maitra C, Jaisawal GK, Paul B (2013) Timing and spectral properties of Be/X-ray pulsar EXO 2030+375 during a Type I outburst. ApJ 764:158. https://doi.org/10.48550/arXiv.1301.1112

National Aeronautics and Space Administration Goddard Space Flight Center. https://heasarc.gsfc.nasa.gov/cgi-bin/W3Browse/w3browse.pl (2023). Accessed 11 Dec 2023

Nesci R (2017) Historic light curve of Swift J0243.6+6124. The Astronomer's Telegram 10989. https://ui.adsabs.harvard.edu/abs/2017ATel10989....1N/abstract. Accessed 14 Oct 2024

Okazaki AT, Negueruela I (2001) A natural explanation for periodic X-ray outbursts in Be/X-ray binaries. A&a 377:161–174. https://doi.org/10.1051/0004-6361:20011083

Pacciani L, Costa E, Del Monte E, Donnarumma I, Evangelista Y, Feroci M, Lapshov I, Lazzarotto F, Rapisarda M, Soffitta P, Argan A, Trois A, Tavani M, Piano G, Pucella G, D'ammando F, Sabatini S, Vittorini V, Bulgarelli A, Gianotti F, Trifoglio M, Di Cocco G, Labanti C, Fuschino

F, Marisaldi M, Galli M, Chen A, Vercellone S, Giuliani A, Mereghetti S, Perotti F, Fornari F, Fiorini M, Caraveo P, Pellizzoni A, Barbiellini G, Longo F, Moretti E, Vallazza E, Picozza P, Morselli A, Prest M, Lipari P, Zanello D, Rappoldi A, Pittori C, Verrecchia F, Santolamazza P, Preger B, Giommi P, Salotti L (2009) SuperAGILE detects a flare from the binary source 3A 0114+650. The Astronomer's Telegram 1917. https://www.astronomerstelegram.org/?read=1917. Accessed 14 Oct 2024

van Paradijs J, Kuiper L (1987) Optical photometry of massive X-ray binaries—SMCX-1/Sk 160. A&A 138, 71–76. https://adsabs.harvard.edu/full/1984A%26A...138...71V

Rangelov B, Pavlov GG, Kargaltsev O, Durant M, Bykov AM, Krassilchtchikov A (2016) First detection of a pulsar bow shock nebula in far-UV: PSR J0437–4715. Astrophys J 831:129–129. https://doi.org/10.3847/0004-637X/831/2/129

Reig P, Coe MJ X-ray spectral properties of the pulsar EXO 2030+375 during an outburst. MNRAS 302:700–706. https://doi.org/10.1046/j.1365-8711.1999.02179.x

Reig P, Coe MJ (1998) Timing properties of the X-ray pulsar EXO 2030+375 during an X-ray outburst. MNRAS 294:118–126. https://doi.org/10.1046/j.1365-8711.1998.01252.x

Reig P, Nersesian A, Zezas A, Gkouvelis L, Coe MJ (2016) Long-term optical variability of high-mass X-ray binaries. II. Spectroscopy. Astron Astrophys 590A. https://doi.org/10.1051/0004-6361/201628271

Romano P, Sidoli L, Krimm HA, Cusumano G, Kuulkers E, Vercellone S, Burrows DN, Gehrels N, Kennea JA, La Parola V, Mangano V (2009) Swift and INTEGRAL observations of activity in XTEJ1739–302/IGRJ17391–3021. The Astronomer's Telegram 1961. 2009ATel.1961....1R. Accessed 14 Oct 2024

RXTE Guest Observer Facility. https://heasarc.gsfc.nasa.gov/docs/xte/ASM/sources.html (2017). Accessed 8 July 2017

Scaled Map Transient Analysis for SMC X-1 https://swift.gsfc.nasa.gov/results/transients/SMC X-1/. (2023). Accessed 14 Oct 2024

Stanek KZ, Kochanek CS, Thompson TA, Shappee BJ, Holoien TWS, Prieto JL, Dong S (2017) ASAS-SN optical light curve of swift J0243.6+6124 shows long term variability. The Astronomer's Telegram 10811. https://ui.adsabs.harvard.edu/abs/2017ATel10811....1S/abstract. Accessed 14 Oct 2024

Stella L, White NE, Rosner R (1986) Intermittent stellar wind acceleration and the long-term activity of population I Binary systems containing an X-Ray pulsar. ApJ 308:669. https://doi.org/10.1086/164538

Townsend LJ, Charles PA (2020) Orbital and superorbital periods in ULX pulsars, disc-fed HMXBs, Be/X-ray binaries, and double-periodic variables. MNRAS 495:L139–L143. https://doi.org/10.1093/mnrasl/slaa078

Vela X-1 History from Apr 28, 2024 to Oct 11, 2024. https://gammaray.nsstc.nasa.gov/gbm/science/pulsars/lightcurves/velax1.html (2024). Accessed 14 Oct 2024

Wilson CA, Finger MH, Coe MJ, Laycock S, Fabregat J (2002) A decade in the life of EXO 2030+375: a multiwavelength study of an accreting X-ray pulsar. ApJ 570:287. https://doi.org/10.48550/arXiv.astro-ph/0201227

Wilson CA, Finger MH, Camero-Arranz A (2008) Outbursts large and small from EXO 2030+375. ApJ 678 1263–1272. https://doi.org/10.48550/arXiv.0804.1375

x0115+634 19.6327° +63.7400°. http://xte.mit.edu/asmlc/srcs/x0115+634.html (2023). Accessed 8 Dec 2023

xper 58.8457° + 31.0460°. http://xte.mit.edu/asmlc/srcs/xper.html. Accessed 7 Jul 2017

Chapter 33
Enhanced 1D Analytical Approach for Modeling the Effect of Transverse Cracking on Slender Structural Elements Displacement

Denis Imamović and Matjaž Skrinar

Abstract According to the elementary static bending theory, transverse and axial displacements for non-cracked straight structural elements are not interdependent. However, the occurrence of a transverse single-sided crack causes a redistribution of stresses within the non-cracked section and changes the deflection behavior mechanism along the element's axis due to the shift of the neutral axis deeper into the non-cracked cross-section. Consequently, both transverse and axial loading simultaneously induce transverse and axial (static and kinematic) quantities on transversely cracked structural elements. This paper presents a new analytical two-parameter model with an eccentric substitute rotational spring, focusing on more precise modeling of this phenomenon for slender structural elements. In this study, the crack is still modeled with a rotational spring, but displaced from the neutral axis of the non-cracked part of the element to its new position. This approach involves solving the differential equations for the transverse and axial displacements of non-cracked segments of the beam using the Euler–Bernoulli bending theory, where determining the integration constants is based on the continuity and boundary conditions. The model was initially validated using numerical examples of a cantilever beam with a crack, and the obtained results were compared to those of a more precise 2D model in the commercial software SAP2000. The results obtained show that the analytical model consistently predicts the transverse and axial displacements of the beam, achieving exceptionally low average relative errors. Additional analyses for different crack depths further confirmed that the simplified model's results align closely with the discrete results of the 2D model. The key finding of this research is that the new analytical model enables accurate prediction of the mechanical responses of beams with cracks, which is crucial for the safe design as well as maintenance of structures.

D. Imamović (✉) · M. Skrinar
University of Maribor, FGPA, Maribor, Slovenia
e-mail: denis.imamovic@um.si

M. Skrinar
e-mail: matjaz.skrinar@um.si

A. Öchsner and H. Altenbach (eds.), *Engineering Design Applications VII*, Advanced Structured Materials 230, https://doi.org/10.1007/978-3-031-84346-4_33

Keywords Transverse crack · Discrete spring model · Analytical model · Static bending analysis · Transverse and axial displacements

33.1 Introduction

For bending analyses, slender structural elements such as beams and columns are typically treated as one-dimensional (1D) continua based on the Euler–Bernoulli (EB) theory (Andersen and Nielsen 2008). These models facilitate not ideally accurate and fast yet still highly reliable response calculations. For non-cracked EB elements, it is well-known that under pure bending, the cross-section rotates about the neutral axis. This results in the exclusive transverse displacement of the neutral axis, which, in the case of pure bending, coincides with the centroid of the cross-section. Conversely, axial forces only cause displacements of the cross-section along the element's axis. Since transverse and axial displacements in non-cracked beams are independent in the classical engineering theory, they can be analyzed separately for straight, non-cracked elements, as reflected in numerous studies and practical applications (Reddy 1994).

However, in structural members, the presence of a single-sided local anomaly, such as a transverse crack (which can also be considered a notch) not only leads to a reduction in bending stiffness but also significantly alters the stress and strain state in its immediate vicinity. Normal stresses (σ_x) due to inner axial forces (N_x) and/or bending moments (M_z) increase and redistribute within the neighboring non-cracked section area (A), causing a local shift of the neutral axis, as depicted in Fig. 33.1. This results in the cross-section rotating about its new position, leading to longitudinal fiber on the neutral axis of the non-cracked part of the element moving not only transversely (as in non-cracked elements) but also along the longitudinal axis.

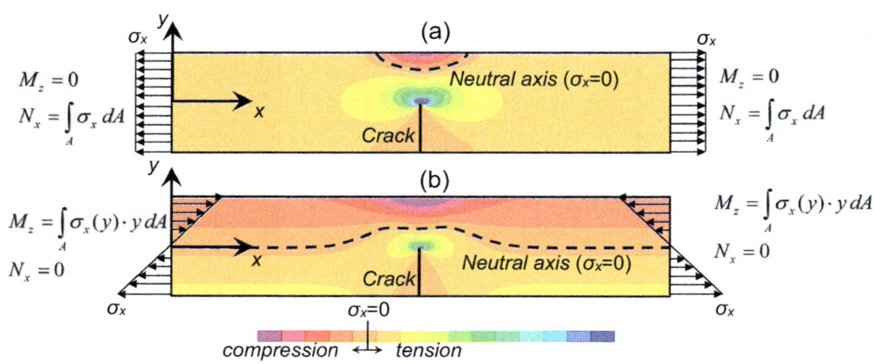

Fig. 33.1 Neutral axis position in a cracked part of element **a** subjected to inner axial force; **b** subjected to inner bending moment

When a detailed stress distribution analysis near the crack is not of interest, a simplified linear computational model can be effectively used for bending analysis. In this model, two adjacent non-cracked elastic segments to the left and right of the crack are connected at the crack location with a substitute discrete rotational spring attached at the cross-section centroid (Okamura et al. 1969). This classical model does not account for the actual shift of the neutral axis in the cracked section, thus not directly incorporating the effects of axial forces on transverse displacements and bending moments on axial displacements.

Previous initial studies (Skrinar 2020; Skrinar and Imamović 2023) using the classical model have considered the influence of axial tensile force (N_x) on bending by incorporating a local inner concentrated moment at the crack location M_N, defined as product of axial force with the distance from the neutral axis (lever arm) of the non-cracked beam to the neutral axis of the cracked section (Fig. 33.2a). The position of the crack, either at the bottom surface (producing a positive bending moment) or at the top surface of the beam (producing a negative bending moment), was in the analysis introduced through the sign of the inner moment M_N, defined by the axial force lever arm. In the initial study (Skrinar 2020), it was simply assumed (without fracture mechanics analysis) that the lever arm for the inner moment equals half the crack depth d, and the definition of the rotary spring K_r was modified accordingly. In the upgraded and improved analysis (Skrinar and Imamović 2023), however, the lever arm was given more realistically as a function of the relative crack depth δ, and this definition was even shown to agree with Okamura's original definition of rotational spring stiffness K_r, unifying the analysis of transverse displacement due to both influences. However, both studies were limited exclusively to observing the axial force's influence on the cracked beam's transverse displacements.

In this new study, an improved mathematical model for the static analysis of beams with a substitute rotational spring at the crack location is presented. This improved new model differs from the existing established model in that now the rotational spring with bending stiffness K_r is being shifted from the position of the neutral axis of the non-cracked section to the (apparent) neutral axis of the cracked section (Fig. 33.2b). This change in the model requires redefining of the lever arm ρ, which now provides a more accurate representation of rotational behavior at the crack location, enhancing the precision and reliability of the model. This novel approach allows for a more comprehensive analysis, accommodating the complex interactions between bending moments and axial forces, thereby improving predictive capabilities and offering greater versatility in practical applications.

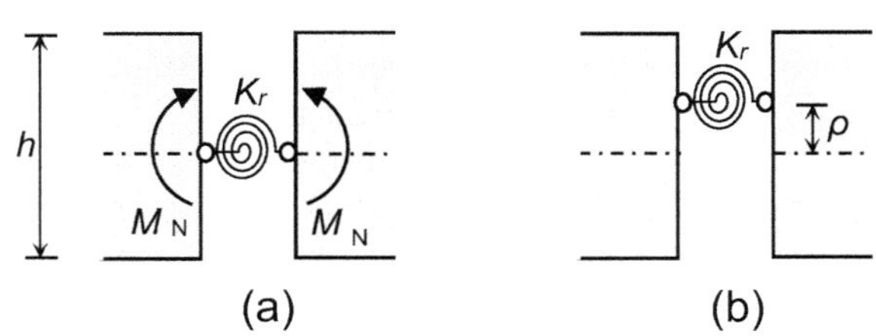

Fig. 33.2 Mathematical model of a crack: **a** with a substitute rotational spring at the cross-section centroid and inner moment M_N; **b** with a substitute eccentric rotational spring

Furthermore, the presented enhanced model now also enables the modeling of the influence of bending on the axial displacement in the crack.

33.2 Novel Two—Parameter Formulation of a Crack with a Simplified Model of an Eccentric Rotational Spring

The distance of the rotational spring (denoted by the lever arm ρ) from the neutral axis of the genuine non-cracked cross-section to its new position deeper within the non-cracked part of the cracked cross-section is generally defined as:

$$\rho = \frac{d}{f_{\rho}(\delta)} \tag{33.1}$$

where $f_{\rho}(\delta)$ represents the compliance function for the lever arm and primarily depends on the relative crack depth $\delta = d/h$, with d being the depth of the transverse crack and h being the height of the cross-section.

For the rotational stiffness of the crack K_r, many definitions have been presented (Ostachowicz and Krawczuk 1990; Liang et al. 1992; Sundermeyer and Weaver 1995), all stemming from the same basis:

$$K_r = \frac{EI}{h \cdot f_{Kr}(\delta)} \tag{33.2}$$

where EI is the bending stiffness of the genuine non-cracked section at the crack location, defined as the product of the elastic modulus E and the second moment of area I_z. $f_{Kr}(\delta)$ represents the compliance function for the rotational stiffness, which also primarily depends on the relative crack depth δ.

Since the stress distribution at the crack location due to the bending moment M_z differs from that caused by the axial force N_x, the definitions of the lever arm ρ and the rotational stiffness K_r also differ. The normal stress distribution σ_x under a bending moment is linear across the non-cracked cross-section, with maximum normal stress occurring at the outer fibers and zero stress at the neutral axis. In contrast, the stress distribution under an axial force is uniform across the entire non-cracked cross-section. These fundamental differences in stresses' distributions clearly necessitate distinct definitions and considerations for the lever arm and rotational stiffness. Therefore, three different formulations of the simplified mathematical crack model were separately presented.

The primary advancement of these formulations, in addition to a more realistic kinematic description, lies in their ability to account for the coupled effects of bending moments and axial displacements due to the presence of an eccentric rotational spring at the crack location. This novel approach significantly improves the accuracy of predicting mechanical responses in cracked beams, as it allows for the calculation of both transverse and axial displacements simultaneously. The enhanced model, entitled two-parameter simplified crack model (TPSCM), achieves a high level of precision while requiring significantly less computational effort compared to more complex numerical methods.

33.2.1 Mathematical Model with an Eccentric Rotational Spring Due to Inner Bending Moment M_z

In this formulation, as shown in Fig. 33.3, due to the bending moment M_z in the simplified model, the total rotation of the rotational spring $\Delta\varphi_{,M}$ occurs as the difference between the oppositely directed rotations $\varphi_{1,M}$ and $\varphi_{2,M}$ of the left and right segments and is defined as the ratio between the moment M_z and the corresponding rotational stiffness $K_{r,M}$:

$$\Delta\varphi_M = \varphi_{2,M} - \varphi_{1,M} = \frac{M_z}{K_{r,M}} \tag{33.3}$$

The rotations $\varphi_{1,M}$ and $\varphi_{2,M}$ (due to the eccentricity of the rotational spring) cause axial displacements $u_{1,\mathrm{M}}$ and $u_{2,\mathrm{M}}$ at the neutral axis location of both non-cracked segments. The total axial displacement Δu_M is defined, according to the small displacement theory, as:

$$\Delta u_M = u_{2,M} - u_{1,M} = \Delta\varphi_M \cdot \rho_M \tag{33.4}$$

This model already allows the calculation of the axial displacement due to the bending moment, thus overcoming the limitations of the previous computational model with the concentrated inner moment M_N.

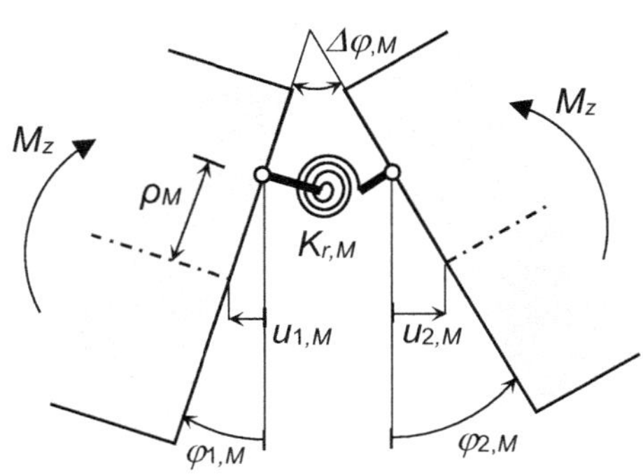

Fig. 33.3 Mathematical model formulation with an eccentric rotational spring due to the action of inner bending moment M_z

33.2.2 Mathematical Model with Rotational and Translational Springs Due to Inner Axial Force N_x

In this case, the tensile axial force N_x at the crack location causes non-uniform transverse stretching of the longitudinal fibers. This results in increased local rotation due to the moment M_N, which arises from the eccentricity ρ_φ relative to the neutral axis, as shown in Fig. 33.4.

Due to the bending moment $M_N = N_x \cdot \rho_\varphi$, resulting from the eccentricity of the axial force, in this simplified model, the total rotation $\Delta\varphi_{,N}$ (given as the difference between the rotations $\varphi_{1,N}$ and $\varphi_{2,N}$ of the left and right segments), is defined as:

$$\Delta\varphi_N = \varphi_{2,N} - \varphi_{1,N} = \frac{N_x \cdot \rho_\varphi}{K_{r,\varphi}} \tag{33.5}$$

The rotations $\varphi_{1,N}$ and $\varphi_{2,N}$ (due to the eccentricity of the rotational spring) also cause axial displacements $u_{1,\varphi}$ and $u_{2,\varphi}$. At the neutral axis location of both segments, additional displacements $u_{1,x}$ and $u_{2,x}$ occur due to the translational spring. The total axial displacement Δu_N is thus defined as the sum of both contributions:

$$\Delta u_N = u_{2,\varphi} - u_{1,\varphi} + u_{2,x} - u_{1,x} = \Delta\varphi_N \cdot \rho_\varphi + \Delta u_x \tag{33.6}$$

where the axial displacement Δu_x is defined as the ratio between the axial force N_x and the translational stiffness K_x:

$$\Delta u_x = \frac{N_x}{K_x} \tag{33.7}$$

Although this model leans on the previous computational model in the part where it models the influence of axial force onto transverse displacements, it still presents a step forward as it also allows modeling the effect of the crack on the axial displacements due to the axial force. However, the drawback of this model is that, in addition

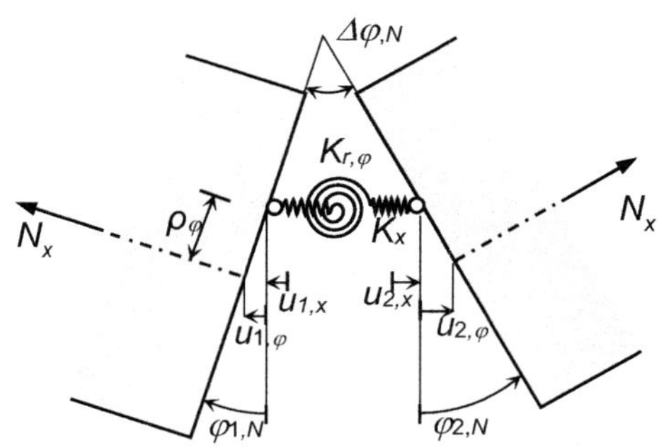

Fig. 33.4 Formulation of the mathematical model with eccentric rotational spring and translational spring due to the action of axial force N_x

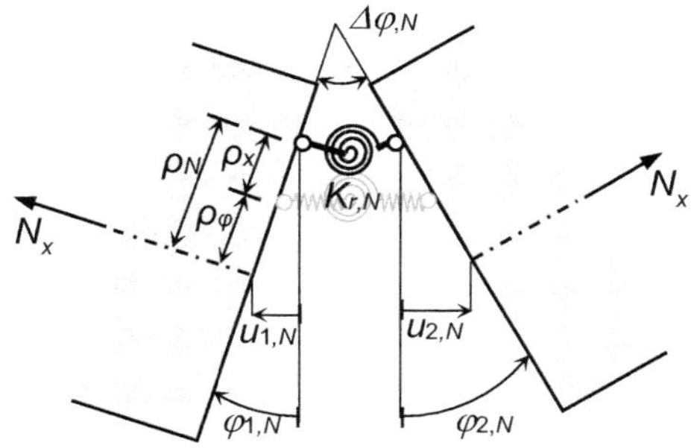

Fig. 33.5 Formulation of the equivalent mathematical model with eccentric rotational spring due to the action of axial force N_x

to the lever arm ρ_φ and rotational stiffness $K_{r,\varphi}$, the translational stiffness K_x also needs to be defined.

To reduce the number of model parameters, an evolution of the previous model was carried out. The axial displacement Δu_x caused by the extension of the translational spring was introduced in the model by extending (an additional shifting of) the lever arm ρ_φ by $\rho_x = \frac{\Delta u_x}{\Delta \varphi_N}$. This results in a new, adequate definition of the total lever arm $\rho_N = \rho_\varphi + \rho_x$ and the corresponding rotational spring $K_{r,N}$, as shown in Fig. 33.5.

The total rotation $\Delta\varphi_N$ and axial displacement Δu_N are now redefined as:

$$\Delta\varphi_N = \varphi_{2,N} - \varphi_{1,N} = \frac{N_x \cdot \rho_N}{K_{r,N}} \tag{33.8}$$

$$\Delta u_N = u_{2,N} - u_{1,N} = \Delta\varphi_N \cdot \rho_N \tag{33.9}$$

By implementing this model, the need to define a special translational stiffness K_x is avoided. Therefore, the definition of the eccentric crack model due to inner axial force N_x that includes only two parameters ($K_{r,N}$ and ρ_N) instead of three ($K_{r,\varphi}$, K_x and ρ_φ) were used in further analysis.

33.3 Mathematical Formulation and Derivation of New Definitions of the Lever Arm and Rotational Stiffness of the Rotational Spring

The derivations of definitions for the lever arm and stiffness of the rotational spring were in this study conducted on a cantilever beam of length L with a transverse crack on the bottom surface at a distance L_1 from the clamped end. The cantilever beam was chosen for this study because it is an ideal choice, allowing for creating a stress state in the vicinity of the crack solely due to bending moments or axial forces. Additionally, this choice facilitates the straightforward examination and further derivation of both the lever arm and rotational stiffness definitions. To derive the lever arm ρ_M and rotational stiffness $K_{r,M}$ exclusively due to the inner moment M_z, the beam was

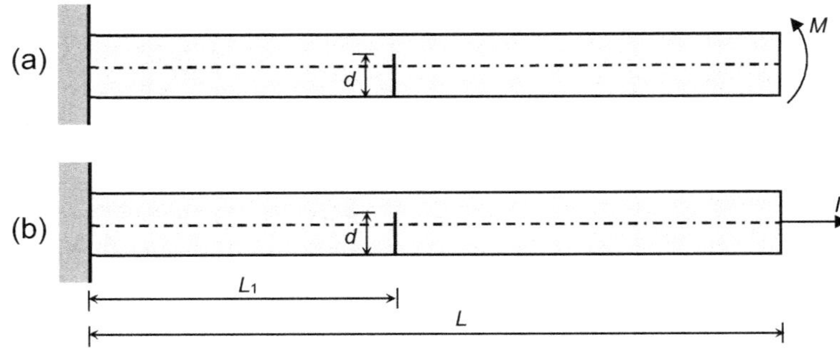

Fig. 33.6 Cantilever beam with a transverse crack: **a** loaded with a concentrated moment M for defining lever arm ρ_M and rotational stiffness $K_{r,M}$; **b** loaded with a concentrated axial force N for defining lever arm ρ_N and rotational stiffness $K_{r,N}$

loaded at the free end with a concentrated moment M (Fig. 33.6a). Consequently, the inner moment M_z is, therefore, the only static quantity presented in the beam and at the location of the crack. For deriving the lever arm ρ_N and rotational stiffness $K_{r,N}$ due to the inner axial force N_x, the beam was loaded at the free end with a concentrated tensile axial force N (Fig. 33.6b). Consequently, the inner axial force N_x is the only inner static quantity present in the beam.

33.3.1 *Formulation of a Simplified Cantilever Beam Model with Eccentric Rotational Spring*

For the first loading case (with the concentrated moment M), the cantilever beam was modeled using an eccentric equivalent rotational spring with stiffness $K_{r,M}$ and lever arm ρ_M (Fig. 33.7a). For the second loading case (with the concentrated axial force N_x), the cantilever beam was modeled using an eccentric equivalent rotational spring with stiffness $K_{r,N}$ and lever arm ρ_N (Fig. 33.7b). Since both loading cases fundamentally represent the same models, differing only in the lever arm and rotational stiffness definitions, the derivation was generalized to $K_{r,i}$ and ρ_i ($i = M$ and N) for further analysis.

Considering only the deformation of the rotational spring, the right segment rotates due to the inner moment in the spring as a rigid body around the point located at the spring's position with lever arm ρ_i from the beam's neutral axis. For both loading cases, the transverse displacement of any point on the right segment was defined as the product of rotation with the perpendicular distance from the centre of rotation to the selected point on the neutral axis of the right segment. The discrete absolute total displacement (v) at the centroid of the transverse section's far-right point, as shown in Fig. 33.7, thus has horizontal and vertical components (v_x in v_y), which are generally defined for both loading cases ($i = M, N$) as:

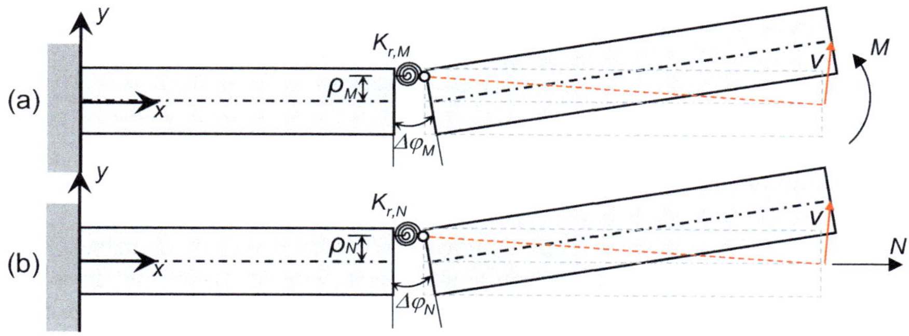

Fig. 33.7 Cantilever beam with eccentric equivalent rotational spring: **a** loaded with concentrated moment M to define lever arm ρ_M and rotational stiffness $K_{r,M}$; **b** loaded with concentrated axial force N to define lever arm ρ_N and rotational stiffness $K_{r,N}$

$$v_{x,i} \cong \Delta\varphi_i \cdot \rho_i \tag{33.10}$$

$$v_{y,i} \cong \Delta\varphi_i \cdot (L - L_1) \tag{33.11}$$

Conversely, using Eqs. (33.10) and (33.11), the displacements ($v_{x,i}$ in $v_{y,i}$) can be used to define the required lever arm as ($i = M, N$):

$$\rho_i = (L - L_1) \cdot \frac{v_{x,i}}{v_{y,i}} \tag{33.12}$$

and the rotation $\Delta\varphi_i$ ($i = M, N$) at the crack location as:

$$\Delta\varphi_i = \frac{v_{y,i}}{(L - L_1)} \tag{33.13}$$

From Eq. (33.13), rotational stiffnesses of the springs $K_{r,i}$ ($i = M, N$) were further determined using Eqs. (33.3) and (33.8), respectively:

$$K_{r,M} = \frac{M \cdot (L - L_1)}{v_{y,M}} \tag{33.14}$$

$$K_{r,N} = N \cdot v_{x,N} \cdot \left(\frac{L - L_1}{v_{y,N}}\right)^2 \tag{33.15}$$

Finally, based on Eqs. (33.1) and (33.2), for both loading cases, the compliance functions for lever arm and stiffness were expressed as ($i = M, N$):

$$f_{\rho,i}(\delta) = \frac{d}{\rho_i} \tag{33.16}$$

$$f_{Kr,i}(\delta) = \frac{EI}{h \cdot K_{r,i}} \tag{33.17}$$

Application of Betti-Maxwell's Reciprocal Theorem

Applying Betti-Maxwell's reciprocal theorem to the depicted cantilever beam (Fig. 33.7), it was found that the horizontal displacement $v_{x,M}$ at the end of the cantilever due to moment M equals the rotation $\Delta\varphi_N$ due to an equally sized force N:

$$v_{x,M} \cong \Delta\varphi_N \tag{33.18}$$

By considering Eq. (33.10) in Eq. (33.18), it was obtained:

$$\Delta\varphi_M \cdot \rho_M \cong \Delta\varphi_N \tag{33.19}$$

In Eq. (33.19), Eqs. (33.3) and (33.8) were further used, with implemented unit bending moment M_z and unit axial force N_x. Thus, the following relation was obtained:

$$\frac{\rho_M}{K_{r,M}} = \frac{\rho_N}{K_{r,N}} \tag{33.20}$$

Equation (33.20) was rearranged to obtain the ratio:

$$\overline{\alpha} = \frac{\rho_N}{\rho_M} = \frac{K_{r,N}}{K_{r,M}} \tag{33.21}$$

This implies that the ratio $\overline{a}$ between the lever arms ρ_M and ρ_N, equals the ratio between the rotational stiffnesses of the springs $K_{r,M}$ and $K_{r,N}$.

33.3.2 Formulation of a Planar Model for Identification of Model's Parameters

With a planar or spatial model, using a sufficiently dense mesh of 2D or 3D finite elements (FE), the crack can be discretely modeled, which further allows to determine displacements accurately. While 2D and 3D models generally do not yield significant differences in results, a spatial model is necessary when a 2D model cannot be applied, such as in the modeling of cracked beams with varying widths.

Determining displacements ($v_{x,i}$ in $v_{y,i}$) at the free end of the cantilever beam for both loading cases ($i = M, N$) requires three steps.

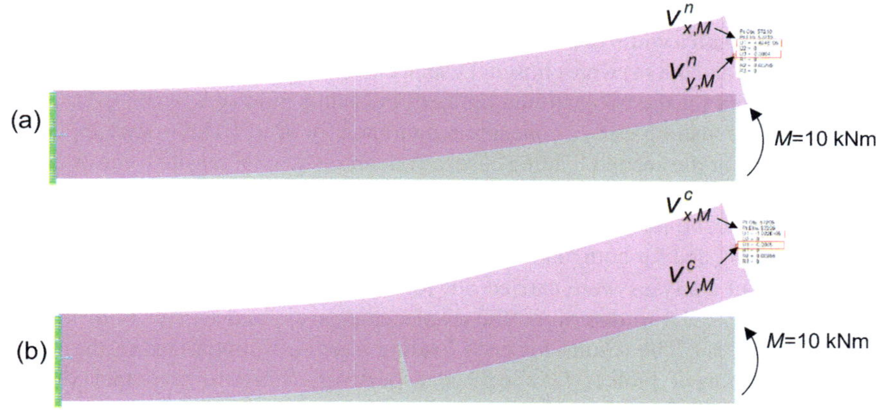

Fig. 33.8 Deformed cantilever beam loaded with concentrated moment $M = 10$ kNm modeled in SAP2000: **a** non-cracked case to determine the $v^n_{x,M}$ and $v^n_{y,M}$ values; b cracked case to determine the $v^n_{x,M}$ and $v^n_{y,M}$ values

- Step 1: Calculate the displacements ($v^n_{x,i}$ and $v^n_{y,i}$) at the free edge of the non-cracked cantilever beam, as shown in Fig. 33.8a for the loading case with a concentrated moment M.
- Step 2: Determine the displacements ($v^c_{x,i}$ and $v^c_{y,i}$) at the same cantilever beam's free edge with the modeled transverse crack, as shown in Fig. 33.8b for the loading case with a concentrated moment M (with $\delta = 0.5$). These displacements are influenced by the stress–strain state of the entire beam, not just near the crack.
- Step 3: Subtract the values of displacements between cracked and non-cracked beams to obtain final values of transverse and axial displacements ($v_{x,i} = v^c_{x,i} - v^n_{x,i}$ and $v_{y,i} = v^c_{y,i} - v^n_{y,i}$), which are caused solely by the stress–strain state in the vicinity of the crack.

These obtained values of displacements ($v_{x,i}$ and $v_{y,i}$) from the planar or spatial model were further used in Eq. (33.12) to define lever arms ρ_i for both loading cases ($i = M, N$). Using Eqs. (33.14) and (33.15), rotational stiffnesses $K_{r,M}$ and $K_{r,N}$ were then determined. Finally, compliance functions for lever arms ($f_{\rho,M}$ in $f_{\rho,N}$) and rotational stiffness ($f_{Kr,M}$ in $f_{Kr,N}$) were determined using Eqs. (33.16) and (33.17), respectively.

33.3.3 Determination of Compliance Function Definitions for a Prismatic Beam with Rectangular Cross-Section

A cantilever beam with a length of $L = 6$ m and a transverse crack at a distance of $L_1 = 3$ m was modeled in SAP2000. The model used a high-density 2D finite element mesh with up to 61,587 elements for the numerical studies. A rectangular cross-section with a width of $b = 0.4$ m was selected and three different height values (0.2 m, 0.4 m, and 0.8 m) were chosen. Displacements at the free end of the cantilever were calculated for the two extreme limits of Poisson's ratio ($\nu = 0$ and 0.5).

For the first loading case, a concentrated moment of $M = 10$ kNm was applied to the free end. For the second loading case, a concentrated axial tensile force of $N = 1$ MN was applied at the free end. Initial analyses were carried out for a non-cracked beam ($\delta = 0$). The results of the displacements $v^n_{x,i}$ and $v^n_{y,i}$ at the free end of the cantilever beam and for both loading cases ($i = M, N$) are summarized in Table 33.1.

Subsequent analyses were carried out for beams with cracks of relative depths $\delta = 0.1, 0.2, \ldots 0.5$ to determine the displacements $v^c_{x,i}$ and $v^c_{y,i}$ for both loading cases ($i = M, N$). The results for each loading case are summarized in the fourth and fifth columns of Tables 33.2 and 33.3, respectively. Since the non-cracked beam does not axially deform due to the concentrated moment M ($v^n_{x,M} = 0$), it follows that $v_{x,M} = v^c_{x,M}$, as considered in the 4th column of Table 33.2. Similarly, in the 4th column of Table 33.3, it is taken into account that the non-cracked beam does not transversely deform due to the concentrated axial force N ($v^n_{y,N} = 0$), hence $v_{y,N} = v^c_{y,N}$.

The remaining displacements $v_{y,M}$ and $v_{x,N}$, obtained by subtracting the displacements of the cracked and non-cracked beams, are given in the sixth column of Tables 33.2 and 33.3, respectively. In addition, columns seven to ten give the derived definitions of the lever arms and rotational stiffnesses as well as the corresponding and compliance functions for the eccentric rotational spring model.

The results from Tables 33.2 and 33.3 for both loading cases indicate that the compliance functions for both lever arms ($f_{\delta,M}$ and $f_{\delta,N}$ in the columns before penultimate) and rotational stiffnesses ($f_{Kr,M}$ and $f_{Kr,N}$ in last columns) are significantly influenced only by the relative crack depth δ, while the height of the beam h and Poisson's ratio ν have negligible effects. For all five considered relative depths, average discrete values of compliance functions for lever arms as well as rotational stiffness were computed from six obtained results (for three different heights and

Table 33.1 The results of the non-cracked ($\delta = 0$) beam displacements $v^n_{x,i}$ and $v^n_{y,i}$ for both loading cases of the 2D FE model

H [m]	ν [/]	Loading case			
		M = 10 kNm		N = 1 MN	
		$v^n_{x,M}$ [mm]	$v^n_{y,M}$ [mm]	$v^n_{x,N}$ [mm]	$v^n_{x,N}$ [mm]
0.2	0.0	0.0000	22.504	2.4993	0.0000
0.2	0.5	0.0000	22.484	2.4966	0.0000
0.4	0.0	0.0000	2.8143	1.2493	0.0000
0.4	0.5	0.0000	2.8100	1.2466	0.0000
0.8	0.0	0.0000	0.3525	0.6245	0.0000
0.8	0.5	0.0000	0.3517	0.6218	0.0000

Table 33.2 The results of the cracked beam for loading case $M = 10$ kNm

δ [/]	h [m]	ν [/]	$v_{x,M} = v^{c}_{x,M}$ [10^{-3} mm]	$v^{c}_{y,M}$ [mm]	$v_{y,M}$ [10^{-3} mm]	ρ_M [mm]	$f_{\rho,M}$ [/]	$K_{r,M}$ [kNm]	$f_{Kr,M}$ [/]
0.1	0.2	0.0	0.9043	22.579	75.000	36.171	0.5529	400.00	0.1000
0.1	0.2	0.5	0.9120	22.560	75.700	36.144	0.5533	396.30	0.1009
0.1	0.4	0.0	0.4767	2.8341	19.740	72.447	0.5521	1519.8	0.1053
0.1	0.4	0.5	0.4811	2.8299	19.910	72.489	0.5518	1506.8	0.1062
0.1	0.8	0.0	0.2297	0.3572	4.7640	144.64	0.5530	6297.2	0.1016
0.1	0.8	0.5	0.2311	0.3565	4.7920	144.70	0.5528	6260.4	0.1022
0.2	0.2	0.0	3.9909	22.807	303.00	39.514	1.0123	99.010	0.4040
0.2	0.2	0.5	4.0074	22.788	304.20	39.521	1.0121	98.619	0.4056
0.2	0.4	0.0	1.9955	2.8901	75.760	79.018	1.0124	395.99	0.4041
0.2	0.4	0.5	2.0037	2.8861	76.050	79.041	1.0121	394.48	0.4056
0.2	0.8	0.0	1.0027	0.3715	19.030	158.07	1.0122	1576.5	0.4060
0.2	0.8	0.5	1.0061	0.3708	19.089	158.12	1.0119	1571.6	0.4072
0.3	0.2	0.0	10.204	23.210	706.60	43.323	1.3849	42.457	0.9421
0.3	0.2	0.5	10.252	23.193	709.60	43.341	1.3844	42.277	0.9461
0.3	0.4	0.0	5.1020	2.9910	176.66	86.641	1.3850	169.82	0.9422
0.3	0.4	0.5	5.1258	2.9874	177.39	86.686	1.3843	169.12	0.9461
0.3	0.8	0.0	2.5927	0.3973	44.802	173.61	1.3824	669.61	0.9558
0.3	0.8	0.5	2.5993	0.3966	44.903	173.66	1.3820	668.11	0.9579
0.4	0.2	0.0	22.245	23.898	1394.8	47.845	1.6721	21.508	1.8597
0.4	0.2	0.5	22.326	23.883	1399.1	47.872	1.6711	21.442	1.8655
0.4	0.4	0.0	11.122	3.1630	348.70	95.692	1.6720	86.036	1.8597
0.4	0.4	0.5	11.163	3.1598	349.80	95.741	1.6712	85.768	1.8655

(continued)

Table 33.2 (continued)

δ [/]	h [m]	ν [/]	$v_{x,M} = v^c_{x,M}$ [10^{-3} mm]	$v^c_{y,M}$ [mm]	$v_{y,M}$ [10^{-3} mm]	ρ_M [mm]	$f_{\rho,M}$ [/]	$K_{r,M}$ [kNm]	$f_{Kr,M}$ [/]
0.4	0.8	0.0	5.6245	0.4405	88.000	191.68	1.6694	340.80	1.8779
0.4	0.8	0.5	5.6370	0.4399	88.200	191.74	1.6689	340.15	1.8815
0.5	0.2	0.0	45.62	25.078	2.5752	53.141	1.8818	11.650	3.4336
0.5	0.2	0.5	45.76	25.066	2.5822	53.166	1.8809	11.618	3.4429
0.5	0.4	0.0	22.81	3.4581	0.6438	106.28	1.8818	46.599	3.4335
0.5	0.4	0.5	22.88	3.4555	0.6455	106.33	1.8809	46.473	3.4429
0.5	0.8	0.0	11.50	0.5146	0.1621	212.83	1.8795	185.08	3.4579
0.5	0.8	0.5	11.52	0.5141	0.1624	212.90	1.8788	184.75	3.4642

Table 33.3 The results of the cracked beam for loading case $N = 1$ MN

δ [/]	h [m]	ν [/]	$v^{c}_{x,N}$ [mm]	$v_{y,N} = v^{c}_{y,N}$ [mm]	$v_{x,N}$ [10^{-1} mm]	ρ_N [mm]	$f_{\rho,N}$ [/]	$K_{r,N}$ [kNm]	$f_{Kr,N}$ [/]
0.1	0.2	0.0	2.5025	0.27129	0.0327	36.161	0.5531	399.89	0.1000
0.1	0.2	0.5	2.4999	0.27361	0.0330	36.183	0.5527	396.72	0.1008
0.1	0.4	0.0	1.2527	0.14301	0.0346	72.582	0.5511	1522.6	0.1051
0.1	0.4	0.5	1.2501	0.14433	0.0349	72.544	0.5514	1507.9	0.1061
0.1	0.8	0.0	0.6278	0.06891	0.0333	144.80	0.5529	6304.1	0.1015
0.1	0.8	0.5	0.6252	0.06934	0.0335	144.84	0.5523	6266.4	0.1021
0.2	0.2	0.0	2.5151	1.1974	0.1583	39.665	1.0084	99.388	0.4025
0.2	0.2	0.5	2.5125	1.2022	0.1590	39.677	1.0081	99.009	0.4040
0.2	0.4	0.0	1.2651	0.5986	0.1583	79.330	1.0084	397.55	0.4025
0.2	0.4	0.5	1.2625	0.6011	0.1590	79.353	1.0081	396.03	0.4040
0.2	0.8	0.0	0.6404	0.3008	0.1591	158.66	1.0084	1582.3	0.4045
0.2	0.8	0.5	0.6378	0.3018	0.1597	158.69	1,0082	1577.3	0.4058
0.3	0.2	0.0	2.5438	3.0612	0.4456	43.669	1.3740	42.796	0.9347
0.3	0.2	0.5	2.5414	3.0754	0.4479	43.691	1.3733	42.619	0.9385
0.3	0.4	0.0	1.2938	1.5306	0.4456	87.338	1.3740	171.18	0.9347
0.3	0.4	0.5	1.2914	1.5377	0.4479	87.382	1.3733	170.48	0.9385
0.3	0.8	0.0	0.6699	0.7778	0.4538	175.02	1.3713	675.05	0.9481
0.3	0.8	0.5	0.6673	0.7798	0.4551	175.07	1.3708	673.54	0.9502
0.4	0.2	0.0	2.6071	6.6734	1.0784	48.479	1.6502	21.793	1.8354
0.4	0.2	0.5	2.6049	6.6977	1.0829	48.505	1.6493	21.726	1.8411
0.4	0.4	0.0	1.3571	3.3367	1.0784	96.958	1.6502	87.175	1.8354
0.4	0.4	0.5	1.3549	3.3488	1.0829	97.010	1.6493	86.904	1.8411
0.4	0.8	0.0	0.7338	1.6873	1.0925	194.26	1.6474	345.36	1.8531
0.4	0.8	0.5	0.7313	1.6911	1.0953	194.31	1.6468	344.71	1.8566
0.5	0.2	0.0	2.7461	13.685	2.4682	54.108	1.8481	11.861	3.3722
0.5	0.2	0.5	2.7443	13.728	2.4774	54.137	1.8472	11.830	3.3812
0.5	0.4	0.0	1.4961	6.8424	2.4682	108.23	1.8481	47.447	3.3722
0.5	0.4	0.5	1.4943	6.8643	2.4774	108.27	1.8472	47.32	3.3812
0.5	0.8	0.0	0.87371	3.4497	2.4920	216.71	1.8458	188.46	3.3959
0.5	0.8	0.5	0.87164	3.4571	2.4982	216.79	1.8451	188.12	3.4020

both extreme values of Poisson's ratio). These averages were subsequently used to determine polynomial compliance functions for lever arm and rotational stiffness, defined over the range $0 \le \delta \le 0.5$, for the loading case with an applied concentrated moment (M):

$$f_{\rho,M}(\delta)=6.0297\,\delta - 5.3104\,\delta^2 + 3.4146\,\delta^3 - 6.5383\,\delta^4 + 5.6050\,\delta^5 \qquad (33.22)$$

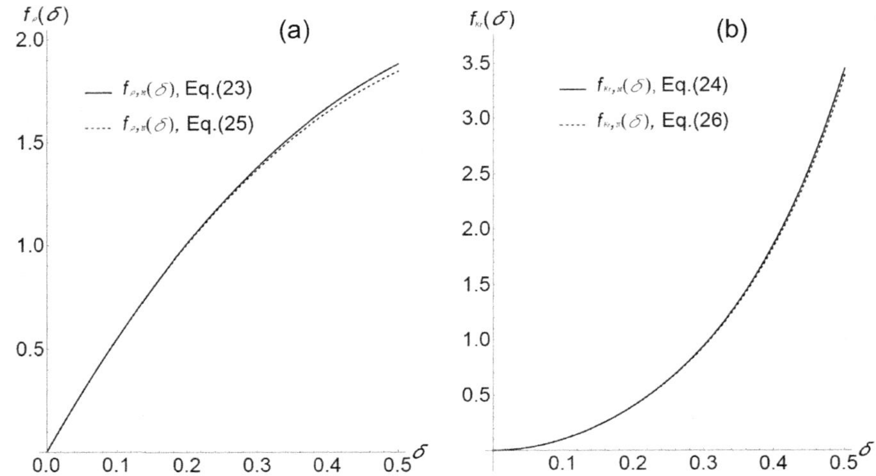

Fig. 33.9 Compliance functions: **a** for lever arms ρ; **b** for rotational stiffness K_r

$$f_{Kr,M}(\delta) = 0.01543\delta + 10.424\delta^2 - 3.5254\delta^3 + 0.5525\delta^4 + 39.624\delta^5 \quad (33.23)$$

Similarly, polynomial compliance functions for lever arm and stiffness were determined for the loading case with axial force, defined over the range $0 \leq \delta \leq 0.5$:

$$f_{\rho,N}(\delta) = 6.0354\delta - 5.4041\delta^2 + 3.3280\delta^3 - 7.1031\delta^4 + 6.6615\delta^5 \quad (33.24)$$

$$f_{Kr,N}(\delta) = 0.02091\delta + 10.323\delta^2 - 2.9389\delta^3 - 1.6574\delta^4 + 40.436\delta^5 \quad (33.25)$$

All four obtained compliance functions are presented in Fig. 33.9. It is evident from Fig. 33.9a that the definitions of the compliance functions for lever arms from Eqs. (33.22) and (33.23) do not differ significantly. The same applies to the compliance functions for rotational stiffness shown in Fig. 33.9b from Eqs. (33.24) and (33.25). Deviations in compliance functions for both parameters (ρ and K_r) gradually increase with the relative depth δ, reaching a maximum at relative depth $\delta = 0.5$, with discrepancies of approximately 2% for both loading cases.

33.4 Analytical Formulation of Static Bending Problem

The linear static analysis was chosen to validate the effectiveness of the proposed two-parameter model. Although this is the simplest possible analysis, this choice still allows for a clear and direct test of the new approach under controlled conditions, which is crucial for the initial validation of the model before applying it to more complex scenarios.

The proposed new two-parameter model, utilizing an eccentric equivalent rotational spring, allows for analytical analysis of the static bending of a straight beam with a crack in the x–y plane using EB theory. This approach requires solving differential equations for transverse displacements $v_i(x)$ and axial displacements $u_i(x)$ in both undamaged segments ($i = 1, 2$), which are generally expressed as:

$$\frac{d^2}{dx^2}\left(EI(x)\cdot\frac{d^2v_i(x)}{dx^2}\right) = q_y(x) \qquad \begin{matrix} i=1 & for & 0 \le x \le L_1 \\ i=2 & \text{for} & L_1 \le x \le L \end{matrix} \tag{33.26}$$

$$\frac{d}{dx}\left(EA(x)\cdot\frac{du_i(x)}{dx}\right) = q_x(x) \qquad \begin{matrix} i=1 & for & 0 \le x \le L_1 \\ i=2 & \text{for} & L_1 \le x \le L \end{matrix} \tag{33.27}$$

where $EA(x)$ is the axial stiffness defined as the product of the elastic modulus E and the area of the non-cracked cross-section $A(x)$, $q_y(x)$ is the distributed transverse load, and $q_x(x)$ is the distributed axial load on the beam.

The corresponding differential relations for bending moment $M_z(x)$, transverse force $V_y(x)$, and axial force $N_x(x)$, considering the continuous load along the entire beam, are:

$$M_z(x) = EI(x)\cdot\frac{d^2v_i(x)}{dx^2} \qquad (0 \le x \le L) \tag{33.28}$$

$$V_y(x) = \frac{d}{dx}\left(EI(x)\cdot\frac{d^2v_i(x)}{dx^2}\right) \qquad (0 \le x \le L) \tag{33.29}$$

$$N_x(x) = EA(x)\cdot\frac{du_i(x)}{dx} \qquad (0 \le x \le L) \tag{33.30}$$

By integrating Eq. (33.26) for both elastic segments, eight integration constants for the transverse displacement functions were initially obtained. An additional four constants for the axial displacement functions were derived by integrating Eq. (33.27), resulting in a total of twelve integration constants. Six of them are obtained from boundary conditions, while the remaining six continuity conditions are imposed at the crack location, including four standard conditions identical to those applicable to the undamaged state of the beam. These standard continuity conditions are expressed as:

$$v_1(L_1) = v_2(L_1) \tag{33.31}$$

$$\left.\frac{d^2v_1(x)}{dx^2}\right|_{x=L_1} = \left.\frac{d^2v_2(x)}{dx^2}\right|_{x=L_1} \tag{33.32}$$

$$\frac{d}{dx}\left(EI(x)\frac{d^2v_1(x)}{dx^2}\right)\bigg|_{x=L_1} = \frac{d}{dx}\left(EI(x)\frac{d^2v_2(x)}{dx^2}\right)\bigg|_{x=L_1} \tag{33.33}$$

$$\frac{du_1(x)}{dx}\bigg|_{x=L_1} = \frac{du_2(x)}{dx}\bigg|_{x=L_1} \tag{33.34}$$

The remaining two »continuity« conditions manifest as differences in rotation and axial displacement, which are interconnected due to the eccentricity of the rotational spring. The difference between the rotations of both segments at the crack location equals the sum of their total rotations $\Delta\varphi_M$ and $\Delta\varphi_N$ due to the simultaneous action of both inner forces (M and N_x):

$$\Delta\varphi = \Delta\varphi_M + \Delta\varphi_N = \frac{dv_2(x)}{dx}\bigg|_{x=L_1} - \frac{dv_1(x)}{dx}\bigg|_{x=L_1} \tag{33.35}$$

Similarly, the difference between the axial displacements of both segments at the crack location equals the superposition of their axial displacements Δu_M and Δu_N due to the simultaneous action of both inner forces (M and N_x):

$$\Delta u = \Delta u_M + \Delta u_N = u_2(L_1) - u_1(L_1) \tag{33.36}$$

Considering Eqs. (33.3), (33.4), (33.8) and (33.9), along with the differential relationships from Eqs. (33.28) to (33.30), both "cracked" continuity conditions from Eqs. (33.35) and (33.36) can finally be expressed in their ultimate form as:

$$\begin{aligned}&\left(EI(x)\frac{d^2v_1(x)}{dx^2}\right)\bigg|_{x=L_1}\cdot\frac{1}{K_{r,M}} + \left(EA(x)\cdot\frac{du_1(x)}{dx}\right)\bigg|_{x=L_1}\\ &\cdot\frac{\rho_N}{K_{r,N}} = \frac{dv_2(x)}{dx}\bigg|_{x=L_1} - \frac{dv_1(x)}{dx}\bigg|_{x=L_1}\end{aligned} \tag{33.37}$$

$$\begin{aligned}&\left(EI(x)\cdot\frac{d^2v_1(x)}{dx^2}\right)\bigg|_{x=L_1}\cdot\frac{\rho_M}{K_{r,M}} + \left(EA(x)\cdot\frac{du_1(x)}{dx}\right)\bigg|_{x=L_1}\\ &\cdot\frac{\rho_N^2}{K_{r,N}} = u_2(L_1) - u_1(L_1)\end{aligned} \tag{33.38}$$

The final solution of the TPSCM's governing differential equation (GDE) is further determined by boundary conditions, which depend on the support conditions on both sides of each considered structure.

The significant advantage of this comprehensive unified model is that it allows not only the calculation of transverse displacements, inner bending moments, and inner transverse forces due to axial forces but also, conversely, the calculation of axial displacements and inner axial forces in a cracked beam due to bending moments.

33.5 Presented Model's Verification with Numerical Examples

This section presents the verification of the TPSCM's GDE solutions using two numerical examples. The first example involves a cantilever beam with a crack, and the second example features a clamped–clamped beam with multiple cracks. These examples demonstrate the model's ability to accurately predict transverse and axial displacements, as well as inner static quantities.

33.5.1 First Example—Cantilever Beam with a Crack

A simple cantilever beam of length $L = 6$ m with a rectangular cross-section of height $h = 0.5$ m and width $b = 0.25$ m was considered (Fig. 33.10) to demonstrate the new TPSCM's GDE solutions for transverse and axial displacements. The values for the elastic modulus and Poisson's ratio were $E = 30$ GPa and $\nu = 0.3$, respectively. A crack with a relative depth of $\delta = 0.5$ was located at a distance of $L_1 = 0.9$ m from the fixed end on the left side. The stiffness values of the equivalent rotational springs and lever arms from Eqs. (33.1) and (33.2), taking into account Eqs. (33.22)–(33.25) for this crack depth were $K_{r,M} = 22{,}672.2$ kNm and $\rho_M = -0.132936$ m, and $K_{r,N} = 23{,}116.5$ kNm and $\rho_N = -0.135361$ m, respectively. The beam was subjected to two separately applied loading cases (l.c.) as follows:

- 1st loading case: uniformly distributed transverse load q_y = -2 kN/m along the entire beam
- 2nd loading case: uniformly distributed axial load q_x = 10 kN/m along the entire beam
- 3rd loading case: uniformly distributed transverse load q_y = -2 kN/m and axial load q_x = 10 kN/m along the entire beam, representing the superposition of previous two load cases.

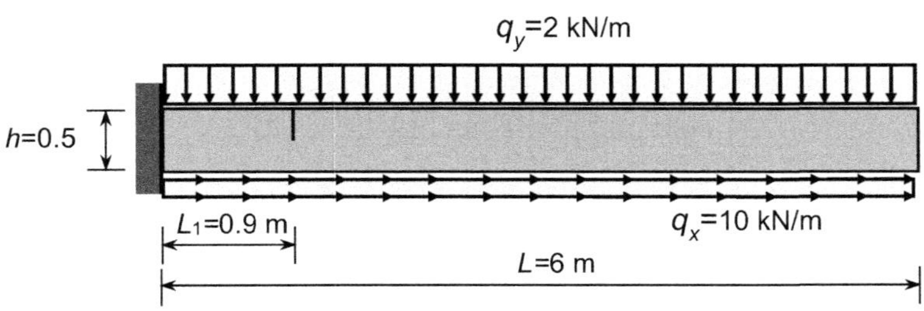

Fig. 33.10 Cantilever beam with a crack loaded with both loading cases presented together

The GDE solutions of the TPSCM for all loading cases were verified against a higher-quality solution, modeled in this example with 63,640 2D shell finite elements in SAP2000, which required solving a system of 193,611 linear equations.

Results and Discussion

The GDE solutions, Eqs. (33.26) and (33.27), were obtained subject to the continuity conditions, Eqs. (33.31)–(33.38), along with six specific boundary conditions for the cantilever beam. For both loading cases ($i = 1, 2$), these boundary conditions can generally be written as: $v_1^i(0) = 0$, $v_1^{i\prime}(0) = 0$, $u_1^i(0) = 0$, $M_z^i(L) = 0$, $V_y^i(L) = 0$ in $N_x^i(L) = 0$.

For the 1st loading case ($i = 1$), the final solutions for transverse and axial displacements functions are:

$$v_1^1(x) = -2.304 \cdot 10^{-4}x^2 + 2.560 \cdot 10^{-5}x^3 - 1.067 \cdot 10^{-6}x^4 (0 \le x \le 0.9\,\text{m})$$

$$v_2^1(x) = 5.162 \cdot 10^{-4} - 5.736 \cdot 10^{-4}x - 2.304 \cdot 10^{-4}x^2 + 2.560 \cdot 10^{-5}x^3 - 1.067 \cdot 10^{-6}x^4 (0.9\,\text{m} \le \text{x} \le 6.\,\text{m})$$

$$u_1^1(x) = 0 \qquad (0 \le x \le 0.9\,\text{m})$$

$$u_2^1(x) = 7.625 \cdot 10^{-5} \qquad (0.9\,\text{m} \le x \le 6.\,\text{m})$$

For the 2nd loading case ($i = 2$), the final solutions for transverse and axial displacement functions are:

$$v_1^2(x) = 0 \qquad (0 \le x \le 0.9\,\text{m})$$

$$v_2^2(x) = 1.346 \cdot 10^{-4} - 1.495 \cdot 10^{-4}x \qquad (0.9\,\text{m} \le x \le 6.\,\text{m})$$

$$u_1^2(x) = 1.600 \cdot 10^{-5}x - 1.333 \cdot 10^{-6}x^2 \qquad (0 \le x \le 0.9\,\text{m})$$

$$u_2^2(x) = 2.024 \cdot 10^{-5} + 1.600 \cdot 10^{-5}x - 1.333 \cdot 10^{-6}x^2 \qquad (0.9\,\text{m} \le x \le 6.\,\text{m})$$

The TPSCM's GDE solutions, as well as the 63,640 2D FE analyses of the third loading case, confirmed that it is a superposition of the first and second loading cases. Therefore, the following discussion is focused primarily on the first and second loading cases. The discrete results for transverse and axial displacements at the free end of the cantilever beam, calculated with the TPSCM's GDE and the 2D FE model,

Table 33.4 Transverse and axial displacements at the free end of the cantilever beam

Load Case (i)	Transverse displacements			Axial displacements		
	GDE [mm]	36,640 2D FE [mm]	Error [%]	GDE [mm]	36,640 2D FE [mm]	Error [%]
1	–7.0726	–7.0977	0.354	0.07625	0.07641	0.207
2	–0.7625	–0.7633	0.102	0.06824	0.06822	0.033
3	–7.8351	–7.8610	0.329	0.14449	0.14463	0.097

are for all loading cases presented in Table 33.4. In addition to the displacement values, the table presents the percentage of relative errors of the GDE solutions compared to the much more computational effort requiring 2D FE model.

For transverse displacements, the TPSCM's GDE results are very close to the 2D FE model's results. The largest (relative) error of the GDE solutions occurs in 1st load case, amounting to 0.354%, while the smallest error is in 2nd load case, amounting to 0.102%. The average error of the GDE solutions is very small, indicating good agreement with the 2D FE model in the analysis of transverse displacements.

Even for axial displacements, the TPSCM's GDE results are even slightly better to the 2D FE model's results. The largest error is in 1st load case with a value of 0.207%, while the smallest error appears in 2nd load case with a value of 0.033%. The average error in the analysis of axial displacements is also low, further confirming the great correspondence of the GDE solutions with the 2D FE model.

Overall, the differences between the TPSCM's GDE solutions and 2D FE models' results are minimal, indicating that the GDE solutions of the presented two-parameter model accurately predict the transverse and axial displacements of the cantilever beam. Low error values, all below 0.4%, demonstrate the high precision of the model considered.

Further analyses for several crack depths were conducted for 1st and 2nd loading cases. Figures 33.11 and 33.12 show transverse and axial displacements along the beam for the crack located at $L_1 = 0.9$ m from the fixed end of the beam and five considered relative crack depths ($\delta = 0.1, 0.2, 0.3, 0.4$, and 0.5). The displacement functions from the GDE solutions of the TPSCM are presented alongside the discrete displacements from the 63,640 2D FE model. The only noteworthy exceptions occur for axial displacements in the first load case. In the TPSCM, a discrete "jump" is observed near the crack, whereas in the 63,640 2D FE model, axial displacements increase steeply near the crack.

33.5.2 Second Example—Clamped–Clamped Beam with Multiple Cracks

A slender clamped–clamped beam with a length of $L = 8$ m, a rectangular cross-section with a height of $h = 0.2$ m, and a width of $b = 0.1$ m was analyzed to validate

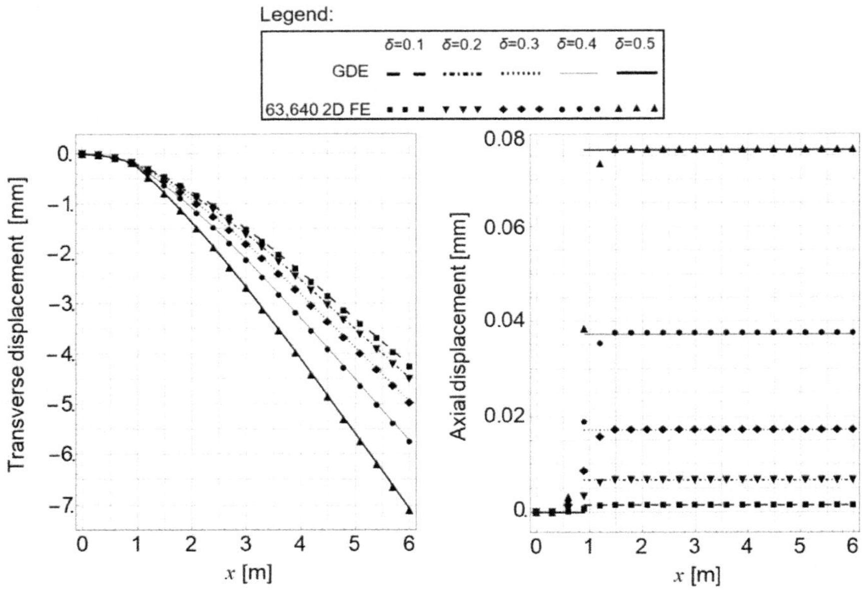

Fig. 33.11 Transverse and axial displacements of the cantilever beam for the 1st load case

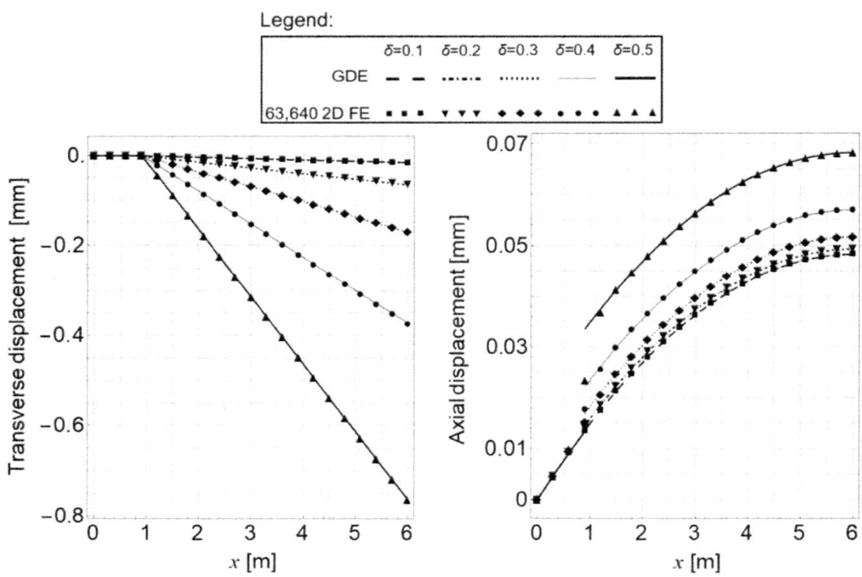

Fig. 33.12 Transverse and axial displacements of the cantilever beam for the 2nd load case

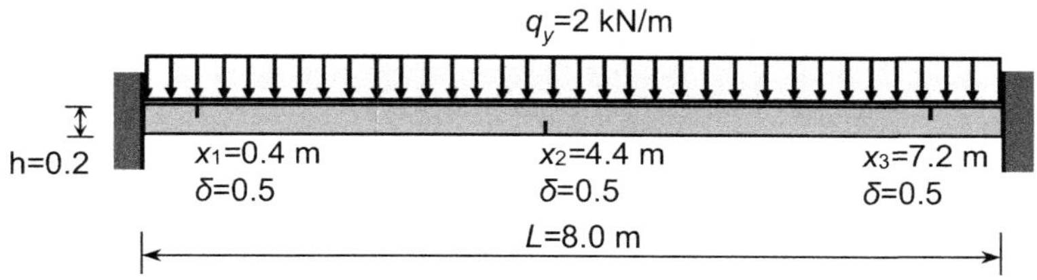

Fig. 33.13 Cracked clamped–clamped beam under uniform transverse load

the presented model further, Fig. 33.13. The beam was subjected to a uniform transverse load $q_y = 2$ kN/m directed downward. The slender beam was specifically chosen to minimize the influence of shear deformations on the transverse displacement. The material properties selected were an elastic modulus $E = 30$ GPa and Poisson's ratio $\nu = 0.2$.

Three transverse cracks, each with a relative depth $\delta = 0.5$, were introduced at three locations. The cracks, located at distances 0.4 m and 7.2 m from the left end, were located on the top side of the beam. The crack, located at a distance of 4.4 m, was located on the bottom side of the beam. The rotational stiffness values at all crack locations were the same and, according to Eqs. (33.2) and taking into account Eqs. (33.23) and (33.25), were $K_{r,M} = 2902.04$ kNm and $K_{r,N} = 2954.99$ kNm, respectively. The lever arm values for the cracks at distances 0.4 m and 7.2 m on the top side, according to Eqs. (33.1) and taking into account Eqs. (33.22) and (33.24), were $\rho_M = $ -0.053174 m and $\rho_N = $ -0.054144 m, respectively. For the crack at distance 4.4 m appearing on the bottom side, the lever arm values were the same in magnitude but positive: $\rho_M = 0.053174$ m and $\rho_N = 0.054144$ m.

The results of the GDE solutions of the TPSCM were verified against a higher-quality solution, modeled in this study with 64,000 2D shell finite elements in SAP2000, which required solving a system of 196,857 linear equations.

33.6 Results and Discussion

First, the general TPSCM's GDE solutions for transverse and axial displacements ($v_i(x)$ and $u_i(x)$) of the model were obtained for the four non-cracked segments using Eqs. (33.26) and (33.27). The 24 integration constants were determined by applying the continuity conditions at the crack locations, given by Eqs. (33.31)-(33.38), and the six boundary conditions at the clamped ends of the beam, which were expressed as $v_1(0) = 0$, $v_1'(0) = 0$, $u_1(0) = 0$, $v_4(L) = 0$, $v_4\prime(L) = 0$ and $u_4(L) = 0$.

The final solutions for transverse and axial displacements for all four segments are thus:

$$v_1(x) = -2.406 \cdot 10^{-3}x^2 + 6.540 \cdot 10^{-4}x^3 - 4.167 \cdot 10^{-5}x^4 \qquad (0 \leq x \leq 0.4\,\text{m})$$

$$v_2(x) = 7.745 \cdot 10^{-4} - 1.936 \cdot 10^{-3}x - 2.406 \cdot 10^{-3}x^2 + 6.540 \cdot 10^{-4}x^3 - 4.167 \cdot 10^{-5}x^4 (0.4\,\text{m} \leq x \leq 4.4\,\text{m})$$

$$v_3(x) = -6.081 \cdot 10^{-3} - 3.783 \cdot 10^{-4}x - 2.406 \cdot 10^{-3}x^2 + 6.540 \cdot 10^{-4}x^3 - 4.167 \cdot 10^{-5}x^4 (4.4\,\text{m} \leq x \leq 7.2\,\text{m})$$

$$v_4(x) = 3.685 \cdot 10^{-3} - 1.735 \cdot 10^{-3}x - 2.406 \cdot 10^{-3}x^2 + 6.540 \cdot 10^{-4}x^3 - 4.167 \cdot 10^{-5}x^4 (7.2\,\text{m} \leq x \leq 8.0\,\text{m})$$

$$u_1(x) = -3.211 \cdot 10^{-5}x \qquad (0 \leq x \leq 0.4\,\text{m})$$

$$u_2(x) = 1.0262 \cdot 10^{-4} - 3.211 \cdot 10^{-5}x \qquad (0.4\,\text{m} \leq x \leq 4.4\,\text{m})$$

$$u_3(x) = 1.851 \cdot 10^{-4} - 3.211 \cdot 10^{-5} \qquad (4.4\,\text{m} \leq x \leq 7.2\,\text{m})$$

$$u_4(x) = 2.569 \cdot 10^{-4} - 3.211 \cdot 10^{-5} \qquad (7.2\,\text{m} \leq x \leq 8.0\,\text{m})$$

Figure 33.14 shows excellent agreement between the TPSCM's GDE and the 2D FE models at all points. The relative error in transverse displacement at the middle of the beam is merely 0.21% when compared to the 2D FE model. However, noticeable "jumps" in axial displacements in the GDE solutions at the crack locations contrast with a steep increase observed in the 2D FE model. Nevertheless, the overall error remains low, indicating the high accuracy of the GDE solutions of the TPSCM in predicting both types or this type of displacements. In addition, the figure, just for comparison, includes transverse displacements of the non-cracked beam (marked with a dashed line), clearly highlighting the influence of the cracks. Meanwhile, axial displacements do not exist in the non-cracked scenario and, therefore, are not presented in the figure.

Next, the inner static quantities' functions were determined directly from the transverse and axial displacement functions using Eqs. (33.28)–(33.30), which are:

$$\text{M}_z(x) = -9.623 + 7.847x - 1.\,\text{x}^2 \qquad (0. \leq x \leq 8.0\,\text{m})$$

$$\text{V}_y(x) = 7.847\text{x} - 2.\text{x} \quad (0. \leq x \leq 8.0\,\text{m})$$

$$\text{N}_x(x) = -19.2683\text{kN} \quad (0. \leq x \leq 8.0\,\text{m})$$

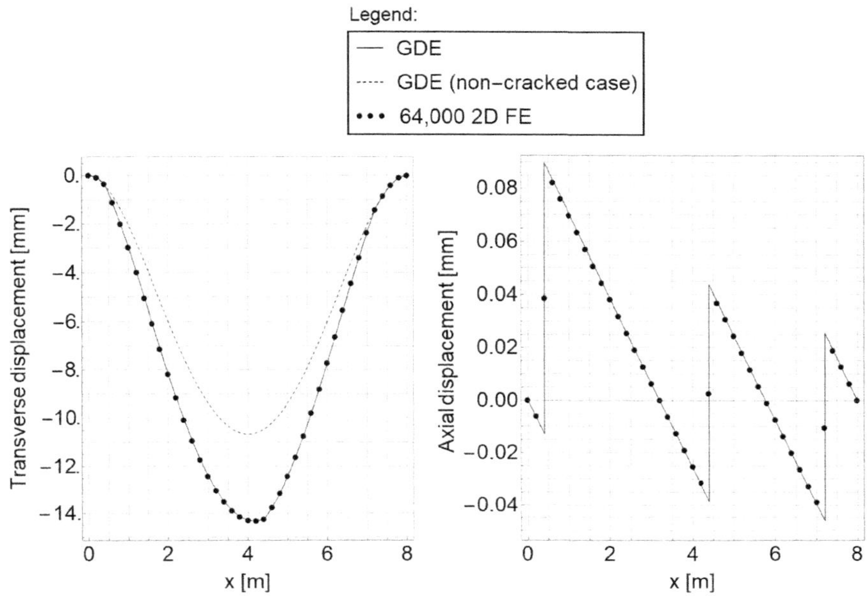

Fig. 33.14 Transverse and axial displacements of the clamped–clamped beam

Figure 33.15 thus shows a comparison of the inner static quantities (M_z, V_y and N_x) between the TPSCM's GDE solutions and the 2D FE model, demonstrating excellent agreement at all points examined. The presence of cracks induces a constant compressive axial force throughout the beam, which is 19.27 kN in the GDE solutions and 18.99 kN in the 2D FE model at all nodes. Notably, the relative error of the axial force values between the two models is minimal at only 1.45%.

The dashed line in the figure represents the bending moments and shear forces of the non-cracked clamped–clamped beam. It is evident that the influence of the transverse cracks on the bending moments M_z and transverse forces V_y is not very noticeable. In contrast, the crack effect on axial forces N_x is significant, as axial forces do not exist in the non-cracked scenario and are, therefore, not shown in the figure. This example underscores the major advantages of the new two-parameter model: it enables the simultaneous calculation of both axial displacements and axial forces in a transversely loaded cracked beam, a capability that previous models have lacked.

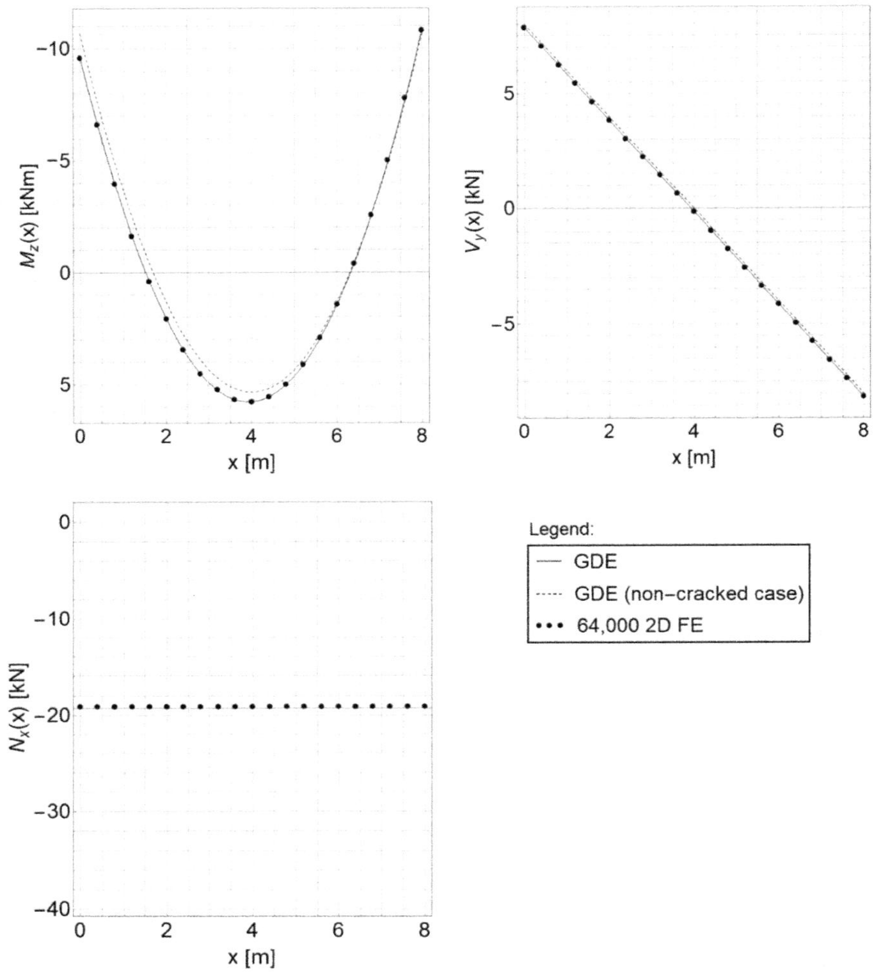

Fig. 33.15 Inner static quantities of the clamped–clamped beam

33.7 Conclusions

This study developed a novel analytical approach of the enhanced unified model to solve the problem of static bending of a straight beam with a crack, modeled with an eccentric equivalent rotational spring. Initially, the differential equations for transverse and axial displacements for both segments of the beam were solved, leading to the determination of 12 integration constants. At the crack location, six continuity conditions were introduced, including additional conditions for rotation and axial displacement, which now account for the eccentricity of the rotational spring.

The developed analytical model was verified with two numerical examples. The results of the model's analytical examples (presented in the form of GDE solutions) were compared with those of a more accurate 2D model in SAP2000. For transverse and axial displacements, the GDE solutions of the TPSCM closely matched the 2D FE model results, with a minor negligible relative error for both examples. Overall, the comparison with the 2D FE model underscores the high accuracy and reliability of the two-parameter model in linking transverse and axial displacements, as well as inner bending moments and axial forces, in the presence of transverse cracks. The minor discrepancy in their values highlights the model's robustness in practical engineering applications, undoubtedly ensuring reliable performance in structural analysis. Although both presented examples to analyze cracked prismatic beams, the expressions given in the article are prepared for the analysis of beams of non-constant cross-section.

Based on the obtained results, it is evident that the computational model, in conjunction with the analytical solutions of GDEs, consistently predicts transverse and axial displacements and proves highly beneficial for analyzing beams with cracks. To thoroughly validate the effectiveness of the new two-parameter model, this study was deliberately limited to the fundamental linear static analysis. Given the model's exceptional efficiency, expanding its application to other, more complex structures and analyses is logical. In order to simplify the modeling of complex frame geometries and enable the integration of the new two-parameter model into computer programs as well, the development of new finite elements incorporating this two-parameter model will be the next logical step. This advanced crack's modeling directs us towards further refining the model, including advanced dynamic analyses and considering the influence of axial forces on transverse displacements according to the second-order theory. On the other hand, separate studies should be dedicated to considering the proposed model's application for analyses of other types of local stiffness reductions.

Acknowledgements The second author acknowledges partial financial support from the Slovenian Research Agency (research core funding No. P2-0129 (A) "Development, modelling and optimization of structures and processes in civil engineering and traffic").

References

Andersen L, Nielsen SR (2008) Elastic beams in three dimensions. Aalborg University

Liang RY, Hu J, Choy F (1992) Theoretical study of crack-induced eigenfrequency changes on beam structures. J Eng Mech 118:384–396. https://doi.org/10.1061/(ASCE)0733-9399(1992)118:2(384)

Okamura H, Liu HW, Chu C-S, Liebowitz H (1969) A cracked column under compression. Eng Fract Mech 1(3):547–564. https://doi.org/10.1016/0013-7944(69)90011-3

Ostachowicz WM, Krawczuk M (1990) Vibration analysis of a cracked beam. Comput Struct 36:245–250. https://doi.org/10.1016/0045-7949(90)90123-J
Reddy JN (1994) An introduction to the finite element method, 2nd edn. McGraw-Hill
Skrinar M (2020) Transverse displacements of transversely cracked beams with a linear variation of width due to axial tensile forces. In: Öchsner A, Altenbach H (eds) Engineering design applications III. Advanced Structured Materials, vol 124. Springer, Cham
Skrinar M, Imamović D (2023) Enhanced transverse displacements analysis of transversely cracked beams with a linear variation of width due to axial tensile forces. Mater Werkst 54:482–495. https://doi.org/10.1002/mawe.202200287
Sundermeyer JN, Weaver RL (1995) On crack identification and characterization in a beam by non-linear vibration analysis. J Sound Vib 183:857–871. https://doi.org/10.1006/jsvi.1995.0290

Chapter 34
Simulation of Energy Harvesting by Using Piezoelectric Materials

Mahmoud A. Essam, Noha M. Abdeltawab, Ahmed M. Balata, Ahmed Y. Shash, and Mostafa M. El-Sayed

Abstract Over the years, piezoelectric devices have found particular use in the energy harvesting sector. Researchers in a variety of technical applications are drawn to these generators because of their simplicity. The objective of this work is to provide a rigorous process for estimating the potential for electrical energy generation from piezoelectric devices while taking various piezoelectric dimensions into account. To calculate the amount of energy that can be extracted from the constant load of 6 MPa on the upper and lower surfaces of a piezoelectric device, a finite element model of the device is created. Four distinct piezoelectric models with varying diameters were studied. The stresses along the vertical axis, the displacement in the x-direction, the electrical displacement along the polarization direction, and the output electrical energy were all examined.

Keywords Piezoelectric · Electrical energy · Electric Flux · FEA · Energy harvesting · Electrical displacement

---

M. A. Essam
Mechanical Engineering Department, Higher Technological Institute (HTI), 10Th of Ramadan City, Egypt
e-mail: Mahmoud.elsefy@hti.edu.eg

N. M. Abdeltawab · A. M. Balata · A. Y. Shash (✉)
Faculty of Engineering and Materials Science, German University in Cairo, Cairo, Egypt
e-mail: ahmed.shash@cu.edu.eg; ahmed.shash@guc.edu.eg

N. M. Abdeltawab
e-mail: noha.abdeltawab@guc.edu.eg

A. Y. Shash
Faculty of Engineering, Mechanical Design and Production Engineering Department, Cairo University, Giza, Egypt

M. M. El-Sayed
Manufacturing and Production Engineering Department, Modern Academy for Engineering and Technology, Cairo, Egypt

A. Öchsner and H. Altenbach (eds.), *Engineering Design Applications VII*, Advanced Structured Materials 230, https://doi.org/10.1007/978-3-031-84346-4_34

34.1 Introduction

These days, the general tendency in the development of electronic devices is miniaturization, multifunctionality, portability, flexibility, high processing capabilities, and low power communication (Sezer and Koç 2021). Wireless sensor networks, mobile electronics, wearable and implanted biomedical devices, and electromagnetic radiation in the form of light and radio waves (RF) can all be powered cleanly by harnessing the energies found in the surrounding environment, such as heat, fluid flows, mechanical vibrations, and in vivo energies (Sezer and Koç 2021). Electrochemical batteries have traditionally been used to power these devices. But battery life is finite and frequently less than that of electronic devices (Song 2019). It thereby incurs additional costs for replacement or recharging (Song 2019).

Since the piezoelectric effect is solely based on the intrinsic polarization of the material and does not require a separate voltage source, magnetic field, or contact with another material as in the case of electrostatic, electromagnetic, and turboelectric energy harvesting, respectively (Xie and Wang 2015), piezoelectric energy harvesting is a very convenient mechanism for capturing ambient mechanical energy and converting it into electric power (Wang et al. 2018a).

In comparison to other energy harvesting techniques, piezoelectric generators are robust, dependable, more sensitive to small strains, and have around 3–5 times higher density power output and higher voltage output (Chen et al. 2019). Furthermore, piezoelectric generators are easily integrated into microelectromechanical systems and can be produced in smaller sizes (Fei et al. 2018). Furthermore, environmental elements like humidity have little effect on them (Wu et al. 2017). As a result, the most promising ambient energy harvesting technology is piezoelectric transduction, which has found use in a wide range of industries, including wearable and implantable biomedical devices, transportation, wireless electronics, microelectromechanical systems, and structures (Siang et al. 2018; Madinei et al. 2016; Lu et al. 2018).

To create piezoelectric generators that may be used in a variety of areas, a variety of piezoelectric materials, including single crystals, ceramics, polymers, composites, and bio-inspired materials, have been generated in the form of nanostructures, thin-films, and layer stacks (Song 2019; Aslam et al. 2018). Wireless sensor networks are one of the most common uses for piezoelectric generators. It has become easier to harvest ambient energy for the sensor nodes' sustainable power supply due to the steady increase in the number of sensor nodes deployed in a variety of fields and the notable decrease in node size and power requirements (Cahill et al. 2018).

Potential fields of application for piezoelectric energy harvesters include transportation, avionics, defense, infrastructure, environmental monitoring, networking, electronics, and healthcare, among many other industries. For example, sensors powered by piezoelectric energy harvesters can be wirelessly communicated to check tire pressure. Tire pressure sensors are powered by piezoelectric generators built into automobile wheels, a technology that Snyder patented (Snyder 1985).

The driver was notified of aberrant tire (Parvez Mahmud et al. 2018) pressure via a low-power radio link, and the generator was fueled by the wheel vibrations experienced while driving. The prediction of rail wheel bearing failure using wireless communication of piezoelectric sensors, which transform rolling stock vibration into electrical power, is another example of energy harvesting in transportation (Siang et al. 2018).

Expanding the realms of use for piezoelectric technology has been made possible by recent research and development activities focused on micro- and nano-scale piezoelectric materials and production techniques (Chen et al. 2019). Emerging fields like wireless sensor networks, the internet of things, wearable electronics, and implantable biomedical devices have favored the use of newly developed piezoelectric materials like ceramics, polymers, composites, and bio-inspired natural materials in the form of various nanostructures and thin films with satisfactory physical properties like high piezoelectric coefficient, flexibility, stretch-ability, and durability (Xie and Wang 2015). As a result, there is increasing interest in piezoelectric energy harvesting across the globe.

In the ambient environment, mechanical energy is the most abundant and adaptable type of energy (Ahasan et al. 2018). Through mechanical-to-electrical transduction, motion, flow, and vibration of a source can be captured and transformed into electrical power (Lu et al. 2018). The most popular electromechanical energy harvesting technique, known as inertial energy harvesting, depends on a mass's resistance to acceleration (Tomasi 2017).

Because its crystal structure lacks a center of symmetry, ferroelectrics and other materials that undergo mechanical deformation (strain) shift in polarization, a phenomenon known as piezoelectricity (Butt et al. 2017). Active materials known as piezoelectric materials produce electricity in reaction to tiny mechanical forces. The kind of material used in piezoelectric energy harvesting greatly affects how well it works. As a result, a variety of materials have been studied for piezoelectric energy harvesting, including inorganic, organic, and composite materials (Safaei et al. 2019).

Large dielectric and piezoelectric coefficients, electromechanical coupling factors, and high energy conversion rates are characteristics of piezoceramics. But because of their extreme brittleness, they cannot withstand heavy stresses without breaking. Conversely, piezopolymers have a low electromechanical coupling factor and a great degree of flexibility (Ansari and Karami 2017). Another family of materials that combines the advantages of ceramics and polymers needed for certain applications is piezoelectric composites (Nestorović et al. 2013).

The crystal structures of the piezoelectric materials utilized in energy harvesting are perovskite or quartzite. Generally speaking, materials with a perovskite structure perform better piezoelectrically than those with a quartzite structure (Wu et al. 2017). However, in order to induce piezoelectric behavior, materials with a perovskite structure need to undergo an extra poling step. In order to give the generator the necessary flexibility and shield the piezoelectric layer from external elements like humidity, piezoelectric generators are usually made by sandwiching prepared piezoelectric materials between elastomeric substrates. In order to gather the created electric charge and ultimately send it to the external load, the top and bottom electrodes are linked (Butt et al. 2017).

Applications for harvesting include the following: square root of the mechanical–electrical energy conversion efficiency, mechanical quality factor "Q," which indicates the degree of damping (lower value indicates higher damping), dielectric constant which indicates the material's capacity to store charge, and piezoelectric strain constant "d," which indicates polarization per unit stress applied, and voltage constant "g," which indicates electric field per unit stress applied, electromechanically Generally speaking, inorganic piezoelectric materials have values of d, k, and ε significantly higher than piezoelectric polymers. Since polymers have substantially lower dielectric constants than inorganic materials, their g constants are larger (Song 2019).

Fabricating materials that are advantageous for energy applications has become possible with the development of nano- and micro-scale materials and manufacturing techniques (Luo et al. 2018). Many piezoelectric generators have been designed with desired features unique to their intended uses, including high piezoelectric performance, stretchability, durability, biocompatibility, and biodegradability. Because of this, research teams from a wide range of disciplines have looked at the possibility of using generators built of various materials to harvest ambient energy at the nano- to mesoscale (Qiu et al. 2017). To give readers an idea of the wide range of applications for piezoelectric generators, this section reviews research on the use of piezoelectric energy harvesting in a variety of fields, such as transportation, structures, aerial applications, in water applications, smart systems, microfluidics, biomedical, wearable and implantable electronics, and tissue regeneration (Fei et al. 2018; Li et al. 2018).

Road load mechanical energy can be used for a variety of things, like driving traffic signals, checking the condition of the roads structurally, powering self-contained car weighing systems, and more. The tracking wheel-pressure test of a beam piece was utilized by Li et al. (2018) to ascertain the road pavement's capacity to generate piezoelectricity. 65.2 V, the maximum voltage, was attained. 0.23 MJ of electric energy were created by the main wheel rolling impact. It is possible to generate daily electric capabilities of 0.8 kW/h, which is enough to run traffic signal lights. The market has witnessed the implementation of piezoelectric energy harvesting in the asphalt industry, Moure et al. (2016) use several modules featuring piezoelectric elements to extract vibrational energy from asphalt in the ambient.

Vibration energy harvesters are a useful tool for converting the energy of bridge vibrations—caused by passing cars—into electrical power (Lu et al. 2018; Karimi et al. 2016). This energy can be used for a number of things, like illumination and bridge structural health monitoring. For example, Karimi et al. (2016) harvested energy from bridge vibrations using a cantilever beam-type piezoelectric generator. At the generator's resonance frequency of 13.5 Hz, 200 kΩ was found to be the ideal load resistance, translating into a maximum output of about 35 μW. The literature on energy harvesting technologies in bridges was examined by Wang et al. (2018b).

It was determined that the generated power output for a single piezoelectric transducer under one vehicle pass on the bridge is typically low in piezoelectric energy harvesting. Therefore, additional sensor arrays under repeated traffic loading are needed to create more electric power (Li et al. 2018). One way that piezoelectric generators might generate electricity is by utilizing the torsional vibration that internal combustion engines produce. A rotating shaft had a cantilever beam construction fixed to its surface. It produced 14 μW of electrical power with a vibrating torque of 30 Nm. At an engine rotation speed of 2000 rpm, the output power may power wireless transducers, like a shaft torque transducer (Xie and Wang 2015).

A P(VDF-TrFE)/BT nanocomposite-based flexible and stretchable kerygma-structured piezoelectric generator was additively built by Zhou et al. (2020) and may be put on a sock to capture foot-stamping energy and function as an independent gait sensor. Three energy sources from walking activity can be employed to create electricity: compression force from an individual's weight acting on the shoe, acceleration from leg swing, and acceleration pulses from heel striking. Fan et al. (2017) created a piezoelectric generator that was installed within a shoe and was made up of a ferromagnetic ball, a crossbeam, and a piezoelectric cantilever beam in order to capture energy from these three distinct excitations of walking activity. When walking speed varied from 2 km/h to 8 km/h, output voltage and power rose from 1.06 V to 3.55 V and 0.03–0.35 MW, respectively. Nia et al. (2017) provide a thorough analysis of walking energy harvesting with piezoelectric materials.

34.2 Methodology

Our objectives are to examine additional material qualities and optimize the thickness at which the maximum electric potential is reached. We are using a combination of piezoelectric material and aluminum for our simulation, which is being run using the Abaqus simulation program as explained in Fig. 34.1.

The geometry is a piezoelectric material and aluminum composite structure. The thickness of the aluminum component in this design is 1 mm, although it is only nominally significant. In contrast, careful attention is paid to the variation in thickness of the piezoelectric material, which ranges from 0.18 to 0.25 mm with an increase of around 0.02 mm in each model that was studied.

The geometry is assembled by combining two different piezoelectric elements with aluminum. To ensure excellent contact and structural integrity, this painstaking assembly involves precisely aligning the piezoelectric elements with the aluminum substrate. Tables 34.1–34.3 provide a reference for the associated mechanical parameters of this material, as provided by the supplier.

The discretization of the computational domain into smaller geometric parts for the purpose of enabling precise numerical analysis is known as mesh generation. Figure 34.2 shows the mesh generation. In order to create the mesh for this project, we carefully divided the computational area into finite elements. That being said, it is noteworthy that in comparison to the piezoelectric material, the number of elements assigned to the aluminum component is somewhat less relevant. This choice is influenced by the diverse material qualities and structural details, which call for a

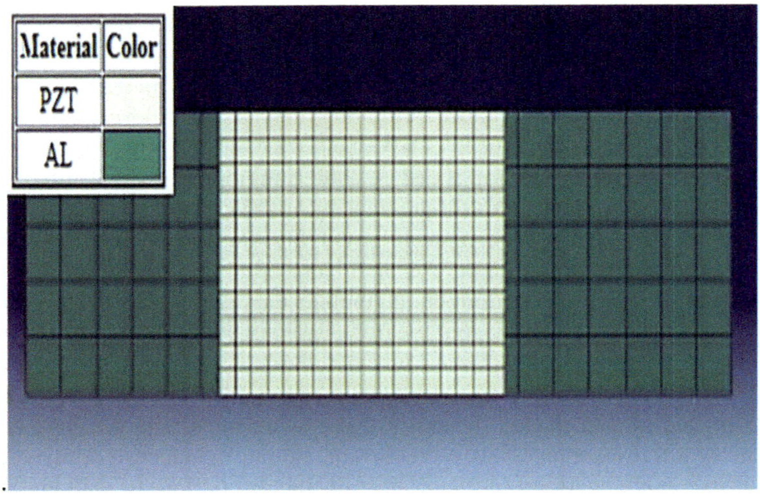

Fig. 34.1 PZT structure

Table 34.1 Dielectric constants

Dielectric constants (c/vm)		
D11	D22	D33
6.6E–006	5.8E–009	0

Table 34.2 Piezoelectric constants

Piezoelectric constants (c/m^2)			
D1111	D1122	D2222	D1212
13,700,000,000	74,100,000,000	112,000,000,000	25,500,000,000

Table 34.3 Elastic constants

Elastic constants (N/m^2)		
E1 12	E2 11	E2 22
12.9	− 5.4	15.8

different mesh density in order to properly represent the behavior of every component in the composite structure.

Figure 34.3 explains the boundary conditions that are derived using experimental data, such as force–displacement measurements and material parameters that were taken from an actual scenario. This entails precisely modeling material behavior, applying realistic loading conditions based on vehicle forces, and securing the base with 28 fixed supports. The program's fixations and boundary conditions are based

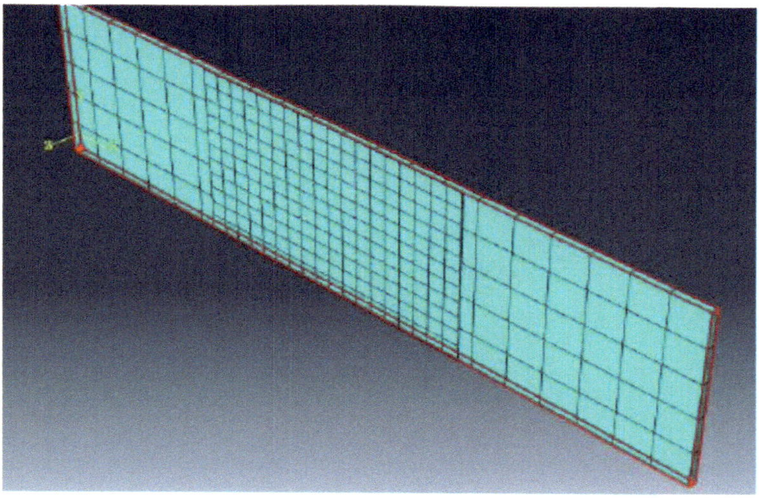

Fig. 34.2 Mesh generation

on an actual speed bump; the boundaries are fixed in the translation and rotation directions.

Surfaces that interact with one another are represented in contact analysis in Abaqus using the phrases "slave surface" and "master surface." The primary surface that regulates contact behavior is the master surface, whereas the slave surface complies with the master's contact requirements. To determine which of the two

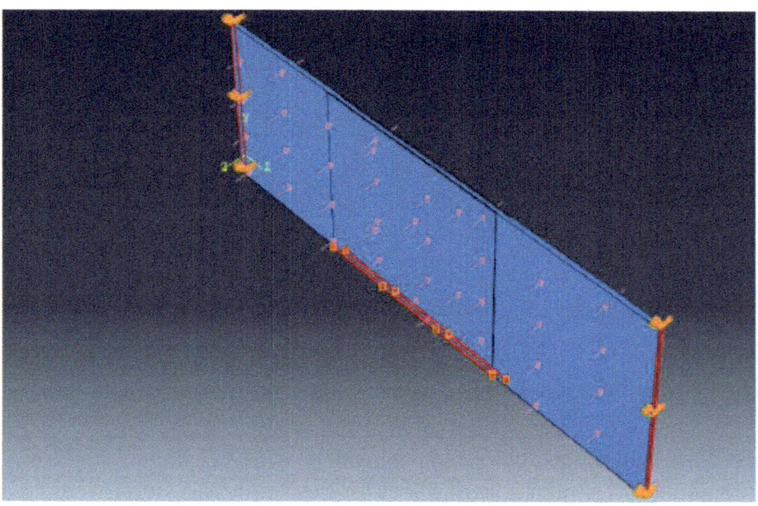

Fig. 34.3 Boundary conditions

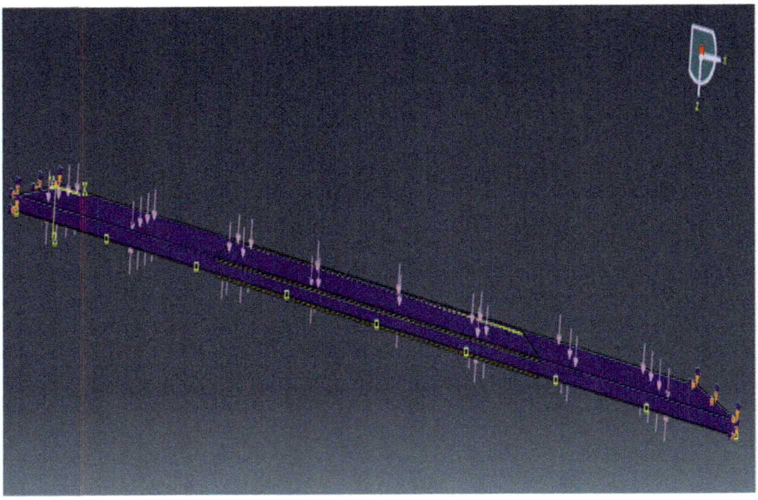

Fig. 34.4 Applying load

is the master and which is the slave, boundary conditions can be used. Boundary requirements for a master surface are often set as fixed supports or specified displacements that control its motion. These boundary conditions illustrate how the master surface moves, changes, or deforms in relation to the slave surface.

However, a slave is typically subjected to restrictions in order to conform to those set by their interactions with a master. These limitations restrict the second object's movement in relation to the first based on forces that are specified through contacts, such as frictional forces, adhesive forces, and so on. For instance, additional events like pressure applied at points where they touch may occur. The piezoelectric material in this instance serves as the master surface, and aluminum serves as the slave surface.

The application of our composite is based on a speed bump, and the stresses are based on experimental data. The loads are evenly distributed; load 1 is the force that is applied when an object is pressed, and load 2 is the reaction force, which is the same force exerted in the opposite direction, Fig. 34.4 explains the applying load.

34.3 Results and Discussions

34.3.1 Von Mises Stress

The von Mises stress, also known as equivalent stress, examines stress in relation to the combined X, Y, and Z axes. Plots for the von Mises stress were created by following a route through each model, which was carefully selected based on areas that had the greatest chance of fluctuation. As a result of its tendency to reveal notable

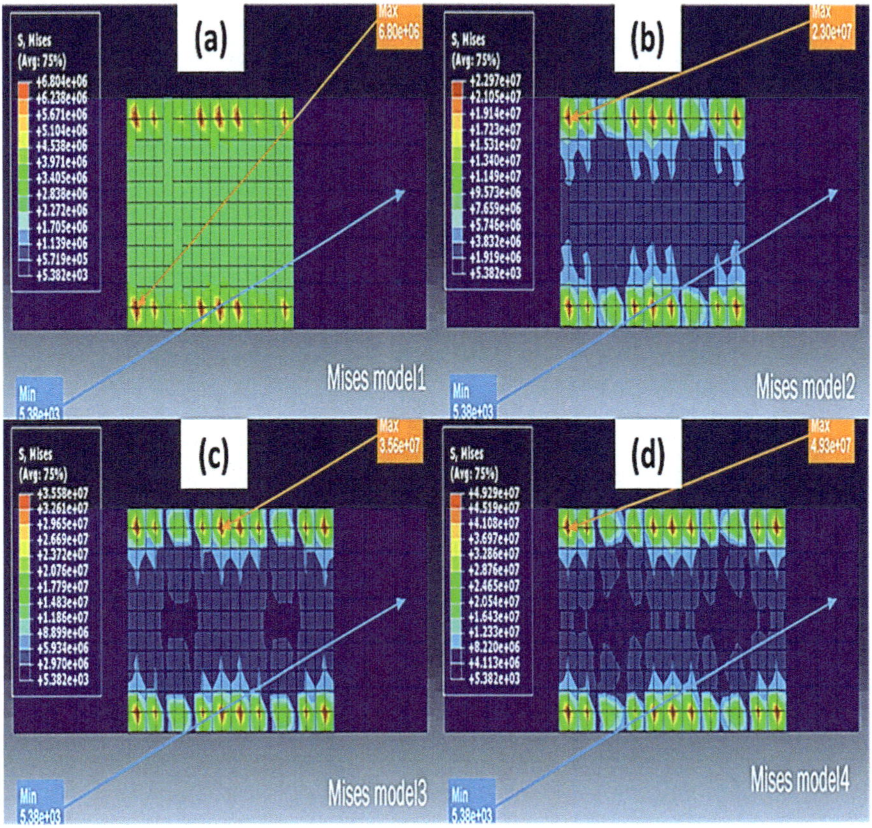

Fig. 34.5 Mises stresses **a** model 1 **b** model 2 **c** model 3 and **d** model 4

variances, "path1," which is located at the bottom of each model, was chosen in this instance. The path vs. stress results are displayed as follows in Fig. 34.5.

The normal stress in X-direction (S11) stress increases as thickness increases, and the stress is progressively increased on the piezoelectric part rather than the aluminum as the thickness is increased as shown in Fig. 34.6.

34.3.2 Results of Displacement

This problem deals with a piezoelectric material's shear deformation. The application of an electric field perpendicular to the direction of polarization causes shear strain in the Z-direction of the piezoelectric material. Figure 34.7 depicts the deformed shape of a 1 mm thick strip and a 10–4 mm thick strip obtained by Abaqus.

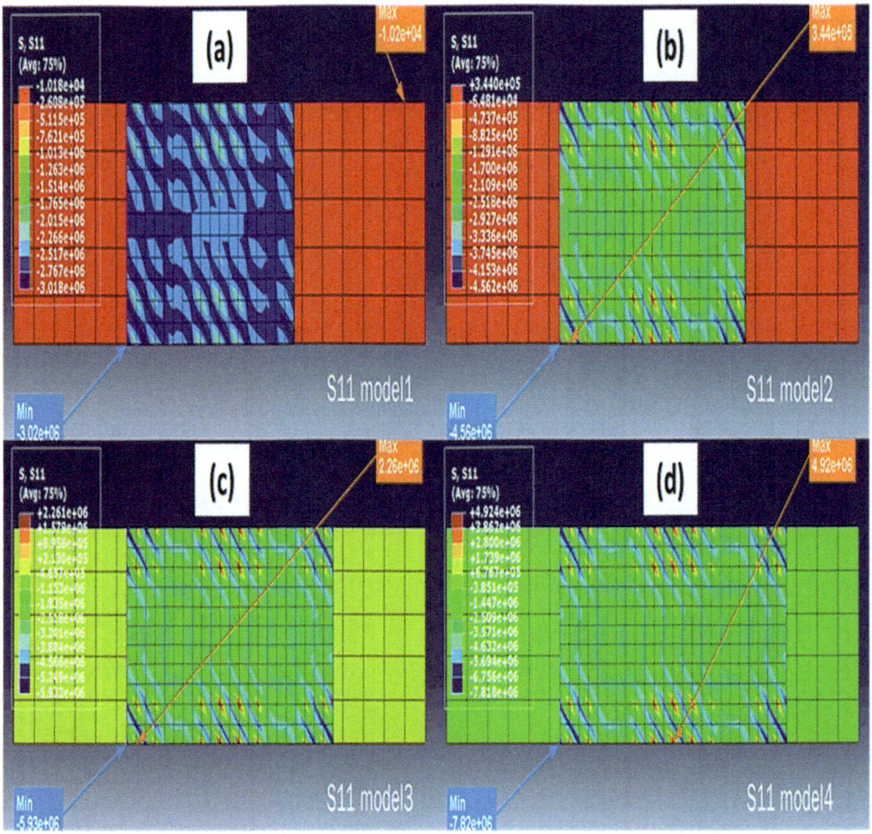

Fig. 34.6 Normal stresses along X-direction **a** model 1 **b** model 2 **c** model 3 and **d** model 4

34.3.3 Electric Flux

Plots based on selected routes and Abaqus visualization show that altering a model's thickness affects electric current, with an increase in thickness causing an increase in current in the x-direction. The straight lines in this particular feature are caused by several electric current-producing points at a single node or point. Fig. 34.8 demonstrates the electric flux (EFLUX).

34.3.4 Electric Potential

It can be seen from Fig. 34.9 that the blue region on the final job of PZT 5A depicts the minimum electrical potential at the same load under a changing frequency ranging

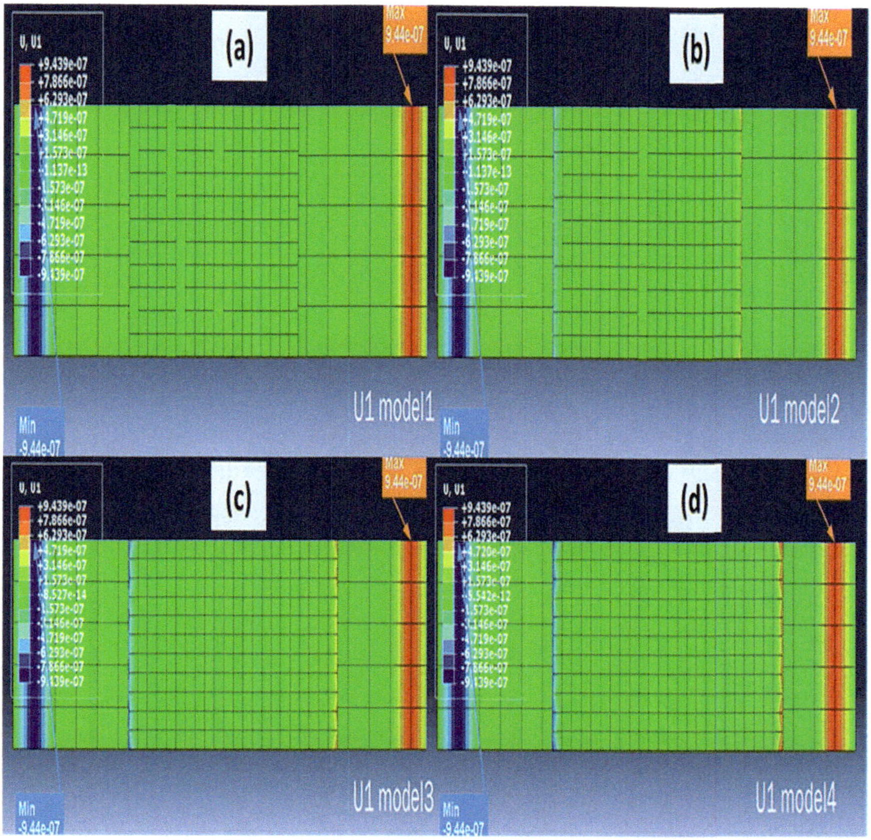

Fig. 34.7 Displacement along X-directions **a** model 1 **b** model 2 **c** model 3 and **d** model 4

from 1 to 70 Hz, and the red region on the final job of PZT 5A shows the maximum electrical potential (V) at constant load of 20 KN. The data presented in Fig. 34.9 indicate that the output voltage of a piezoelectric material drops as frequency increases at varying loading conditions. From the electric potential (EPOT) results, it is clear that, of all the properties examined, a change in part thickness has the greatest impact on the property electric potential (EPOT) because of its obvious influence.

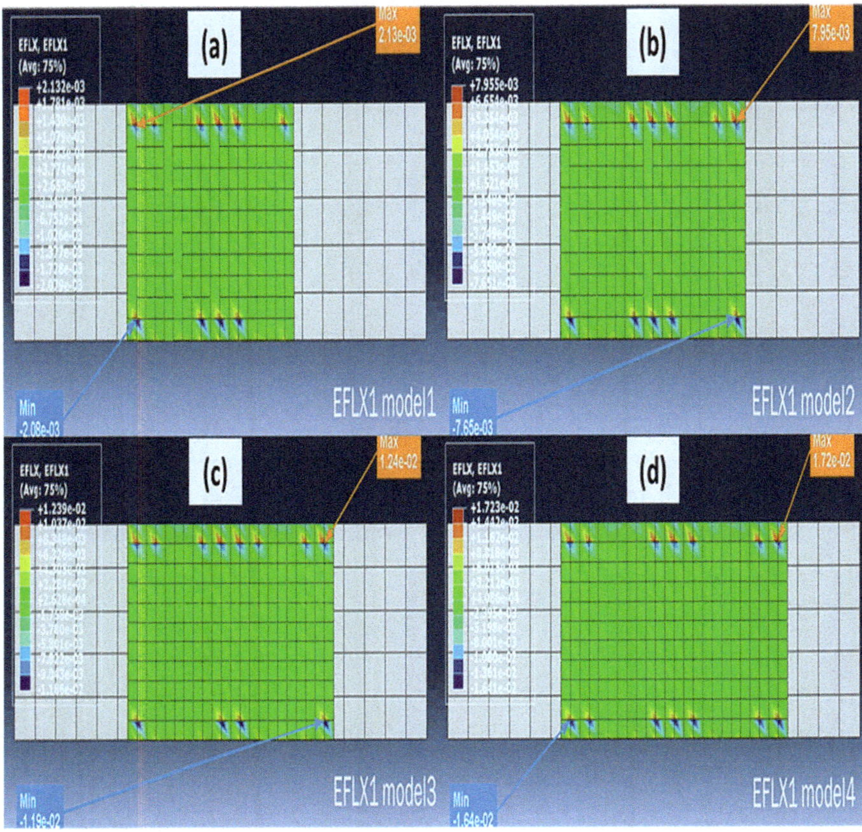

Fig. 34.8 Electric displacement along X-direction **a** model 1 **b** model 2 **c** model 3 and **d** model 4

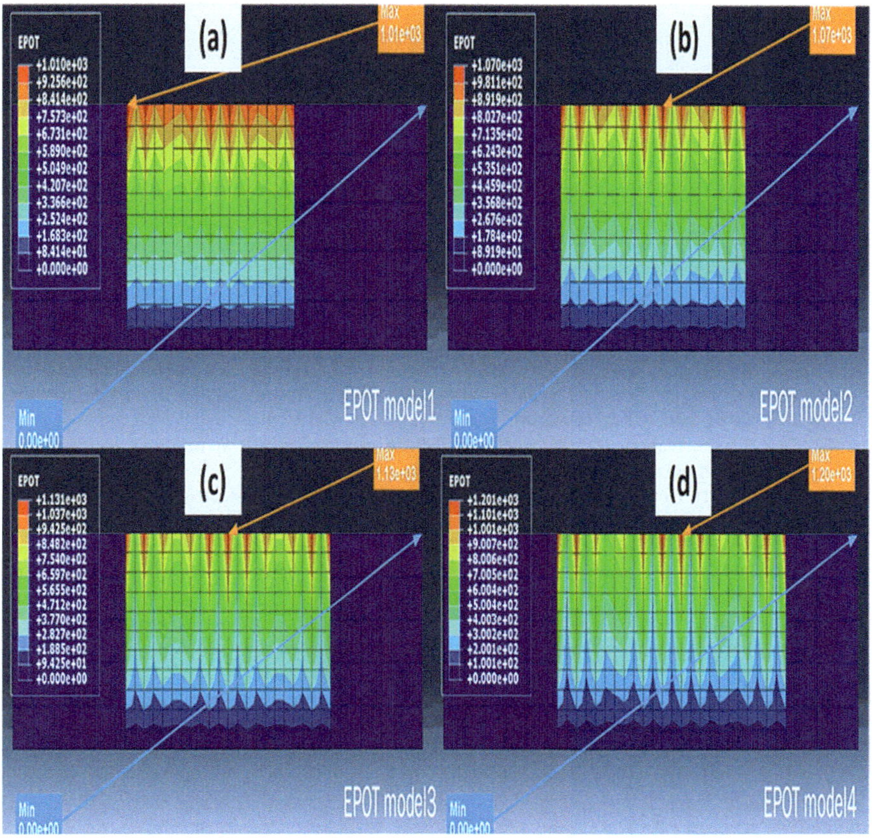

Fig. 34.9 Electrical potential transformation **a** model 1 **b** model 2 **c** model 3 and **d** model 4

34.4 Conclusion

In this work, Abaqus software was utilized to analyze four distinct forms of piezoelectric solids. The following conclusions can be drawn:

- As the piezoelectric device's dimensions increased, the output electrical voltage EPOT increased as well. For example, the model number 4's EPOT is equivalent to 1.2×10^3, while the model number 1's is equal to 1.1×10^3.
- Compared to model 1, the electrical displacement rose by 18.81% for model 4.
- Additionally, the mechanical displacement went up from the first model to the fourth model.

References

Ahasan MF, Hassan MF, Mondal D (2018) Numerical analysis of energy harvesting on a wind turbine blade by using piezoelectric material. Energy 2:1

Ansari M, Karami MA (2017) Experimental investigation of fan-folded piezoelectric energy harvesters for powering pacemakers. Smart Mater Struct 26:065001

Aslam M, Nagarajan P, Remanan M (2018) Analysis of piezoelectric solids using finite element method. In: IOP conference series: materials science and engineering, p 012009

Butt Z, Rahman SU, Pasha RA, Mehmood S, Abbas S, Elahi H (2017) Characterizing barium titanate piezoelectric material using the finite element method. Trans Electr Electron Mater 18:163–168

Cahill P, Hazra B, Karoumi R, Mathewson A, Pakrashi V (2018) Vibration energy harvesting based monitoring of an operational bridge undergoing forced vibration and train passage. Mech Syst Signal Process 106:265–283

Chen J, Oh SK, Nabulsi N, Johnson H, Wang W, Ryou J-H (2019) Biocompatible and sustainable power supply for self-powered wearable and implantable electronics using III-nitride thin-film-based flexible piezoelectric generator. Nano Energy 57:670–679

Fan K, Liu Z, Liu H, Wang L, Zhu Y, Yu B (2017) Scavenging energy from human walking through a shoe-mounted piezoelectric harvester. Appl Phys Lett 110(14)

Fei C, Liu X, Zhu B, Li D, Yang X, Yang Y, Zhou Q (2018) AlN piezoelectric thin films for energy harvesting and acoustic devices. Nano Energy 51:146–161

Karimi M, Karimi A, Tikani R, Ziaei-Rad S (2016) Experimental and theoretical investigations on piezoelectric-based energy harvesting from bridge vibrations under travelling vehicles. Int J Mech Sci 119:1–11

Li R, Yu Y, Zhou B, Guo Q, Li M, Pei J (2018) Harvesting energy from pavement based on piezoelectric effects: Fabrication and electric properties of piezoelectric vibrator. J Renew Sustain Energy 10(5)

Lu Q, Liu L, Scarpa F, Leng J, Liu Y (2018) A novel composite multi-layer piezoelectric energy harvester. Compos Struct 201:121–130

Luo S, Zhao J, Zou J, He Z, Xu C, Liu F, Huang Y, Dong L, Wang L, Zhang H (2018) Self-standing polypyrrole/black phosphorus laminated film: promising electrode for flexible supercapacitor with enhanced capacitance and cycling stability. ACS Appl Mater Interfaces 10:3538–3548

Madinei H, Khodaparast HH, Adhikari S, Friswell M (2016) Design of MEMS piezoelectric harvesters with electrostatically adjustable resonance frequency. Mech Syst Signal Process 81:360–374

Moure A, Rodríguez MI, Rueda SH, Gonzalo A, Rubio-Marcos F, Cuadros DU, Pérez-Lepe A, Fernández J (2016) Feasible integration in asphalt of piezoelectric cymbals for vibration energy harvesting. Energy Convers Manage 112:246–253

Nestorović T, Shabadi S, Marinković D, Trajkov M (2013) Modeling of piezoelectric smart structures by implementation of a user defined shell finite element. Facta Univ-Ser Mech Eng 11:1–12

Nia EM, Zawawi NAWA, Singh BSM (2017) A review of walking energy harvesting using piezoelectric materials. In: IOP conference series: materials science and engineering, p. 012026

Parvez Mahmud M, Huda N, Farjana SH, Asadnia M, Lang C (2018) Recent advances in nanogenerator-driven self-powered implantable biomedical devices. Adv Energy Mater 8:1701210

Qiu M, Sun Z, Sang D, Han X, Zhang H, Niu C (2017) Current progress in black phosphorus materials and their applications in electrochemical energy storage. Nanoscale 9:13384–13403

Safaei M, Sodano HA, Anton SR (2019) A review of energy harvesting using piezoelectric materials: state-of-the-art a decade later (2008–2018). Smart Mater Struct 28:113001

Sezer N, Koç M (2021) A comprehensive review on the state-of-the-art of piezoelectric energy harvesting. Nano Energy 80:105567

Siang J, Lim M, Salman Leong M (2018) Review of vibration-based energy harvesting technology: mechanism and architectural approach. Int J Energy Res 42:1866–1893

Snyder DS (1985) Piezoelectric reed power supply for use in abnormal tire condition warning systems. ed: Google Patents
Song Y (2019) Finite-element implementation of piezoelectric energy harvesting system from vibrations of railway bridge. J Energy Eng 145:04018076
Tomasi JT (2017) Method for the prediction of the energy harvested by vibrational piezoelectric devices applied to civil infrastructure. Lehigh University
Wang D, Mo J, Wang X, Ouyang H, Zhou Z (2018a) Experimental and numerical investigations of the piezoelectric energy harvesting via friction-induced vibration. Energy Convers Manage 171:1134–1149
Wang H, Jasim A, Chen X (2018b) Energy harvesting technologies in roadway and bridge for different applications–A comprehensive review. Appl Energy 212:1083–1094
Wu N, Wang Q, Xie X (2017) Ocean wave energy harvesting with a piezoelectric coupled buoy. ed: Google Patents
Xie X, Wang Q (2015) Energy harvesting from a vehicle suspension system. Energy 86:385–392
Zhou X, Parida K, Halevi O, Liu Y, Xiong J, Magdassi S, Lee PS (2020) All 3D-printed stretchable piezoelectric nanogenerator with non-protruding kirigami structure. Nano Energy 72:104676

Chapter 35
Research of Nonlinear Output Regulation for Systems Described by TakagiSugeno Fuzzy Descriptor Models and LMI for the Optimization and Implementation of a Clamp Mechanism

Luis Alberto Poblete-Salinas, Juan Alfonso Beltrán-Fernández, Mauricio González-Rebattú y González, Juan Carlos Hermida-Ochoa, Alejandro Tonatiu Velázquez-Sánchez, Erik Omar Alvarado-Alcántara, Karen Pamela Vázquez-Thierry, Pablo Moreno-Garibaldi, and Cassandra Anahí Hernández-Copca

Abstract The primary objective of output regulation is to control the system's output effectively, ensuring asymptotic tracking of a desired reference trajectory while rejecting disturbances originating from an exo-system, all while maintaining closed-loop stability. Designing controllers to achieve this regulation typically requires

---

L. A. Poblete-Salinas (✉) · J. A. Beltrán-Fernández · A. T. Velázquez-Sánchez · E. O. Alvarado-Alcántara · K. P. Vázquez-Thierry · P. Moreno-Garibaldi · C. A. Hernández-Copca
Instituto Politécnico Nacional—Escuela Superior de Ingeniería Mecánica y Eléctrica—Sección de Estudios de Posgrado e Investigación Edificio 5, 2Do Piso, Unidad Profesional Adolfo López Mateos "Zacatenco" Col. Lindavista, Ciudad de México, México
e-mail: lpobletes2300@alumno.ipn.mx

J. A. Beltrán-Fernández
e-mail: jbeltran@ipn.mx

A. T. Velázquez-Sánchez
e-mail: atvelazquezs@ipn.mx

E. O. Alvarado-Alcántara
e-mail: eomar_1@yahoo.com.mx

K. P. Vázquez-Thierry
e-mail: kappa.thierry@gmail.com

P. Moreno-Garibaldi
e-mail: pmg170588@gmail.com

C. A. Hernández-Copca
e-mail: cassandracopca@gmail.com

A. Öchsner and H. Altenbach (eds.), *Engineering Design Applications VII*, Advanced Structured Materials 230, https://doi.org/10.1007/978-3-031-84346-4_35

solving a set of nonlinear system of equations. However, solving these partial differential equations can be numerically challenging, particularly when electromechanical systems are modeled using the Euler–Lagrange formulation, leading to the development of nonlinear descriptor systems.

Keywords Output regulation · Steady-state mapping · Biomechanics · Linear matrix inequalities

35.1 Introduction

In control systems, the problem of ensuring the system's output asymptotically tracks a desired reference, rejects undesired disturbances, and maintains the stability of the closed-loop system is frequently addressed due to its wide applicability in areas such as mechanical systems, robotics, and biomechanics. This task is referred to as the output regulation problem (Isidori and Byrnes 1990). Francis established and demonstrated that this problem can be solved as a particular case of the nonlinear problem (Francis 1977), providing necessary and sufficient conditions through a set of nonlinear partial differential equations known as the Francis-Isidori-Byrnes (FIB) equations.

Solving these conditions can, however, present significant numerical challenges. Some numerical solutions have been proposed, most notably in terms of Takagi–Sugeno (TS) fuzzy models. Recently, a lineal matrix inequalities (LMI) solution for the nonlinear mappings has become available. Nonlinear descriptor systems, which may arise when the Euler–Lagrange formalism is employed, have not yet been extensively addressed in the context of mechanical, electromechanical, and biomechanical system modeling.

This study aims to provide a numerical solution to the nonlinear output regulation problem for descriptor systems using Takagi–Sugeno models and convex optimization techniques. Singular linear systems are considered, alongside the representation of descriptor models, where the input matrix remains constant, as noted by Meda-Campaña et al. (2009). The goal is to derive a solution using lineal matrix inequalities for the nonlinear mappings through optimization, applying this methodology

M. González-Rebattú y González
Hospital Regional "1 de Octubre" del ISSSTE, Av. Instituto Politécnico Nacional 1669, Revolución IMSS, Gustavo A. Madero, Ciudad de México, México
e-mail: mauriciorebattu@yahoo.com.mx

J. C. Hermida-Ochoa
Centro de Investigación y Laboratorio Biomecánico (CILAB), Chimalistac, Ciudad de México, México
e-mail: jhermida@gmail.com

to develop a control law for a clamp mechanism in-hand manipulation. This biomechanical system, intended for patients who have undergone phalangeal amputations, seeks to create a stable and functional mechanism.

The methodology of this work consists of four main steps:

1. Representation: The clamp mechanism, composed of the thumb and index fingers, is mathematically modeled.
2. Parametrization: The system's mathematical model is obtained through parameter identification.
3. Control Law: A control law, incorporating a nonlinear stabilizer and combined regulator, is derived from the mathematical model.
4. Implementation: The control law is implemented within a biomechanical system.

35.2 Methodology

35.2.1 Representation

In order to derive the dynamic equations governing the hand's grasping mechanism, it is necessary to represent it through a mathematical model. In this study, the Lagrangian formalism is utilized to derive the equations of motion from the system's kinetic and potential energies. A schematic representation of the grasping mechanism, which consists of the thumb and index finger, is presented in Fig. 35.1. The motion of these components is considered to contribute to the system's total energy.

The kinetic energy of the proximal phalanx of the thumb finger is

$$E_{k1} = \frac{M_1 \dot{\tilde{x}}^2}{2} \tag{35.1}$$

And the kinetic energy of the distal phalanx is

$$E_{k2} = \frac{m_1 \dot{\tilde{x}}^2{}_G}{2} + \frac{m_1 \dot{y}_G^2}{2} + \frac{I\dot{\theta}^2}{2} \tag{35.2}$$

where: $x_G = \tilde{x} + lsin(\theta)$

$$y_G = lcos(\theta)$$

The total kinetic energy in the thumb finger is

$$E_k = E_{k1} + E_{k2} = \frac{1}{2}(M + m)\dot{\tilde{x}}^2 + ml\dot{\tilde{x}}\dot{\theta}cos(\theta) + \frac{1}{2}\left(I + ml^2\right)\dot{\theta}^2 \tag{35.3}$$

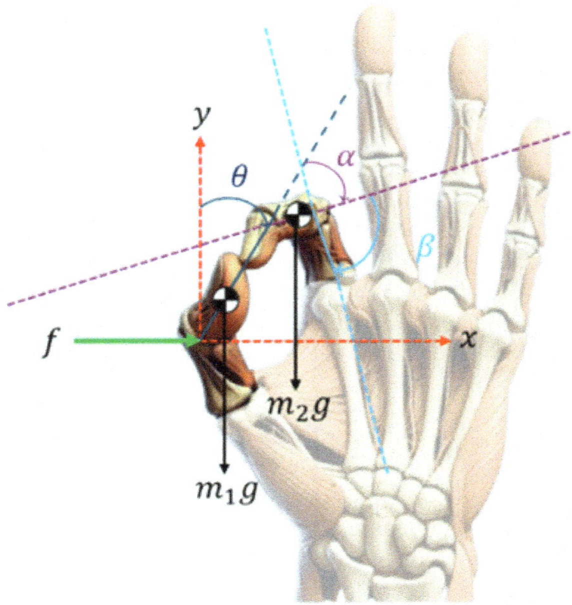

Fig. 35.1 Representation of the clamp mechanism

The kinetic energy of the proximal phalanx of the index finger is

$$E_{k3} = \frac{M_2\dot{\tilde{x}}^2}{2} \tag{35.4}$$

And the kinetic energy of the medium phalanx is

$$E_{k4} = \frac{m_2\dot{\tilde{x}}^2{}_G}{2} + \frac{m_2\dot{y}_G^2}{2} + \frac{I\dot{\beta}^2}{2} \tag{35.5}$$

where: $x_G = \tilde{x} + l_{med}sin(\beta)$

$$y_G = l_{med}cos(\beta)$$

The last kinetic energy, is of the distal phalanx

$$E_{k5} = \frac{m_{2d}\dot{\tilde{x}}^2{}_G}{2} + \frac{m_{2d}\dot{y}_G^2}{2} + \frac{I\dot{\alpha}^2}{2} \tag{35.6}$$

where: $x_G = \tilde{x} + l_{dis}sin(\alpha)$

$$y_G = l_{dis}cos(\alpha)$$

The total kinetic energy in the index finger is

$$E_k = E_{k3} + E_{k4} + E_{k5}$$

$$= \frac{1}{2}(M_2 + m_2 + m_{2d})\dot{\bar{x}}^2 + m_2 l\dot{\bar{x}}\dot{\beta}cos(\beta) + \frac{1}{2}\left(I + m_2 l^2\right)\dot{\beta}^2$$
$$+ m_{2d} l\dot{\bar{x}}\dot{\alpha}cos(\alpha) + \frac{1}{2}\left(I + m_{2d} l^2\right)\alpha^2$$

The potential energy in the system is

$$E_p = m_2 gl((cos(\beta) + \cos(\alpha)) - 1) \tag{35.7}$$

The Lagrange function is represented by

$$L_a = E_k - E_p$$

$$L_a = \frac{1}{2}(M_2 + m_2 + m_{2d})\dot{\bar{x}}^2 + m_2 l\dot{\bar{x}}\dot{\beta}cos(\beta) + \frac{1}{2}\left(I + m_2 l^2\right)\dot{\beta}^2 + m_{2d} l\dot{\bar{x}}\dot{\alpha}cos(\alpha)$$
$$+\frac{1}{2}\left(I + m_{2d} l^2\right)\alpha^2 - m_2 gl((cos(\beta) + \cos(\alpha)) - 1)n \tag{35.8}$$

The corresponding equations of motion are derived using the Lagrange equation

$$\frac{d}{dt}\left(\frac{\partial L_a}{\partial \dot{q}}(q, \dot{q})\right) - \frac{\partial L_a}{\partial q}(q, \dot{q}) = \tau \tag{35.9}$$

Were $q = (q_1, q_2, \ldots, q_n)^T$ represents the generalized variables, one for each degree of freedom of the system, $\tau = (\tau_1, \tau_2, \ldots, \tau_n)^T$ denote the external forces applied to the system. For this example, the generalized variables are x and α, as $q = (x, \alpha)^T$ therefore, one gets

$$\frac{\partial L_a}{\partial \dot{X}} = (M_2 + m_{2d})\dot{\bar{x}}^2 + m_2 l\dot{\bar{x}}\dot{\alpha}cos(\alpha) \tag{35.10}$$

$$\frac{\partial L_a}{\partial X} = 0 \tag{35.11}$$

$$\frac{\partial L_a}{\partial \dot{\alpha}} = m_2 l\dot{\bar{x}}\dot{\alpha}cos(\alpha) + \frac{1}{2}\left(I + m_2 l^2\right)\dot{\alpha}^2 \tag{35.12}$$

$$\frac{\partial L_a}{\partial \alpha} = m_2 glsin(\alpha) - m_2 l\dot{\bar{x}}\dot{\alpha}sin(\alpha) \tag{35.13}$$

The system is considered underactuated, with the medium phalanx connected to a motor, while the distal phalanx is modeled as a pendulum. The proximal phalanx is treated as a fixed base, oriented vertically to simplify the analysis and facilitate modeling of the system as underactuated. The dynamic equations of the system are derived, and their final form is represented by:

$$M(q)\ddot{q} + C_o(q, \dot{q})\dot{q} + G(q) = \tau \tag{35.14}$$

Were $q \in \mathbb{R}^{n_q}$ is the generalized coordinate, τ is the torque, $M(q) \in \mathbb{R}^{n_q \times n_q}$ is the inertia matrix, $C_o(q, \dot{q}) \in \mathbb{R}^{n_q}$ is the centrifugal forces and Coriolis matrix and $G(q) \in \mathbb{R}^{n_q}$ is known as the gravitational torque vector. In this example, it is assumed that there are no dissipative forces or friction. Substituting the values into the Lagrange Eq. (35.9), we obtain that (10) and (12) corresponding to the inertia matrix because it contains the term $\dot{\tilde{x}}$, in the Eq. (35.13) the second term corresponds to the Coriolis matrix because it contains $\dot{\alpha}$. For the gravitational torque vector, see the first term of (13) This assumption was considered because it accounts for gravity. Finally, the system, with a motor connected to the proximal phalanx, is considered to have a torque represented by f.

$$q = \begin{bmatrix} \tilde{x} \\ \alpha \end{bmatrix} M(q) = \begin{bmatrix} M_2 + m_{2d} & m_{2d} l cos(\alpha) \\ m_{2d} l cos(\alpha) & m_{2d} l^2 \end{bmatrix}$$

$$C_o(q, \dot{q}) = \begin{bmatrix} 0 & -m_{2d} l sin(\alpha)\dot{\alpha} \\ 0 & 0 \end{bmatrix}$$

$$G(q) = \begin{bmatrix} 0 \\ -m_{2d} g l sin(\alpha) \end{bmatrix} \tau = \begin{bmatrix} f \\ 0 \end{bmatrix}$$

35.2.2 Parametrization

The system is considered in descriptor form

$$E(x)\dot{x} = f(x) + g(x)u, \; y = h(x)$$

where $E(x) = \begin{bmatrix} I & 0 \\ 0 & M(\bullet) \end{bmatrix}$, $u = \tau$. The output is considered by $y = q$, is held in state space $y = Cx$ with $C = \begin{bmatrix} I & 0 \end{bmatrix}$. Considering the assignment of the variables of state $x_1 = \tilde{x}$, $x_2 = \dot{\tilde{x}}$, $x_3 = \alpha$ y $x_4 = \dot{\alpha}$, is obtained the following representation in state space in descriptor form:

$$\begin{bmatrix} 1 & 0 & 0 & 0 \\ 0 & M_2 + m_{2d} & 0 & m_{2d}lcosx_3 \\ 0 & 0 & 1 & 0 \\ 0 & m_{2d}lcosx_3 & 0 & m_{2d}l^2 \end{bmatrix} \begin{bmatrix} \dot{x}_1 \\ \dot{x}_2 \\ \dot{x}_3 \\ \dot{x}_4 \end{bmatrix} = \begin{bmatrix} x_2 \\ m_{2d}lsinx_3{x_4}^2 + u \\ x_4 \\ m_{2d}lgsinx_3 \end{bmatrix} \tag{35.15}$$

To find the output regulation it is important to consider the convex model in descriptor form of the non linear system, Taniguchi (Taniguchi et al. 1999, 2000).

$$\sum_{i=1}^{r} \mu_i(z)E_i\dot{x} = \sum_{i=1}^{r} \mu_i(z)(A_ix + B_iu),\ y = \sum_{i=1}^{r} \mu_i(z)C_ix \tag{35.16}$$

where $z \in \mathrm{R}^{\mathrm{p}}$ is the premises vector, r is the number of rules, $(A_i, B_i, C_i, E_i), i \in \{1, 2, \ldots, r\}$ are the vertex matrices of appropriate dimensions associated with each rule and $\mu_i(z), i \in \{1, 2, \ldots, r\}$ are the membership functions that satisfy the convex sum property.

$$0 \le \mu_i(z) \le 1, \sum_{i=1}^{r} \mu_i(z) = 1. \tag{35.17}$$

The model (15) can be an exact representation of the nonlinear system in descriptor form. The non-constant terms must be identified and grouped in the premise vector, where each entry of this vector is bounded $\Omega_x \subset R^n$, i.e., $z_i(x) \in \left[z_i^0, z_i^1\right]$, were z_i^0 and z_i^1 are the minimum and maximum of $z\_i(x)$. Thus, each term can be expressed as convex sums of its bounds, that is

$$z_i(x) = \mu_0^i(z_i)z_i^0 + \mu_1^i(z_i)z_i^1 \tag{35.18}$$

where

$$\mu_0^i(z_i) = \frac{z_i^1 - z_i(x)}{z_i^1 - z_i^0},\ \mu_1^i(z_i) = 1 - \mu_0^i(z_i), \tag{35.19}$$

are called weighting functions and fulfill the convex sum property in $\forall x \in \Omega_x$, i.e., $\mu_0^i(z_i) + \mu_1^i(z_i) = 1$ and $\mu_0^i, \mu_1^i \in [0,1]$. Then, the so-called membership functions can be defined

$$\mu_i(z) = \prod_{j=1}^{p} \mu_{i_j}^p(z_j),\ i_j = \{0,1\} \tag{35.20}$$

they also hold the convex sum property: $0 \le \mu_i(z) \le 1$ and $\sum_{i=1}^{r} \mathrm{m}\mu_i(z) = 1$. Finally, a Takagi–Sugeno model for the system can be expressed as in Eq. (35.13). In this work, we assume that the exosystem is linear, this is,

$$\dot{w}(t) = Sw(t), \quad S \in R^{s \times s}, \tag{35.21}$$

with S having its eigenvalues on the imaginary axis.

35.2.3 Control Law

In (Meda-Campaña et al. 2009) it is established that the nonlinear system, with a Takagi–Sugeno representation, performs output regulation if the input distribution matrix $B(x)$ and the steady-state mapping $\pi(w)$ are constant.

While $\gamma(w)$ has a convex form. Based on this and taking advantage of the structure in the convex form, it is defined as $\pi(w) = \Pi w, \Pi \in R^{n\times s}$ and $\gamma(w) = \sum_{i=1}^{r} \mathrm{m}\mu_i(z)\Gamma_i w, \Gamma_i \in R^{m\times s}, i \in \{1, 2, \ldots, r\}$. Propose the following adapted r-Fransis Isidori Byrnes equations for descriptor systems as mentioned in Poblete-Salinas et al. (2024).

$$E\_i\Pi S = A\_i\Pi + B\Gamma\_i + D\_i \tag{35.22}$$

$$0 = C\Pi - Q \tag{35.23}$$

where the r set of linear equations must hold simultaneously for Γ_i, $i \in \{1, 2, \ldots, r\}$ and a single Π. Thanks to the structure of the descriptor systems, there are no crossed-products between Γ_i and B_i, However, the nonlinear control law that performs output regulation is

$$u = K_{\mu}(x - \Pi w) + \Gamma_{\mu} w \tag{35.24}$$

With K_μ designed such that the origin of the nonlinear descriptor system is asymptotically stable. To design the stabilizer, based on the descriptor convex model, the augmented system *redundancy form* is used, $\dot{x} = \dot{x}$ and $0 \times \ddot{x} = A_\mu x + Bu - E_\mu \dot{x}$, then

$$\overline{E}\dot{\overline{x}} = \overline{A}_\mu \overline{x} + \overline{B}u$$

$$y = \overline{C}\overline{x}$$

where $\overline{x} = \begin{bmatrix} x \\ \dot{x} \end{bmatrix}$,

$$\overline{E} = \begin{bmatrix} I & 0 \\ 0 & 0 \end{bmatrix},$$

$$\overline{A}_\mu = \begin{bmatrix} 0 & I \\ A_\mu & E_\mu \end{bmatrix},$$

$$B = \begin{bmatrix} 0 \\ B \end{bmatrix}$$

and $\overline{C} = \left[C \; 0 \right]$. The nonlinear control law under design is a PDC (parallel distributed compensator)

$$u = \sum\nolimits_{i=1}^{r} \mu_i(z) K_i x = \left[K_\mu \; 0 \right] \begin{bmatrix} x \\ \dot{x} \end{bmatrix} = \overline{K_\mu} \overline{x} \tag{35.25}$$

where $K_i \in R^{m \times n}$, $i \in \{1, 2, \ldots, r\}$ are the controller gains to be designed.

To this end consider the following Lyapunov function candidate $V(\overline{x}) = \overline{x}^T \overline{E}^T \overline{P} x$, $\overline{E^T P} = \overline{P^T E} \geq 0$ were $\overline{P} = \begin{bmatrix} P_1 & 0 \\ P_3 & P_4 \end{bmatrix}$, $P_i \in R^{n \times n}.i \in \{1, 3, 4\}$ such that $P_1 = P_1^T > 0$ and P_4 being invertible. Indeed, guaranteeing $\dot{V} < 0$ ensures the stability of the closed-loop system; it is considered $\dot{V} < 0$ if

$$\overline{A}_\mu^T \overline{P} + \overline{K}_\mu^T \overline{B}^T \overline{P} + \overline{P}^T \overline{A}_\mu + \overline{P}^T \overline{BK}_\mu < 0 \tag{35.26}$$

Now, $X = \begin{bmatrix} X_1 & 0 \\ X_3 & X_4 \end{bmatrix}$, with $X_1 = P_1^{-1}$, $X_4 = P_4^{-1}$, $X_3 = -P_4^{-1} P_3 P_1^{-1}$ and multiplying (26) by X^T on the left side and X by the right side, yields

$$X^T \overline{A}_\mu^T + \overline{M}_\mu^T \overline{B}^T + \overline{A}_\mu X + \overline{BM}_\mu < 0, \tag{35.27}$$

with $\overline{M}_\mu = \overline{K}_\mu X$. The previous inequality guarantees that $\dot{V} < 0$, thus asymptotic stability at the origin is achieved.

Nevertheless, in practical cases it is important to guaranteeing a maximum speed convergence $\varphi > 0$ while holding a bound on the control input $||u(t)|| \leq \beta$, $\beta > 0$. The former is guaranteed if $\dot{V} \leq -2\varphi$ holds, in terms of LMIs this is implied by

$$\begin{bmatrix} X_3 + X_3^T + 2\varphi X_1 & (*) \\ A_i X_1 + BM_i - E_i X_3 + X_4^T & -E_i X_4 - X_4^T E_i^T \end{bmatrix} < 0, \tag{35.28}$$

hold for all $i \in \{1, 2, \ldots, r\}$. If feasible, the controller gains are computed as $K_i = M_i X_1^{-1}$, $i \in \{1, 2, \ldots, r\}$. As for bounding the control law it is obtained

$$\begin{bmatrix} 1 & x^T(0) \\ x(0) & X_1 \end{bmatrix} \geq 0 y \begin{bmatrix} X & M_i^T \\ M_i & \beta^2 \end{bmatrix} \geq 0, i \in \{1, 2, \ldots, r\} \tag{35.29}$$

where $x(0)$ is an initial condition.

Based on the results in Bernal et al.(2012), the conditions (22) and (23) can be rewritten as a LMI (linear matrix inequalities) problem, thus the design of Π and $\gamma(w(t))$ can be considered as an optimization problem.

It is important let $N(x)$ and $R(x)$ be continuously differentiable linear matrix functions and x as the decision vector. And the approximated solution for $N(x) - R(x) = 0$ can be settled as a minimization problem, that is,

$\min\ \in > 0 : - \in < N(x) - R(x) <, <$ stands for element-wise ordinary lower than, in other words:

$$\min \in > 0 : \begin{matrix} N(x) - R(x) - \in < 0 \\ N(x) - R(x) + \in > 0 \end{matrix} \tag{35.30}$$

Considered the conditions (22) and (23), it can be expressed as the following LMI optimization problem:

$$\min \epsilon > 0 : \begin{bmatrix} A_i\Pi + B\Gamma_i + D_i - E_i\Pi S & 0 \\ 0 & C\backslash Pi - Q \end{bmatrix} \prec \epsilon \tag{35.31}$$

with ϵ being an arbitrarily small scalar.

35.3 Results

Consider the underactuated system in Fig. 35.1. The clamp mechanism dynamical equations can be represented from the Lagrange equation of motion as $M(q)\ddot{q} + C_o(q, \dot{q})\dot{q} + G(q) = \tau$ where $q = [x\ \alpha]^T$ is a vector of generalized coordinates and x is the distance of the phalanx and θ, α and β are the angles from the vertical of each phalanx. The exact TS representation is:

$$M(q) = \begin{bmatrix} M_2 + m_{2d} & m_{2d}lcos(\alpha) \\ m_{2d}lcos(\alpha) & m_{2d}l^2 \end{bmatrix}, G(q) = \begin{bmatrix} 0 \\ -m_{2d}glsin(\alpha) \end{bmatrix},$$

$$C_{o(q,\dot{q})} = \begin{bmatrix} 0 & -m_{2d}lsin(\alpha)\dot{\alpha} \\ 0 & 0 \end{bmatrix} and\, \tau = \begin{bmatrix} f \\ 0 \end{bmatrix} \tag{35.32}$$

Selecting $x_1 = q_1, x_2 = \dot{q}_1, x_3 = q_2$ and $x_4 = \dot{q}_2$ as state variables, the following state-space representation is obtained in the descriptor form (15).

Rewriting the nonlinear descriptor system as a convex sum in a compact set $\Omega_x = \{|x_3| \leq 17^\circ, |x_4| \leq 1.2\text{rad/s}\}$ with their nonlinearities and their bounds settle in the Table 35.1, one obtains an exact TS representation:

$$\sum_{i=1}^{32} \mu(z)E_i\dot{x} = \sum_{i=1}^{32} \mu(z)(A_i x + Bu) \tag{35.33}$$

were $\mu_i(z) = \mu^1_{i_1}(z)\mu^2_{i_2}(z)\mu^3_{i_3}(z)\mu^4_{i_4}(z)$ and the indexes $[i_1 i_2 i_3 i_4]$ are the digit binary representation of $(i-1), i \in \{1, 2, \ldots, 16\}$ due to space reasons, some of the 32 matrices A_i, E_i and $B = [0\,1\,0\,0]^T$ are given below:

Table 35.1 Nonlinearities in the clamp system and their boundaries

Nonlinearity	Definition	Lower bounds	Upper bounds
z_1	$\sin x_3$	−0.2588	0.2582
z_2	$\sin x_3/x_3$	0.9886	1
z_3	x_4	−1.5708	1.5702
z_4	$\cos x_3$	0.9659	1

$$A_1 = \begin{bmatrix} 0 & 1 & 0 & 0 \\ 0 & 0 & 0 & 0.024 \\ 0 & 0 & 0 & 1 \\ 0 & 0 & 0.582 & 0 \end{bmatrix}, A_8 = \begin{bmatrix} 0 & 1 & 0 & 0 \\ 0 & 0 & 0 & -0.024 \\ 0 & 0 & 0 & 1 \\ 0 & 0 & 0.588 & 0 \end{bmatrix}$$

$$A_{16} = \begin{bmatrix} 0 & 1 & 0 & 0 \\ 0 & 0 & 0 & -0.024 \\ 0 & 0 & 0 & 1 \\ 0 & 0 & 0.588 & 0 \end{bmatrix}, A_{32} = \begin{bmatrix} 0 & 1 & 0 & 0 \\ 0 & 0 & 0 & 0.024 \\ 0 & 0 & 0 & 1 \\ 0 & 0 & 0.588 & 0 \end{bmatrix}$$

$$E_1 = \begin{bmatrix} 1 & 0 & 0 & 0 \\ 0 & 1.6 & 0 & 0.058 \\ 0 & 0 & 1 & 0 \\ 0 & 0.058 & 0 & 0.009 \end{bmatrix}, E_8 = \begin{bmatrix} 1 & 0 & 0 & 0 \\ 0 & 1.6 & 0 & 0.058 \\ 0 & 0 & 1 & 0 \\ 0 & 0.058 & 0 & 0.009 \end{bmatrix}$$

$$E_{16} = \begin{bmatrix} 1 & 0 & 0 & 0 \\ 0 & 1.6 & 0 & 0.058 \\ 0 & 0 & 1 & 0 \\ 0 & 0.058 & 0 & 0.009 \end{bmatrix}, E_{32} = \begin{bmatrix} 1 & 0 & 0 & 0 \\ 0 & 1.6 & 0 & 0.058 \\ 0 & 0 & 1 & 0 \\ 0 & 0.058 & 0 & 0.009 \end{bmatrix}$$

For this system, the task is that distal phalanx of the thumb finger follows a sinusoidal reference, that is, x_3 asymptotically tracks ω_1 generated by the exosystem $\dot{\omega}_1 = \omega_2$, $\dot{\omega}_2 = -\omega_1$. The LMIs are found feasible, for illustration purposes some of the gains are shown

$$K_1 = [135.7 \quad 235.4 \quad 628.5 \quad 166.8],$$

$$K_8 = [135.0 \quad 235.6 \quad 629.1 \quad 167.0],$$

$$K_{16} = [135.5 \quad 235.4 \quad 628.5 \quad 166.9],$$

$$K_{32} = [135.0 \quad 235.6 \quad 629.1 \quad 166.9],$$

the minimum ϵ is 9.0757 x 10^{-16} the common steady state is

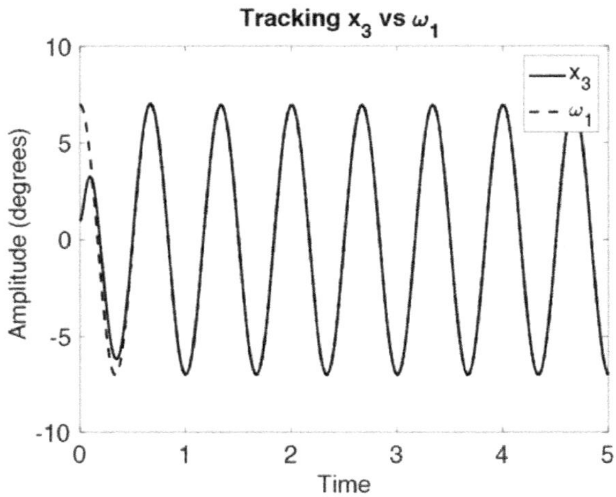

Fig. 35.2 Output x_3 versus reference ω_1 for the clamp mechanism system with decay rate and input constraints

$$\Pi = \begin{bmatrix} 1.784 & 0 \\ 0 & 1.784 \\ 1 & 0 \\ 0 & 1 \end{bmatrix}$$

and some of steady state inputs Γ_i related to i subsystems:

$\Gamma_1 = \begin{bmatrix} 0.323 \; 0.232 \end{bmatrix}$, $\Gamma_8 = \begin{bmatrix} -1.010 \; -1.100 \end{bmatrix}$, $\Gamma_{13} = \begin{bmatrix} 1.656 \; -1.100 \end{bmatrix}$ and $\Gamma_{32} = \begin{bmatrix} 0.323 \; -2.434 \end{bmatrix}$. Simulation results have been performed for initial conditions $x(0) = [10^\circ \; 0.05 \; 0.1 \; 0.05]^T$, $w(0) = [5 \; 0]^T$, the behavior closed-loop trajectories are depicted in Fig. 35.2. As it can be observed, the tracking error goes to zero as time increases. The rest of the states and the control input are shown in Fig. 35.3. Notice that at the beginning, the control law requires a large amount of energy.

35.4 Discussion of Results

The proposed nonlinear output regulation approach for the clamp mechanism, modeled using Takagi–Sugeno fuzzy descriptor systems, demonstrated significant improvements in controlling biomechanical systems. By leveraging linear matrix inequalities, the stabilization and regulation of the system were achieved with greater efficiency compared to traditional nonlinear methods. The inclusion of a nonlinear stabilizer ensured the global attraction of steady-state mappings, which was crucial for the precise control required in the biomechanical application. Additionally, the

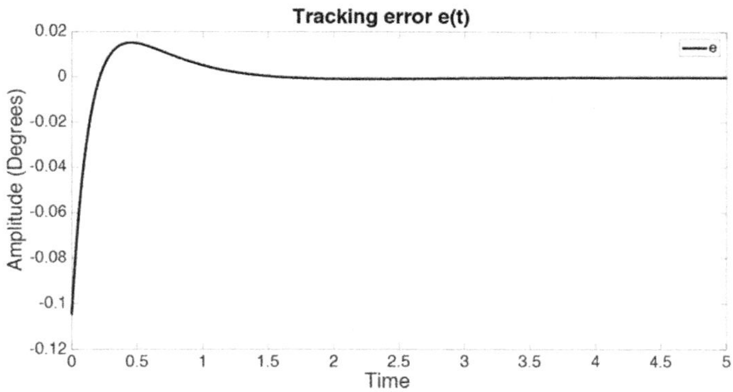

Fig. 35.3 Tracking error $e(t)$ for the clamp mechanism system with decay rate and input constraints

model's ability to handle underactuated dynamics, as seen in the hand's clamp mechanism, further highlights the flexibility and robustness of the proposed methodology. The results show that the implementation of this control law in a practical biomechanical system is feasible, providing accurate motion control and stability for patients who have undergone phalangeal amputations. Future work may focus on refining the LMI-based solutions to handle more complex mechanical interactions, particularly in systems with higher degrees of freedom or multiple interacting actuators.

35.5 Conclusions

In this work, the nonlinear output regulation problem for a hand clamp mechanism in descriptor form has been defined, with sufficient conditions for the controller provided in terms of linear matrix inequalities. The proposed approach consists of five main components:

1. A nonlinear stabilizer that ensures global attraction towards the steady-state by redefining the system's state equations.
2. Steady-state mappings based on the new state equations, which allow accurate regulation.
3. The parametrization of the dynamic model of the clamp mechanism, incorporating biomechanical constraints and underactuated dynamics.
4. The development of a control law that integrates both stabilization and output regulation, ensuring system stability and effective tracking.
5. Implementation of the control strategy in a biomechanical hand mechanism, demonstrating its potential for improving object manipulation for patients with phalangeal amputations.

This comprehensive approach provides an effective framework for addressing the output regulation problem in biomechanical systems.

Acknowledgements • The authors wish to express their gratitude to the National Council of Humanities, Science, and Technology (CONAHCYT) for providing research scholarships at the advanced level.

• Special thanks are also extended to the National Polytechnic Institute for its ongoing support of this work.

• Additionally, the authors acknowledge the support provided by IPN SIP ID: 20,242,695, "Assessment of Tumor Extension in Breast Cancer Using Computed Tomography Scans and 3D Printed Models."

References

Bernal M, Márquez R, Estrada-Manzo V, Castillo-Toledo B (2012) Nonlinear output regulation via Takagi-Sugeno fuzzy mappings: a full-information LMI approach. In: 2012 IEEE international conference on fuzzy systems 1–7. https://doi.org/10.1109/FUZZ-IEEE.2012.6251000

Francis BA (1977) The linear multivariable regulator problem. SIAM J Control Optim 15(3):486–505. https://doi.org/10.1137/0315034

Isidori A, Byrnes CI (1990) Output regulation of nonlinear systems. IEEE Trans Autom Control 35(2):131–140. https://doi.org/10.1109/9.46374

Meda-Campaña JA, Castillo-Toledo B, Chen G (2009) Synchronization of chaotic systems from a fuzzy regulation approach. Fuzzy Sets Syst 160(19):2860–2875. https://doi.org/10.1016/j.fss.2009.04.001

Poblete-Salinas LA, Beltrán-Fernández JA, González-Rebattú M, Hermida-Ochoa JC, Velázquez-Sánchez AT, Alvarado-Alcántara EO, ... Hernández-Copca CA (2024) Research of nonlinear output regulation for systems described by takagi–sugeno fuzzy descriptor models with steady state mapping for the optimization and implementation of a hand rehabilitation. In: Engineering design applications VI: structures, materials and processes 361–373. Cham: Springer Nature Switzerland

Taniguchi T, Yamaguchi T, Ogura T (1999) Fuzzy descriptor systems: stability analysis and design via LMIs. Proc Am Control Conf 3:1827–1831. https://doi.org/10.1109/ACC.1999.786165

Taniguchi T, Tanaka K, Wang HO (2000) Fuzzy descriptor systems and nonlinear model following control. IEEE Trans Fuzzy Syst 8(4):442–452. https://doi.org/10.1109/91.868950

Chapter 36
Bottle Inspection and Classification System by Volume and Colour

Mauricio Aarón Pérez Romero, Alejandra Armenta Molina, Luis Héctor Hernández Gómez, Luis Fernando Franco Guzmán, Armando Josué Piña Díaz, Marcial Margarito Sánchez Sánchez, and Tanya Nerina Arreola Valles

Abstract A prototype is reported to inspect liquid soap bottles and their classification by colour. The process is carried out in the following way. The bottles are centered at the input of a conveyor belt with a positioning mechanism. Then, they are taken to the inspection system. Their images are obtained, and it is verified that the bottle satisfies the filling requirements. The bottles that satisfy the required liquid level, continue on the conveyor belt. In the next stage, they are separated according to the colour of their content. If a bottle does not satisfy the requirements, it will continue to the conveyor belt output. The system has a Pixy camera (CMUcam5). The image is obtained and processed by an Arduino Mega. It activates the bottle

M. A. Pérez Romero · L. F. Franco Guzmán · A. J. Piña Díaz
Escuela Superior de Ingeniería Mecánica y Eléctrica, Ingeniería en Control y Automatización, Instituto Politécnico Nacional, Ciudad de México, México
e-mail: maperezro@ipn.mx

L. F. Franco Guzmán
e-mail: lfrancog@ipn.mx

A. J. Piña Díaz
e-mail: apinad@ipn.mx

A. Armenta Molina (✉) · L. H. Hernández Gómez · T. N. Arreola Valles
Escuela Superior de Ingeniería Mecánica y Eléctrica, Sección de Estudios de Posgrado e Investigación, Instituto Politécnico Nacional, Ciudad de México, México
e-mail: aarmentamo@ipn.mx

L. H. Hernández Gómez
e-mail: lhhernandez@ipn.mx

T. N. Arreola Valles
e-mail: tarreolav1700@alumno.ipn.mx

M. M. Sánchez Sánchez
Escuela Superior de Ingeniería Mecánica y Eléctrica, Ingeniería en Comunicaciones y Electrónica, Instituto Politécnico Nacional, Ciudad de México, México
e-mail: mmsanchezs@ipn.mx

A. Öchsner and H. Altenbach (eds.), *Engineering Design Applications VII*, Advanced Structured Materials 230, https://doi.org/10.1007/978-3-031-84346-4_36

classification system (by liquid level and colour). In addition, it counts the inspections carried out and displays the information on the HMI. The interface is a module with a touch screen that allows navigation among the different menus. To reduce the space required by the system, a Raspberry Pi 3 minicomputer was used. The system prototype was successfully implemented. It inspects 19 bottles per minute.

36.1 Introduction

Vision systems were used only to validate a manufactured product without providing any additional information. Currently, a vision system is necessary to optimize product quality and reduce production costs. In addition, they facilitate tasks where certain specific characteristics are required. Despite this, some activities are carried out mechanically and manually. Visual systems are not used in small production systems due to their high cost. This is the case of product inspection. A process inspection system has many advantages, such as improving the performance and quality of the process. Manually performed activities can cause visual fatigue for the operator.

Computers with a combination of hardware and software apply methods to obtain, process, and analyze real-world images. The purpose is to produce information that can be processed. Regarding an artificial vision system, it has been mentioned in Khan et al. (2021) that "it is an autonomous system that replaces some of the tasks of the human vision system. It obtains data from the environment to perform the previously defined tasks or activities. It can be related to identifying and inspecting objects and recognizing characters in an image of a three-dimensional interpretation of complicated scenes."

An analogy can be made with the human senses. People use their eyes to perceive the world around them. In this way, artificial vision tries to produce such an effect. They interpret an image or sequence of images and make decisions. They achieve this by decomposing the image into small fragments (pixels) and subsequently studying them (Fig. 36.1). In (Cognex: Visión Artificial. 2023) establishes that "the main components of a machine vision system include lighting, lenses, image sensor, vision processor and communications" (Fig. 36.2).

The bottles to be inspected are illuminated. Features that stand out and are easily captured on camera, are searched. The lens captures the image and transmits it as a ray of light to a sensor, which converts it into a digital image. Subsequently, it is analyzed with a vision processor. Its algorithms interpret the image, execute the inspection, and classify it. Finally, communication is carried out using an input–output communication protocol.

There are two main variants of artificial vision systems: (1) Artificial vision applied to industry, which covers computing, optics, mechanical engineering, and industrial automation. They integrate digital image capture systems, input/output devices, and computer networks for the control of equipment intended for production processes, supported by robotic arms and conveyor belts. On the other hand, (2) academic

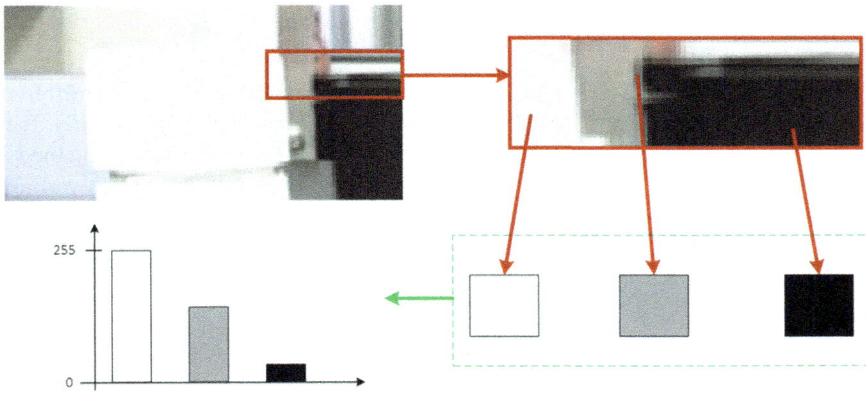

Fig. 36.1 Development of an image under the intensity of light

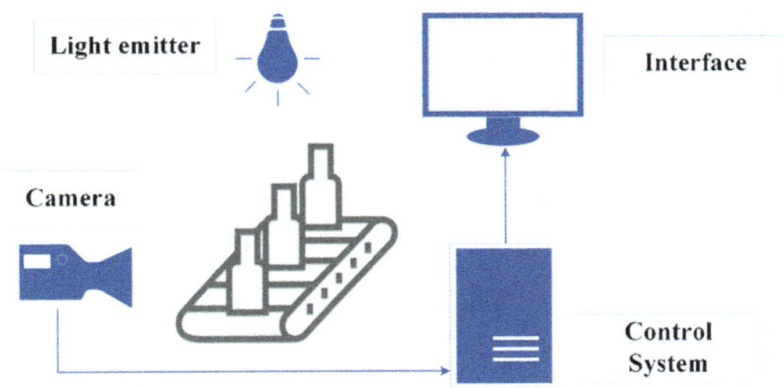

Fig. 36.2 Main components of the artificial vision systems

artificial vision, in which the systems rely on image processing. Technology has had important advances in the information processing units used in this field (Wiley and Lucas 2023).

The use of microcontrollers and PLCs satisfies the needs of the industrial market. However, the facilities offered by open-source platforms and the development of programmable electronic cards for this type of software have created controllers that can replace a PLC in specific tasks. Specifically, where a robust controller is not required. It is mentioned in Alves and Morris (2018) that the selection of one of these options is a function of the requirements. Currently, the most common cards that use open source play an important role in the development of new prototypes. They replace classic controllers (microcontrollers). Its implementation in the industry is proving to be a success. It is mentioned in Yang et al. (2020) that "the microcontroller has already eclipsed even the Industrial Revolution."

36.2 Statement of the Problem

The system consists of a conveyor belt. It carries liquid soap bottles of three different colors. These will be centered using a positioning mechanism. Subsequently, they will be inspected with a camera in an inspection booth. The information obtained is processed with a microcontroller. The volume and color of liquid soap are verified. Bottles that do not meet the established requirements are removed. Figure 36.3 shows a diagram of the process, and a schematic arrangement is illustrated in Fig. 36.4.

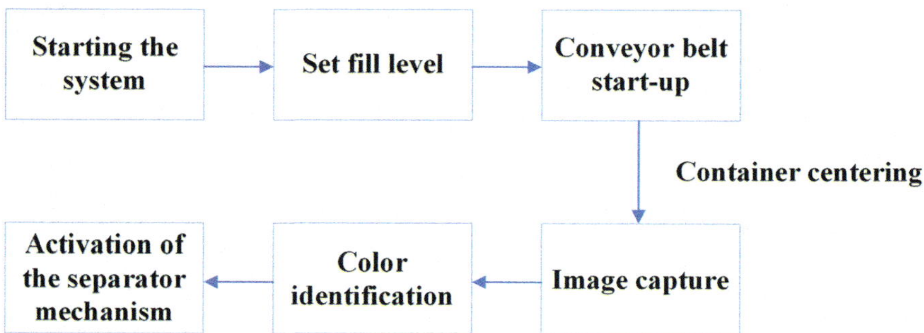

Fig. 36.3 Schematic diagram of the process

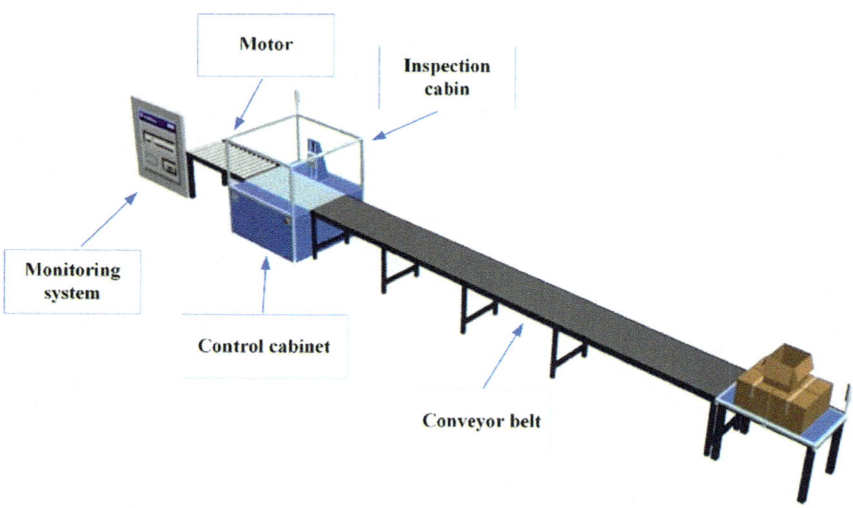

Fig. 36.4 Isometric view of the setup

36.3 Architecture of the Bottle Inspection and Classification System

A minicomputer will be used, which gives many advantages over a conventional computer. Computers have a general purpose. They are designed to cover a wide range of needs and minicomputers are designed for specific tasks. This represents a big difference in terms of costs, since a computer, whether for common or industrial use, is much more expensive. Figure 36.5 shows the block diagram that represents the architecture of the bottle inspection and classification system.

A minicomputer, which was linked by serial communication, was used. At the same time, it was connected to an image sensor to perform image acquisition. Such a minicomputer has an application software. It is an interface between the user and the system. A diagram with the selected components is shown in Fig. 36.6.

The importance of defining data-sending and reading functions lies in the possibility of manipulating a certain number of bytes, in a set time. This prevents reading errors in the controller and the user interface. The main functions, used in this project, are mentioned below.

The *Void sendMessage function ()* sends a set of data that was previously obtained with the bottle counters and the liquid level detection. With an "if" structure, the current time is compared with the scheduled time, and the data-sending function is called.

The *millis function ()* defines the programmed time and replaces the delay function. The microcontroller stops when the latter is executed. This would affect the correct execution of the program.

The *Function void readSBus()* evaluates whether Serial is receiving information. If so, the read information is saved in a variable. The different data obtained from the interface are saved by going through an array.

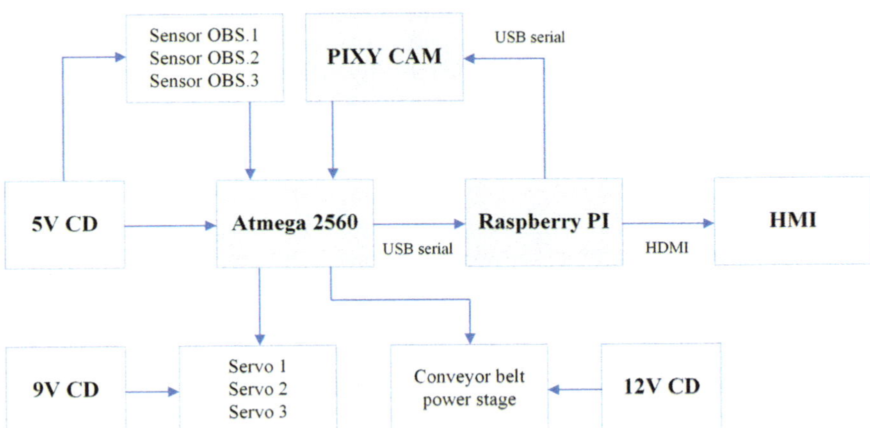

Fig. 36.5 Architecture of the bottle inspection and classification system

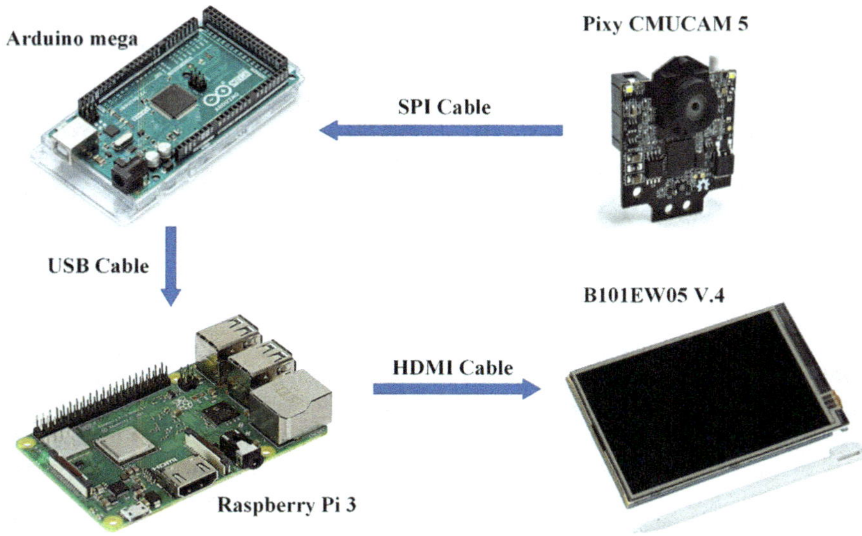

Fig. 36.6 Main elements of the proposed setup

The complete logical process, which is followed by the inspection and classification system by level and color, is shown in Figs. 36.7, 36.8, and 36.9.

36.4 Development of the Setup

The system verifies that the bottles passing through the conveyor belt meet the desired volume. If the bottle satisfies the desired liquid level, it will be classified according to the color of the bottle's contents, orange, purple, or blue. For this purpose, an infrared sensor coupled with a servomotor is used. Firstly, colors must be assigned to each signature that will be used in the program. It is done with the PixyMon software, selecting the area occupied by the color to be inspected. This step is done with each color. The system should identify bottles that do not meet the level, as shown in Fig. 36.10.

The color intensity and brightness, with which the images are detected, are configured with the PixyMon software. The goal is to have complete detection. As an example, an orange bottle is passed through the positioning mechanism. This performs its function. Subsequently, the data of the bottle is obtained by the Pixy Cam (Fig. 36.11) and the image is processed. Tests were carried out with various bottles. The results were consistent.

Communication is relevant. The devices are interconnected, and it is easy to collect information. The tests validated that the program and the communication between the system components work properly. Communication between the intelligent image

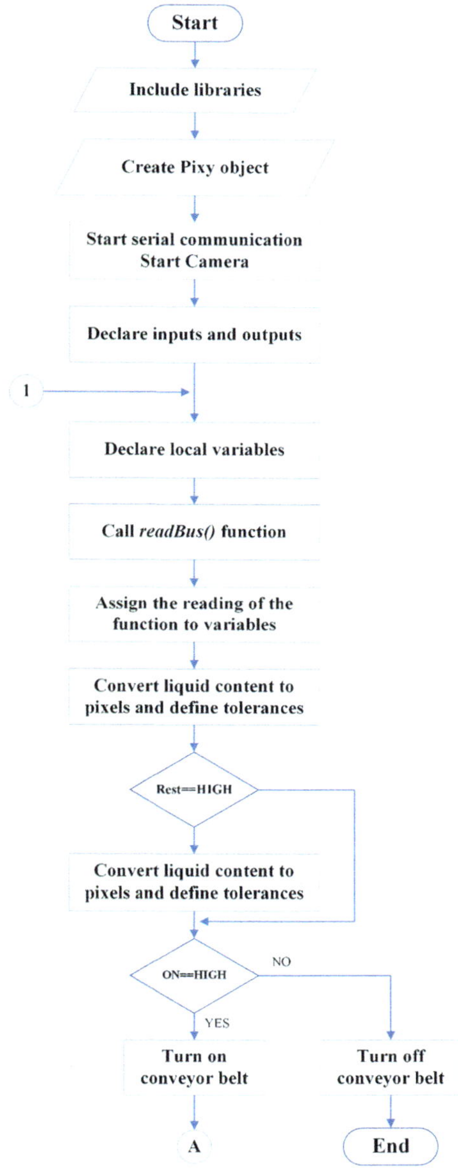

Fig. 36.7 Logical process of the Inspection and classification system from START to node A

sensor (Pixy Cam) and the PixyMon Software is quickly verified with the USB cable that connects to the Raspberry. The Pixy side is connected with micro-USB. It is verified that the camera projects an image, which is shown in Fig. 36.12.

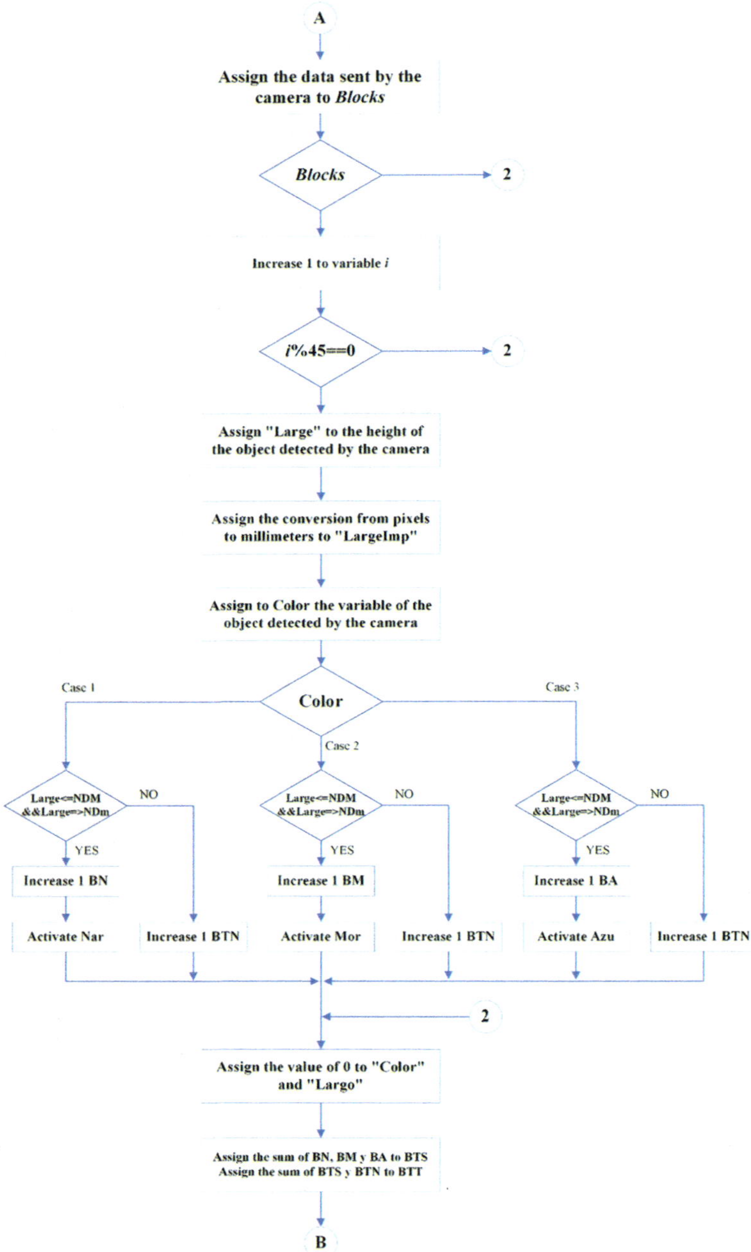

Fig. 36.8 Logical process of the Inspection and classification system from node A to node B

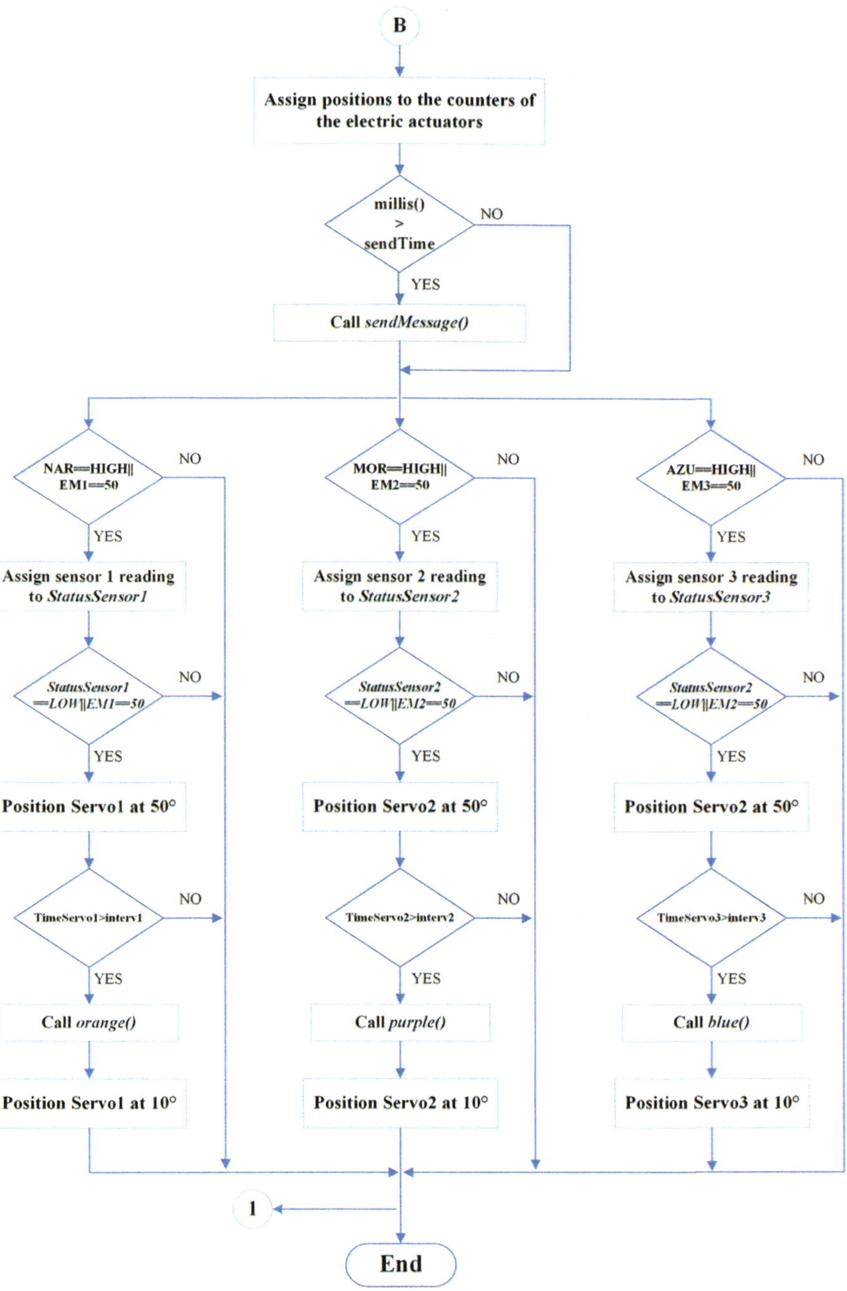

Fig. 36.9 Logical process of the Inspection and classification system from node B to END

Fig. 36.10 Bottles with incorrect liquid level (non-compliant)

Fig. 36.11 Orange bottle detected by Pixy Cam

Communication between the Arduino and the Pixy Cam is done through a 6-pin SPI cable. To verify its operation, the height attribute provided by any signature is received and displayed on the Arduino Serial Port (Fig. 36.13).

Fig. 36.12 Communication between Pixy Cam and PixyMon

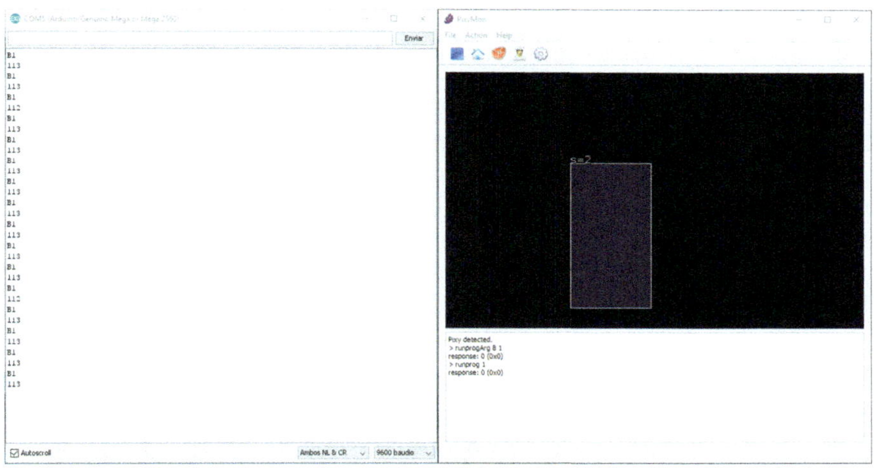

Fig. 36.13 Pixy Cam and Arduino MEGA communication

36.5 Results

The tests were carried out with 100 bottles. Each one included the colors orange, purple, and blue in a random order. Figure 36.14 summarizes the results of the first test. The numbers 1, 2, and 3 on the vertical axis are related to the colors blue, orange, and purple, respectively. The bottles that do not satisfy the liquid level are related to zero. This is the case of bottles 10, 30, 71, and 85. The results showed that the vision system works properly.

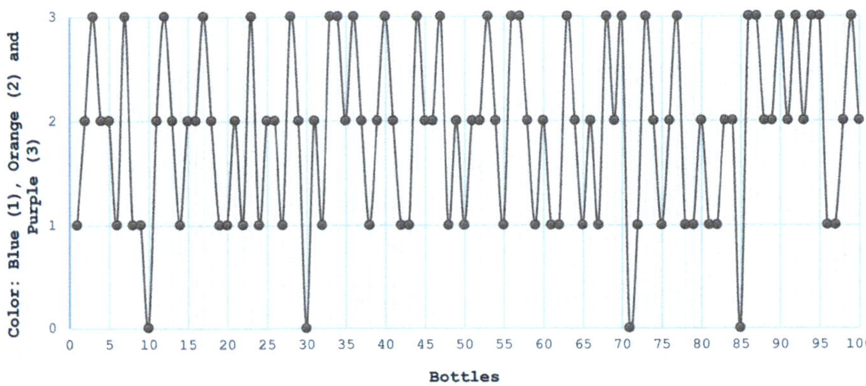

Fig. 36.14 Classification of bottles by color and soap content

The HMI programming was carried out in an open-source language based on Java. It creates designs of interface components from written instructions. The design was made friendly and intuitive for the operator, thus making its manipulation easier. Another great advantage of HMI is its compatibility with other open-source platforms, such as Arduino.

The operation time was compared with the results obtained by Galicia et al. (2016). They were manually and semi-automatically inspected and classified. It takes 12 s for an individual to perform this activity. A semi-automatic system performs the same activity in 3.33 s. Table 36.1 summarizes these evaluations for different periods. The proposed system optimizes this process. It takes 3.15 s to inspect and classify a bottle. In this case, the distance between bottles is 70 cm on the conveyor belt and the belt has a speed of 13 m/min.

Table 36.1 Number of bottles inspected in different periods with manual, semi-automatic, and proposed systems

Time	Manually inspected bottles	Bottles inspected with a semi-automatic system	Bottles inspected with the developed system
1 min	5 bottles	18 bottles	19 bottles
60 min	300 bottles	1,080 bottles	1140 bottles
8 h	2,400 bottles	8,640 bottles	9,120 bottles
7 days	16,800 bottles	60,480 bottles	63,840 bottles

36.6 Conclusions

The main component of the proposed system is the microcontroller, the Arduino Mega board. It communicates to the other components using the Serial protocol. It is a byte-by-byte communication. Sending and receiving a larger amount of data requires special handling. This project shows the procedure to perform such tasks.

A user interface was developed with the processing program. It has a structured code programming mode that is like an Arduino. This software is widely used for the development of user interfaces. It has compatibility with the selected microcontroller.

The developed system has advantages for the process in which it was implemented. The operator performance is optimized. The inspection and classification are monitored. The quality of the product is increased, as well as the productivity of the process.

Acknowledgements The authors kindly acknowledge the grant awarded by the Instituto Politécnico Nacional and COFAA for the development of this research.

References

Alves T, Morris T (2018) OpenPLC: an IEC 61,131–3 compliant open source industrial controller for cyber security research. Comp & Sec 78:364–379. https://doi.org/10.1016/j.cose.2018.07.007

Cognex: Visión Artificial (2023). Accessed 22 Nov 2023. https://www.cognex.com/es-ar/what-is/machine-vision/components

Galicia A, Jiménez M, Martínez J (2016) Development of a semi-automatic system for identifying and separating liquid soap containers. Dissertation, Instituto Politécnico Nacional

Khan AA, Laghari AA, Awan SA (2021) Machine learning in computer vision: a review. EAI End Trans Scal Inform Sys 8(32):1–11. https://doi.org/10.4108/eai.21-4-2021.169418

Wiley V, Lucas T (2023) Computer vision and image processing: a paper review. Int J Art Int Res 2(1):29–36. https://doi.org/10.29099/ijair.v2i1.42

Yang J, Wang C, Jiang B, Song H, Meng Q (2020) Visual perception enabled industry intelligence: state of the art, challenges and prospects. IEEE Trans Ind Inf 17(3):2204–2219 (2020). https://doi.org/10.1109/TII.2020.2998818

Chapter 37
Solution of the Swanson Reaction-Diffusion Model via Split and Symmetries

Jorge Luiz de Mello Caurio Junior, Claudio Zen Petersen, Fernanda Tumelero, Fernanda Krüger Tomaschewski, and Aquiles Almeida Ribeiro

Abstract The present work aims to propose an exact solution through split and symmetries for a model used to simulate the growth rate of cancer cells in a specific region of the human body. The model in question is the Swanson model, which has been numerically solved. In this study, we are excluding the term related to BNCT treatment and considering a constant diffusion coefficient per region. Attainment of an exact solution for the Swanson model holds potential significance in the treatment landscape of select cancer types, owing to the discernible benefits derived from the utilization of a refined model. Specifically, the deployment of a more precise model correlates with concomitant mitigation in the deleterious effects on healthy cells adjacent to neoplastic counterparts. Subsequent endeavors aim to refine the methodology by incorporating boundary and initial conditions into the comprehensive model, thereby encompassing elements pertinent to chemotherapy and radiotherapy interventions.

J. L. de Mello Caurio Junior (✉)
Universidade Federal do Rio Grande, Rio Grande, RS, Brazil
e-mail: juniorcaurio@gmail.com

C. Z. Petersen · F. K. Tomaschewski
Instituto de Física e Matemática, Universidade Federal de Pelotas, Capão do Leão, RS, Brazil
e-mail: claudio.petersen@ufpel.edu.br

F. K. Tomaschewski
e-mail: fernanda.tomaschewski@ufpel.edu.br

F. Tumelero
Instituto de Matemática, Estatística e Física, Universidade Federal do Rio Grande, Rio Grande, RS, Brazil
e-mail: fernandatumelero@furg.br

A. A. Ribeiro
Universidade Federal de Pelotas, Capão do Leão, RS, Brazil

A. Öchsner and H. Altenbach (eds.), *Engineering Design Applications VII*, Advanced Structured Materials 230, https://doi.org/10.1007/978-3-031-84346-4_37

37.1 Introduction

Cancer represents a heterogeneous spectrum of more than 100 diseases affecting diverse anatomical sites within the human body, characterized by dysregulated and unrestrained cellular proliferation. Malignant cells demonstrate diminished functional specialization relative to their healthy counterparts. Concurrently, the gradual displacement of normal cells by malignant counterparts leads to a progressive decline in the physiological integrity of afflicted tissues (Information 2025).

Within the spectrum of cancer types, gliomas stand out as highly invasive primary brain tumors that diffuse extensively throughout the brain. Approximately 50% of brain tumors fall under this category, with the most aggressive subtype known as glioblastoma multiforme (GBM), leading to a life expectancy of 6–12 months (Hatzikirou et al. 2005). Faced with the resistance and nonspecificity of treatments against these tumors, the primary challenge lies in developing strategies that are more specific to tumor cells and less aggressive to patients. In response to this challenge emerges the concept of *neutron capture therapy* (NCT), based on the observation that certain nuclides, whether radioactive or stable, possess the ability to absorb neutrons, generating a lethal amount of energy for cells through nuclear fission. NCT involves a binary treatment system for cancer, wherein a drug is initially administered to the patient, followed by the incidence of a collimated beam of low-energy neutrons.

With this problem at hand, the objective of this study is to find an exact solution for the Swanson reaction-diffusion model (Swanson et al. 2009). This model describes the diffusion and proliferation of tumor cells and the impact of chemotherapy and radiotherapy treatments, aiming to obtain an estimate of tumor concentration in space-time produced by a glioma-type brain tumor. In this work, we consider the model without the influence of chemotherapy and radiotherapy treatments, utilizing symmetry-based methods (Beck 2009; Sperotto 2007).

37.2 Swanson Model

The equation of the Swanson model without therapy is given by

$$\frac{\partial C}{\partial t} = \nabla \cdot (D(x)\nabla C) + \rho C, \tag{37.1}$$

where $x \in B,\ t \geq 0,\ C(x, 0) = C_0,\ n \cdot \nabla C$ on ∂B, where $C = C(x, t)$ represents the concentration of cancerous tumor cells per mm^3 at point x and time t, the domain B (Brain) is closed and bounded, D is the spatial diffusion coefficient with units of mm^2 per year, ρ represents the cell proliferation rate per year, C_0 is the initial distribution of tumor cells, and $n \cdot \nabla C$ is a zero-flux boundary condition preventing cells from leaving the domain B on its boundary ∂B. Thus, the term $\nabla(D(x)\nabla C)$ describes the net diffusion of glioma cells, and ρC represents the proliferation rate of glioma cells. Given that this model is a reaction-diffusion differential equation describing the net diffusion and proliferation rate of cancer cells, considering that in our study D is constant, and the geometry is Cartesian and bidimensional, the model can be rewritten as

$$\frac{\partial C}{\partial t} = D\left(\frac{\partial^2 C}{\partial x^2} + \frac{\partial^2 C}{\partial y^2}\right) + \rho C. \tag{37.2}$$

To employ the proposed methodology in the next section, we perform the substitution $C = \bar{C}e^{\rho t}$ to ensure the convergence of the geometric series. In this case, we have

$$\frac{\partial \bar{C}}{\partial t}e^{\rho t} + \rho\bar{C}e^{\rho t} = D\left(\frac{\partial^2 \bar{C}}{\partial x^2} + \frac{\partial^2 \bar{C}}{\partial y^2}\right)e^{\rho t} + \rho\bar{C}e^{\rho t}. \tag{37.3}$$

Simplifying the terms $\rho\bar{C}e^{\rho t}$, dividing both sides of the equation by $e^{\rho t}$ and considering that D is constant throughout the domain, Eq. (37.2) becomes

$$\frac{\partial \bar{C}}{\partial t} = D\left(\frac{\partial^2 \bar{C}}{\partial x^2} + \frac{\partial^2 \bar{C}}{\partial y^2}\right). \tag{37.4}$$

To solve Eq. (37.4) exactly, we seek methodologies for obtaining a solution. Among the methodologies applied for this purpose are symmetry-based methods, a class of techniques used to solve partial differential equations (PDEs). They are based on the idea that the presence of symmetries in a PDE can be used to simplify the solution. However, not all PDEs possess symmetries, and not all symmetries lead to exact solutions. It is important to note that these methods are a valuable tool but not a universal solution for solving PDEs.

All variable transformations that turn exact solutions of a particular differential equation into new solutions or equivalent solutions are called Lie symmetries admitted by the equation, auto-Bäcklund transformations, or simply symmetries. Various exact techniques can be utilized to obtain symmetries (Dattoli et al. 1998; Ibragimov et al. 1995; Zwillinger 1992), such as the direct substitution method (Bluman and Kumei 1989). However, at least one solution is necessary to initiate the process, which can be obtained, among other ways, through methods based on split.

37.3 Symmetry-Based Method

In the pursuit of exact solutions for Eq. (37.4), it is expressed as

$$Lf = 0, \tag{37.5}$$

where L is a linear operator decomposed as $L = AB$. Here, A and B are also linear differential operators that commute, i.e., $AB = BA$. This results in the equation:

$$Af = Bf. \tag{37.6}$$

Applying the inverse operator of A, denoted as A^1, to the left side of both members of Eq. (37.6) yields

$$A^{-1}Af = A^{-1}Bf. \tag{37.7}$$

Thus, Eq. (37.5), after adding the null space of the operator A, becomes

$$f = A^{-1}Bf + h(A). \tag{37.8}$$

where $h(A)$ is the null space of A, represented by the set of solutions to the equation $Ah(A) = 0$. Rearranging terms containing the function f results in

$$\left[I - A^{-1}B\right] f = h(A), \tag{37.9}$$

solving for f by using the inverse $\left[I - A^{-1}B\right]^{-1}$ gives

$$f = \left[I - A^{-1}B\right]^{-1} h(A) + h(I - A^{-1}B). \tag{37.10}$$

If we observe closely, we can rewrite the inverse operator appearing in Eq. (37.10) as a geometric series:

$$\frac{1}{I - A^{-1}B} = \sum_{k=0}^{\infty} \left(A^{-1}B\right)^k. \tag{37.11}$$

Considering the relative constraint on the norm of the operator $A^{-1}B$ concerning the convergence of the series, i.e., $\|A^{-1}B\| < 1$, the solution is obtained in the form:

$$f = \sum_{k=0}^{\infty} \left(A^{-1}B\right)^k h(A) + h(I - A^{-1}B). \tag{37.12}$$

To obtain a solution for the proposed equation, it is necessary to choose a function that serves as a proposal for converting the series, in other words, a $\eta_0 \in h(A) \cap h(I - A^{-1}B)$, where $h(I - A^{-1}B)$ is the null space of $A^{-1}B$. Note that the term $h(I - A^{-1}B)$ is redundant, as it comes from:

$$(A^{-1}B)h(I - A^{-1}B) = 0. \tag{37.13}$$

Once this η_0 is chosen, in order to convert the solution given by the series into a finite sum, we obtain the functions belonging to the null space of a finite power n of the operator $A^{-1}B$, as a truncated series:

$$f = \sum_{k=0}^{n} \left(A^{-1}B\right)^k \eta_0, \tag{37.14}$$

this represents an exact solution, as all subsequent terms are automatically null. As a result, no concerns regarding convergence arise during the development of the proposed formulation.

Moreover, the chosen η_0 will contain enough arbitrary constants to satisfy the boundary and initial conditions of the problem. Ideally, the process would commence with a known particular solution from the specified null spaces, thereby facilitating the generation of superior results within a limited number of iterations.

37.4 Finding the Structure of η_0 Using the Split Method

To discover a particular solution for Eq. (37.4) and consequently, the foundational function for the structure of η_0, we apply a mathematical technique known as the split method. The fundamental concept of this method involves breaking down the original PDE into various sub-PDEs, each of which can be solved independently. This decomposition is achieved by carefully choosing how to split the differential operator of the original equation into terms of its different components. In essence, it involves breaking down the target equation into a system containing, generally, two auxiliary differential equations. This allows the development of resolution algorithms that require only basic knowledge of calculus, primarily partial differentiation and integration.

Our aim is to integrate the split method seamlessly into the pursuit of exact solutions for the proposed model. To identify a suitable η_0 that meets predetermined criteria, we reformulate the operators A and B as follows:

$$\begin{cases} A = \dfrac{\partial(\cdot)}{\partial t} \\ B = D\left(\dfrac{\partial^2(\cdot)}{\partial x^2} + \dfrac{\partial^2(\cdot)}{\partial y^2}\right) \end{cases} \tag{37.15}$$

and, consequently,

$$A^{-1} = \int (\cdot)dt. \tag{37.16}$$

The resolution involves equating each half of the original equation to a source, initially constant or even zero, and then solving the resulting system. To break down the differential equation into a system of two equations, we start by grouping its terms, making it take the form $Af = Bf$, where A and B are differential operators. One of them is chosen to be easily invertible, and its null space is known. The criterion for separation is intentionally simplified here, suitable for desirable conditions allowing a split without overly restricting the resulting solution space. The split is based on applying the solution of the obtained system in one step (k) as a source for the next step ($k + 1$). Thus, we can obtain various exact solutions using a process like:

$$\begin{cases} Af_{k+1} = f_k \\ Bf_{k+1} = f_k, \end{cases} \tag{37.17}$$

where functions f_k represent exact solutions of the original equation. To apply the iterative process to the equations generated by the split, it is necessary to verify the commutativity between A and B, i.e., if $[A, B] = 0$. According to the choice made in Eq. (37.15), it is evident that $[A, B] = 0$. Thus, as the coefficients of A and B commute, for $k = 0$, we can have

$$\begin{cases} Af_1 = f_0 \\ Bf_1 = f_0, \end{cases} \tag{37.18}$$

Choosing $f(0) = 0$. Noting that operator A is invertible, and its null space is known, it is possible to find a solution to the equation related to the operator A by integrating it with respect to time and summing the corresponding null space. Therefore, we have a function f_1 defined by

$$f_1 = g_1(x, y). \tag{37.19}$$

The next step is to apply the solution represented in Eq. (37.19) to the equation related to the operator B, resulting in

$$D\left(\frac{\partial^2}{\partial x^2} g_1(x, y) + \frac{\partial^2}{\partial y^2} g_1(x, y)\right) = 0. \tag{37.20}$$

To solve this equation, we apply a new split:

$$\begin{cases} \dfrac{\partial^2}{\partial x^2} g_1(x, y) = 0 \\[2ex] \dfrac{\partial^2}{\partial y^2} g_1(x, y) = 0. \end{cases} \tag{37.21}$$

Integrating the first equation of the system in Eq. (37.21) twice with respect to x and solving for $g_1(x, y)$, we obtain an expression of the form:

$$g_1(x, y) = a_1(y)x + b_1(y), \tag{37.22}$$

Substituting Eq. (37.22) into the second equation of the system in Eq. (37.21), we get

$$a_1''(y)x + b_1''(y) = 0. \tag{37.23}$$

The only way to satisfy this equation is by setting $a_1''(y) = 0$ e $b_1''(y) = 0$. Thus, these functions become

$$a_1(y) = c_1 y + c_2 \qquad e \qquad b_1(y) = c_3 y + c_4, \tag{37.24}$$

providing a solution of the form:

$$f_1 = g_1(x, y) = c_1 xy + c_2 x + c_3 y + c_4. \tag{37.25}$$

Once this solution is found, the process is applied again, now introducing solution Eq. (37.25) as a new source. Thus, for $k = 1$, we have

$$\begin{cases} \dfrac{\partial}{\partial t} f_2 = c_1 xy + c_2 x + c_3 y + c_4 \\[2ex] D\left(\dfrac{\partial^2}{\partial x^2} f_2 + \dfrac{\partial^2}{\partial y^2} f_2\right) = c_1 xy + c_2 x + c_3 y + c_4. \end{cases} \tag{37.26}$$

Integrating the first equation of Eq. (37.26) with respect to t, we obtain

$$f_2 = g_2(x, y, t) = c_1 xyt + c_2 xt + c_3 yt + c_4 t + g_3(x, y). \tag{37.27}$$

In order to solve the second equation of Eq. (37.26), a new split is performed, resulting in the system

$$\begin{cases} D\left(\dfrac{\partial^2}{\partial x^2} f_2\right) = \dfrac{1}{2}\left(c_1 xy + c_2 x + c_3 y + c_4\right) \\[2ex] D\left(\dfrac{\partial^2}{\partial y^2} f_2\right) = \dfrac{1}{2}\left(c_1 xy + c_2 x + c_3 y + c_4\right), \end{cases} \tag{37.28}$$

which, upon rearranging the terms, becomes

$$\begin{cases} \dfrac{\partial^2}{\partial x^2} f_2 = \dfrac{1}{2D}\left(c_1 xy + c_2 x + c_3 y + c_4\right) \\[2ex] \dfrac{\partial^2}{\partial y^2} f_2 = \dfrac{1}{2D}\left(c_1 xy + c_2 x + c_3 y + c_4\right). \end{cases} \tag{37.29}$$

The second derivatives are then expressed as

$$\frac{\partial^2}{\partial x^2} f_2 = \frac{\partial^2}{\partial x^2}\left(g_3(x, y)\right) = \frac{1}{2D}\left(c_1 xy + c_2 x + c_3 y + c_4\right) \tag{37.30}$$

and

$$\frac{\partial^2}{\partial y^2} f_2 = \frac{\partial^2}{\partial y^2}\left(g_3(x, y)\right) = \frac{1}{2D}\left(c_1 xy + c_2 x + c_3 y + c_4\right). \tag{37.31}$$

It is noteworthy that integrating $\frac{\partial^2}{\partial x^2} g_3(x, y)$ twice with respect to x, results in

$$\begin{aligned} g_3(x, y) = \frac{1}{2D}\Big(&\frac{c_1}{6}x^3 y + \frac{c_2}{6}x^3 + \frac{c_3}{2}x^2 y + \frac{c_4}{2}x^2 \\ &+ a_2(y)x + b_2(y)\Big), \end{aligned} \tag{37.32}$$

substituting the value of $g_3(x, y)$ in $\frac{\partial^2}{\partial y^2} g_3(x, y)$ results in the equation

$$\begin{aligned} \frac{\partial^2}{\partial y^2} g_3(x, y) &= \frac{1}{2D}\left(a_2''(y)x + b_2''(y)\right) \\ &= \frac{1}{2D}(c_1 xy + c_2 x + c_3 y + c_4). \end{aligned} \tag{37.33}$$

Equation (37.33) is resolved by setting $a_2''(y) = c_1 y + c_2$, $b_2''(y) = c_3 y + c_4$, and integrating twice with respect to y provides the expressions in

$$\begin{aligned} a_2(y) &= \frac{c_1 y^3}{6} + \frac{c_2 y^2}{2} + c_5 y + c_6 \quad e \\ b_2(y) &= \frac{c_3 y^3}{6} + \frac{c_4 y^2}{2} + c_7 y + c_8. \end{aligned} \tag{37.34}$$

Upon substituting the found values of $a_2(y)$ and $b_2(y)$ into Eq. (37.32), $g_3(x, y)$ takes the form described in

$$\begin{aligned} g_3(x, y) = \frac{1}{2D}\Bigg[&c_1\left(\frac{x^3 y + x y^3}{6}\right) + c_2\left(\frac{x^3 + 3xy^2}{6}\right) \\ &+ c_3\left(\frac{3x^2 y + y^3}{12}\right) c_4\left(\frac{x^2 + y^2}{2}\right) + c_5 xy \\ &+ c_6 x + c_7 y + c_8\Bigg]. \end{aligned} \tag{37.35}$$

Finally, replacing $g_3(x, y)$ obtained in Eq. (37.35) into Eq. (37.27) yields the expression for f_2.

$$f_2 = g_2(x, y, t) = c_1xyt + c_2xt + c_3yt + c_4t + \frac{1}{2D}\left[c_1\left(\frac{x^3y + xy^3}{6}\right) + c_2\left(\frac{x^3 + 3xy^2}{6}\right) + c_3\left(\frac{3x^2y + y^3}{6}\right) + c_4\left(\frac{x^2 + y^2}{2}\right) + c_5xy + c_6x + c_7y + c_8\right]. \tag{37.36}$$

37.5 Results and Discussions

It is evident that each iteration of the split method generates a solution containing a greater number of arbitrary constants than the previous one. For a considerable range of problems, this method can be applied until a sufficient quantity of constants or arbitrary functions is obtained to meet the boundary conditions in subdomains or even to achieve an invariant form of the solution. The higher the number of arbitrary constants present in the solution, the broader the range of boundary conditions that can be satisfied, or the larger the subdomain over which the solution is valid. Now, it is sufficient to observe that Eq. (37.36) is a solution to Eq. (37.4), utilizing the results from Eqs. (37.30) and (37.31), and knowing that

$$\frac{\partial f_2}{\partial t} = c_1xy + c_2x + c_3y + c_4 \tag{37.37}$$

is verified by

$$\frac{\partial f_2}{\partial t} = D\left(\frac{\partial^2 f_2}{\partial x^2} + \frac{\partial^2 f_2}{\partial y^2}\right), \tag{37.38}$$

substituting the terms yields

$$c_1xy + c_2x + c_3y + c_4 = D\left[\frac{1}{2D}(c_1xy + c_2x + c_3y + c_4) + \frac{1}{2D}(c_1xy + c_2x + c_3y + c_4)\right], \tag{37.39}$$

and combining the terms results in

$$c_1xy + c_2x + c_3y + c_4 = D\left[\frac{2}{2D}(c_1xy + c_2x + c_3y + c_4)\right]. \tag{37.40}$$

Thus, we have

$$c_1xy + c_2x + c_3y + c_4 = c_1xy + c_2x + c_3y + c_4. \tag{37.41}$$

Therefore, $f_2 = \bar{C}$ is an exact solution to Eq. (37.4) and correspondingly satisfies Eq. (37.2), as $C = \bar{C}e^{\rho t}$.

$$\begin{aligned} C &= c_1\left(\frac{x^3y + xy^3}{12D} + xyt\right)e^{\rho t} + c_2\left(\frac{x^3 + 3xy^2}{12D} + xt\right)e^{\rho t} \\ &+ c_3\left(\frac{3x^2y + y^3}{12D} + yt\right)e^{\rho t} + c_4\left(\frac{x^2 + y^2}{4D} + t\right)e^{\rho t} \\ &+ c_5xye^{\rho t} + c_6xe^{\rho t} + c_7ye^{\rho t} + c_8e^{\rho t}. \end{aligned} \tag{37.42}$$

Solving for the Partial Derivatives, we obtain

$$\begin{aligned} \frac{\partial C}{\partial t} &= c_1\left[xye^{\rho t} + \left(\frac{x^3y + xy^3}{12D} + xyt\right)\rho e^{\rho t}\right] \\ &+ c_2\left[xe^{\rho t} + \left(\frac{x^3 + 3xy^2}{12D} + xt\right)\rho e^{\rho t}\right] \\ &+ c_3\left[ye^{\rho t} + \left(\frac{3x^2y + y^3}{12D} + yt\right)\rho e^{\rho t}\right] \\ &+ c_4\left[e^{\rho t} + \left(\frac{x^2 + y^2}{4D} + t\right)\rho e^{\rho t}\right] \\ &+ c_5xy\rho e^{\rho t} + c_6x\rho e^{\rho t} + c_7y\rho e^{\rho t} + c_8\rho e^{\rho t}, \end{aligned} \tag{37.43}$$

$$\frac{\partial^2 C}{\partial x^2} = c_1\left(\frac{xy}{2D}\right)e^{\rho t} + c_2\left(\frac{x}{2D}\right)e^{\rho t} + c_3\left(\frac{y}{2D}\right)e^{\rho t} + \frac{c_4}{2D}e^{\rho t} \tag{37.44}$$

and

$$\frac{\partial^2 C}{\partial y^2} = c_1\left(\frac{xy}{2D}\right)e^{\rho t} + c_2\left(\frac{x}{2D}\right)e^{\rho t} + c_3\left(\frac{y}{2D}\right)e^{\rho t} + \frac{c_4}{2D}e^{\rho t}. \tag{37.45}$$

Therefore, we are left with

$$\begin{aligned} & c_1\left[xye^{\rho t} + \left(\frac{x^3y + xy^3}{12D} + xyt\right)\rho e^{\rho t}\right] \\ &+ c_2\left[xe^{\rho t} + \left(\frac{x^3 + 3xy^2}{12D} + xt\right)\rho e^{\rho t}\right] \\ &+ c_3\left[ye^{\rho t} + \left(\frac{3x^2y + y^3}{12D} + yt\right)\rho e^{\rho t}\right] \end{aligned}$$

$$
\begin{aligned}
&+c_4\left[e^{\rho t}+\left(\frac{x^2+y^2}{4D}+t\right)\rho e^{\rho t}\right]\\
&+c_5xy\rho e^{\rho t}+c_6x\rho e^{\rho t}+c_7y\rho e^{\rho t}+c_8\rho e^{\rho t}=\\
&2D\left[c_1\left(\frac{xy}{2D}\right)e^{\rho t}+c_2\left(\frac{x}{2D}\right)e^{\rho t}\right.\\
&\left.+c_3\left(\frac{y}{2D}\right)e^{\rho t}+\frac{c_4}{2D}e^{\rho t}\right]+\rho Ce^{\rho t}.
\end{aligned}
\tag{37.46}
$$

Simplifying and rearranging the terms, we obtain

$$
\begin{aligned}
&c_1\left[xye^{\rho t}+\left(\frac{x^3y+xy^3}{12D}+xyt\right)\rho e^{\rho t}\right]\\
&+c_2\left[xe^{\rho t}+\left(\frac{x^3+3xy^2}{12D}+xt\right)\rho e^{\rho t}\right]\\
&+c_3\left[ye^{\rho t}+\left(\frac{3x^2y+y^3}{12D}+yt\right)\rho e^{\rho t}\right]\\
&+c_4\left[e^{\rho t}+\left(\frac{x^2+y^2}{4D}+t\right)\rho e^{\rho t}\right]\\
&+c_5xy\rho e^{\rho t}+c_6x\rho e^{\rho t}+c_7y\rho e^{\rho t}+c_8\rho e^{\rho t}=\\
&c_1xye^{\rho t}+c_2xe^{\rho t}+c_3ye^{\rho t}+c_4e^{\rho t}+\rho Ce^{\rho t}.
\end{aligned}
\tag{37.47}
$$

This demonstrates that the solution is exact. It is worth noting that the proposed solution incorporates combinations of solutions to Eq. (37.2), meaning that certain multiples of coefficients also represent an exact solution.

It can be observed that structures formed by multiples of coefficients C_n constitute a solution to Eq. (37.4).

C_1 is verified as a solution because,

$$
c_1\left(\frac{x^3y+xy^3}{12D}+xyt\right),
\tag{37.48}
$$

in other words, it is acknowledged as a solution

$$
C=\frac{x^3y+xy^3}{12D}+xyt.
\tag{37.49}
$$

It is noted that the partial derivatives result in

$$
\frac{\partial C}{\partial t}=xy,
\tag{37.50}
$$

$$\frac{\partial^2 C}{\partial x^2} = \frac{6xy}{12D} = \frac{xy}{2D} \tag{37.51}$$

and

$$\frac{\partial^2 C}{\partial y^2} = \frac{6xy}{12D} = \frac{xy}{2D}. \tag{37.52}$$

Therefore,

$$xy = D\left(\frac{xy}{2D} + \frac{xy}{2D}\right) = D\left(\frac{2xy}{2D}\right) = D\left(\frac{2xy}{2D}\right) = xy. \tag{37.53}$$

This indicates that C_1 is a solution.

Analogously, for the combination of C_2 and C_3, we are left with

$$c_2\left(\frac{x^3 + 3xy^2}{12D} + xt\right) + c_3\left(\frac{3x^2y + y^3}{12D} + yt\right), \tag{37.54}$$

so, the solution is given by

$$C = \left(\frac{x^3 + 3xy^2}{12D}\right) + \left(\frac{3x^2y + y^3}{12D}\right) + xt + yt. \tag{37.55}$$

The partial derivatives result in

$$\frac{\partial C}{\partial t} = x + y, \tag{37.56}$$

$$\frac{\partial^2 C}{\partial x^2} = \frac{6x}{12D} + \frac{6y}{12D} = \frac{x + y}{2D} \tag{37.57}$$

and

$$\frac{\partial^2 C}{\partial y^2} = \frac{6x}{12D} + \frac{6y}{12D} = \frac{x + y}{2D}, \tag{37.58}$$

therefore,

$$\begin{aligned} x + y &= D\left(\frac{x + y}{2D} + \frac{x + y}{2D}\right) = D\left(\frac{2(x + y)}{2D}\right) = D\left(\frac{2(x + y)}{2D}\right) \\ &= x + y. \end{aligned} \tag{37.59}$$

The same result is observed for C_4, as

$$c_4\left(\frac{x^2+y^2}{4D}+t\right), \tag{37.60}$$

meaning that the solution appears as

$$C=\frac{x^2+y^2}{4D}+t. \tag{37.61}$$

Its partial derivatives result in

$$\frac{\partial C}{\partial t}=1, \tag{37.62}$$

$$\frac{\partial^2 C}{\partial x^2}=\frac{1}{2D} \tag{37.63}$$

and

$$\frac{\partial^2 C}{\partial y^2}=\frac{1}{2D}. \tag{37.64}$$

Therefore,

$$1=D\left(\frac{1}{2D}+\frac{1}{2D}\right)=D\left(\frac{1}{2D}+\frac{1}{2D}\right)=\frac{2}{2}=1. \tag{37.65}$$

37.6 Conclusions

It can be firmly concluded that the methodology showcased herein not only proves its efficacy but also underscores its indispensability in solving the proposed model. The attainment of an exact solution represents a groundbreaking achievement, particularly considering the longstanding reliance on numerical approaches to tackle similar problems. Thus, the continuation of the study is of great importance, as it is possible to extend the proposed methodology to the complete model, including the term related to treatment, in addition to considering the spatial dependence of biological parameters. Therefore, in this work, an exact solution was obtained, providing an estimate of the concentration of tumor cells without any inherent errors associated with numerical methods or operator approximations. As a future perspective, realistic results for the proposed problem will be generated by considering D, ρ, initial and boundary conditions according to what has been used in the literature.

References

Beck D (2009) Soluções particulares para as equações de Navier-Stokes tridimencionais transientes. Phd thesis, Programa de Pós-Graduação em Engenharia Mecânica, Universidade Federal do Rio Grande do Sul

Bluman GW, Kumei S (1989) Symmetries and differential equations. Springer, New York, NY, in press

Dattoli G, Giannessi L, Quatromini M, Torre A (1998) Exponential operators, operational rules and evolutions problems. Il nuovo cimento 113B:699–710

Hatzikirou H, Deutsch A, Schaller C, Simon A, Swanson K (2005) Mathematical modelling of glioblastoma tumour development: a review. Math Models Methods Appl Sci 15:1779–1794

Ibragimov NH, Aksenov AV, Baikov VA, Chugunov VA, Gazizov RK, Meshkov AG (1995) Applications in engineering and physical sciences. CRC Press, Florida, in press

Information on https://www.gov.br/inca/pt-br/assuntos/cancer

Sperotto FA (2007) Soluções de equações advectivo-difusivas utilizando split, série geométrica e transformação de Bäcklund. Phd thesis, Programa de Pós-Graduação em Engenharia Mecânica, Universidade Federal do Rio Grande d Sul

Swanson KR, Rockne R, Alvord EC Jr, Rockhill JK (2009) A mathematical model for brain tumor response to radiation therapy. J Math Biol 58:561–578

Zwillinger D (1992) The Handbook of integration. Jones and Bartlett Publishers, Boston, MA, in press

Chapter 38
Analysis of the Reliability and Availability of the Algiers Metro Traction System Based on the Pareto and Weibull Three Parameters Methods

Hakim Siguerdjidjene, Abdelkrim Merah, Djamel Saidi, and Amin Houari

Abstract Rail transport by metro is a socio-economic form of transport, and the reliability of traction systems is considered to be one of the key factors required to ensure its smooth operation, safety and public comfort. In addition, it provides rapid mobility for citizens between different areas of a city and reduces the road congestion in urban areas by offering a fast, efficient transport alternative. This research is based on an in-depth analysis of the most frequent operating time data, failures and breakdowns (mechanical, electrical, etc.) of the ELDJAZAIR-Alger metro company's traction systems, with the aim of analysing the reliability and availability of these two series of traction systems, in order to minimize interruptions and maintain a high level of service to meet users' needs. Historical data was collected and thoroughly analyzed. Firstly, we visualized this data in the form of a Pareto chart known as the '20/80% rule' (A, B, C) (Thomas E. Nisonger in The "80/20 Rule" and Core Journals. The Serials Librarian 55(1):62–84, in order to determine the highest priority areas. This makes it easier to process the data. Second, the reliability of vehicle traction systems is analyzed using the three-parameter Weibull probability criterion R (t, γ, η, β) (Vonta Ilia, Mangey Ram. Reliability Engineering:

H. Siguerdjidjene (✉) · A. Merah · A. Houari
Faculty of Technology, M'Hamed Bougara University, Boumerdes, Algeria
e-mail: h.siguerdjidjene@univ-boumerdes.dz

A. Merah
e-mail: abdelkrimerah@univ-boumerdes.dz

A. Houari
e-mail: a.houari@univ-boumerdes.dz

D. Saidi
National Polytechnic School of Algiers (ENP), El Harrach, Algeria
e-mail: djamel.saidi@g.enp.edu.dz

H. Siguerdjidjene · A. Houari
Motor Dynamics and Vibroacoustics Laboratory, Boumerdes, Algeria

A. Merah
Research Unit Materials - Processes and Environment UR-MPE, Boumerdes, Algeria

A. Öchsner and H. Altenbach (eds.), *Engineering Design Applications VII*, Advanced Structured Materials 230, https://doi.org/10.1007/978-3-031-84346-4_38

Theory and Applications. Publisher: CRC Press-Taylor & Francis Group ISBN: 9,780,815,355,175. (2018)), based on the calculation of the time-to-function and mean time between failures and the failure rate λ (t). The results obtained using the Pareto method show that the most critical faults are classified in zone A: where 20% corresponds to 80% of the total number of faults, representing five faults designated by codes as follows: [CLS, MON, CAJ, CAU, VAUX], which are more critical and take priority over emergency intervention by the maintenance department. The results of the analysis show that the correlation coefficient equals to 0.976, which indicates a very strong positive linear relationship between the percentage of variable failures and time and the reliability R (t, γ, η, β) of traction systems is almost in the approximate 50% with its parameters: $\gamma = 0$, $\beta = 3.9776$, $\eta = 1067.62$h and availability A(t) decreases over time, which requires measures to improve the latter by including new techniques for service management, control, preventive and corrective maintenance.

Keywords Failures · Weibull's Law · Reliability R(t) · Availability a(t) · TBF · MTTF · MTBF

38.1 Introduction

Metro transport plays a crucial role in urban mobility throughout the world, helping to reduce road congestion, greenhouse gas emissions and journey times. It is often regarded as one of the most reliable, economical, sustainable and modern public transport systems, offering an efficient solution for quick and easy access to populated areas and business centres, thereby promoting economic development.

In this context, this work looks at the operation and importance of the reliability of Algiers metro traction systems, with particular emphasis on the operator "Algiers Metro Operating Company", using the trade name "Métro El Djazaïr". The operation of line 1 of the Algiers Metro (main section with its three extensions A, B and C), the maintenance of the rolling stock and the maintenance of the equipment, railway infrastructure and engineering structures is entrusted to the operator "Algiers Metro Operating Company".

The metro system currently comprises Line 1 of the Algiers metro, running from north-west to south-east represented in Fig. 38.1. It forks between the "Place of Martyrs" end station and the "El Harrach Centre" and "Aïn Naâdja" end stations. It is approximately 16,200 m long between the termini, divided between:

- Place of Martyrs in Harrach Centre: approximately 13,000 m;
- Hai El Badr in Aïn Naâdja: approximately 3200 m.

It serves nineteen stations, including seventeen underground and two overhead. The stations have two tracks and two side platforms, with the exception of the "Haï El Badr" overhead station, which has four tracks and three central platforms.

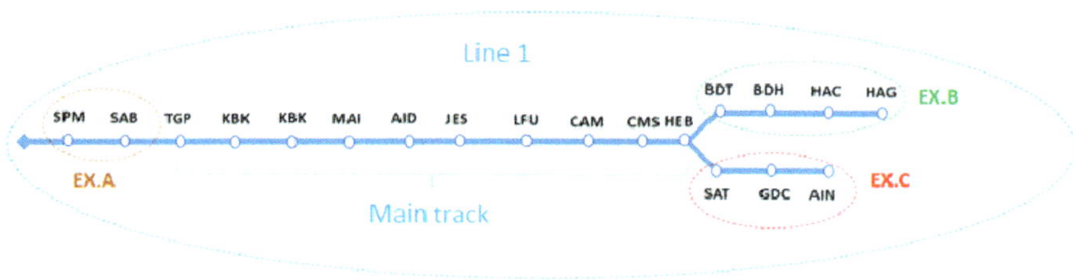

Fig. 38.1 Algiers metro stations

The rolling stock of the Algiers metro consists of two types of trainsets, with a total of 26 iron-wheeled trains powered by 750V DC 3rd rail.

The metro plays an important role in the mobility of the people of Algiers. It should be stressed that this is the only metro system in Algeria. Its smooth operation depends heavily on the reliability and availability of its traction systems. These systems must be serviced and maintained periodically in order to:

- Ensure the smooth running and safety of passengers;
- Increase the lifespan of rolling stock;
- Update efficient planning and management of the maintenance of these traction systems.

Several researches in the field of analysis and reliability of metro transport systems have been carried out by researchers from all over the world among these researches Qin et al. (Qin et al. 2014), Fang et al. (Application, et al. 2019) is motivated to apply reliability-centered maintenance (RCM) to the maintenance of the metro door system. Then a fuzzy FMECA method is proposed to define the hazard degree of 10 failure modes from 5 components with high failure rates to conduct reliability analysis. Cheng et al. 2013 addressed the analysis of metro vehicles. In addition, Rafał et al. (Melnik, et al. 2019) (TRANSCOM) have treated the failure analysis of the two types of metro vehicles and a comparison of their reliability functions.

The research work Yin et al. (Yin et al. 2017) established a selected model by preprocessing (screening, refining, and classification) the original data and statistical analysis, using the A-D test to verify the degree of its fitness to determine the optimal failure distribution model.

38.1.1 Analysis of the Historical Data of the Algiers Metro Train System

Firstly, we will analyse the historical data collected from the operation of the Algiers metro train systems presented in Table1, for the period from 01/01/2022 to 31/12/2022, using the Pareto method, also known as the (ABC) method, based on the principle that 20% of the elements in the area (A) represent 80% of the criterion

Table 38.1 Historical breakdown data for metro traction vehicle systems by code designation

Code	CLS	MON	CAJ	CAU	VAUX	GTA
Designation	Cabin	Air conditioning	Door	Engine (TMC, ESIS)	Traction- brake	Static converter
Nb of failure Year 2022	173	147	143	73	65	30
Code	BOG	PAE	GCC	CAB	TRF	EI
Designation	Box	Auxiliary vehicles	Current capture	Bogie	Communication	Production treatment
Nb of failure Year 2022	29	28	24	21	19	6
Code	COM	ATE	FRE	CBTC	CAT	Sum
Designation	Industrial equipments	Braking direct	Current generation	Hitch	Automatic driving	
Nb of failure	4	3	2	1	0	768

under study, and distinguish the number of faults according to the fault code and determine the faultiest elements from those that are less faulty) to be treated in order to improve the quality of service and increase the safety and mobility of citizens.

38.2 Reliability Analysis of Algiers Metro Vehicle Traction Systems

38.2.1 Reliability Function

The reliability function determines the probability of failure-free operation of the object at a given time. Assuming that the object is new at time $t = 0$ and fails in time T, the reliability function *R(t)* will take the form describe by Tghem (Lucanin 2005):

$$\mathrm{R(t) = P(T > t)\ t \geq 0} \tag{38.1}$$

In other way:

$$R(t) = \int_{t}^{\infty} \mathrm{f}(x)dx = 1 - \mathrm{F}(x) \tag{38.2}$$

The function of the reliability can be calculated as an integral of probability density function of time to failure $f(x)$ or can be equivalently related to the cumulative distribution function $F(x)$ that occurred in the interval [0… t] as following:

$$R(t) = \int_{t}^{\infty} f(x)dx = 1 - F(x) \tag{38.3}$$

$R(t)$ is the component reliability at time t, and $0 < R(t) < 1$

$$R(\infty) = \lim_{t \to \infty} R(t) = 0$$

38.2.2 Study of Reliability

Metro vehicle traction systems are mainly composed of mechanical, electric, electronic and other material components. It is assumed that the life of each component will correctly work.

To this end, the study of the reliability of the traction systems of the Algiers metro is based on the determination of indicators, namely the operation operating time (TBF) and the mean operating time to first failure (MTTF) for non-repairable systems or the Mean Time to Between Failure (MTBF) for repairable systems, Vojkan (Lucanin 2005).

Knowledge of this indicators makes it possible to plan and define adequate maintenance policy for each traction vehicle of the two types of traction systems.

The indicator MTBF, can be defined by the following calculation:

$$MTBF = \int_{0}^{\infty} f(t).dt \tag{38.4}$$

where, $f(t)$ represents the probability of the first failure of the traction vehicle system between $(t, t + \Delta t)$, i.e. the failure density.

The failure rate (λ) at which traction vehicle systems fail is expressed as the number of failures per unit time and can be expressed as follows:

$$\lambda = \frac{N_{failure}}{T_{Total}} \tag{38.5}$$

where, $N_{failure}$ is the total number of failures of the metro traction vehicle systems, T_{total} is the total operating time of the metro traction vehicle systems in hours.

For the repairable system, the working time between two adjacent faults is represented by the mean time between failure (MTBF), the relation between MTBF and failure rate can be described by:

$$\text{MTB} = \frac{1}{\lambda} \tag{38.6}$$

In another way, the mean time between failure, can be defined by the folowing formula:

$$\text{MTBF} = \frac{\sum_{i=1}^{n} \Delta t}{n} \tag{38.7}$$

where Δt is the failure interval, n is the total number of times the fault has occurred in the system.

Moreover, the system failure rate function λ (t) under the exponential distrubution can be expressed as:

$$\lambda(t) = \frac{-\,\text{R}'(\text{t})}{\text{R}(\text{t})} = \lambda \tag{38.8}$$

where λ is the constant failure rate.

In this way, according to EN 13,306: 2001, for this study, several reliability models can be applied, in particular the three-parameter Weibull model which expresses the reliability of a piece of equipment as the probability for it to still be in service at a time (t), Tghem (Teghem 2013).

From the historical data of interventions on the vehicle traction systems of the metro ELDJAZAIR, from Algiers represented in Table 38.2 and assuming that the lifetime of each component of the last ones follows the Weibull distribution with three parameters R (t, γ, η, β)), at the end, we proceed to the calculations and to the analysis of the reliability by a software.

The expression for calculating and analysing the reliability function can then be described as follows:

$$R(t) = e^{-\left(\frac{t-\gamma}{\eta}\right)^{\beta}} \tag{38.9}$$

$$\text{For } \gamma = 0\,,\ R(t) = e^{-(\frac{t}{n})^{\beta}}$$

The associated distribution function (CDF) and the density function for each system traction vehicle components are defined by the following formulas:

$$F(t) = 1 - R(t) = 1 - e^{-(\frac{t-\gamma}{\eta})^{\beta}} = 1 - e^{-\left(\frac{t}{\eta}\right)^{\beta}}\ for\ 1 - e^{-(\frac{t}{\eta})^{\beta}} \tag{38.10}$$

$$f(t, \gamma, \beta, \eta) = \frac{\beta}{\eta^{\beta}}\, t^{\beta-1}\, e^{-\left(\frac{t-\gamma}{\eta}\right)^{\beta}}\ for\ t \geq \gamma \tag{38.11}$$

where:

n — The number of components in the systems;
γ— Position or origin parameter (unit of time);
β— Shape parameter (unitless);
η— Scale parameter (unit of time).

Table 38.2 Historical data on the uptime (TBF) and number of failures (λ) of metro vehicle traction systems per month for the year 2022

Month	Jan-22	Feb-22	Mar-22	Apr-22	May-22	Jun-22	Jul-22	Aug-22	Sep-22	Oct-22	Nov-22	Dec-22
TBF (h)	5149	4650	5149	4526	5028	4951	4249	4241	4704	4917	4525	4693
Nb of Failure	37	43	40	50	61	85	54	69	83	75	84	87
Σ ni	48	80	120	170	231	316	370	439	522	597	681	768
Month	Jan-22	Feb-22	Mar-22	Apr-22	May-22	Jun-22	Jul-22	Aug-22	Sep-22	Oct-22	Nov-22	Dec-22
TBF (h)	5149	4650	5149	4526	5028	4951	4249	4241	4704	4917	4525	4693
Nb of Failure	37	43	40	50	61	85	54	69	83	75	84	87
Σ ni	48	80	120	170	231	316	370	439	522	597	681	768

In the present analysis, it is the operating time of the vehicle traction systems that has been taken into account, and not the mileage. Consequently, the estimated reliability functions are expressed as a function of time *t* in hours for each vehicle traction system (traction system).

The historical overview of work on the vehicle traction systems of the ELDJAZAIR metro (Algiers) is presented in Table 38.2.

38.3 Availability Calculation for Metro Traction Vehicle Systems

The availability of the metro vehicle traction system is the probability that the system will be operational when required for use and can be calculated as follows:

$$\text{Availability} = \text{MTBF}/(\text{MTBF} + \text{MTTR}) \tag{38.12}$$

where MTBF is the mean time between failure and MTTR is the mean time to repair.

The average life means that the equipment is continuously fault-free under the specified conditions, it can be written as:

The influence of railway vehicle reliability on its availability, has been presented in the work research of Vojkan (Lucanin 2005).

$$\text{MTTF} = \text{E(T)} \tag{38.13}$$

Since *E(t)* is finite, we can write:

$$\text{MTTF} = \int_0^\infty \text{R(t)dt} \tag{38.14}$$

$$\text{E(T}crt) = \int_0^\infty \text{t.f(t)dt} \tag{38.15}$$

And it is known from the reliability theory that:

$$\text{f(t)} = \frac{\text{dF(t)}}{\text{dt}} = \frac{-\text{dR(t)}}{\text{dt}} = -\text{R}'\text{(t)} \tag{38.16}$$

$$\text{E(T)} = \int_0^\infty \text{t.f(t)dt} = \int_0^\infty t.[\text{R}'\text{(t)]dt} = \int_0^\infty \text{R(t)dt} \tag{38.17}$$

In our case we take the Weibull method, then the average availability of the systems of traction vehicle is calculating the availability of the traction system by the following formula:

Table 38.3 Pareto damage classification table

Code	CLS	MON	CAJ	CAU	VAUX	GTA	BOG	PAE	GCC	CAB	Other
Nbr of failure	173	147	143	73	65	30	29	28	24	21	35
Percent	22.5	19.1	18.6	9.5	8.5	3.9	3.8	3.6	3.1	2.7	4.6
Cum %	22.5	41.7	60.3	69.9	78.3	82.2	85.9	89.6	92.7	95.4	100

$$A(t) = \frac{E(t)}{T} = \frac{\int_0^\infty R(t)dt}{T} = \frac{\int_0^\infty e^{-\left(\frac{t}{\eta}\right)^\beta} dt}{T} \frac{MTTF}{T} \tag{38.18}$$

38.4 Results and Discussion

38.4.1 Results of Pareto Analysis of Historical Data

From the code data and the failures collected in Table 38.1, we define their categories, then:

We classify them in decreasing order, then we add them together and calculate their cumulative percent presented in damage classification Table 38.3, finally, we determine the appropriate scale for plotting the graph shown in Fig. 38.2.

In addition, the Pareto curve diagram shows that the most critical faults are classified in zone A: 20%, corresponding to 80% of the total number of faults, i.e. five faults designated by the following codes: [CLS, MON, CAJ, CAU, VAUX], which are the most critical and have priority for emergency intervention by maintenance services.

38.4.2 Results of the Analysis of the Reliability and Availability of Algiers Metro System

In this work, the Weibull probability density function with three parameters is expressed as follows:

$$f(t) = \frac{\beta}{\eta^\beta} t^{\beta-1} e^{-\left(\frac{t-\gamma}{\eta}\right)^\beta} \tag{38.19}$$

where γ- Position or origin parameter (unit of time) $\gamma = 0$, and means that no failure are observed before the start of the observation;

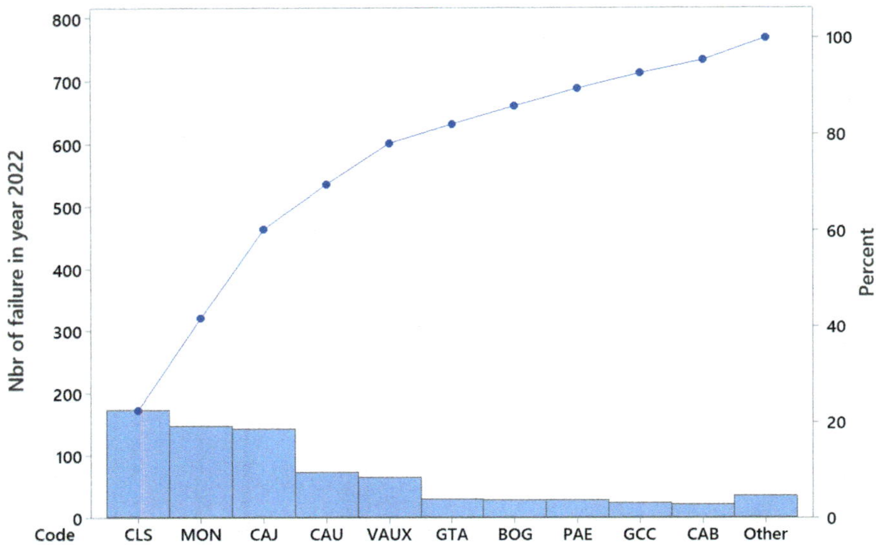

Fig. 38.2 Pareto curve for the analysis of the components of the Algiers metro vehicle system for the year 2022

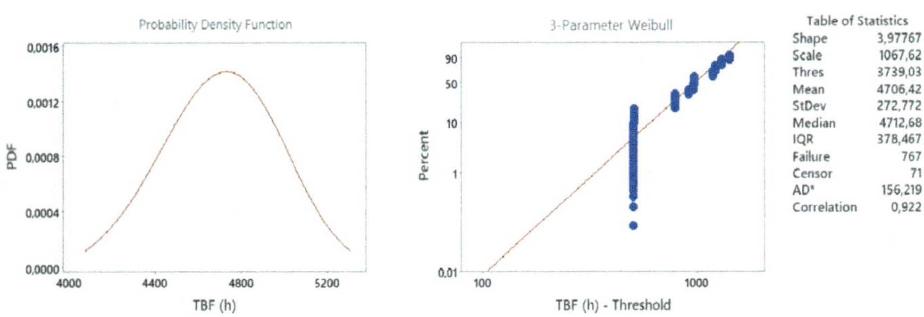

a- Probability density function b- Regression percent in function Time TBF (h)

Fig. 38.3 Distribution overview plot for time(h), of linear squares LSXY- estimates failure censored data

In our case, the reliability function commonly used for metro components is expressed as follows:

$$R(t) = e^{-\left(\frac{t-\gamma}{\eta}\right)^{\beta}}, \gamma = 0, \; R(t) = e^{-\left(\frac{t}{\eta}\right)^{\beta}} \tag{38.20}$$

The cumulative hazard function h(t) is defined as follows.

$$h(t) \; = \; \frac{\beta}{\eta^{\beta}} \, t^{\beta - 1} \tag{38.21}$$

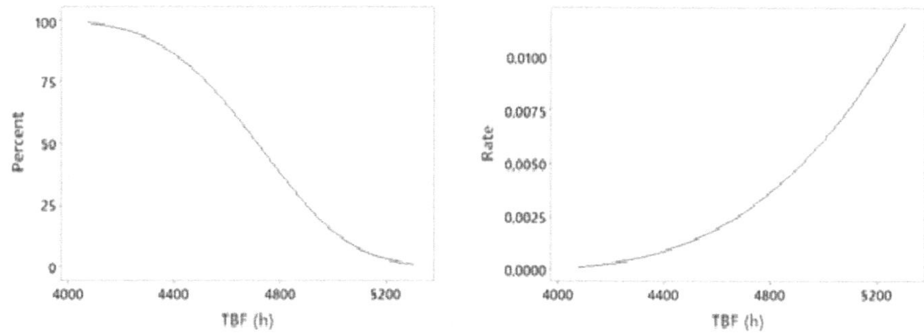

c- Survival function as a function of time

d- Hazard function as a function of time (cumulative failure curve)

Fig. 38.4 Presents the curves of survival and hazard functions in time, by estimation of M. squares XY- full data

In conclusion, the results of the analysis show that the correlation coefficient equals to 0.922, which indicates a very strong positive linear relationship between the percentage of variable failures and time and the reliability R (t, γ, η, β) of traction systems is almost in the approximate 50% with its parameters: $\gamma = 0$, $\beta = 3.9776$, $\eta = 1067.62$h which requires measures to improve the latter by including new techniques for service management, control, preventive and corrective maintenance.

Additionally, the availability of the traction vehicle system of Algiers metro is calculated by the following formula:

$$A(t) = \frac{E(t)}{T} = \frac{\int_0^\infty R(t)dt}{T} = \frac{\int_0^\infty e^{-\left(\frac{t}{\eta}\right)^\beta} dt}{T} \tag{38.22}$$

In order to facilitate the calculation of the availability of the metro vehicle systems *A(t)*, we will proceed to solve the integral Eq. (38.23) above, by developing it using the following formula:

$$\mathrm{A(t)} = \frac{\int_0^\infty e^{-\left(\frac{t}{\eta}\right)^\beta} dt}{T} = \lim_{n\to\infty} -\frac{\frac{1067.62.\Gamma\left(\frac{1}{3.9776} \frac{t^{3.9776}}{9722569711845376}\right)}{3.9776}}{t} + \frac{1067.62\Gamma\left(\frac{1}{3.9776}\right)}{3.9776t} \tag{38.23}$$

The curve in Fig. 38.5 represents the availability *A(t)* of Algiers metro traction vehicle systems in function in time.

The results obtained when calculating the availability of the traction vehicle systems shown in Fig. 38.4, show that its availability $A(t)$ decreases as a function of time (t) in exponential form and this availability $A(t)$ equals 20.6% at mean time to first failure $t = MTTF = 1067.62$h, and tends to 0 over the mean time between failure $t = MTBF$ equals 4706.42 h.

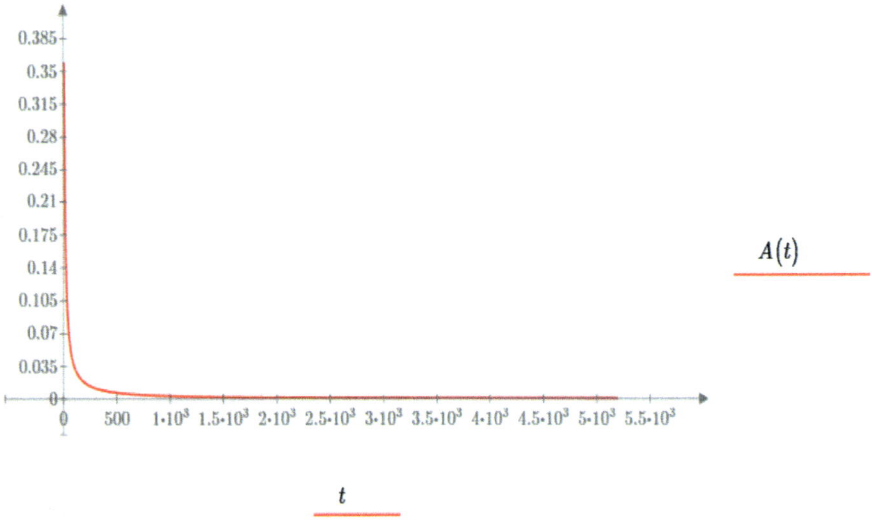

Fig. 38.5 Availability of Algiers metro traction vehicle systems

38.5 Conclusions

The results show that the reliability model and the 3-parameter sequential Weibull method proposed in this paper have enabled us to obtain rich and different reliability indices. Through visual simulation and theoretical calculation, the reliability analysis of the metro traction vehicle system was effectively evaluated.

In this work, EXCEL, Minitab Software, and Mathcad were used to implement the non-linear least squares methods applied for the identification of the parameters of the three-parameter Weibull distribution that models the reliability of the traction vehicle systems of the Algiers metro. The calculations were used to evaluate the main reliability indicators, which are the uptime (TBF) equal to 56,782 h, the mean time to first failure MTTF equal 1067.62h and the mean time between failure MTBF of the two traction systems which is 4706.42 h. The Weibull reliability R (γ, β, η, t) corresponding respectively to MTBF equals 50%.

The availability A(t) of the traction vehicle systems at t = MTTF = 1067.62 h, is equal to 90.6% and begin to decrease to just below 50%, over t = MTBF equals 4706.42 h, it begins to converge towards 0 which requires the implementation of a predictive maintenance policy.

Acknowledgements With this work, i would like to thank all the technical staff of the Algiers metro Company and especially the head of the management direction Mr. R. KERAMEN, Mr. H. SADAT Mr. M. MENACHE for their help and contribution for the realization of this fruitful work, which can bring an added value for the Algiers metro Company, as well as for the scientific research.

References

Feng Fang et al (2019) Application of reliability-centered maintenance in metro door system. IEEE Access 1–1:99. https://doi.org/10.1109/ACCESS.2019.2960521

Cheng X et al (2013) Reliability analysis of metro door system based on FMECA. J Intell Learn Syst Appl 5:216–220

Vonta I, Ram M (2018) Reliability engineering: theory and applications. Publisher: CRC Press-Taylor & Francis Group ISBN: 9780815355175

Lucanin V (2005) Relation between reliability and availability of railway vehicles. FME Transaction 33(3)

Melnik R et al (2019) Reliability analysis of metro vehicles operating in Poland. Transp Res Procedia 40:808–814. https://doi.org/10.1016/j.trpro.2019.07.114

Nisonger TE (2008) The 80/20 Rule and Core Journals. The Serials Librarian 55(1):62–84. https://doi.org/10.1080/03615260801970774

Qin Y et al (2014) Reliability analysis and prediction of Metro Vehicles' Bogie Frame. Appl Mech Mater 590:872–877

Teghem J (2013) Recherche operationnelle. Tome 2. Gestion de production, Modeles aleatoire, Aide multicritere. ISBN 978–2–7298–82020. Ellipses Edition

Yin H et al (2017) Reliability analysis of subway vehicles based on the data of operational failures. EURASIP J Wirel Commun Netw 212 8

The manufacturer's authorised representative in the EU is Springer Nature Customer Service Centre GmbH, Europaplatz 3, 69115 Heidelberg, Germany. If you have any concerns regarding our products, please contact ProductSafety@springernature.com

Printed and bound by CPI Group (UK) Ltd, Croydon, CR0 4YY
30/09/2026
02225586-0019